Modern Experimental Biochemistry

SECOND EDITION

Of related interest from the Benjamin/Cummings Series in the Life Sciences and Chemistry

W. M. Becker and D. W. Deamer
The World of the Cell, second edition (1991)

Benjamin/Maruzen
HGS Molecular Structure Models:
General Chemistry (1969) and *Organic Chemistry Sets* (1988)

I. S. Butler and J. F. Harrod
Inorganic Chemistry: Principles and Applications (1989)

N. A. Campbell
Biology, third edition (1993)

L. E. Hood, I. L. Weissman, W. B. Wood, and J. H. Wilson
Immunology, second edition (1984)

N. Kerner
Chemical Investigations (1986)

H. W. Knoche
Essentials of Organic Chemistry (1986)

G. M. Loudon
Organic chemistry, third edition (1994)

B. Mahan and R. J. Meyers
University Chemistry, fourth edition (1987)

C. K. Mathews and K. E. van Holde
Biochemistry (1990)

G. Sackheim
Introduction to Chemistry for Biology Students, fourth edition (1991)

J. D. Watson, N. H. Hopkins, J. W. Roberts, J. A. Steitz, and A. M. Weiner
Molecular Biology of the Gene, fourth edition (1987)

W. B. Wood, J. H. Wilson, R. M. Benbow, and L. E. Hood
Biochemistry: A Problems Approach (1981)

C. Yoder and C. Schaeffer
Introduction to Multinuclear NMR (1987)

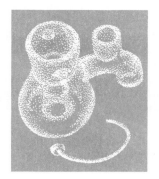

Modern Experimental Biochemistry

SECOND EDITION

Rodney F. Boyer

Hope College

The Benjamin/Cummings Publishing Company, Inc.

Redwood City, California • Menlo Park, California • Reading, Massachusetts
New York • Don Mills, Ontario • Wokingham, U.K. • Amsterdam
Bonn • Sydney • Singapore • Tokyo • Madrid • San Juan

Disclaimer

The experiments in this book have been exhaustively tested for safety and all attempts have been made to use the least hazardous chemicals and procedures possible. However, the author and publisher cannot be held liable for any damage which may occur during the performance of the experiments. It is assumed that before an experiment is initiated, a material safety data sheet for each chemical used will have been studied by the Instructor and students to ensure its safe handling and disposal.

Sponsoring Editor: Robin Heyden
Editorial Assistant: Korinna Sodic
Production Editor: Jean Lake
Art and Design Manager: Michele Carter
Cover Photography and Design: Martucci Studio
Text Design: Eleanor Mennick
Copy Editor: Mary Prescott
Composition and Illustrations: Interactive Composition Corporation
Printing and Binding: R. R. Donnelly and Sons

Library of Congress Cataloging-in-Publication Data

Boyer, Rodney F.
    Modern experimental biochemistry / Rodney F. Boyer. — 2nd ed.
       p. cm. — (Benjamin/Cummings series in the life sciences and chemistry)
    Includes bibliographical references and index.
    ISBN 0-8053-0545-9
    1. Biochemistry—Methodology. 2. Biochemistry—Experiments.
3. Molecular biology—Methodology. 4. Molecular biology—
Experiments. I. Title. II. Series.
QP519.7.B68 1993
574.19′2—dc20   92-35064
                                                      CIP

123456789—D0—9876543

The Benjamin/Cummings Publishing Company, Inc.
390 Bridge Parkway
Redwood City, California 94065

To: H. B., the Putz, and Mäusi

Preface

The second edition of *Modern Experimental Biochemistry* was prompted by many factors—rapid advancement of techniques in experimental biochemistry/molecular biology, encouragement from instructors and students who used the first edition, and the benefit of a sabbatical leave. All of these factors converged to signal that the time was right.

The purpose of the second edition is the same as the first—to provide junior-senior students majoring in biochemistry, chemistry and biology with a modern and complete experience in experimental biochemistry and molecular biology. Biochemistry laboratory has become even more of a prominent fixture in the curriculum during the seven years since the first edition. The American Chemical Society and the American Society for Biochemistry and Molecular Biology have each introduced a recommended curriculum for the bachelor's degree in biochemistry and they both include components of biochemistry laboratory and independent research. The second edition is designed for the recommended one- or two-semester biochemistry major's laboratory.

The Committee on Professional Training (CPT) of the American Chemical Society in early 1992 released a document outlining general techniques recommended for inclusion in an undergraduate biochemistry laboratory. Two categories of techniques were listed:

Important general techniques:

error and statistical analysis of experimental data

spectroscopic methods

electrophoretic techniques

chromatographic separations

isolation and characterization of biological materials

Some selected additional techniques:

use of radioisotopes

enzyme kinetics

immunoassay methods

DNA cloning and sequencing

plasmid isolation and mapping

peptide isolation and sequencing

computer graphics and structure calculations

All of the techniques listed but one (computer graphics and structure calculations) are discussed in Part I or II of the second edition.

As with the first edition, this book is divided into two parts. Instructors and students have expressed positive comments about the usefulness of this organization. Part I, Theory and Experimental Techniques, introduces students to background material that is usually not provided in lecture courses nor readily available in existing laboratory manuals. This part serves as an ideal supplement for instructors who use their own set of lab experiments. Twenty-two experiments representing all areas of biochemistry are included in Part II. One-third of the experiments are new, covering such topics as catalytic RNA, protein-ligand interactions, plasmid DNA, and immobilized enzymes. The remaining two-thirds are thoroughly revised for greater clarity. The experiments are designed for the typical time constraints and equipment inventories at most colleges and universities. All equations used in Parts I and II are highlighted by a triangle (▶). An updated Appendix includes physical properties of biomolecules and buffers and answers to questions at the end of each experiment.

Both Parts I and II have been completely rewritten and reflect the many advances in biochemistry/molecular biology theory and techniques. Especially noteworthy have been the technical advances in chromatography (perfusion, FPLC, bioaffinity), electrophoresis (pulsed gel, capillary, nucleic acid sequencing), spectrophotometry (diode array detectors), and molecular biology (microsequencing of proteins and nucleic acids, blotting, polymerase chain reaction, RNA scissors).

In the development of an experiment, primary consideration was given to the use of modern procedures and techniques. Students will learn procedures that are now used in actual laboratory settings, whether academic or industrial. Each experiment presents a challenging laboratory situation or problem to be solved by the student. In each experiment, a technique is introduced that allows students to obtain and evaluate some property of a biomolecule or biosystem.

The outline for each experiment is as follows.

Introduction and Theory.   In this section it is assumed that a student has studied the general subject in lecture. Only a summary or review of significant aspects of background is included. The general discussion includes theoretical and practical information that is generally not available in biochemistry textbooks. The general thrust of the experiment is explained, and a flow chart of the experiment is presented, if appropriate. This outline of the experiment allows the student to recognize the importance of each part of the experiment to the achievement of the overall objective.

Materials and Supplies.   A complete list of all materials, supplies, and equipment required for the experiment is provided. An Instructor's Manual describing the preparation of all reagents and solutions is available from the publisher.

Experimental Procedure.   This section contains a step-by-step detailed procedure for the experiment. It is divided into logical parts for ease of completion and to facilitate the interruption of an experiment, if necessary. There is an increased level of safety and environmental awareness in the second edition. The introduction of an icon (▲) to alert students and instructors to possible safety concerns is one example of this.

Analysis of Results.   In this section the student is instructed in the proper collection and handling of data from the experiment. Each table or graph that is to be constructed is explained and sample calculations are outlined. Typical data for the experiment may be disclosed to the student but only to aid the student in interpretation of results.

Questions and Problems.   Several questions and problems are provided for student study at the end of each experiment. Some questions will deal with various details of the experiment and numerical problems are emphasized. A triangle (▶) indicates questions that are answered in Appendix IX.

References.   Each experiment ends with a complete list of references that provide either a more detailed theoretical background or an expanded explanation of procedures and techniques.

The majority of the book was written while the author was on an American Cancer Society–funded research sabbatical leave in the laboratory of Thomas R. Cech at the University of Colorado, Boulder. The stimulating environment of the laboratory and the physical beauty of the surroundings all helped to make writing a pleasure. The author encourages comments from instructors and students.

RFB

May, 1992

Holland, Michigan

Acknowledgments

As with the first edition, I had the outstanding secretarial assistance of Mrs. Norma Plasman, who typed all revisions of the manuscript. Her efficiency and organization are exemplary. The expert advice of Tom Cech and the hospitality shown to the author while in his laboratory will long be remembered. Funding from the American Cancer Society and Hope College allowed for a combined research-writing sabbatical leave. Elliott Uhlenhopp (Grinnell College) contributed to the content of Chapter 4 and Experiments 19, 20, and 21.

I greatly appreciate the assistance of many expert reviewers:

First Edition

    Hugh Akers, Lamar University

    John Cronin, Arizona State University

    Patricia Dwyer-Hallquist, Appleton Paper Inc.

    Robert Lindquist, San Francisco State University

    Kenneth Marx, Dartmouth College

    Richard Paselk, Humboldt State University

    William Scovell, Bowling Green State University

    Ev Trip, University of British Columbia

    Dennis Vance, University of British Columbia

    Ronald Watanabe, San Jose State University

Second Edition

    Larry Byers, Tulane University

    Stephen Dahms, San Diego State University

    Anthony Gawienowski, University of Massachusetts, Amherst

    Rebecca Jurtshuk, University of Houston

    Robert Lindquist, San Francisco State University

    Denise Magnuson, Texas A&M University

    Henry Mariani, University of Massachusetts, Boston

    William Meyer, University of Vermont

    Ilka Nemere, University of California, Riverside

    Robert Sanders, University of Kansas

    Anthony Toste, Southwest Missouri State University

    Steven Wietstock, Alma College

Their advice, counsel, and encouragement assisted greatly in the project. Several individuals at the Benjamin/Cummings Publishing Company, including Diane Bowen, David Rogelberg, Robin Heyden, Korinna Sodic, and Jean Lake, provided excellent supervision and assistance with the project.

How could I ever forget the Mäuschen, our blue-point Himalayan, who was a constant writing companion napping under the desk lamp?

Table of Contents

PART II   Experiments

PART I Theory and Experimental Techniques

CHAPTER I

Introduction to the Biochemistry Laboratory

■   Welcome to your biochemistry laboratory course! This is not the first chemistry laboratory course for most of you, but I believe you will find it to be among the most exciting and dynamic of those in which you have enrolled. Most of the experimental techniques and skills that you have acquired over the years will be of great value in this laboratory. However, you will be introduced to several new procedures. Your success in the biochemistry laboratory will depend, to a great extent, on your mastery of these specialized techniques and on your understanding of chemical-biochemical principles.

As you proceed through the schedule of experiments for this term, you will, no doubt, compare your work with previous laboratory experiences. In biochemistry laboratory you will seldom run reactions and isolate products on a gram-size scale as you did in the organic laboratory. Rather, you will work with milligram or even microgram quantities, and in most cases the biomolecules will be dissolved in solution so you never really "see" the materials under study. But, you will observe the dynamic chemical and biological changes brought about by biomolecules. The techniques and procedures introduced in the laboratory will be your "eyes" and will monitor the biochemical events occurring.

This chapter is an introduction to procedures that are of utmost importance for the safe and successful completion of a biochemical project. It is recommended that you become familiar with the following sections before you begin laboratory work.

A. Safety in the Laboratory

The concern for laboratory safety can never be overemphasized. Most students have progressed through at least two years of laboratory work without even a minor accident. This record is, indeed, something to be proud of; however, it should not lead to overconfidence. You must always be aware that chemicals used in the laboratory are potentially toxic, irritating, or flammable. Such chemicals are a hazard, however, only when they are mishandled or improperly disposed of. It is the author's experience that accidents happen least often to those students who come to each laboratory session mentally prepared and with a complete understanding of the experimental procedures to be followed. Since dangerous situations can develop unexpectedly, though, you must be familiar with general safety practices, facilities, and emergency action. Students must have a special concern for the safety of classmates. Carelessness on the part of one student can often cause injury to other students.

The experiments in this book are designed with an emphasis on safety. However, no amount of planning or pretesting of experiments substitutes for awareness and common sense on the part of the student. All chemicals used in the experiments outlined here must be handled with care and respect. The use of chemicals in all U.S. workplaces, including academic research and teaching laboratories, is now regulated by the Federal Hazard Communication Standard, a document written by the Occupational Safety and Health Administration (OSHA).[1] Specifically, the OSHA standard requires all workplaces where chemicals are used to do the following: (1) develop a written hazard communication program, (2) maintain files of Material Safety Data Sheets (MSDS) on all chemicals used in that workplace, (3) label all chemicals with information regarding hazardous properties and procedures for handling, and (4) train employees in the proper use of these chemicals. Several states have passed "right to know" legislation which amends and expands the federal OSHA standard. If you have an interest in or concern about any chemical used in the laboratory, the MSDS for that chemical may be obtained from your instructor or laboratory manager. The actual form of an MSDS for a chemical may vary, but certain specific information must be present. Figure 1.1 is a partial copy of the MSDS for sucrose, common table sugar, a reagent often used in biochemical research. Different systems for labeling chemical reagent bottles are commercially available. One of the most widely used is the Hazardous Materials Identification System (HMIS). A copy of the actual label for sucrose is shown in Figure 1.2A. The health, flammability, reactivity, and personal protection codes are defined in Figure 1.2B.

It is easy to overlook some of the potential hazards of working in a biochemistry laboratory. Students often have the impression that they are working less with chemicals and more with natural biomolecules; therefore, there is less need for caution. However, this is not true, since many reagents used are flammable and toxic. In addition, materials such as fragile glass (disposable pipets), sharp objects (needles), and potentially

---

[1] *Federal Register*, Vol. 48, Nov. 25, 1983, p. 53280; *Federal Register*, Vol. 50, Nov. 27, 1985, p. 48758.

```
                        **SUCROSE**
                        **SUCROSE**
                        **SUCROSE**
                     MATERIAL SAFETY DATA SHEET
      --------------------------------------------------------------

      FISHER SCIENTIFIC                EMERGENCY NUMBER:  (201) 796-7100
      CHEMICAL DIVISION                CHEMTREC ASSISTANCE: (800) 424-9300
      1 REAGENT LANE
      FAIR LAWN NJ 07410
      (201) 796-7100

      THIS INFORMATION IS BELIEVED TO BE ACCURATE AND REPRESENTS THE BEST
      INFORMATION CURRENTLY AVAILABLE TO US.  HOWEVER, WE MAKE NO WARRANTY OF
      MERCHANTABILITY OR ANY OTHER WARRANTY, EXPRESS OR IMPLIED, WITH RESPECT TO
      SUCH INFORMATION, AND WE ASSUME NO LIABILITY RESULTING FROM ITS USE.  USERS
      SHOULD MAKE THEIR OWN INVESTIGATIONS TO DETERMINE THE SUITABILITY OF THE
      INFORMATION FOR THEIR PARTICULAR PURPOSES.

      --------------------------------------------------------------
                         SUBSTANCE IDENTIFICATION

                                                 CAS-NUMBER 57-50-1
      SUBSTANCE: **SUCROSE**

      TRADE NAMES/SYNONYMS:
        SACCHAROSE; SUGAR; CANE SUGAR; BEET SUGAR; CONFECTIONER'S SUGAR;
        BETA-D-FRUCTOTARANOSYL-ALPHA-D-GLUCOPYRANOSIDE; SACCHRUM; GRANULATED SUGAR;
        (ALPHA-D-GLYCOSIDO)-BETA-D-FRUCTOFURANOSIDE; ROCK CANDY;
        ALPHA-D-GLUCOPYRANOSYL BETA-D-FRUCTOFURANOSIDE; NCI-C56597; AMERFOND;
        MICROSE; MANALOX AS; WHITE SUGAR; SUCRALOX; D-(+)-SACCHAROSE; D-(+)-SUCROSE;
        S-5; BP220;

      CHEMICAL FAMILY:
      GLYCOSIDE

      MOLECULAR FORMULA: C12-H22-O11

      MOLECULAR WEIGHT: 342.34

      CERCLA RATINGS (SCALE 0-3):  HEALTH=1  FIRE=U  REACTIVITY=0  PERSISTENCE=0
      NFPA RATINGS (SCALE 0-4):  HEALTH=U  FIRE=U  REACTIVITY=0
      --------------------------------------------------------------
                        COMPONENTS AND CONTAMINANTS

      COMPONENT: SUCROSE                              PERCENT: 100
               CAS# 57-50-1

      OTHER CONTAMINANTS: NONE

      EXPOSURE LIMITS:
      NUISANCE PARTICULATES (NUISANCE DUST):
        5 MG/M3 OSHA TWA (RESPIRABLE DUST); 15 MG/M3 OSHA TWA (TOTAL DUST)
        10 MG/M3 ACGIH TWA (TOTAL DUST) (NO ASBESTOS AND < 1% CRYSTALLINE SILICA)
```

Figure 1.1
Partial MSDS for sucrose. Courtesy of Fisher Scientific.

infectious biologicals (blood, bacteria) must be used and disposed of with caution. The extensive use of electrical equipment including hot plates, stirring motors, and high-voltage power supplies presents special hazards.

Proper disposal of all waste chemicals, sharp objects, and infectious agents is essential not only to maintain safe laboratory working conditions but also to protect the general public and the local environment. Most of the liquid chemical reagents and reaction mixtures from each experiment are relatively safe and may be disposed of in the laboratory drainage system without causing environmental damage. However, special procedures must be followed for the use and disposal of some

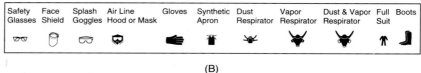

Hazardous Materials Identification System

HAZARD INDEX	PERSONAL PROTECTION INDEX	
4 Severe Hazard	A	G
3 Serious Hazard	B	H
2 Moderate Hazard	C	I
1 Slight Hazard	D	J
0 Minimal Hazard	E	K
	F	X Ask your supervisor for guidance.

PERSONAL PROTECTION SYMBOL DESCRIPTIONS

Safety Glasses	Face Shield	Splash Goggles	Air Line Hood or Mask	Gloves	Synthetic Apron	Dust Respirator	Vapor Respirator	Dust & Vapor Respirator	Full Suit	Boots

(A)

H HEALTH 1
F FLAMMABILITY 0
R REACTIVITY 0
PERSONAL PROTECTION C

(B)

Figure 1.2
(A) Hazardous Materials Identification System (HMIS) for sucrose.   (B) Definition of the hazard index and personal protection index. HMIS is copyrighted by the National Paint and Coatings Association and marketed exclusively through Labelmaster, Chicago, IL.

reagents and materials. Often this means that the instructor will provide detailed information on the proper use procedures. In some cases proper disposal will require the collection of waste materials from each laboratory worker and the institution will be responsible for removal. For each experiment in this book, instructions will be given for the proper handling and disposal of all excess reagents, reaction mixtures, and other materials. At the beginning of each experiment, a precautionary note highlighted by the logo ▲ will appear. Here the dangerous properties of each chemical or procedure will be described along with recommendations for safe handling, first aid, and proper disposal.

So that all students are aware of potential laboratory hazards, it is imperative that a set of rules be developed for each laboratory situation. The following can serve as guidelines:

1. Some form of eye protection is required at all times. Safety glasses with wide side shields are recommended, but normal eyeglasses with safety lenses may be permitted. *Contact lenses should never be worn in the laboratory*. The major problem with contact lenses is that they reduce the rate of self-cleansing of the eye. If a chemical splatters into the eye of a contact lens wearer, he or she will possibly have greater

injury because of less effective irrigation of the eye. Furthermore, the wearer may be reluctant to use an eyewash for fear of losing the contact lens.

2. Never work alone in the laboratory.

3. Be familiar with the physical properties of all chemicals used in the laboratory. This includes their flammability, reactivity, toxicity, and proper disposal. This information may be obtained from your instructor or from an MSDS. Always wear disposable gloves when using potentially dangerous chemicals or infectious agents.

4. Eating, drinking, and smoking in the laboratory are strictly prohibited.

5. Unauthorized experiments are not allowed.

6. Mouth suction should not be used to fill pipets or to start siphons.

7. Become familiar with the location and use of standard safety features in your laboratory. All chemistry laboratories should be equipped with fire extinguishers, eyewashes, safety showers, fume hoods, chemical spill kits, first-aid supplies, and containers for chemical disposal. Any questions regarding the use of these features should be addressed to your instructor or teaching assistant.

Rules of laboratory safety and chemical handling are not designed to impede productivity, nor should they instill a fear of chemicals or laboratory procedures. Rather, their purpose is to create a healthy awareness of potential laboratory hazards, to improve the efficiency of each student worker, and to protect the general public and the environment from waste contamination. The list of references at the end of this chapter includes books and manuals describing proper and detailed safety procedures.

B. The Laboratory Notebook

The biochemistry laboratory experience is not finished when you complete the experimental procedure and leave the laboratory. All scientists have the obligation to prepare written reports of the results of experimental work. Since this record may be studied by many individuals, it must be completed in a clear, concise, and accurate manner. This means that procedural details, observations, and results must be recorded in a laboratory notebook *while* the experiment is being performed. The notebook should be hardbound with quadrille-ruled (gridded) pages and used only for the biochemistry laboratory. This provides a durable, permanent record and the potential for construction of graphs, charts, etc. It is recommended that the first one or two pages of the notebook be used for a constantly updated table of contents. Although your instructor may have his or her own rules for preparation of the notebook, the most readable notebooks are those in which only the right-hand pages are used for record keeping. The left-hand pages may be used for your own notes, reminders and calculations. The outline in Figure 1.3 may be used for each experimental write-up. The basic outline follows that required by most biochemical research journals. Note that Parts I–IIc are labeled Prelab and should be completed before you enter the laboratory.

Details of Experimental Write-Up

I. INTRODUCTION

This section begins with a three- or four-sentence statement of the objective or purpose of the experiment. For preparing this statement, ask yourself, "What are the goals of this experiment?" This statement is followed by a brief discussion of the theory behind the experiment. If a new technique or instrumental method is introduced, give a brief description of the method. Include chemical or biochemical reactions when appropriate.

II. EXPERIMENTAL

Begin this section with a list of all reagents and materials used in the experiment. The sources of all chemicals and the concentrations of solutions should be listed. Instrumentation is listed with reference to company name and model number. A flowchart to describe the stepwise procedure for the experiment should be included after the list of equipment.

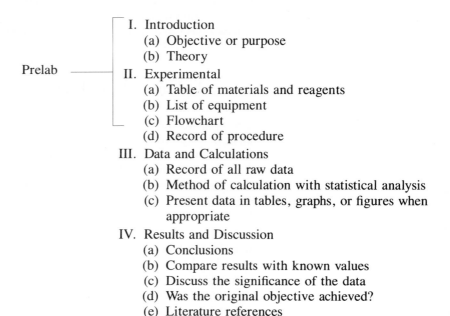

Figure 1.3
Outline of experimental write-up.

Prelab
I. Introduction
   (a) Objective or purpose
   (b) Theory
II. Experimental
   (a) Table of materials and reagents
   (b) List of equipment
   (c) Flowchart
   (d) Record of procedure
III. Data and Calculations
   (a) Record of all raw data
   (b) Method of calculation with statistical analysis
   (c) Present data in tables, graphs, or figures when appropriate
IV. Results and Discussion
   (a) Conclusions
   (b) Compare results with known values
   (c) Discuss the significance of the data
   (d) Was the original objective achieved?
   (e) Literature references

For the early experiments, a flowchart is provided. Flowcharts for later experiments should be designed by the student.

The write-up to this point is to be completed as a Prelab assignment. The experimental procedure followed is then recorded in your notebook as you proceed through the experiment. The detail should be sufficient so that a fellow student can use your notebook as a guide. You should include observations, such as color changes or gas evolution, made during the experiment.

III. DATA AND CALCULATIONS

All raw data from the experiment are to be recorded directly in your notebook, not on separate sheets of paper or paper towels. Calculations involving the data must be included for at least one series of measurements. Proper statistical analysis must be included in this section.

For many experiments, the clearest presentation of data is in a tabular or graphical form. The Analysis of Results section following each experimental procedure in this book describes the preparation of graphs and tables. These must all be included in your notebook.

IV. RESULTS AND DISCUSSION

This is the most important section of your write-up, because it answers the questions, "Did you achieve your proposed goals and objectives?" and "What is the significance of the data?" Any conclusion that you make must be supported by experimental results. It is often possible to compare your data with known values and results from the literature. If this is feasible, calculate percentage error and explain any differences. If problems were encountered in the experiment, these should be outlined with possible remedies for future experiments.

All library references (books and journal articles) that were used to write up the experiment should be listed at the end. The standard format to follow for a book or journal listing is shown at the end of this chapter in the reference section.

Everyone has his or her own writing style, some better than others. It is imperative that you continually try to improve your writing skills. When your instructor reviews your write-up, he or she should include helpful writing tips in the grading. Read the works listed in the last section of references at the end of this chapter for further instructions in scientific writing.

C. Cleaning Laboratory Glassware

The results of your experiments will depend, to a great extent, on the cleanliness of your equipment. There are at least two reasons for this. (1) Many of the chemicals and biochemicals will be used in milligram or microgram amounts. Any contamination, whether on the inner walls of a beaker, in a pipet, or in a glass cuvette, could be a significant percentage of the total experimental sample. (2) Many biochemicals and biochemical processes are sensitive to one or more of the following common contaminants: metal ions, detergents, and organic residues. In fact, the objective of many experiments is to investigate the effect of a metal ion, organic molecule, or other chemical agent on a biochemical process. Contaminated glassware will virtually ensure failure.

Cleaning Plasticware and Glassware

Plastic containers made of polyethylene are increasingly being used in chemical procedures. Polyethylene containers, when used for the first time, should be cleaned in 8 M urea (adjusted to pH 1 with concentrated HCl). After rinsing it with distilled water, wash the plasticware with 1 N KOH, distilled water again, and then 10^{-3} M EDTA to remove metal ion contamination. Finally, thoroughly rinse with glass-distilled water. After this initial treatment, a wash with 0.5% detergent solution followed by thorough glass-distilled water rinsing is recommended after each use. The base-urea-EDTA treatment may be repeated as desired.

Many contaminants, including organics and metal ions, adhere to the inner walls of glass containers. Washing glassware, including pipets, with dilute detergent (0.5% in water) followed by extensive water rinsing is probably sufficient for most purposes. The final rinse should be with distilled or deionized water. Metal ion contamination can be removed from glassware by rinsing with concentrated nitric acid followed by extensive rinsing with purified water.

Dry equipment is required for most processes carried out in biochemistry laboratory. When you needed dry glassware in organic laboratory, you probably rinsed the piece of equipment with acetone, which rapidly evaporated leaving a dry surface. Unfortunately, that surface is coated with an organic residue consisting of nonvolatile contaminants in the acetone. Since this residue could interfere with your experiment, it is best to refrain from acetone washing. Glassware and plasticware should be rinsed well with purified water and dried in an oven designated for glassware, not one used for drying chemicals. *Never dry cellulose nitrate centrifuge tubes in an oven.* Cellulose nitrate is an explosive. If you are not sure of the type of plastic in a container, do not dry it in an oven.

Cleaning Quartz and Glass Cuvettes

Never clean cuvettes or any optically polished glassware with ethanolic KOH or other strong base, as this will cause etching. All cuvettes should be cleaned carefully with 0.5% detergent solution, using a cotton-tipped wood applicator. These may be purchased or made by twisting a small wad of cotton around a cylindrical wood applicator. Rinse with plenty of glass-distilled water after cleaning.

D. Quantitative Transfer of Liquids

Practical biochemistry is highly reliant on analytical methods. Many analytical techniques must be mastered, but few are as important as the quantitative transfer of solutions. Some type of pipet will almost always be used in liquid transfer. Since students may not be familiar with the many types of pipets and the proper techniques in pipetting, this instruc-

Figure 1.4
Examples of pipets and pipet fillers. (A) Latex bulb courtesy of VWR Scientific Division of Univar. (B) Pipet filler courtesy of Curtin Matheson Scientific, Inc. (C) Mechanical pipet filler courtesy of VWR Scientific, Division of Univar. (D) Pipettor pump courtesy of Cole-Parmer Instrument Co. (E) Pasteur pipet courtesy of VWR Scientific, Division of Univar. (F) Volumetric pipet courtesy of VWR Scientific, Division of Univar. (G) Mohr pipet courtesy of VWR Scientific, Division of Univar. (H) Serological pipet courtesy of VWR Scientific, Division of Univar.

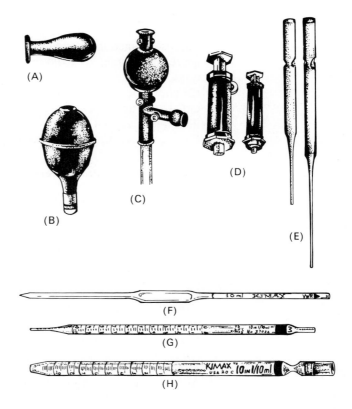

tion is included here. Figure 1.4 illustrates the various types of pipets and fillers. Following is a discussion of the use of each.

Filling a Pipet

The use of any pipet requires some means of drawing reagent into the pipet. *Liquids should never be drawn into a pipet by mouth suction on the end of the pipet.* Small latex bulbs are available for use with disposable pipets (Figure 1.4A). For volumetric and graduated pipets, two types of bulbs are available. One type (Figure 1.4B) features a special conical fitting that accommodates common sizes of pipets. To use these, first place the pipet tip below the surface of the liquid. Squeeze the bulb with your left hand (if you are a right-handed pipettor) and then hold it tightly to the end of the pipet. Slowly release the pressure on the bulb to allow liquid to rise to 2 or 3 cm above the top graduated mark. Then, remove the bulb and quickly grasp the pipet with your index finger over the top end of the pipet. The level of solution in the pipet will fall slightly, but not below the top graduated mark. If it does fall too low, use the bulb to refill.

Mechanical pipet fillers (sometimes called safety pipet fillers, pro-pipets, or pi-fillers) are more convenient than latex bulbs (Figure 1.4C,D). Equipped with a system of hand-operated valves, these fillers can be used for the complete transfer of a liquid. The use of a safety pipet filler is outlined in Figure 1.5. *Never allow any solvent or solution to enter the pipet bulb.* To avoid this, two things must be kept in mind: (1) always maintain careful control while using valve S to fill the pipet, and (2) never use valve S unless the pipet tip is below the surface of the liquid. If the tip moves above the surface of the liquid, air will be sucked into the pipet and solution will be flushed into the bulb. Other pipet fillers are used in a similar fashion.

Disposable Pasteur Pipets

Often it is necessary to perform a semiquantitative transfer of a small volume (1 to 10 mL) of liquid from one vessel to another. Since pouring is not efficient, a **Pasteur pipet** with a small latex bulb may be used (Figure 1.4A,E). Pasteur pipets are available in two lengths (15 cm and 23 cm) and hold about 2 mL of solution. These are especially convenient for the transfer of nongraduated amounts to and from test tubes. Typical recovery while using a Pasteur pipet is 90 to 95%. If dilution is not a problem, rinsing the original vessel with a solvent will increase the transfer yield.

Calibrated Pipets

Although most quantitative transfers are now done with automatic pipetting devices, which are described later in the chapter, instructions will be

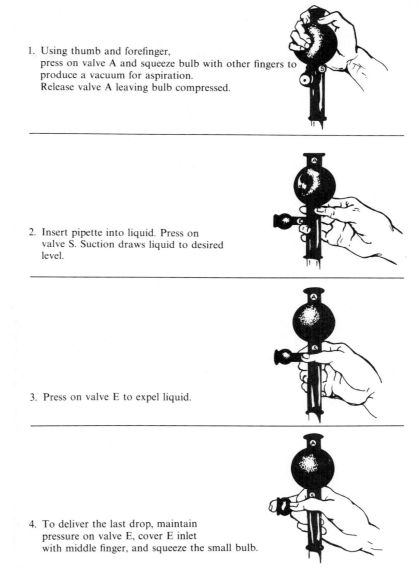

1. Using thumb and forefinger,
   press on valve A and squeeze bulb with other fingers to
   produce a vacuum for aspiration.
   Release valve A leaving bulb compressed.

2. Insert pipette into liquid. Press on
   valve S. Suction draws liquid to desired
   level.

3. Press on valve E to expel liquid.

4. To deliver the last drop, maintain
   pressure on valve E, cover E inlet
   with middle finger, and squeeze the small bulb.

Figure 1.5
How to use a Spectroline safety
pipet filler. Courtesy of Spec-
tronics Corporation, Westbury,
NY 11590.

given for the use of all types of pipets. If a quantitative transfer of a
specific and accurate volume of liquid is required, some form of calibrated pipet must be used. **Volumetric pipets** (Figure 1.4F) are used for
the delivery of liquids required in whole milliliter amounts (1, 2, 3, 4, 5,
10, 15, 20, 25, 50, and 100 mL). To use these pipets, draw liquid with a
latex bulb or mechanical pipet filler to a level 2 to 3 cm above the "fill
line." Wipe the outside of the pipet with a lint-free tissue. Touch the tip
of the pipet to the inside of the glass wall of the container from which it

was filled. Release liquid from the pipet until the meniscus is directly on the fill line. Transfer the pipet to the inside of the second container and release the liquid. Hold the pipet vertically, allow the solution to drain until the flow stops, and then wait an additional 5 to 10 seconds. Touch the tip of the pipet to the inside of the container to release the last drop from the outside of the tip. Remove the pipet from the container. Some liquid may still remain in the tip. Most volumetric pipets are calibrated as "TD" (to deliver), which means the intended volume is transferred *without final blow-out,* i.e., the pipet delivers the correct volume.

Fractional volumes of liquid are transferred with **graduated pipets**, which are available in two types—Mohr and serological. **Mohr pipets** (Figure 1.4G) are available in long- or short-tip styles. Long-tip pipets are especially attractive for transfer to and from vessels with small openings. Virtually all Mohr pipets are TD and are available in many sizes (0.1 to 10.0 mL). The marked subdivisions are usually 0.01 or 0.1 mL, and the markings end a few centimeters from the tip. Selection of the proper size of pipet is especially important. For instance, do not try to transfer 0.2 mL with a 5 or 10 mL pipet. Use the smallest pipet that is practical. Class A pipets have tolerance limits ranging from $\pm$ 0.005 mL for a 0.1 mL pipet to $\pm$ 0.02 mL for a 10 mL pipet.

The use of a Mohr pipet is similar to that of a volumetric pipet. Draw the liquid into the pipet with a pipet filler to a level about 2 cm above the "0" mark. Clean the tip with a tissue, and lower the liquid level to the 0 mark. Remove the last drop from the tip by touching it to the inside of the glass container. Transfer the pipet to the receiving container and release the desired amount of solution. The solution should not be allowed to move below the last graduated mark on the pipet. Touch off the last drop. If an additional transfer of the same solution is to be done, the same pipet may be used, but be sure to wipe the tip with a tissue before inserting it into the reagent.

Serological pipets (Figure 1.4H) are similar to Mohr pipets, except that they are graduated downward to the very tip and are designed for blow-out. Their use is identical to that of a Mohr pipet except that the last bit of solution remaining in the tip must be forced out into the receiving container with a rubber bulb. This final blow-out should be done after 15 to 20 seconds of draining. Serological pipets are used less often than Mohr pipets in biochemical procedures. Tolerance limits are in the same range as for Mohr pipets.

Automatic Pipetting Systems

For most quantitative transfers, including many identical small-volume transfers, a mechanical microliter pipettor (Eppendorf type) is ideal. This allows accurate, precise, and rapid dispensing of fixed volumes from 1 to 5000 μL (5 mL). The pipet's push-button system can be operated with one hand, and it is fitted with detachable polypropylene tips (Figure 1.6).

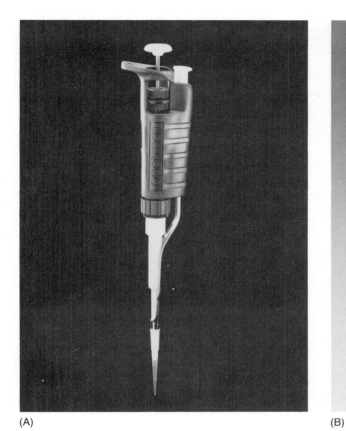

(A)

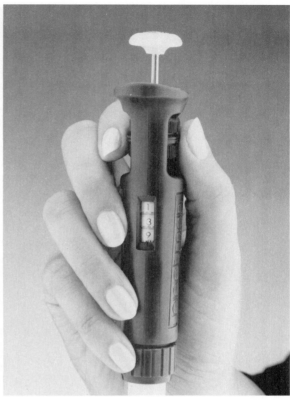

(B)

Figure 1.6
How to use an adjustable pipetting device (A).   (B) Set the digital micrometer to
the desired volume using the adjustment knob. Attach a new disposable tip to the
shaft of the pipet. Press on firmly with a slight twisting motion.   (C) Depress the
plunger to the first positive stop, immerse the disposable tip into the sample liquid
to a depth of 2 to 4 mm, and allow the pushbutton to return slowly to the up position and wait 1 to 2 seconds.   (D) To dispense sample, place the tip end against
the side wall of the receiving vessel and depress the plunger slowly to the first stop.
Wait 2 to 3 seconds, and then depress the plunger to the second stop to achieve
final blow-out. Withdraw the device from the vessel carefully with the tip sliding
along the inside wall of the vessel. Allow the plunger to return to the up position.
Discard the tip by depressing the tip ejector button. Photos courtesy of Rainin Instrument Company, Inc., Woburn, MA. Pipetman is a registered trademark of Gilson
Medical Electronics. Exclusive license to Rainin Instrument Company, Inc.

The advantage of polypropylene tips is that the reagent film remaining in
the pipet after delivery is much less than for glass tips. Mechanical pipet-
tors are available in up to 25 different sizes. Newer models offer continuous volume adjustment, so a single model can be used for delivery of
specific volumes within a certain range.

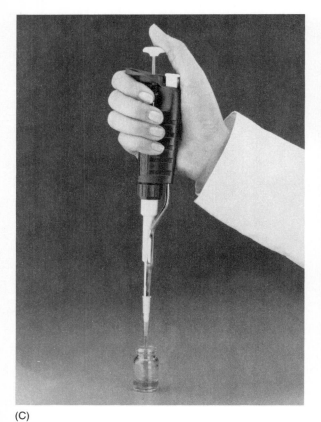
(C)

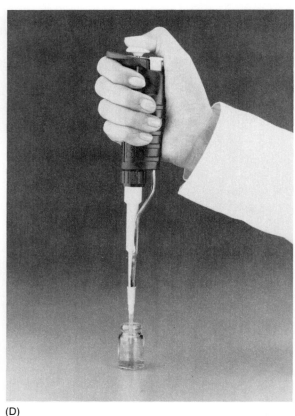
(D)

Figure 1.6   continued

To use the pipettor, choose the proper size and place a polypropylene pipet tip firmly onto the cone as shown in Figure 1.6. Tips for pipets are available in several sizes, for $1-20$ μL, $20-250$ μL, $200-1000$ μL, and $1000-5000$ μL capacity. Details of the operation of an adjustable pipet are given in Figure 1.6.

For rapid and accurate transfer of volumes greater than 5 mL, automatic repetitive dispensers are commercially available. These are particularly useful for the transfer of corrosive materials. The dispensers, which are available in several sizes, are simple to use. The volume of liquid to be dispensed is mechanically set; the syringe plunger is lifted for filling and pressed downward for dispensing. Hold the receiving container under the spout while depressing the plunger. Touch off the last drop on the inside wall of the receiving container. The set volumes are dispensed to an accuracy of $\pm 1\%$ and with a reproducibility of $\pm 0.1\%$.

Cleaning and Drying Pipets

Special procedures are required for cleaning glass pipets. Immediately after use, every pipet should be placed, tip up, in a vertical cylinder containing a dilute detergent solution (less than 0.5%). The pipet must be completely covered with solution. This ensures that any reagent remaining in the pipet is forced out through the tip. If reagents are allowed to dry inside a pipet, the tips can easily become clogged and are very difficult to open. After several pipets have accumulated in the detergent solution, the pipets should be transferred to a pipet rinser. Pipet rinsers continually cycle fresh water through the pipets. Immediately after detergent wash, tap water may be used to rinse the pipets, but distilled water should be used for the final rinse. Pipets may then be dried in an oven.

E. Preparation and Storage of Solutions

Water Quality

The most common solvent for solutions used in the biochemical laboratory is water. Tap water should never be used for the preparation of any reagent solutions. Purified water (deionized or distilled) should be used for all procedures. The water quality necessary will depend on the solution to be prepared and on the biochemical procedure to be investigated. Most laboratories are equipped with "in-house" purified water, which is prepared by ion exchange, reverse osmosis, or distillation. Water that is purified by ion exchange will be low in metal ion concentration but may contain certain organics that are washed from the ion-exchange resin. These contaminants will increase the ultraviolet absorbance properties of water. If sensitive ultraviolet absorbance measurements are to be made, distilled water is better than deionized.

Distilled water may be produced in a metal still or one in which the water comes in contact only with glass (glass-distilled water). Clearly, water in contact with metals will contain mineral ions that may interfere with reagents or biomaterials. This water will, however, be very low in organic contamination if it is distilled at a relatively slow rate.

For most procedures in the biochemistry laboratory, glass-distilled water is the choice. Several varieties of glass stills are commercially available. A glass still should operate only on deionized or distilled water. Use of tap water will cause a buildup of solid salts, which leads to "bumping" in the boiling flask. The still must be drained and cleaned with 10% hydrochloric acid almost daily if the tap water has a high mineral content. Organic contaminants can be removed during glass distillation of water by maintaining a slow flow rate of cooling water through the condenser coils. The organics are "steam distilled" and are carried out in the vapor escaping from the condenser. The instructions for glass stills give the most efficient flow rate for entry of water into the still and for condenser-cooling water.

All water becomes contaminated upon storage. Ultraviolet-absorbing organics are leached from some plastic containers, and metal ions may be present on glass surfaces. Bacterial contamination of glass-distilled water is not uncommon. A good practice is to tightly cap the purified water and store it in a refrigerator. If there is obvious bacterial growth (turbidity) in stored water, discard it.

Ultrapure water is often required for absorbance or fluorescence measurements. Distillation over acid permanganate (1 liter of water, 1 g of potassium permanganate, and 1 mL of 85% phosphoric acid in a glass still) will destroy organic contaminants by oxidation.

In summary, it is difficult, if not impossible, to remove all contaminants from large volumes of water. Compromises will have to be made in the choice of water. When deciding the method of water purification and storage, one must determine which type of contaminant has the least potential for interference. For most procedures described in this book, glass-distilled water is of sufficient purity.

Reagent and Solution Preparation and Storage

With only a few exceptions, reagents and chemicals used in this manual can be obtained from commercial sources and used without further purification. If a specific grade of chemical is required for an experiment, this will be stated in the instructions for solution preparation.

The concentrations for most biochemical solutions are in units of molarity, % (weight or volume), or weight per volume (mg/mL, g/liter, etc.). Solids should be weighed on weighing paper or plastic weighing boats, using an analytical or top-loading balance. Liquids are more conveniently dispensed by volumetric techniques; however, this assumes that the density is known. If a small amount of a liquid is to be weighed, it should be added to a tared flask by means of a disposable Pasteur pipet with a latex bulb. The hazardous properties of all materials should be known before use and the proper safety precautions obeyed.

The storage conditions of reagents and solutions are especially critical. Although some will remain stable indefinitely at room temperature, it is good practice to store all solutions in a closed container and in a refrigerator if they are prepared in advance. This inhibits bacterial growth and decomposition of the reagents. Some solutions may require storage below 0°C. If these are aqueous solutions or others that will freeze, be sure there is room for expansion inside the container.

All stored containers, whether at room temperature, 4°C, or below freezing, must be properly sealed. This reduces contamination by bacteria and vapors in the laboratory air (CO_2, NH_3, HCl, etc.). Volumetric flasks, of course, have glass stoppers, but test tubes, Erlenmeyers, bottles, and other containers should be sealed with screw caps, corks, or hydrocarbon foil (Parafilm). Remember that hydrocarbon foil, a wax, is dissolved by solutions containing nonpolar organic solvents.

Bottles of pure chemicals and reagents should also be properly stored. Many manufacturers now include the best storage conditions for a reagent on the label. The common conditions are: store at room temperature, store at 0–4°C, store below 0°C, and store in a desiccator at 0–4°C or below 0°C. Many biochemical reagents form hydrates by taking up moisture from the air. If the water content of a reagent increases, the molecular weight and purity of the reagent change. For example, when NAD^+ is purchased, the label usually reads: "Anhydrous molecular weight = 663.5; when assayed, contained 3 H_2O per mole." The actual molecular weight that should be used for solution preparation is $663.5 + (18)(3) = 717.5$. However, if this reagent is stored in a moist refrigerator or freezer outside of a dessicator, the moisture content will increase to an unknown value.

If the proper method of storage for a specific reagent is in doubt, it is usually safe to refrigerate it at 4°C in a dessicator containing a drying agent.

F. Statistical Analysis of Experimental Measurements

The purpose of each laboratory exercise in this book is to observe and measure characteristics of a biomolecule or a biological system. These measured characteristics may be the molecular weight of a protein, the pH of a buffer solution, the absorbance of a colored solution, the rate of an enzyme-catalyzed reaction, or the radioactivity associated with a molecule. If you measure a single characteristic many times under identical conditions, a slightly different result will most likely be obtained each time. For example, if a radioactive sample is counted twice under identical experimental and instrumental conditions, the second measurement immediately following the first, the probability is very low that the numbers of counts will be identical. If the absorbance of a solution is determined several times at a specific wavelength, the value of each measurement will surely vary from the others. Which measurements, if any, are correct? Before this question can be answered, you must understand the source and treatment of variations in experimental measurements.

Analysis of Experimental Data

An **error** in an experimental measurement is defined as a deviation of an observed value from the true value. There are two types of errors, **determinate** and **indeterminate**. Determinate errors are those that can be controlled by the experimenter and are associated with malfunctioning equipment, improperly designed experiments, and variations in experimental conditions. These are sometimes called human errors because they can be corrected or at least partially alleviated by careful design and performance of the experiment. Indeterminate errors are random or accidental and cannot be controlled by the experimenter. Specific examples of indeterminate errors are variations in radioactive counting and small differences in the successive measurements of glucose in a serum sample.

Two statistical terms involving error analysis that are often used and misused are **accuracy** and **precision**. Precision refers to the extent of agreement among repeated measurements of an experimental value. Accuracy is defined as the difference between the experimental value and the true value for the quantity. Since the true value is seldom known, accuracy is better defined as the difference between the experimental value and the accepted true value. Several experimental measurements may be precise (that is, in close agreement with each other) without being accurate.

If an infinite number of identical measurements could be made on a biosystem, this series of numerical values would constitute a **statistical population**. The average of all of these numbers would be the **true value** of the measurement. It is obviously not possible to achieve this in practice. The alternative is to obtain a relatively small **sample of data**, which is a subset of the infinite population data. The significance and precision of these data are then determined by statistical analysis.

This section explores the mathematical basis for the statistical treatment of experimental data. Most measurements required for the completion of the experiments can be made in duplicate, triplicate, or even quadruplicate, but it would be impractical and probably a waste of time and materials to make numerous determinations of the same measurement. Rather, when you perform an experimental measurement in the laboratory, you will collect a small sample of data from the population of infinite values for that measurement. To illustrate, imagine that an infinite number of experimental measurements of the pH of a buffer solution are made, and the results are written on slips of paper and placed in a container. It is not feasible to calculate an average value of the pH from all of these numbers, but it is possible to draw five slips of paper, record these numbers, and calculate an average pH. By doing this, you have collected a sample of data. By proper statistical manipulation of this small sample, it is possible to determine whether it is representative of the total population and the amount of confidence you should have in these numbers.

The data analysis will be illustrated here primarily with the counting of radioactive materials, although it is not limited to such applications. Any replicate measurements made in the biochemistry laboratory can be analyzed by these methods.

Determination of the Mean, Sample Deviation, and Standard Deviation

Radioactive decay with emission of particles is a random process. It is impossible to predict with certainty when a radioactive event will occur. Therefore, a series of measurements made on a radioactive sample will result in a series of different count rates, but they will be centered around an **average** or **mean** value of counts per minute. Table 1.1 contains such

Table 1.1
The Observed Counts and Sample Deviation
from a Typical Radioactive Sample

Counts per Minute	Sample Deviation $x_i - \overline{x}$
1243	+21
1250	+28
1201	−21
1226	+4
1220	−2
1195	−27
1206	−16
1239	+17
1220	−2
1219	−3
Mean = 1222	

a series of count rates obtained with a scintillation counter on a single radioactive sample. A similar table could be prepared for other biochemical measurements, including the rate of an enzyme-catalyzed reaction or the protein concentration of a solution as determined by the Bradford method. The arithmetic average or mean of the numbers is calculated by totaling all the experimental values observed for a sample (the counting rates, the velocity of the reaction, or protein concentration) and dividing the total by the number of times the measurement was made. The mean is defined by Equation 1.1.

$$\blacktriangleright \quad \overline{x} = \frac{\sum\limits_{i}^{n} x_i}{n} \qquad\qquad \text{(Equation 1.1)}$$

where

$\overline{x}$ = arithmetic average or mean

x_i = the value for an individual measurement

n = the total number of experimental determinations

The mean counting rate for the data in Table 1.1 is 1222. If the same radioactive sample were again counted for a series of ten observations, that series of counts would most likely be different from those listed in the table, and a different mean would be obtained. If we were able to make an infinite number of counts on the radioactive sample, then a **true mean** could be calculated. The true mean would be the actual amount of radioactivity in the sample. Although it would be desirable, it is not possible experimentally to measure the true mean. Therefore, it is necessary to use the average of the counts as an approximation of the true mean and to use statistical analysis to evaluate the precision of the measurements (that is, to assess the agreement among the repeated measurements).

Since it is not usually practical to observe and record a measurement many times as in Table 1.1, what is needed is a means to determine the reliability of an observed measurement. This may be stated in the form of a question. How close is the result to the true value? One approach to this analysis is to calculate the **sample deviation**, which is defined as the difference between the value for an observation and the mean value, $\bar{x}$ (Equation 1.2). The sample deviations are also listed for each count in Table 1.1.

▶   Sample deviation $= x_i - \bar{x}$ (Equation 1.2)

A more useful statistical term for error analysis is **standard deviation**, a measure of the spread of the observed values. Standard deviation, s, for a sample of data consisting of n observations may be estimated by Equation 1.3.

▶ $$s = \sqrt{\frac{\sum (x_i - \bar{x})^2}{n - 1}}$$ (Equation 1.3)

It is a useful indicator of the probable error of a measurement. Standard deviation is often transformed to **standard deviation of the mean** or **standard error**. This is defined by Equation 1.4, where n is the number of measurements.

▶ $$s_m = \frac{s}{\sqrt{n}}$$ (Equation 1.4)

It should be clear from this equation that as the number of experimental observations becomes larger, s_m becomes smaller, or the precision of a measurement is improved.

Standard deviation may also be illustrated in graphical form (Figure 1.7). The shape of the curve in Figure 1.7 is closely approximated by the **Gaussian distribution** or **normal distribution curve**. This mathematical treatment is based on the fact that a plot of relative frequency of a given event yields a dispersion of values centered about the mean, $\bar{x}$. The value of $\bar{x}$ is measured at the maximum height of the curve. The normal distribution curve shown in Figure 1.7 defines the spread or dispersion of the data. The probability that an observation will fall under the curve is unity or 100%. By using an equation derived by Gauss, it can be calculated that for a single set of sample data, 68.3% of the observed values will occur within the interval $\bar{x} \pm s$, 95.5% of the observed values within $\bar{x} \pm 2s$, and 99.7% of the observed values within $\bar{x} \pm 3s$. Stated in other terms, there is a 68.3% chance that a single observation will be in the interval $\bar{x} \pm s$.

For many experiments, a single measurement is made so a mean value, $\bar{x}$, is not known. In these cases, error is expressed in terms of s but is defined as the **percentage proportional error**, $\%E_x$, in Equation 1.5.

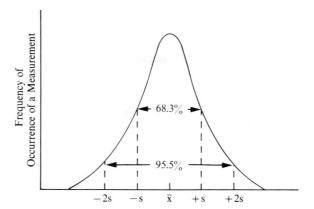

Figure 1.7
The normal distribution curve.

$$%E_x = \frac{100k}{\sqrt{n}}$$

(Equation 1.5)

The parameter k is a proportional constant between E_x and the standard deviation. The percent proportional error may be defined within several probability ranges. **Standard error** refers to a confidence level of 68.3%; that is, there is a 68.3% chance that a single measurement will not exceed the $%E_x$. For standard error, $k = 0.6745$. **Ninety-five hundredths error** means there is a 95% chance that a single measurement will not exceed the $%E_x$. The constant k then becomes 1.45.

The previous discussion of standard deviation and related statistical analysis placed emphasis on estimating the reliability or precision of experimentally observed values. However, standard deviation does not give specific information about how close an experimental mean is to the true mean. Statistical analysis may be used to estimate, within a given probability, a range within which the true value might fall. The range or confidence interval is defined by the experimental mean and the standard deviation. This simple statistical operation provides the means to determine quantitatively how close the experimentally determined mean is to the true mean. **Confidence limits** (L_1 and L_2) are created for the sample mean as shown in Equations 1.6 and 1.7.

$$L_1 = \overline{x} + (t)(s_m)$$

(Equation 1.6)

$$L_2 = \overline{x} - (t)(s_m)$$

(Equation 1.7)

where

t = a statistical parameter that defines a distribution between a sample mean and a true mean

The parameter t is calculated by integrating the distribution between percent confidence limits. Values of t are tabulated for various confidence limits. Such a table is in Appendix VIII. Each column in the table refers to a desired confidence level (0.05 for 95%, 0.02 for 98%, and 0.01 for 99% confidence). The table also includes the term **degrees of freedom**, which is represented by $n - 1$, the number of experimental observations minus 1. The values of $\bar{x}$ and s_m are calculated as previously described in Equations 1.1 and 1.4.

Statistical Analysis in Practice

The equations for statistical analysis that have been introduced in this chapter are of little value if you have no understanding of their practical use, meaning, and limitations. A set of experimental data will first be presented, and then several statistical parameters will be calculated using the equations. This example will serve as a summary of the statistical formulas and will also illustrate their application.

Ten identical protein samples were analyzed by the Bradford method for protein analysis. The following values for protein concentration were obtained.

Observation Number	Protein Concentration (mg/mL), x
1	1.02
2	0.98
3	0.99
4	1.01
5	1.03
6	0.97
7	1.00
8	0.98
9	1.03
10	1.01

SAMPLE MEAN

$$\bar{x} = \frac{\sum x}{n} = \frac{10.02}{10} = 1.00 \text{ mg/mL}$$

SAMPLE DEVIATION

Sample deviation $= x_i - \bar{x}$

Observation	$x_i - \bar{x}$
1	+0.02
2	−0.02
3	−0.01
4	+0.01
5	+0.03
6	−0.03
7	0.00
8	−0.02
9	+0.03
10	+0.01

Calculation of the sample deviation for each measurement gives an indication of the precision of the determinations.

STANDARD DEVIATION

$$s = \sqrt{\frac{\Sigma\,(x_i - \bar{x})^2}{n - 1}}$$

$$s = 0.02$$

The mean can now be expressed as $\bar{x} \pm s$ (for this specific example, 1.00 ± 0.02 mg/mL). The probability of a single measurement falling within these limits is 68.3%. For 95.5% confidence ($2s$), the limits would be 1.00 ± 0.04 mg/mL.

STANDARD ERROR OF THE MEAN

$$s_m = \frac{s}{\sqrt{n}}$$

$$s_m = \frac{0.02}{\sqrt{10}}$$

$$s_m = 0.006$$

This statistical parameter can be used to gauge the precision of the experimental data.

CONFIDENCE LIMITS

The desired confidence limits will be set at the 95% confidence level. Therefore we will choose a value for t from Appendix VIII in the column labeled $t_{0.05}$ and $n - 1 = 9$.

$$L_1 = \overline{x} + (t_{0.05})(s_m)$$

$$L_2 = \overline{x} - (t_{0.05})(s_m)$$

$$s_m = 0.006$$

$$t_{0.05} = 2.262$$

$$\overline{x} = 1.00$$

$$L_1 = 1.00 + (2.262)(0.006)$$

$$L_1 = 1.01$$

$$L_2 = 0.99$$

We can be 95% confident that the true mean falls between 0.99 and 1.01 mg/mL.

References

Error Analysis

R. Anderson, *Practical Statistics for Analytical Chemistry* (1987), Van Nostrand Reinhold (New York).

J. Brewer, A. Pesce, and R. Ashworth, *Experimental Techniques in Biochemistry* (1974), Prentice-Hall (Englewood Cliffs, NJ), pp. 10–31, 305–310. An excellent treatment of error analysis by Gaussian distribution. Also, an application of statistics to radioactive counting.

T. Cooper, *The Tools of Biochemistry* (1977), John Wiley & Sons (New York), pp. 108–113. A brief discussion of the statistics of radioactive counting.

J. Lee and T. Lee, *Statistics and Numerical Methods in BASIC for Biologists* (1982), Van Nostrand Reinhold (New York).

J. C. Miller and J. N. Miller, *Statistics for Analytical Chemistry*, 2nd ed. (1988), Ellis Horwood (Chichester).

J. Robyt and B. White, *Biochemical Techniques: Theory and Practice* (1987), Waveland Press (Prospect Heights, IL), pp. 5–17. Statistical analysis of data.

W. Schefler, *Statistics for the Biological Sciences*, 2nd ed. (1980), Addison-Wesley (Reading, MA). Emphasis on quantitative measurements in biology.

Safety

Biosafety in the Laboratory: Prudent Practices for the Handling and Disposal of Infectious Materials (1989), National Research Council, National Academy Press (Washington, DC).

J. Dux and R. Stalzer, *Managing Safety in the Chemical Laboratory* (1988), Van Nostrand Reinhold (New York).

G. Lowry and R. Lowry, *Lowry's Handbook of Right-to-Know and Emergency Planning* (1990), Lewis Publishers (Chelsea, MI).

W. Mahn, *Fundamentals of Laboratory Safety: Physical Hazards in the Academic Laboratory* (1991), Van Nostrand Reinhold (New York). An excellent introduction for students and instructors.

M. Pitt and E. Pitt, *Handbook of Laboratory Waste Disposal* (1985), Halsted Press (New York).

N. Steere, Editor, *Handbook of Laboratory Safety,* 2nd ed. (1988), CRC Press (Boca Raton, FL).

J. Young, Editor, *Improving Safety in the Chemical Laboratory—A Practical Guide* (1987), Wiley-Interscience (New York).

Writing Laboratory Reports

R. Day, *How to Write and Publish a Scientific Paper*, 2nd ed. (1983), ISI Press (Philadelphia).

J. Dodd, Editor, *The ACS Safety Guide: A Manual for Authors and Editors* (1986), American Chemical Society (Washington, DC).

H. Kanare, *Writing the Laboratory Notebook* (1985), American Chemical Society (Washington, DC).

J. Walker, *Biochem. Educ.* **19**, 31–32 (1991). "A Student's Guide to Practical Write-ups."

CHAPTER 2

General Laboratory Procedures

■ All biochemical laboratory activities, whether in teaching, research, or industry, are replete with techniques that must be carried out almost on a daily basis. This chapter outlines the theoretical and practical aspects of some of these general and routine procedures, including use of buffers, pH and other electrodes, dialysis, membrane filtration, lyophilization, and methods for protein and nucleic acid measurement.

A. pH Measurements, Buffers, and Electrodes

Most biological processes in the cell take place in a water-based environment. Water is an **amphoteric** substance; that is, it may serve as a proton donor (acid) or a proton acceptor (base). Equation 2.1 shows the ionic equilibrium of water.

▶ $$H_2O \rightleftharpoons H^+ + OH^-$$ (Equation 2.1)

In pure water, $[H^+] = [OH^-] = 10^{-7}$; in other words, the pH or $-\log [H^+]$ is 7. Acidic and basic molecules, when dissolved in water in a biological cell or test tube, react with either H^+ or OH^- to shift the equilibrium of Equation 2.1 and result in a pH change of the solution.

Biochemical processes occurring in cells and tissues depend on strict regulation of the hydrogen ion concentration. Biological pH is maintained at a constant value by natural buffers. When biological processes are studied *in vitro,* artificial media must be prepared that mimic the cell's natural environment. Because of the dependence of biochemical reactions on pH, the accurate determination of hydrogen ion concentration has always been of major interest. Today, we consider the measurement and control of pH to be a simple and rather mundane activity. However, an inaccurate pH measurement or a poor choice of buffer can lead to failure in the biochemistry laboratory. You should become familiar with several aspects of pH measurement, electrodes, and buffers.

Measurement of pH

A pH measurement is usually taken by immersing a glass combination electrode into a solution and reading the pH directly from a meter. At one time, pH measurements required two electrodes, a pH-dependent glass electrode sensitive to H^+ ions and a pH-independent calomel reference electrode (Figure 2.1A,C). The potential difference that develops between the two electrodes is measured as a voltage as defined by Equation 2.2.

$$\blacktriangleright \quad V = E_{\text{constant}} + \frac{2.303RT}{F}\text{pH} \qquad \text{(Equation 2.2)}$$

where

$\qquad V =$ voltage of the completed circuit

$\qquad E_{\text{constant}} =$ potential of reference electrode

$\qquad R =$ the gas constant

$\qquad T =$ the absolute temperature

$\qquad F =$ the Faraday constant

A pH meter is standardized with buffer solutions of known pH before a measurement of an unknown solution is taken. It should be noted from Equation 2.2 that the voltage depends on temperature. Hence, pH meters must have some means for temperature correction. Older instruments usually have a knob labeled "temperature control," which is adjusted by the user to the temperature of the measured solution. Newer pH meters automatically display a temperature-corrected pH value.

Most pH measurements today are obtained using a single **combination electrode** (Figure 2.1B). Both the reference and the pH-dependent electrode are contained in a single glass tube. Although these are more expensive than dual electrodes, they are much more convenient to use, especially for smaller volumes of solution. Using a pH meter with a combination electrode is relatively easy, but certain guidelines must be followed. A pH meter not in use is left in a "standby" position. Before use, check the level of saturated KCl in the electrode (Figure 2.1B). If it is low, check with your instructor for the filling procedure. Turn the temperature control, if available, to the temperature of the standard calibration buffers and the test solutions. Be sure the function dial is set to pH. Lift the electrode out of the storage solution, rinse it with distilled water from a wash bottle, and *gently* clean and dry the electrode with a tissue. Im-

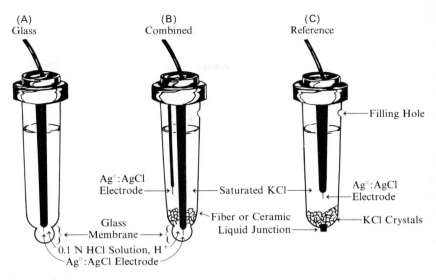

Figure 2.1
pH electrodes.   (A) pH-dependent glass electrode.   (B) Combination electrode.
(C) Calomel reference electrode. From T. G. Cooper, *The Tools of Biochemistry.*
Copyright 1977, Wiley-Interscience.

merse the electrode in a standard buffer. Common standard buffers are
pH 4, 7, and 10 with accuracy of ±0.02 pH unit. The standard buffer
should have a pH within two pH units of the expected pH of the test solution. The bulb of the electrode must be completely covered with solution.
Turn the pH meter to "on" or "read" and adjust the meter with the
"calibration dial" (sometimes called "intercept") until the proper pH of
the standard buffer is indicated on the dial. Turn the pH meter to standby
position. Remove the electrode and again rinse with distilled water and
carefully blot dry with tissue. Immerse the electrode in a standard buffer
of different pH and turn the pH meter to "read." The dial should read
within ±0.05 pH unit of the known value. If it does not, adjust to the
proper pH and again check the first standard pH buffer. Clean the electrode and immerse it in the test solution. Record the pH of the test solution.

As with all delicate equipment, the pH meter and electrode must receive proper care and maintenance. All glass electrodes should be stored
in 3 *M* KCl and kept clean. Normally, rinsing with distilled water and
careful blotting with tissue paper is sufficient. Glass electrodes are fragile
and expensive, so they must be handled with care. If pH measurements
of protein solutions are often taken, a protein film may develop on the
electrode; it can be removed by soaking in 5% pepsin in 0.1 *N* HCl for 2
hours and rinsing well with water.

Measurements of pH are always susceptible to experimental errors.
Some common problems are:

1. The Sodium Error   Most glass combination electrodes are sensitive to Na^+ as well as H^+. The sodium error can become quite significant at high pH values, where $0.1 \ N \ Na^+$ may decrease the measured pH by 0.4 to 0.5 unit. Several things may be done to reduce the sodium error. Some commercial suppliers of electrodes provide a standard curve for sodium error correction. Newer glass electrodes that are virtually Na^+ impermeable are now commercially available. If neither a standard curve nor a special glass electrode is available, potassium salts may be substituted for sodium salts.

2. Concentration Effects   The pH of a solution varies with the concentration of buffer ions or other salts in the solution. This is because the pH of a solution depends on the *activity* of an ionic species, not on the concentration. **Activity**, you may recall, is a thermodynamic term used to define species in a nonideal solution. At infinite dilution, the activity of a species is equivalent to its concentration. At finite dilutions, however, the activity of a solute and its concentration are not equal.

It is common practice in biochemical laboratories to prepare concentrated "stock" solutions and buffers. These are then diluted to the proper concentration when needed. Because of the concentration effects described above, it is important to adjust the pH of these solutions *after* dilution.

3. Temperature Effects   The pH of a buffer solution is influenced by temperature. This effect is due to a temperature-dependent change of the dissociation constant (pK_a) of ions in solution. The pH of the commonly used buffer Tris is greatly affected by temperature changes, with a $\Delta pK_a/C°$ of -0.031. This means that a pH 7.0 Tris buffer made up at 4°C would have a pH of 5.95 at 37°C. The best way to avoid this problem is to prepare the buffer solution at the temperature at which it will be used and to standardize the electrode with buffers at the same temperature as the solution you wish to measure.

Biochemical Buffers

Buffer ions are used to maintain solutions at constant pH values. The selection of a buffer for use in the investigation of a biochemical process is of critical importance. Before the characteristics of a buffer system are discussed, we will review some concepts in acid-base chemistry.

Weak acids and bases do not completely dissociate in solution but exist as equilibrium mixtures (Equation 2.3).

$$\blacktriangleright \quad HA \; \underset{k_2}{\overset{k_1}{\rightleftharpoons}} \; H^+ + A^- \qquad\qquad \text{(Equation 2.3)}$$

HA represents a weak acid and A^- represents its conjugate base; k_1 represents the rate constant for dissociation of the acid and k_2 the rate constant for association of the conjugate base and hydrogen ion. The equilibrium constant, K_a, for the weak acid HA is defined by Equation 2.4.

▶ $\quad K_a = \dfrac{k_1}{k_2} = \dfrac{[H^+][A^-]}{[HA]}$ (Equation 2.4)

which can be rearranged to define $[H^+]$ (Equation 2.5).

▶ $\quad [H^+] = \dfrac{K_a[HA]}{[A^-]}$ (Equation 2.5)

The $[H^+]$ is often reported as pH, which is $-\log[H^+]$. In a similar fashion, $-\log K_a$ is represented by pK_a. Equation 2.5 can be converted to the $-\log$ form by substituting pH and pK_a:

▶ $\quad pH = pK_a + \log\dfrac{[A^-]}{[HA]}$ (Equation 2.6)

Equation 2.6 is the familiar Henderson-Hasselbalch equation, which defines the relationship between pH and the ratio of acid and conjugate base concentrations. The Henderson-Hasselbalch equation is of great value in buffer chemistry because it can be used to calculate the pH of a solution if the molar ratio of buffer ions ($[A^-]/[HA]$) and the pK_a of HA are known. Also, the molar ratio of HA to A^- that is necessary to prepare a buffer solution at a specific pH can be calculated if the pK_a is known.

A solution containing both HA and A^- has the capacity to resist changes in pH; i.e., it acts as a **buffer**. If acid (H^+) were added to the buffer solution, it would be neutralized by A^- in solution:

▶ $\quad H^+ + A^- \longrightarrow HA$ (Equation 2.7)

Base (OH^-) added to the buffer solution would be neutralized by reaction with HA:

▶ $\quad OH^- + HA \longrightarrow A^- + H_2O$ (Equation 2.8)

The most effective buffering system contains equal concentrations of the acid, HA, and the conjugate base, A^-. According to the Henderson-Hasselbalch equation (2.6), when $[A^-]$ is equal to $[HA]$, pH equals pK_a. Therefore, the pK_a of a weak acid-base system represents the center of the buffering region. The effective range of a buffer system is generally two pH units, centered at the pK_a value (Equation 2.9).

▶ $\quad \begin{array}{c}\text{Effective pH range}\\ \text{for a buffer}\end{array} = pK_a \pm 1$ (Equation 2.9)

Selection of a Biochemical Buffer

Virtually all biochemical investigations must be carried out in buffered aqueous solutions. The natural environment of biomolecules and cellular organelles is under strict pH control. When these components are extracted from cells, they are most stable if maintained in their normal pH range of 6 to 8. An artificial buffer system is found to be the best substitute for the natural cell milieu. It should also be recognized that many biochemical processes (especially some enzyme processes) produce or consume hydrogen ions. The buffer system neutralizes these solutions and maintains a constant chemical environment.

Although most biochemical solutions require buffer systems effective in the pH range 6 to 8, there is occasionally a need for buffering over the pH range 2 to 12. Obviously, no single acid–conjugate base pair will be effective over this entire range, but several buffer systems are available that may be used in discrete pH ranges. Figure 2.2 compares the effective buffering ranges of common biological buffers. It should be noted that some buffers (phosphate, succinate, and citrate) have more than one pK_a value, so they may be used in different pH regions. Many buffer sys-

Figure 2.2
Effective buffering ranges of several common buffers.

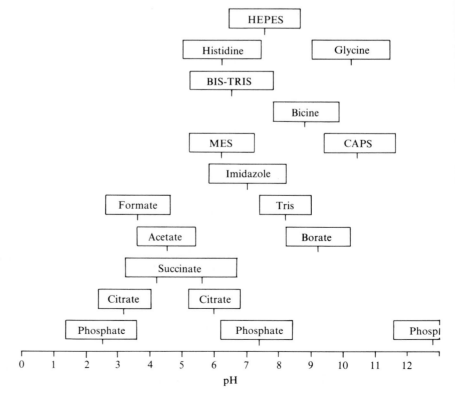

tems are effective in the usual biological pH range (6 to 8); however, there may be major problems in their use. Several characteristics of a buffer must be considered before a final selection is made.

The molecular weights and pK values of several common buffer compounds are listed in Appendix II. Following is a discussion of the advantages and disadvantages of the commonly used buffers (Gueffroy, 1975).

Phosphate Buffers

The phosphates are among the most widely used buffers. These solutions have high buffering capacity and are very useful in the pH range 6.5 to 7.5. Because phosphate is a natural constituent of cells and biological fluids, its presence affords a more "natural" environment than many buffers. Sodium or potassium phosphate solutions of all concentrations are easy to prepare. The major disadvantages of phosphate solutions are (1) precipitation or binding of common biological cations (Ca^{2+} and Mg^{2+}) and (2) inhibition of some biological processes, including some enzymes.

Tris Buffer

The use of the synthetic buffer Tris [tris(hydroxymethyl)aminomethane] is now probably greater than that of phosphate. It is useful in the pH range 7.5 to 8.5. Tris is available in a basic form as highly purified crystals, which makes buffer preparation especially convenient. To prepare solutions, the appropriate amount of Tris base is weighed and dissolved in water. For 1 L of a 0.1 M solution, 12.11 g (0.1 mole) of Tris base is weighed and dissolved in 950 to 975 mL of distilled water. The pH is adjusted by addition of acid (concentrated hydrochloric if Tris-HCl is desired), with stirring, until the appropriate pH is attained. Water is added to a final volume of 1 L and a final pH check is made. Although Tris is a primary amine, it causes minimal interference with biochemical processes and does not precipitate calcium ions. However, Tris has several disadvantages, including (1) pH dependence on concentration, since the pH decreases 0.1 pH unit for each 10-fold dilution; (2) interference with some pH electrodes; and (3) a large $\Delta pK_a/C°$ compared to most other buffers. Most of these drawbacks can be minimized by (1) adjusting the pH *after* dilution to the appropriate concentration, (2) purchasing electrodes that are compatible with Tris, and (3) preparing the buffer at the temperature at which it will be used.

Carboxylic Acid Buffers

The most widely used buffers in this category are acetate, formate, citrate, and succinate. This group is useful in the pH range 3 to 6, a region that offers few other buffer choices. All of these acids are natural metabolites, so they may interfere with the biological processes under investigation. Also, citrate and succinate may interfere by binding metal ions (Fe^{3+}, Zn^{2+}, Mg^{2+}, etc.). Formate buffers are especially useful because they are volatile and can be removed by evaporation under reduced pressure.

Borate Buffers

Buffers of boric acid are useful in the pH range 8.5 to 10. Borate has the major disadvantage of complex formation with many metabolites, especially carbohydrates.

Amino Acid Buffers

The most commonly used amino acid buffers are glycine (pH 2 to 3 and 9.5 to 10.5), histidine (pH 5.5 to 6.5), glycine amide (pH 7.8 to 8.8), and glycylglycine (pH 8 to 9). These provide a more "natural" environment to cellular components and extracts; however, they may interfere with some biological processes, as do the carboxylic acid and phosphate buffers.

Zwitterionic Buffers (Good's Buffers)

In the mid-1960s, N. E. Good and his colleagues recognized the need for a set of buffers specifically designed for biochemical studies (Good and Izawa, 1972). He and others noted major disadvantages of the established buffer systems. Good outlined several characteristics essential in a biological buffer system:

1. pK_a between 6 and 8.
2. Highly soluble in aqueous systems.
3. Exclusion or minimal transport by biological membranes.
4. Minimal salt effects.
5. Minimal effects on dissociation due to ionic composition, concentration, and temperature.
6. Buffer–metal ion complexes nonexistent or soluble and well defined.
7. Chemically stable.
8. Insignificant light absorption in the ultraviolet and visible regions.
9. Readily available in purified form.

Good investigated a large number of synthetic zwitterionic buffers and found many of them to meet these criteria. Table 2.1 lists several of these buffers and their properties. Good's buffers are widely used, but their main disadvantage is high cost. Some Good's buffers, such as Tris,

Table 2.1

Structures and Properties of Several Synthetic Zwitterionic Buffers

Buffer Name	Abbreviation	Structure (All structures shown in salt form)	pK_a (20°C)	Useful pH Range	$\Delta pK_a/C°$	Concentration of a Saturated Solution (M, 0°C)
N-2-Acetamido-2-amino-ethanesulfonic acid	ACES	$H_2NCOCH_2\overset{+}{N}H_2CH_2CH_2SO_3^-$	6.9	6.4–7.4	–0.020	0.22
N-2-Acetamidoiminodiacetic acid	ADA	$H_2NCOCH_2\overset{+}{N}H(CH_2COO^-)CH_2COO^-$	6.6	6.2–7.2	–0.011	—
N,N-Bis(2-hydroxyethyl)-2-aminoethanesulfonic acid	BES	$(HOCH_2CH_2)_2\overset{+}{N}HCH_2CH_2SO_3^-$	7.15	6.6–7.6	–0.016	3.2
N,N-Bis(2-hydroxyethyl)-glycine	Bicine	$(HOCH_2CH_2)_2\overset{+}{N}HCH_2COO^-$	8.35	7.5–9.0	–0.018	1.1
3-(Cyclohexylamino)-propanesulfonic acid	CAPS	 cyclohexyl–$NHCH_2CH_2CH_2SO_3^-$	10.4	10.0–11.0	–0.009	0.85
Cyclohexylaminoethane-sulfonic acid	CHES	 cyclohexyl–$\overset{+}{N}HCH_2CH_2SO_3^-$	9.5	9.0–10.0	–0.009	0.85
Glycylglycine	Gly-Gly	$H_3\overset{+}{N}CH_2CONHCH_2COO^-$	8.4	7.5–9.5	–0.028	1.1
N-2-Hydroxyethyl-piperazine-N'-2-ethanesulfonic acid	HEPES	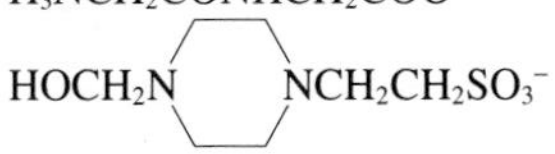 $HOCH_2N$ (piperazine ring) $\overset{+}{N}CH_2CH_2SO_3^-$	7.55	7.0–8.0	–0.014	2.25
2-(N-Morpholino)ethane-sulfonic acid	MES	 morpholine–$\overset{+}{N}HCH_2CH_2SO_3^-$	6.15	5.8–6.5	–0.011	0.65
3-(N-Morpholino)-propanesulfonic acid	MOPS	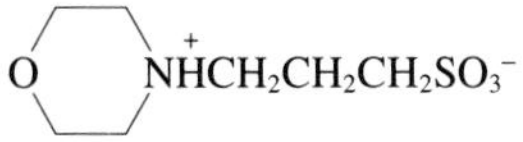 morpholine–$\overset{+}{N}HCH_2CH_2CH_2SO_3^-$	7.20	6.5–7.9	—	—
Piperazine-N,N'-bis-2-ethanesulfonic acid	PIPES	 $^-O_3SCH_2CH_2\overset{+}{N}$ (piperazine ring) $\overset{+}{N}HCH_2CH_2SO_3^-$	6.8	6.4–7.2	–0.0085	—
N-Tris(hydroxymethyl)-methyl-2-aminoethane-sulfonic acid	TES	$(HOCH_2)_3\overset{+}{N}HCH_2CH_2SO_3^-$	7.5	7.0–8.0	–0.020	2.6
N-Tris(hydroxymethyl)-methylglycine	Tricine	$(HOCH_2)_3C\overset{+}{N}H_2CH_2COO^-$	8.15	7.5–8.5	–0.021	0.8
Tris(hydroxymethyl)-aminomethane	Tris	$(HOCH_2)_3C\overset{+}{N}H_3$	8.3	7.5–9.0	–0.031	2.4

HEPES, and PIPES, have been shown to produce free radicals under a variety of experimental conditions, so they should be avoided if biological redox processes or free radical–based reactions are being studied. Radicals are not produced from MES or MOPS buffers (Grady et al., 1988).

The Oxygen Electrode

Second in popularity only to the pH electrode is the oxygen electrode. This device is a polarographic electrode system that can be used to measure oxygen concentration in a liquid sample or to monitor oxygen uptake or evolution by a chemical or biological system. The only known interfering materials are H_2S and SO_2.

The basic electrode was developed by L. C. Clark, Jr. (Clark et al., 1953). The popular Clark-type oxygen electrode with biological oxygen monitor is shown in Figure 2.3. The electrode system, which consists of a platinum cathode and silver anodes, is molded into an epoxy block and surrounded by a Lucite holder. A thin Teflon membrane is stretched over the electrode end of the probe and held in place with an O-ring. The membrane separates the electrode elements, which are bathed in an electrolyte solution (saturated KCl), from the test solution in the sample chamber. The membrane is permeable to oxygen and other gases, which then come into contact with the surface of the platinum cathode. If a suitable polarizing voltage (about 0.8 volt) is applied across the cell, oxygen is reduced at the cathode.

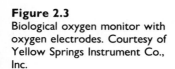

Figure 2.3
Biological oxygen monitor with oxygen electrodes. Courtesy of Yellow Springs Instrument Co., Inc.

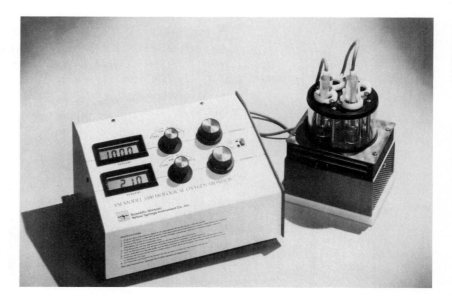

$$O_2 + 2e^- + 2H_2O \longrightarrow [H_2O_2] + 2OH^-$$

$$[H_2O_2] + 2e^- \longrightarrow 2OH^-$$

Total $\quad O_2 + 4e^- + 2H_2O \longrightarrow 4OH^-$

The reaction occurring at the anode is

$$4Ag^\circ + 4Cl^- \longrightarrow 4AgCl + 4e^-$$

The overall electrochemical process is

$$4Ag^\circ + O_2 + 4Cl^- + 2H_2O \longrightarrow 4AgCl + 4OH^-$$

The current flowing through the electrode system is, therefore, directly proportional to the amount of oxygen passing through the membrane. According to the laws of diffusion, the rate of oxygen flowing through the membrane depends on the concentration of dissolved oxygen in the sample. Therefore, the magnitude of the current flow in the electrode is directly related to the concentration of dissolved oxygen. The function of the electrode can also be explained in terms of oxygen pressure. The oxygen concentration at the cathode (inside the membrane) is virtually zero because oxygen is rapidly reduced. Oxygen pressure in the sample chamber (outside the membrane) will force oxygen to diffuse through the membrane at a rate that is directly proportional to the pressure (or concentration) of oxygen in the test solution. The current generated by the electrode system is electronically amplified and transmitted to a meter, which usually reads in units of % oxygen saturation.

Many biochemical processes consume or evolve oxygen. The oxygen electrode is especially useful for monitoring such changes in the concentration of dissolved oxygen. If oxygen is consumed, for example, by a suspension of mitochondria or by an oxygenase enzyme system, the rate of oxygen diffusion through the probe membrane will decrease, leading to less current flow. A linear relationship exists between the electrode current and the amount of oxygen consumed by the biological system. Measurements of these changes in O_2 concentration are most conveniently made with a biological oxygen monitor equipped with a recorder.

The use of an oxygen electrode is not free of experimental and procedural problems. Probably the most significant source of trouble is the Teflon membrane. A torn, damaged, or dirty membrane is easily recognized by recorder noise or electronic "spiking." The membrane must not be touched with the fingers, contaminated with chemicals, or creased when it is stretched over the electrode. The membrane can also become coated with chemical reagents and biomolecules (especially proteins) present in the sample chamber. It is good practice to inspect the membrane routinely for damage and contamination with a low-power microscope. The silver anode should be cleaned frequently with dilute ammonium hydroxide in order to remove silver sulfide and maintain a bright surface.

Temperature control of the sample chamber and electrode is especially important because the rate of oxygen diffusion through the membrane depends on temperature. Commercial sample chambers are equipped with a bath assembly for temperature control. The bath assembly also contains a magnetic stirrer. It is essential that the solution in the sample chamber be stirred constantly to prevent oxygen depletion at the cathode.

Air bubbles in the sample chamber or in the KCl electrolyte solution inside the membrane lead to considerable error in measurements. When an air bubble (which may contain up to 20 times more oxygen than air-saturated water) is in the vicinity of the probe, erratic readings result from this rapid increase in oxygen concentration. If bubbles are present in the electrolyte solution inside the membrane, the membrane must be removed and carefully replaced to avoid bubbles. Air bubbles are easily removed from the sample chamber because the tightly fitting Lucite plunger is slanted and contains a small groove or access slot. When the plunger is slowly pushed into the sample, air bubbles can be directed out this small groove. Reactants can also be added to the sample chamber through the access slot without removing the probe.

Ion-Selective Electrodes and Biosensors

Although electrodes that respond to H^+ and O_2 are the most widely used, other ions and gases may be measured with specific electrodes and potentiometric methods. Table 2.2 lists some of the electrodes that are valuable in biochemical measurements. Each type of electrode is made from a specially prepared glass that has increased permeability toward a single type of ion. Ion-specific electrodes are sensitive (some are useful in the parts-per-billion range), require simple, inexpensive equipment (may be hooked to a pH meter set in the millivolt mode), and save time (after electrode calibration, an analysis takes about 1 minute). A major difficulty that has slowed the adoption of ion-selective electrodes is interference by other ions. In most cases, the serious interfering ions are known (see Table 2.2), and procedures are available for masking or eliminating some competing ions.

Electrodes or electrode-like devices are now being developed for the specific measurement of physiologically important molecules such as urea, carbohydrates, enzymes, antibodies, and metabolic products. This type of device, now referred to as a **biosensor**, is defined as "an analytical tool or system consisting of an immobilized biological material (such as an enzyme, antibody, whole cell, organelle or combinations thereof) in intimate contact with a suitable transducer device which will convert the biochemical signal into a quantifiable electric signal" (Gronow, 1984). The important components of a biosensor as shown in Figure 2.4

Table 2.2
Ion-Specific Electrodes

Ions	Concentration Range (M)	Interferences[1]
Ammonium	10^0 to 10^{-6}	Volatile amines
Bromide	10^0 to 5×10^{-6}	S^{2-}, I^-
Cadmium	10^0 to 10^{-7}	Ag^+, Hg^{2+}, Cu^{2+}, Pd^{2+}, Fe^{3+}
Carbon dioxide	10^{-2} to 10^{-4}	Sr^{2+}, Mg^{2+}, Ba^{2+}, Ni^{2+}, volatile weak acids
Chloride	10^0 to 8×10^{-6}	ClO_4^-, I^-, NO_3^-, SO_4^{2-}, Br^-, OH^-, OAc^-, HCO_3^-, F^-
Cupric	Saturated to 10^{-8}	S^{2-}, Hg^{2+}, Ag^+
Cyanide	10^{-2} to 10^{-6}	S^{2-}, I^-, Br^-, Cl^-
Divalent cations	10^0 to 6×10^{-6}	Na^+, Cu^{2+}, Zn^{2+}, Fe^{2+}, Ni^{2+}, Sr^{2+}, Bi^{2+}
Fluoride	Saturated to 10^{-6}	OH^-
Iodide	10^0 to 2×10^{-5}	S^{2-}
Lead	10^0 to 10^{-7}	Ag^+, Hg^{2+}, Cu^{2+}, Cd^{2+}, Fe^{3+}
Nitrate	10^0 to 6×10^{-6}	I^-, Br^-, NO_2^-
Nitrite	10^{-2} to 5×10^{-7}	CO_2
Potassium	10^0 to 10^{-5}	Cs^+, NH_4^+, H^+
Silver/sulfide	10^0 to $\times 10^{-7}$	Hg^{2+}
Sodium	Saturated to 10^{-6}	Cs^+, Li^+, K^+
Thiocyanate	10^0 to 5×10^{-6}	OH^-, Br^-, Cl^-

[1]The presence of these ions does not rule out the use of a specific-ion electrode, since many ions interfere only at relatively high concentrations.

Figure 2.4
Components of a biosensor system for determining the identity and concentration of biomolecule A. A is stoichiometrically converted to B by a specific enzyme in the reaction center. The concentration of B is measured by the sensor.

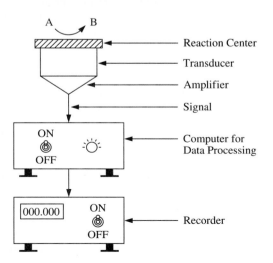

are (1) a reaction center consisting of a membrane or gel containing the biochemical system to be studied, (2) a transducer, (3) an amplifier, and (4) a system for data acquisition and processing. When biomolecules in the reaction center interact, a physicochemical change occurs. This change in the molecular system, which may be a modification of concentration, absorbance, mass, conductance, or redox state, is converted into an electrical signal by the transducer. The signal is then amplified and displayed on a recorder or by digital readout. Each biosensor is specifically designed for a type of molecule or biological interaction; therefore, details of construction and function vary.

B. Dialysis and Ultrafiltration

One of the oldest procedures applied to the purification and characterization of biomolecules is **dialysis**, an operation used to separate dissolved molecules on the basis of their molecular size (McPhie, 1971). The technique involves sealing an aqueous solution containing both macromolecules and small molecules in a porous membrane. The sealed membrane is placed in a large container of low-ionic-strength buffer (Figure 2.5). The membrane pores are too small to allow diffusion of macromolecules of molecular weight greater than about 10,000. Smaller molecules diffuse freely through the openings (Figure 2.6). The passage of small molecules continues until their concentrations inside the dialysis tubing and outside in the large volume of buffer are equal. Thus, the concentration of small molecules inside the membrane is reduced. Equilibrium is reached after 4 to 6 hours. Of course, if the outside solution (dialysate) is replaced with fresh buffer after equilibrium is reached, the concentration of small molecules inside the membrane will be further reduced by continued dialysis.

Dialysis membranes are available in a variety of materials and sizes. The most common materials are collodion, cellophane, and cellulose.

Figure 2.5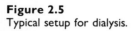
Typical setup for dialysis.

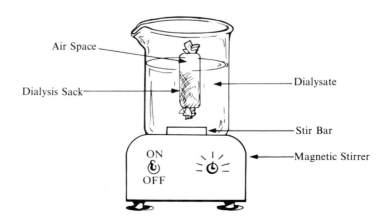

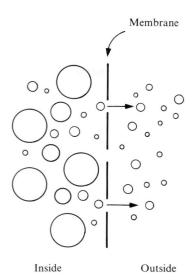

Figure 2.6
Diffusion of smaller molecules through dialysis membrane pores.

Recent modifications in membrane construction make a range of pore sizes available. Spectrum Medical Industries sells Spectropor cellulose tubings with complete molecular weight cutoff ranging from 1000 to 50,000.

Applications of Dialysis

Dialysis is most commonly used to remove salts and other small molecules from solutions of macromolecules. During the separation and purification of biomolecules, small molecules are added to selectively precipitate or dissolve the desired molecule. For example, proteins are often precipitated by addition of organic solvents or salts such as ammonium or sodium sulfate. Since the presence of organics or salts usually interferes with further purification and characterization of the molecule, they must be removed. Dialysis is a simple, inexpensive, and effective method for removing all small molecules, ionic or nonionic.

Dialysis is also useful for removing small ions and molecules that are weakly bound to biomolecules. Protein cofactors such as NAD, FAD, and metal ions can often be dissociated by dialysis. The removal of metal ions is facilitated by the addition of a chelating agent (EDTA) to the dialysate.

Several precautions should be observed during dialysis setup. Commercial dialysis tubing is usually contaminated with glycerol, heavy-metal ions, and traces of sulfur-containing compounds and occasionally contaminated with hydrolytic enzymes (proteinases and nucleases), which must be removed before use. Typical preparation begins with cutting the tubing into convenient lengths (6 to 8 inches), soaking it in 1% acetic acid for 1 hour, and then transferring it to glass-distilled water. To remove metal ions, the tubing is then boiled in basic EDTA solution (1% sodium carbonate, 10^{-3} M EDTA) for 1 hour. This step is repeated with a fresh portion of basic EDTA. The EDTA solution is poured off and heating is repeated about five times with distilled water. Recommended storage is in distilled water at 4°C with a small amount of sodium azide or a few drops of chloroform added as a bacterial-growth inhibitor. The treated tubing should not be handled with bare fingers, since it may become contaminated by hydrolytic enzymes. Because biomolecules are usually less stable at elevated temperatures, dialysis is done in a refrigerated chamber at 3 to 4°C. The dialysate containing the sealed tube should be stirred very gently so that the small molecules are evenly dispersed throughout the solution and not concentrated in the vicinity of the dialysis bag. It is good practice to leave a small air bubble (about 10% of the total volume) in the dialysis sack when tying the final end. This ensures that the sack will float and not be damaged by the rotating stir bar.

Applications of Ultrafiltration

Although dialysis is still used occasionally as a purification tool, it has been largely replaced by gel filtration (see Chapter 3) and ultrafiltration techniques. The major disadvantage of dialysis that is overcome by the newer methods is that it may take several days of dialysis to attain a suitable separation. The other methods require 1 to 2 hours or less.

Ultrafiltration involves the separation of molecular species on the basis of size, shape, and/or charge. The solution to be separated is forced through a membrane under the influence of high pressure or centrifugal force. Membranes may be chosen for optimum flow rate, molecular specificity, and molecular weight cutoff. Two applications of membrane filtration are obvious: (1) desalting buffers or other solutions and (2) clarification of turbid solutions by removal of micron- or submicron-sized particles. Other applications are discussed below.

Membrane filters are divided into two major classes, *depth* and *screen*. **Depth filters**, which may be composed of paper, cotton, or fiberglass, function by trapping particles primarily within the "depths" of the filter matrix. The interior of these filters, as shown in Figure 2.7, is a random arrangement of fiber material forming tiny channels. Particles that are larger than the passages are retained in the filter by entrapment in the matrix. Since they are thick, depth filters have a high load capacity, retaining particles both on the surface and within the matrix. In addition, they have relatively high flow rates, are inert to most solvents, and are inexpensive. However, they have several disadvantages, including (1) ill-defined and variable pore sizes due to a random matrix, (2) extensive absorption and loss of liquid filtrate, and (3) loss of filter fragments that contaminate the filtrate.

Many of these disadvantages are overcome by **screen filters**, which have uniform pore size. The screen filters function by retaining particles

Figure 2.7
Electron micrograph of a microporous membrane. Particles greater than 0.1 μm in diameter are trapped on its surface or within pores. Courtesy of Amicon Corporation.

on their surfaces rather than within the matrix (Figure 2.8). The most widely used screen-type filters are composed of polycarbonate and cellulose esters (cellulose nitrate and acetate). Membrane filters of these materials can be manufactured with a predetermined and accurately controlled pore size. Filters are available with a mean pore size ranging from 0.025 to 15 μm. These filters clog more readily than do depth filters and require suction, pressure, or centrifugal force for liquid flow. A typical flow rate for the commonly used 0.45-μm membrane is 57 mL min^{-1} cm^{-2} at 10 psi. Clogging can be reduced by combining depth and screen filters. The depth filter serves as a "prefilter" to remove particles that would rapidly clog the screen filter.

Ultrafiltration devices are available for macroseparations (up to 50 L) or for microseparations (milli- to microliters). For solutions larger than a few milliliters, gas-pressurized cells or suction-filter devices are used. For concentration and purification of samples in the milli- to microliter range, disposable filters are available. These devices, often called microconcentrators, offer the user simplicity, time saving, and high recovery. The sample is placed in a reservoir above the membrane and centrifuged in a fixed-angle rotor. The time and centrifugal force required depend on

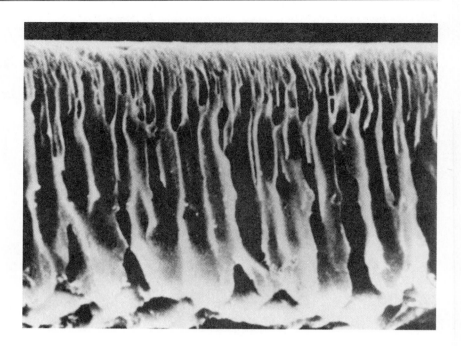

Figure 2.8
Electron micrograph of an ultrafiltration membrane composed of a thin dense skin and larger voids. Any species capable of passing through the skin can freely pass through the membrane. Courtesy of Amicon Corporation.

the membrane, with spin times varying from 30 minutes to 2 hours and forces from $1000 \times g$ to $7500 \times g$. Figure 2.9 outlines the use of a centrifuge microfilter. These are available from a variety of sources, including Schleicher and Schuell, Bio-Rad, Pierce, Amicon, and Millipore.

The principles behind ultrafiltration are sometimes misunderstood. The nomenclature implies that separations are the result of physical trapping of the particles and molecules by the filter. With polycarbonate and fiberglass filters, separations are made primarily on the basis of physical size. Other filters (cellulose nitrate, polyvinylidene fluoride, and to a lesser extent cellulose acetate) trap particles that cannot pass through the pores, but also retain macromolecules by adsorption. In particular, these materials have protein and nucleic acid binding properties. Each type of membrane displays a different affinity for various molecules. For protein, the relative binding affinity is polyvinylidene fluoride > cellulose nitrate > cellulose acetate. We can expect to see many applications of the "affinity membranes" in the future as the various membrane surface chemistries are altered and made more specific. Obvious applications are purification of macromolecules and quantitative binding assays. Others are described on pages 48–49.

CLARIFICATION OF SOLUTIONS

Because of low solubility and denaturation, solutions of biomolecules or cellular extracts are often turbid. This is a particular disadvantage if spectrophotometric analysis is desired. The transmittance of turbid solutions can be greatly increased by passage through a membrane filter system. A typical filter setup is shown in Figure 2.10. Since the pore sizes

Figure 2.9
Use of a centrifuge microfilter. Courtesy of Amicon Corporation.

1. Pipette up to 2 ml of sample into the top reservoir.

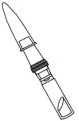

2. Cover with the cone-shaped cap and spin in a centrifuge with a fixed-angle rotor.

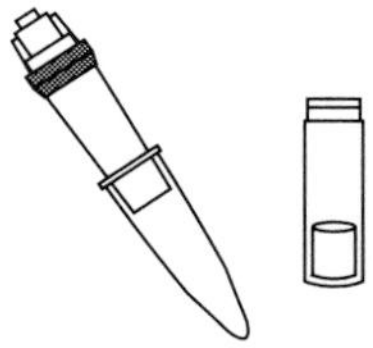

3. Detach and invert top half.

4. Centrifuge 1 to 2 minutes to spin the concentrate into the cone-shaped cap.

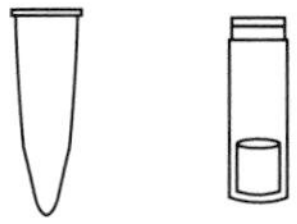

5. Concentrate is now completely accessible for further analysis. Both concentrate and filtrate can be stored in collection cups.

Figure 2.10
Assembly for membrane filtration. Courtesy of Millipore Corporation.

are so small, liquids will not enter the filter materials by gravity; suction (water aspirator) or pressure must be applied.

This simple technique may also be applied to the sterilization of nonautoclavable materials such as protein and nucleic acid solutions or heat-labile reagents. Bacterial contamination can be removed from these solutions by passing them through filter systems that have been sterilized by autoclaving. Most membrane filters can be autoclaved, but it is a good idea to read the manufacturer's literature before doing so.

COLLECTION OF PRECIPITATES FOR ANALYSIS

The collection of small amounts of very fine precipitates is the basis for many chemical and biochemical analytical procedures. Membrane filtration is an ideal method for sample collection. This is of great advantage in the collection of radioactive precipitates. Cellulose nitrate and fiberglass filters are often used to collect radioactive samples because they can be analyzed by direct suspension in an appropriate scintillation cocktail (see Chapter 6).

HARVESTING OF BACTERIAL CELLS FROM FERMENTATION BROTHS

The collection of bacterial cells from nutrient broths is typically done by batch centrifugation. This time-consuming operation can be replaced by membrane filtration. Filtration is faster than centrifugation, and it allows extensive cell washing.

CONCENTRATION OF BIOMOLECULE SOLUTIONS

Protein or nucleic acid solutions obtained from extraction or various purification steps are often too dilute for further investigation. Since they cannot be concentrated by high-temperature evaporation of solvent, gentler methods have been developed. One of the most effective is the use of ultrafiltration pressure cells as shown in Figure 2.11. A membrane filter is placed in the bottom and the solution is poured into the cell. High pressure, exerted by compressed nitrogen (air could cause oxidation and denaturation of biomolecules), forces the flow of small molecules, including solvent, through the filter. Membranes are available in a number of sizes allowing a large variety of molecular weight cutoffs. Larger molecules that cannot pass through the pores are, of course, concentrated in the sample chamber. This method of concentration is rapid and gentle and can be performed at cold temperatures to ensure minimal inactivation of the molecules. One major disadvantage is clogging of the pores, which reduces the flow rate through the filter; this can be lessened by constant but gentle stirring of the solution.

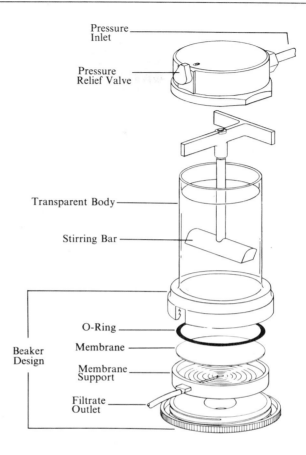

Pressure Inlet

Pressure Relief Valve

Transparent Body

Stirring Bar

Beaker Design

O-Ring

Membrane

Membrane Support

Filtrate Outlet

Figure 2.11
Schematic of an ultrafiltration cell. Courtesy of Amicon Corporation.

BINDING ASSAYS

The applications of membrane filters in binding assays are numerous. As described earlier in this section, the surface chemistry of membranes can be altered to allow selectivity in the collection of macromolecules. Membrane filters have been applied to drug-hormone binding assays, ligand-protein binding assays, RNA-DNA hybrid formation, and antigen-antibody interactions.

C. Lyophilization (Freeze-Drying)

Although ultrafiltration is being used more and more for the concentration of biological solutions, the older technique of lyophilization is still used. There are some situations (storing or transporting biological materials) in which lyophilization is preferred. The technique of freeze-drying involves the removal of solvent from a frozen sample. This is one of the most effective methods for drying or concentrating heat-sensitive materials. In practice, a biological solution to be concentrated is "shell-frozen" on the walls of a round-bottom or freeze-drying flask. Freezing of the biological material is accomplished by placing the flask (half full with sample) in a dry ice–acetone bath and slowly rotating it as it is held at a 45° angle. The biological material freezes in layers on the wall of the flask. This provides a large surface area for evaporation of water. The flask is then connected to the lyophilizer, which consists of a refrigeration unit and a vacuum pump (see Figure 2.12). The combined unit maintains the sample at −40°C for stability of the biological materials and applies a vacuum of approximately 5 to 25 mm on the sample. Ice formed from the aqueous solution sublimes and is pumped from the sample vial. In fact, all materials that are volatile under these conditions (−40°C, 5 to 25 mm) will be removed, and nonvolatile materials (proteins, buffer salts, nucleic acids, etc.) will be concentrated into a light, fluffy precipitate. Most freeze-dried biological materials are stable for long periods of time and some remain viable for many years.

As with any laboratory method, there are precautions and limitations of lyophilization that must be understood. Only aqueous solutions should be lyophilized. Organic solvents lower the melting point of aqueous solutions and increase the chances that the sample will melt and become denatured during freeze-drying. There is also the possibility that organic vapors will pass through the cold trap into the vacuum pump, where they may cause damage.

Freeze-drying of buffered solutions involves special problems. All nonvolatile materials in the aqueous solution, including inorganic or buffer salts, crystallize during the freeze-drying. Biological materials

Figure 2.12
Schematic showing the components of a freeze dryer. Courtesy of FTS Systems, Inc., Stone Ridge, NY 12484.

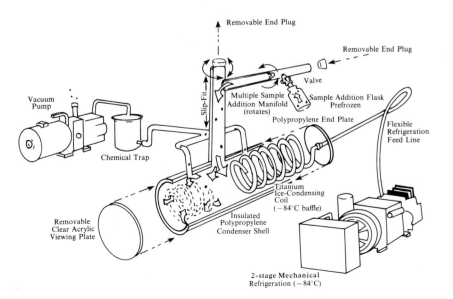

generally can tolerate high salt concentrations; however, freezing of a buffered solution may cause large changes in pH. When pH 7.0 phosphate buffer is frozen, disodium phosphate crystallizes before monosodium phosphate. The pH of the solution just before freezing is about 3.5. Problems associated with buffered solutions can be minimized by very rapid freezing of the solution before lyophilization.

Some biological samples, especially dilute protein solutions, form a very light and fluffy material upon lyophilization. Unless there is some barrier (a small piece of filter paper or glass wool) between the sample and the vacuum hookup, sample may be lost by suction.

Freeze dryers are now available in several models and styles. The most versatile is a compact bench-top unit that contains a refrigerator and has the capability to freeze-dry up to 12 samples. More details of the theory and techniques of lyophilization are available in Everse and Stolzenbach (1971).

D. Measurement of Protein Solutions

Biochemical research often requires the measurement of protein concentrations in solutions. Several techniques have been developed; however, most have limitations because either they are not sensitive enough or they are based on reactions with specific amino acids in the protein. Since the amino acid content varies from protein to protein, no single assay will be suitable for all proteins. In this section we discuss five assays: three older, classical methods that are occasionally used today and two newer methods that are widely used. In four of the methods, chemical reagents are added to protein solutions to develop a color whose intensity is measured in a spectrophotometer. A "standard protein" of known concentration is also treated with the same reagents and a calibration curve is constructed. The other assay relies on a direct spectrophotometric measurement. None of the methods is perfect because each is dependent on the amino acid content of the protein. However, each will provide a satisfactory result if the proper experimental conditions are used and/or a suitable standard protein is chosen. Other important factors in method selection include the sensitivity and accuracy desired, the presence of interfering substances, and the time available for the assay. The various methods are compared in Table 2.3.

The Biuret and Lowry Assays

When substances containing two or more peptide bonds react with the biuret reagent, alkaline copper sulfate, a purple complex is formed. The colored product is the result of coordination of peptide nitrogen atoms with Cu^{2+}. The amount of product formed depends on the concentration of protein.

Table 2.3

Methods for Protein Measurement

Method	Sensitivity	Time	Principle	Interferences	Comments
Biuret	Low 1–20 mg	Moderate, 20–30 min	Peptide bonds + alkaline $Cu^{2+} \rightarrow$ purple complex	Good's buffers, Tris, some amino acids	Fast but not sensitive. Similar color with all proteins.
Lowry	High $\sim$5 μg	Slow, $\sim$40–60 min	(1) Biuret reaction (2) Reduction of phosphomolybdate-phosphotungstate by Tyr and Trp	Ammonium sulfate, glycine, Good's buffers, mercaptans	Time-consuming. Color varies with proteins. Critical timing of procedure.
Bradford	High, $\sim$1 μg	Rapid, $\sim$15 min	λ_{max} of Coomassie dye shifts from 465 nm to 595 nm when protein-bound	Strongly basic buffers, Triton X-100, SDS	Excellent method, only slight interferences, stable color which varies with proteins. Reagents commercially available.
BCA	High, 1 μg	Slow, $\sim$60 min	(1) Biuret reaction (2) Copper complex with BCA; λ_{max} = 562 nm	EDTA, dithiothreitol, ammonium sulfate	Compatible with detergents. Reagents commercially available.
Spectrophotometric (Warburg-Christian)	Moderate 50–100 μg	Rapid, 5–10 min	Absorption of 280-nm light by Tyr and Trp residues	Purines, pyrimidines, nucleic acids	Useful for monitoring column eluents. Nucleic acid absorption can be corrected.

In practice, a calibration curve must be prepared by using a standard protein solution. The best protein to use as a standard is a purified preparation of the protein to be assayed. Since this is rarely available, the experimenter must choose another protein as a relative standard. A relative standard must be selected that provides a similar color yield. An aqueous solution of bovine serum albumin is an appropriate standard. Various known amounts of this solution are treated with the biuret reagent and the color is allowed to develop. Measurements of absorbance at 540 nm (A_{540}) are made against a blank containing biuret reagent and buffer or water. The A_{540} data are plotted versus protein concentration (mg/mL). Unknown protein samples are treated with biuret reagent and the A_{540} is measured after color development. The protein concentration is determined from the standard curve. Few substances interfere with the biuret test. There is abnormal color development with amino acid buffers, Good's buffers, and Tris buffer. The biuret assay has several advantages including speed, similar color development with different proteins, and few interfering substances. Its primary disadvantage is its lack of sensitivity. Its lower limit of sensitivity is about 1 mg.

The Lowry protein assay is one of the most sensitive and, hence, widely used. The Lowry procedure can detect protein levels as low as 5 μg. The principle behind color development is identical to that of the biuret assay except that a second reagent (Folin-Ciocalteu) is added to increase the amount of color development. Two reactions account for the intense blue color that develops: (1) the coordination of peptide bonds with alkaline copper (biuret reaction) and (2) the reduction of the Folin-Ciocalteu reagent (phosphomolybdate-phosphotungstate) by tyrosine and tryptophan residues in the protein.

The procedure is similar to that of the biuret assay. A standard curve is prepared with bovine serum albumin or other pure protein, and the concentration of unknown protein solutions is determined from the graph.

The obvious advantage of this assay is its sensitivity, which is up to 100 times greater than that of the biuret assay; however, more time is required for the Lowry assay. The Lowry assay, unfortunately, is greatly affected by the presence of interfering substances, and critical timing is necessary for addition of reagents. Good's buffers result in a high blank absorbance and may not be present in the assay. Other interfering substances are ammonium sulfate (concentration greater than 0.15%), glycine (greater than 0.5%), and mercaptans. Since proteins have varying contents of tyrosine and tryptophan, the amount of color development changes with different proteins, including the bovine serum albumin standard. Because of this, the Lowry protein assay should be used only for measuring changes in protein concentration, not absolute values of protein concentration. This does not decrease the value of the Lowry assay, because for many biochemical procedures (such as protein purification) relative measurements of protein concentration changes are suitable.

The Bradford Assay

The many limitations of the biuret and Lowry assays have encouraged researchers to seek better methods for quantitation of protein solutions. A procedure based on protein binding of a dye provides numerous advantages over other methods (Bradford, 1976; Bio-Rad, 1984).

The binding of Coomassie Brilliant Blue dye (Figure 2.13) to protein in acidic solution causes a shift in wavelength of maximum absorption (λ_{max}) of the dye from 465 nm to 595 nm. The absorption at 595 nm is directly related to the concentration of protein.

In practice, a calibration curve is prepared using bovine plasma gamma globulin or bovine serum albumin as a standard (Figure 2.14). The assay requires only a single reagent, an acidic solution of Coomassie Brilliant Blue G-250. After addition of dye solution to a protein sample, color development is complete in 2 minutes and the color remains stable for up to 1 hour. The sensitivity of the Bradford assay rivals and may surpass that of the Lowry assay. With a microassay procedure, the Bradford assay can be used to determine proteins in the range of 1 to 20 μg. The Bradford assay shows significant variation with different proteins, but this also occurs with the Lowry assay. The Bradford method is not only rapid but also has very few interferences by nonprotein components. The only known interfering substances are detergents, Triton X-100, and sodium dodecyl sulfate. The many advantages of the Bradford assay have led to its wide adoption in biochemical research laboratories.

The BCA Assay

The BCA protein assay is based on chemical principles similar to those of the biuret and Lowry assays (Smith et al., 1985). The protein to be analyzed is reacted with Cu^{2+} (which produces Cu^+) and bicinchoninic acid (BCA). The Cu^+ is chelated by BCA, which converts the apple-green color of the free BCA to the purple color of the copper-BCA complex. Samples of unknown protein and relative standard are treated with the reagent, the absorbance is read in a spectrophotometer at 562 nm, and concentrations are determined from a standard calibration curve. This as-

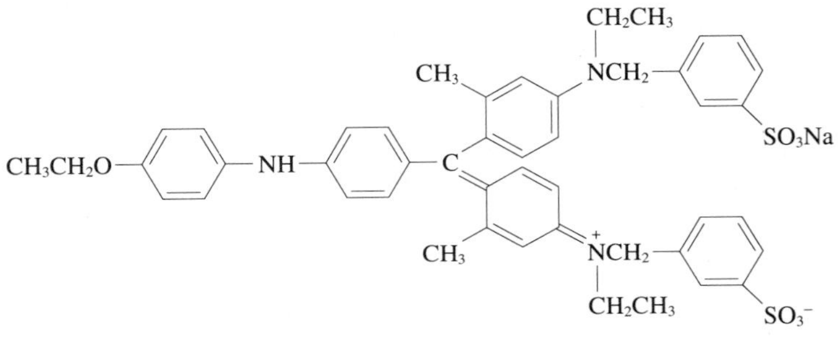

Figure 2.13
Coomassie Brilliant Blue G-250 dye.

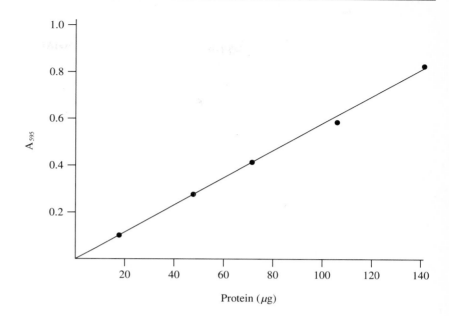

Figure 2.14
Standard calibration curve for the Bradford assay using bovine gamma globulin as standard protein.

say has the same sensitivity level as the Lowry and Bradford assays. Its main advantages are its simplicity and its usefulness in the presence of 1% detergents such as Triton or sodium dodecyl sulfate (SDS).

The Spectrophotometric Assay

Most proteins have relatively intense ultraviolet light absorption centered at 280 nm. This is due to the presence of tyrosine and tryptophan residues in the protein. However, the amount of these amino acid residues varies in different proteins, as was pointed out earlier. If certain precautions are taken, the A_{280} value of a protein solution is proportional to the protein concentration. The procedure is simple and rapid. A protein solution is transferred to a quartz cuvette and the A_{280} is read against a reference cuvette containing the protein solvent only (buffer, water, etc.).

Cellular extracts contain many other compounds that absorb in the vicinity of 280 nm. Nucleic acids, which would be common contaminants in a protein extract, absorb strongly at 280 nm ($\lambda_{max} = 260$). Early researchers developed a method to correct for this interference by nucleic acids. Mixtures of pure protein and pure nucleic acid were prepared, and the ratio A_{280}/A_{260} was experimentally determined (Table 2.4). The following empirical equation may be used for protein solutions containing up to 20% nucleic acids (Layne, 1957):

$$\text{Protein concentration (mg/mL)} = 1.55A_{280} - 0.76A_{260}$$

Table 2.4

Protein and Nucleic Acid Estimation by UV Spectrophotometry[1]

A_{280}/A_{260}	% Nucleic Acid	A_{280}/A_{260}	% Nucleic Acid
1.75	0.00	0.85	5.55
1.63	0.25	0.82	6.00
1.52	0.50	0.80	6.50
1.40	0.75	0.78	7.00
1.36	1.00	0.77	7.50
1.30	1.25	0.75	8.00
1.25	1.50	0.73	9.00
1.16	2.00	0.71	10.00
1.09	2.50	0.67	12.00
1.03	3.00	0.64	14.00
0.95	3.50	0.62	17.00
0.94	4.00	0.60	20.00
0.87	5.00		

[1] Adapted from E. Layne, in *Methods in Enzymology,* Vol. III, S. P. Colowick and N. O. Kaplan, Editors, (1957), p. 447. Copyright 1957 Academic Press.

Table 2.4 can also be used to estimate the amount of nucleic acid in a protein solution.

Although the spectrophotometric assay of proteins is fast, relatively sensitive, and requires only a small sample size, it is nothing more than an estimate of protein concentration. It has certain advantages over the colorimetric assays in that most buffers and ammonium sulfate do not interfere. The spectrophotometric assay is particularly suited to the rapid measurement of protein elution from a chromatography column, where only protein concentration changes are required.

E. Measurement of Nucleic Acid Solutions by Spectrophotometry

In the previous section on protein assays, it was noted that the concentration of nucleic acid in a protein solution can be estimated from the ratio A_{280}/A_{260} (see Table 2.4). This method is suitable if one wishes to know the approximate amount of nucleic acid contaminating a protein solution or vice versa. But what if one requires the concentration of pure RNA or DNA? Solutions of nucleic acids strongly absorb ultraviolet light with a λ_{max} of 260 nm. The intense absorption is due primarily to the presence of aromatic rings in the purine and pyrimidine bases. The concentration of nucleic acid in a solution can be calculated if one knows the value of A_{260} of the solution. A solution of double-stranded DNA at a concentration of 50 μg/mL in a 1-cm quartz cuvette will give an A_{260} reading of 1.0. Therefore, the concentration (X) of a double-stranded DNA solution with A_{260} of 0.15 is

$$\frac{1.0}{50\ \mu\text{g/mL}} = \frac{0.15}{\text{X}\ \mu\text{g/mL}}, \qquad \text{X} = 7.5\ \mu\text{g/mL}$$

A solution of single-stranded DNA or RNA that has an A_{260} of 1.0 in a cuvette with a 1-cm path length has a concentration of 40 μg/mL.

This spectrophotometric assay for nucleic acids is simple, rapid, nondestructive, and very sensitive. The minimum concentration of nucleic acid that can be accurately determined is 2.5–5.0 μg/mL.

References

Amicon Division, W. R. Grace and Co., *Laboratory Products* (1990), Beverly, MA. "Ultrafiltration, a Review."

A. Athani and U. Banakar, *BioPharm.* January 23, pp. 23–30 (1990). "The Development and Potential Uses of Biosensors."

Bio-Rad, Bio-Rad Protein Assay, Bulletin 1069 (1984). 2200 Wright Avenue, Richmond, CA 94804.

M. Bradford, *Anal. Biochem.* **72,** 248–254 (1976). "A Rapid and Sensitive Method for the Quantitation of Microgram Quantities of Protein Utilizing the Principle of Protein-Dye Binding."

L. C. Clark, Jr., R. Wold, G. Granger, and Z. Taylor, *J. Appl. Physiol.* **6,** 189 (1953). "Continuous Recording of Blood Oxygen Tensions."

B. Danielsson and K. Mosbach, in *Methods in Enzymology,* Vol. 137, K. Mosbach, Editor (1988), Academic Press (San Diego), pp. 3–14. "Analytical Applications with Emphasis on Biosensors."

N. C. den Boer, C. van der Heiden, B. Leijnse, and J. H. M. Souverijn, Editors, *Clinical Chemistry—An Overview* (1989), Plenum Publishing (New York), pp. 113–128. Biosensors for biochemical and medical use.

J. Everse and F. Stolzenbach, in *Methods in Enzymology,* Vol. XXII, W. Jakoby, Editor (1971), Academic Press (New York), pp. 33–39. "Lyophilization."

D. C. Fork, in *Methods in Enzymology,* Vol. XXIV, A. San Pietro, Editor (1972), Academic Press (New York), pp. 113–122. "Oxygen Electrode."

N. E. Good and S. Izawa, in *Methods in Enzymology,* Vol. XXIV, A. San Pietro, Editor (1972), Academic Press (New York), pp. 53–68. "Hydrogen Ion Buffers for Photosynthesis Research."

J. K. Grady, N. D. Chasteen, and D. C. Harris, *Anal. Biochem.* **173,** 111–115 (1988). "Radicals from Good's Buffers."

M. Gronow, *Trends Biochem. Sci.* **9,** 336–340 (1984). "Biosensors."

D. E. Gueffroy, Editor, *A Guide for the Preparation and Use of Buffers in Biological Systems* (1975), Calbiochem, 10933 N. Torrey Pines Road, La Jolla, CA 92037.

R. Hughes, A. Ricco, M. Butler, and S. Martin, *Science* **254,** 74–80 (1991). "Chemical Microsensors."

Labconco Corporation, *A Guide to Freeze Drying for the Laboratory* (1987), 811 Prospect, Kansas City, MO 64132.

E. Layne, in *Methods in Enzymology,* Vol. III, S. P. Colowick and N. O. Kaplan, Editors (1957), Academic Press (New York), pp. 447–454. "Spectrophotometric and Turbidimetric Methods for Measuring Protein."

V. Linek, V. Vacek, J. Sinkule, and P. Benes, *Measurement of Oxygen by Membrane-Covered Probes* (1988) Ellis Horwood (Chichester). Theory and applications of the oxygen electrode.

O. Lowry, N. Rosebrough, A. Farr, and R. Randall, *J. Biol. Chem.* **193,** 265–275 (1951). "Protein Measurement with the Folin Phenol Reagent."

E. Marques, Jr., F. Spencer Netto, and J. Kima Filho, *Biochem. Educ.* **19,** 77–78 (1991). "A Low Cost Oxygen Meter."

P. McPhie, in *Methods in Enzymology,* Vol. XXII, W. Jakoby, Editor (1971), Academic Press (New York), pp. 23–32. "Dialysis."

P. Smith, R. Krohn, G. Hermanson, A. Mallia, F. Gartner, M. Provenzano, E. Fujimoto, N. Goeke, B. Olson, and D. Klenk, *Anal. Biochem.* **150,** 76–85 (1985). "Measurement of Protein Using Bicinchoninic Acid."

Yellow Springs Instrument Co., *Instructions for YSI Model 53 Biological Oxygen Monitor* (1976), Yellow Springs, OH 45387.

CHAPTER **3**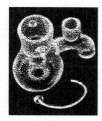

Separation and Purification of Biomolecules by Chromatography

■ The molecular details of a biochemical process cannot be fully elucidated until the reacting molecules have been isolated and characterized. Therefore, our understanding of biochemical principles has increased at about the same pace as the development of techniques for the separation and identification of biomolecules. Chromatography has been and will continue to be the most effective technique for isolating and purifying all types of biomolecules. In addition, it is widely used as an analytical tool.

A. Introduction to Chromatography

All types of chromatography are based on a very simple principle. The sample to be examined (called the **solute**) is allowed to interact with two physically distinct entities—a **mobile phase** and a **stationary phase** (see Figure 3.1). The mobile phase, which may be a gas or liquid, moves the sample through a region containing the solid or liquid stationary phase called the **sorbent**. The stationary phase will not be described in detail at this time, since it varies from one chromatographic method to another. However, it may be considered as having the ability to "bind" some types of solutes. The sample, which may contain one or many molecular components, comes into contact with the stationary phase. The components distribute themselves between the mobile and stationary phases. If some of the sample components are preferentially bound by the stationary phase, they spend more time in the stationary phase and, hence, are retarded in their movement through the chromatography system. Molecules that show weak affinity for the stationary phase spend more time with the mobile phase and are more rapidly removed or **eluted** from

59

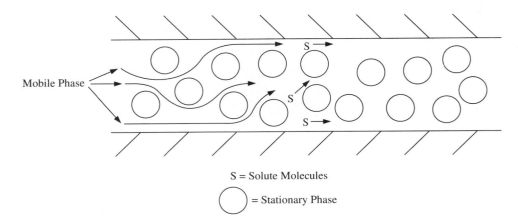

S = Solute Molecules

= Stationary Phase

Figure 3.1
A representation of the principles of chromatography.

the system. The many interactions that occur between solute molecules and the stationary phase bring about a separation of molecules because of different affinities for the stationary phase. The general process of moving a solute mixture through a chromatographic system is called **development**.

The mobile phase can be collected as a function of time at the end of the chromatographic system. The mobile phase, now called the **effluent**, contains the solute molecules. If the chromatographic process has been effective, fractions or "cuts" that are collected at different times will contain the different components of the original sample. In summary, molecules are separated because they differ in the extent to which they are distributed between the mobile phase and the stationary phase.

Throughout this chapter and others, biochemical techniques will be designated as **preparative** or **analytical**, or both. A preparative procedure is one that can be applied to the purification of a relatively large amount of a biological material. The purpose of such an experiment would be to obtain purified material for further characterization and study. Analytical procedures are used most often to determine the purity of a biological sample; however, they may be used to evaluate any physical, chemical, or biological characteristic of a biomolecule or biological system.

Partition versus Adsorption Chromatography

Chromatographic methods are divided into two types according to how solute molecules bind to or interact with the stationary phase. **Partition** chromatography is the distribution of a solute between two liquid phases. This may involve direct extraction using two liquids, or it may use a liquid immobilized on a solid support as in the case of paper, thin-layer, and gas-liquid chromatography. For partition chromatography, the stationary phase in Figure 3.1 consists of inert solid particles coated with liquid adsorbent. The distribution of solutes between the two phases is based primarily on solubility differences. The distribution may be quantified by using the **partition coefficient**, K_D (Equation 3.1).

▶ $$K_D = \frac{\text{concentration of solute in stationary phase}}{\text{concentration of solute in mobile phase}}$$ (Equation 3.1)

Adsorption chromatography refers to the use of a stationary phase or support, such as an ion-exchange resin, that has a finite number of relatively specific binding sites for solute molecules. There is not a clear distinction between the processes of partition and adsorption. All chromatographic separations rely, to some extent, on adsorptive processes. However, in some methods (paper, thin-layer, and gas chromatography) these specific adsorptive effects are minimal and the separation is based primarily on nonspecific solubility factors. Adsorption chromatography relies on relatively specific interactions between the solute molecules and binding sites on the surface of the stationary phase. The attractive forces between solute and support may be ionic, hydrogen bonding, or hydrophobic interactions. Binding of solute is, of course, reversible.

Because of the different interactions involved in partition and adsorption processes, they may be applied to different separation problems. Partition processes are the most effective for the separation of small molecules, especially those in homologous series. Partition chromatography has been widely used for the separation and identification of amino acids, carbohydrates, and fatty acids. Adsorption techniques, represented by ion-exchange chromatography, are most effective when applied to the separation of macromolecules including proteins and nucleic acids.

In the rest of the chapter, various chromatographic methods will be discussed. You should recognize that no single chromatographic technique relies solely on adsorption or partition effects. Therefore, little emphasis will be placed on a classification of the techniques; instead, theoretical and practical aspects will be discussed.

B. Paper and Thin-Layer Chromatography

Because of the similarities in the theory and practice of these two procedures, they will be considered together. Both are examples of partition chromatography. In paper chromatography, the cellulose support is extensively hydrated, so distribution of the solutes occurs between the immobilized water (stationary phase) and the mobile developing solvent. The initial stationary liquid phase in thin-layer chromatography (TLC) is the solvent used to prepare the thin layer of adsorbent. However, as developing solvent molecules move through the stationary phase, polar solvent molecules may bind to the immobilized support and become the stationary phase.

Preparation of the Stationary Support

The support medium may be a sheet of cellulose or a glass or plastic plate covered with a thin coating of silica gel, alumina, or cellulose. Large sheets of cellulose chromatography paper are available in different porosities. These may be cut to the appropriate size and used without further treatment. The paper should never be handled with bare fingers, especially if amino acids are to be analyzed. Although thin-layer plates can easily be prepared, it is much more convenient to purchase ready-made plates. These are available in a variety of sizes, materials, and thicknesses of stationary support. They are relatively inexpensive and have a more uniform support thickness than hand-made plates.

Figure 3.2 outlines the application procedure. A very small drop of sample is spotted onto the plate with a disposable microcapillary pipet

Figure 3.2
The procedure of paper and thin-layer chromatography.   (A) Application of the sample.   (B) Setting plate in solvent chamber.   (C) Movement of solvent by capillary action.   (D) Detection of separated components and calculation of R_f.

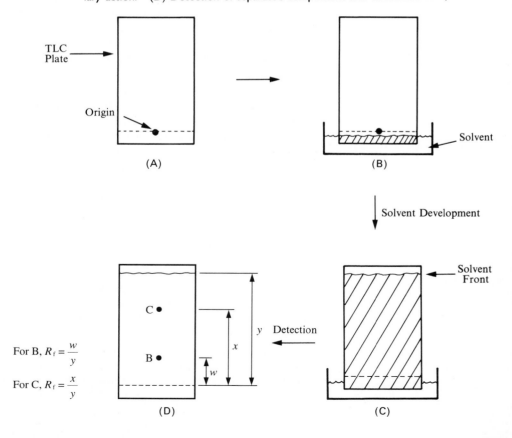

For B, $R_f = \dfrac{w}{y}$

For C, $R_f = \dfrac{x}{y}$

and allowed to dry; then the spotting process is repeated by superimposing more drops on the original spot. The exact amount of sample applied is critical. There must be enough sample so the developed spots can be detected, but overloading will lead to "tailing" and lack of resolution. Finding the proper sample size is a matter of trial and error. It is usually recommended that two or three spots of different concentrations be applied for each sample tested. Spots should be applied along a very faint line drawn with a pencil and ruler. TLC plates should not be heavily scratched or marked. Very faint marks may be made every 2 to 4 cm along the edge, but be sure the sample is not applied on or near these marks.

Solvent Development of the Support

A wide selection of solvent systems is available in the biochemical literature. If a new solvent system must be developed, a preliminary analysis must be done on the sample with a series of solvents. Solvents can be rapidly screened by developing several small chromatograms (2 × 6 cm) in small sealed bottles containing the solvents. For the actual analysis, the sample should be run on a larger plate with appropriate standards in a development chamber (Figure 3.3). The chamber must be airtight and saturated with solvent vapors. Filter paper on two sides of the chamber, as shown in Figure 3.3, enhances vaporization of the solvent.

Paper chromatograms may be developed in either of two types of arrangements—ascending or descending solvent flow. Descending solvent flow leads to faster development because of assistance by gravity, and it can offer better resolution for compounds with small R_f values because the solvent can be allowed to run off the paper. R_f values cannot

Figure 3.3
A typical chamber for paper and thin-layer chromatography.

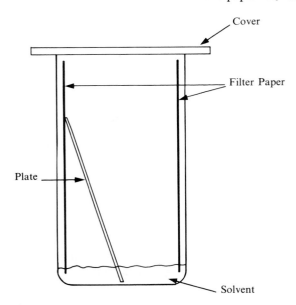

Cover

Filter Paper

Plate

Solvent

be determined under these conditions, but it is useful for quantitative separations.

Two-dimensional chromatography is used for especially difficult separations. The chromatogram is developed in one direction by a solvent system, air dried, turned 90°, and developed in a second solvent system.

Detection and Measurement of Components

Unless the components in the sample are colored, their location on a chromatogram will not be obvious after solvent development. Several methods can be used to locate the spots, including fluorescence, radioactivity, and treatment with chemicals that develop colors. Substances that are highly conjugated may be detected by fluorescence under a UV lamp. Chromatograms may be treated with different types of reagents to develop a color. **Universal reagents** produce a colored spot with any organic compound. When a solvent-developed plate is sprayed with concentrated H_2SO_4 and heated at 100°C for a few minutes, all organic substances appear as black spots. A more convenient universal reagent is I_2. The solvent-developed chromatogram is placed in an enclosed chamber containing a few crystals of I_2. The I_2 vapor reacts with most organic substances on the plate to produce brown spots. The spots are more intense with unsaturated compounds.

Specific reagents react with a particular class of compound. For example, rhodamine B is often used for visualization of lipids, ninhydrin for amino acids, and aniline phthalate for carbohydrates.

The position of each component of a mixture is quantified by calculating the distance traveled by the component relative to the distance traveled by the solvent. This is called **relative mobility** and symbolized by R_f. In Figure 3.2D, the R_f values for components B and C are calculated. The R_f for a substance is a constant for a certain set of experimental conditions. However, it varies with solvent, type of stationary support (paper, alumina, silica gel), temperature, humidity, and other environmental factors. R_f values are always reported along with solvent and temperature.

Applications of Paper and Thin-Layer Chromatography

Thin-layer chromatography is now more widely used than paper chromatography. In addition to its greater resolving power, TLC is faster and plates are available with several sorbents (cellulose, alumina, silica gel).

Partition chromatography as described in this section may be applied to two major types of problems: (1) identification of unknown samples and (2) isolation of the components of a mixture. The first application is, by far, the more widely used. Paper chromatography and TLC require only a minute sample size, the analysis is fast and inexpensive, and detection is straightforward. Unknown samples are applied to a plate along with appropriate standards, and the chromatogram is developed as a single experiment. In this way any changes in experimental conditions (temperature, humidity, etc.) affect standards and unknowns to the same extent. It is then possible to compare the R_f values directly.

Purified substances can be isolated from developed chromatograms; however, only tiny amounts are present. In paper chromatography, the spot may be cut out with a scissors and the piece of paper extracted with an appropriate solvent. Isolation of a substance from a TLC plate is accomplished by scraping the solid support from the region of the spot with a knife edge or razor blade and extracting the sorbent with a solvent. "Preparative" thin-layer plates with a thick coating of sorbent (up to 2 mm) are especially useful because they have higher sample capacity.

C. Gas Chromatography (GC)

When the mobile phase of a chromatographic system is gaseous and the stationary phase is a liquid coated on inert solid particles, the technique is **gas-liquid chromatography**, or simply gas chromatography. Separation by GC methods is based primarily on partitioning processes. The stationary phase, inert particles coated with a thin layer of liquid, is confined to a long stainless steel or glass tube, called the **column**, which is maintained at a suitable (usually elevated) temperature. A gaseous mobile phase under high pressure is continuously swept through the column. The sample to be analyzed is vaporized, introduced into the warm gaseous phase, and swept through the stationary phase. The vaporous chemical constituents in the sample then distribute themselves between the mobile phase and the stationary liquid film on the solid support. Components of the sample mixture that have affinity for the stationary phase are retarded in their movement through the column. Ideally, each component will have a different partition coefficient, K_D (see Equation 3.1), and each will pass through the column at a different rate.

Instrumentation

The essential components of a gas chromatography system are shown in Figure 3.4. The mobile phase (called the carrier gas) is inert, usually helium, nitrogen, or argon. The gas is directed past an injection port, the entry point of the sample. The sample, dissolved in a solvent, is injected with a syringe through a rubber septum into the injection port. The column, injection port, and detector are in individual ovens maintained at elevated temperatures so that the sample components remain vaporized throughout their residence time in the system.

Two types of columns are widely used. A **packed column** is one filled with inert, solid particles coated with a liquid stationary phase. Standard tubing is about 0.5 cm in diameter, with lengths ranging from 1 m to 20 m; however, columns for large-scale preparative work may be up to 5 cm in diameter and several meters long. Commonly used solid sup-

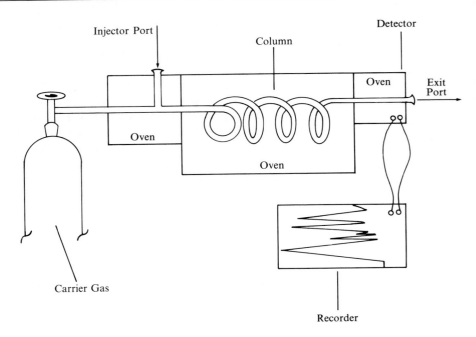

Figure 3.4
The essential components of a gas chromatography system.

ports are diatomaceous earth, Teflon powder, and glass beads. The stationary liquid must be chosen on the basis of the compounds to be analyzed. A more recently developed type of column is the **open-tubular** or **capillary column**. This is prepared by coating the inner wall of the column with the stationary liquid phase. The inside diameter of a typical capillary tube is 0.25 mm, and the length ranges from 10 to 100 meters. The length of a column depends on the degree of resolution required. Clearly, the longer the column, the better the separation; however, a long column increases the pressure differential from the beginning to the end of the column. A large pressure change hampers analysis by increasing retention time and causing peak broadening. Stainless steel columns are still in widespread use, but they cannot be used with sensitive compounds. Many biochemical substances decompose when they contact warm metal surfaces. For example, derivatives of amino acids can be analyzed only in gas chromatographs with glass injection ports and columns.

After passage through the column, the carrier gas and separated components of the mixture are directed through a detector. Several types of detectors are available, but the two most commonly used for bioanalytical purposes are the **thermal conductivity cell (TC cell)** and the **flame ionization detector (FID)**. The TC cell functions by measuring the temperature-dependent electrical resistance of a hot wire. The detector unit consists of a platinum or platinum-alloy wire through which an electric current is passed. The hot wire is cooled as the carrier gas passes over it. The extent of cooling depends on the gas flow rate and the thermal conductivity of the gas. Since the rate of carrier gas flow remains constant during a typical GC analysis, only changes in the thermal conductivity of the vapor will change the temperature and, therefore, the resistance of the wire. Organic vapors usually have lower thermal conductivities than the carrier gas. When carrier gas containing an organic vapor exits the column and passes over the wire, the electrical resistance of the wire decreases. The changing current in the wire is amplified and monitored by a recorder. The extent of resistance change depends on the amount of organic vapor in the carrier gas. Therefore, the size of the recorder signal is a measure of the amount of that chemical constituent in the sample. The thermal conductivity cell has a poor level of sensitivity (about 5 μg), but it is widely used because it responds to all organic compounds. Since it does not destroy the sample, a TC cell can be used when samples are to be collected from the column.

In contrast to the TC detector, the FID is sensitive to about 10^{-5} μg. A flame supported by hydrogen gas and air is used to burn organic vapors as they leave the column. Electrons and ionic fragments are produced upon combustion of the organics. A wire loop (called the ion collector) collects the charged particles and produces an electrical current, which is fed into a recorder. The FID is sensitive only to oxidizable compounds and does not respond to water vapor or CO_2. Since the detector response is proportional to the amount of organic material in the carrier gas, quantitative analysis can be done. Samples cannot be collected after passage through the FID; however, it is possible to split the carrier gas flow before it enters the detector. Only a small fraction of the carrier gas with organic vapor is directed through the detector, and the rest is collected for further analysis.

The electrical signal from a detector is amplified and fed into a strip chart recorder. A typical recorder trace is shown in Figure 3.5. Each peak represents a component in the original mixture. A peak is identified by a **retention time**, the time lapse between injection of the sample and the maximum signal from the recorder. This number is a constant for a particular compound under specified conditions of the carrier gas flow rate; temperature of the injector, column, and detector; and type of column. Retention time in GC analysis is analogous to the R_f value in thin-layer or paper chromatography.

Selection of Operating Conditions

Each type of gas chromatography has its own set of operating instructions, but general experimental conditions are appropriate for all instruments. Three important factors must always be considered when a GC analysis is to be completed: (1) selection of the proper column; (2) choice of temperatures for injector, oven, and detector; and (3) adjustment of

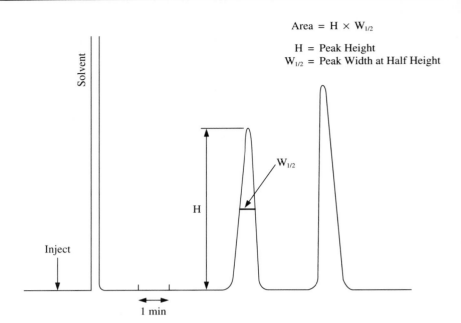

Area = H × W$_{1/2}$

H = Peak Height
W$_{1/2}$ = Peak Width at Half Height

Figure 3.5
A recorder trace from gas chromatography.

gas flow. Because hundreds of stationary phases are available, it is impossible to outline the characteristics of each. Selecting the stationary phase requires some knowledge of the nature of the sample to be analyzed. Books and journal articles on gas chromatography and commercial catalogs of GC supplies are the best sources of information to help select the right column.

The operating temperature of the column oven is critical, as it greatly affects the resolving ability of the column. If the temperature is too high, low-boiling components are swept rapidly through the column without equilibration with the stationary phase. A temperature that is too low causes condensation of some organic compounds in the liquid stationary phase, resulting in very extended retention times. A column temperature set near the boiling point of the primary component in the sample is often the best. If a sample containing a mixture of compounds with a wide range of boiling points is to be analyzed, a temperature gradient may be required. Research-grade chromatographs are equipped with temperature programmers that gradually increase the column temperature at a preselected rate.

The detector and injector temperatures should be maintained about 10° above the column temperature. This ensures against condensation of vapor and causes rapid vaporization of the sample upon injection.

The separation of a mixture by GC also depends on the flow rate of the carrier gas. A low rate allows extensive equilibration of the sample between the stationary phase and the gaseous phase and leads to better separation; however, the recorder peaks become very broad. A high flow rate greatly decreases column efficiency by decreasing equilibration time. A suitable flow rate for a 0.5-cm packed column is between 50 and 120 mL/min.

Analysis of GC Data

Gas chromatography can be applied to both qualitative and quantitative analyses. There are two general methods for qualitative identification of eluted compounds. A compound may be identified by **retention time** or by **peak enhancement**. If you have some idea of the identity of an unknown peak or if you know it is one of three or four compounds, then you can inject the known compounds into the GC under conditions identical to those for the unknown sample. Compounds with the same retention times (± 2 or 3%) can generally be considered identical. Alternatively, you can add pure sample of a suspected component to the unknown sample and inject the mixture into the GC. If a recorder peak is increased in size, this also provides evidence for the identity of an unknown. The latter technique is called **peak enhancement** or **spiking**.

Identification of unknown compounds by retention times or peak enhancement is not conclusive or absolute proof of identity. It is possible for two different substances to have identical retention times under the same experimental conditions. For positive identification, the sample must be collected at the exit port and characterized by mass spectrometry, infrared, nuclear magnetic resonance, or chemical analysis.

The response from most recorders and detectors is proportional to the amount of compound in the original sample. In other words, the area of each recorder peak is a relative measure of the concentration of that substance in the sample. GC then becomes an important tool for quantitative analysis. Methods for measuring peak area range from sophisticated to primitive. The most ideal method is to use an electronic integrator, which automatically measures peak area and can be programmed to yield % composition of each component. Alternatively, one can use the method of triangulation. Here the peaks are assumed to be triangles with an area calculated according to Equation 3.2, where H is height and $W_{1/2}$ is the width at one-half height.

▶ $$\text{Area} = HW_{1/2} \qquad\qquad \text{(Equation 3.2)}$$

This method gives acceptable results if the peaks are symmetrical. If the area of each peak is measured, the % composition of the sample can be estimated using Equation 3.3.

▶ $$\%x = \frac{\text{Area}_x}{\text{Area}_a + \text{Area}_b + \text{Area}_x} \times 100 \qquad\qquad \text{(Equation 3.3)}$$

The percent of x in the sample is equal to the area represented by component x divided by the sum of all peak areas. Area_a and Area_b refer to other components in the mixture.

These methods for quantitative analysis have a common limitation. Most detectors do not respond with equal sensitivity to all components in the sample. If samples that are isomeric or in a homologous series are analyzed, detector response is probably similar. However, if dissimilar substances are to be measured, response correction factors (RCFs) must be determined. The procedure for measuring RCFs is outlined in Experiment 8.

Advantages and Limitations of GC

It is difficult to imagine another analytical technique that has all the advantages of gas chromatography. It provides excellent separation of most organic molecules; it is simple, versatile, and rapid; it is highly sensitive; and, by most standards, it is relatively inexpensive.

The major limitation of GC is the requirement for heat stability and volatility of the sample. Obviously, compounds that decompose at elevated temperatures (below 250°C) cannot normally be subjected to GC analysis. Many compounds of biochemical interest are not volatile in the useful temperature range of GC (up to about 200–250°C). Such compounds can often be converted to volatile derivatives. Hydroxyl groups in alcohols, carbohydrates, and sterols are converted to derivatives by trimethylsilylation or acetylation. Amino groups can also be converted to volatile derivatives by acetylation and silylation. Fatty acids are transformed to methyl esters for GC analysis, as described in Experiment 8.

D. Column Chromatography

Adsorption chromatography in biochemical applications usually consists of a solid stationary phase and a liquid mobile phase. The most useful technique is column chromatography, in which the stationary phase is confined to a glass tube and mobile phase (a solvent or buffer) is allowed to flow through the solid adsorbent. A small amount of the sample to be analyzed is layered on top of the column. The sample mixture enters the column of adsorbing material and is distributed between the mobile phase and the stationary phase. The various components in the sample have different affinities for the two phases and move through the column at different rates. Collection of the liquid phase emerging from the column yields separate fractions containing the individual components in the sample.

Specific terminology is used to describe various aspects of column chromatography. When the actual adsorbing material is made into a column, it is said to be **poured** or **packed**. Application of the sample to the top of the column is **loading** the column. Movement of solvent through the loaded column is called **developing** or **eluting** the column. The **bed volume** is the total volume of solvent and adsorbing material taken up by the column. The volume taken up by the liquid phase in the column is the **void volume**. The **elution volume** is the amount of solvent

Table 3.1
Adsorbents Useful in Biochemical Applications

Adsorbing Material	Uses
Alumina	Small organics, lipids
Silica gel	Amino acids, lipids, carbohydrates
Fluorisil (magnesium silicate)	Neutral lipids
Calcium phosphate (hydroxyapatite)	Proteins, polynucleotides, nucleic acids

Table 3.2
Mesh Sizes of Adsorbents and Typical Applications

Mesh Size	Applications
20–50	Crude preparative work, very high flow rate
50–100	Preparative applications, high flow rate
100–200	Analytical separations, medium flow rate
200–400	High-resolution analytical separations, low flow rate

required to remove a particular solute from the column. This is analogous to R_f values in thin-layer or paper chromatography or to retention time in GC.

In adsorption chromatography, solute molecules take part in specific interactions with the stationary phase. Herein lies the great versatility of adsorption chromatography. Many varieties of adsorbing materials are available, so a specific sorbent can be chosen that will effectively separate a mixture. There is still an element of trial and error in the selection of an effective stationary phase. However, experiences of many investigators are recorded in the literature and are of great help in choosing the proper system. Table 3.1 lists the most common stationary phases employed in adsorption column chromatography.

Adsorbing materials come in various forms and sizes. The most suitable forms are dry powders or a slurry form of the material in an aqueous buffer or organic solvent. Alumina, silica gel, and fluorisil do not normally need special pretreatment. The size of particles in an adsorbing material is defined by *mesh size*. This refers to a standard sieve through which the particles can pass. A 100 mesh sieve has 100 small openings per square inch. Adsorbing material with high mesh size (400 and greater) is extremely fine and is most useful for very high resolution chromatography. Table 3.2 lists standard mesh sizes and the most appropriate application. For most biochemical applications, 100 to 200 mesh size is suitable.

Operation of a Chromatographic Column

A typical column setup is shown in Figure 3.6. The heart of the system is, of course, the column of adsorbent. In general, the longer the column, the better the resolution of components. However, a compro-

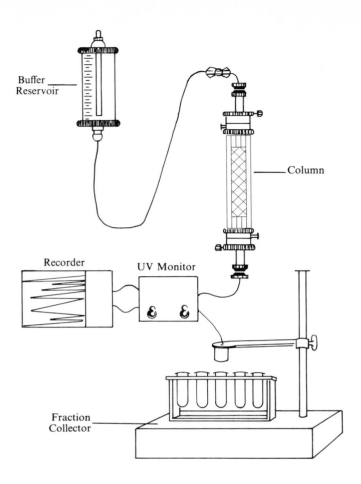

Figure 3.6
Setup for the operation of
a chromatography column.

mise must be made because flow rate decreases with increasing column
length. The actual size of a column depends on the nature of the adsorbing material and the amount of chemical sample to be separated. For
preparative purposes, column heights of 20 to 50 cm are usually sufficient to achieve acceptable resolution. Column inside diameters may vary
from 0.5 to 5 cm. Increasing the inside diameter of a column does not
improve resolution but it does increase the capacity.

Packing the Column

Once the adsorbing material and column size have been selected, the
column is poured. If the glass tube does not have a fritted disc in the bottom, a small piece of glass wool or cotton should be used to support the
column. Most columns are packed by pouring a slurry of the sorbent into
the glass tube and allowing it to settle by gravity into a tight bed. The
slurry is prepared with the solvent or buffer that will be used as the initial developing solvent. The slurry should be thin to avoid trapping air
bubbles in the bed while the column is poured. Pouring of the slurry
must be continuous to avoid formation of sorbent layers. Excess solvent
is eluted from the bottom of the column while the sorbent is settling. The

column must never run dry. Additional slurry is added until the column bed reaches the desired height. The top of the settled adsorbent is then covered with a small circle of filter paper or glass wool to protect the surface while the column is loaded with sample or the eluting solvent is changed.

Sometimes it is necessary to pack a column under pressure (5 to 10 psi). This leads to a tightly packed bed that yields more reproducible results, especially with gradient elution (see below).

Loading the Column

The sample to be analyzed by chromatography should be applied to the top of the column in a concentrated form. If the sample is solid, it is dissolved in a minimum amount of solvent; if already in solution, it may be concentrated by ultrafiltration as described in Chapter 2. After the sample is loaded onto the column with a graduated or disposable pipet, it is allowed to percolate into the adsorbent. A few milliliters of solvent are then carefully added to wash the sample into the column material. The glass column is then filled with eluting solvent.

Eluting the Column

The chromatography column is developed by continuous flow of a solvent. Maintaining the appropriate flow rate is important for effective separation. If the flow rate is set too high, there is not sufficient time for complete equilibration of the sample components with the two phases. Too low a flow rate allows diffusion of solutes, which leads to poor resolution and broad elution peaks. It is difficult to give guidelines for the proper flow rate of a column, but, in general, a column should be adjusted to a rate slightly less than "free flow." Sometimes it is necessary to find the proper flow rate by trial and error. One problem encountered during column development is a changing flow rate. As the solvent height above the column bed is reduced, there is less of a "pressure head" on the column, so the flow rate decreases. This can be avoided by storing the developing solvent in a large reservoir and allowing it to enter the column at the same rate as it is emerging from the column (see Figure 3.6).

Adsorption columns are eluted in one of three ways. All components may be eluted by a single solvent or buffer. This is referred to as **continual elution**. In contrast, **stepwise elution** refers to an incremental change of solvent to aid development. The column is first eluted with a volume of one solvent and then with a second solvent. This may continue with as many solvents or solvent mixtures as desired. In general, the first solvent should be the least polar of any used in the analysis, and each additional solvent should be of greater polarity or ionic strength. Finally, adsorption columns may be developed by **gradient elution** brought about by a gradual change in solvent composition. The composition of the eluting solvent can be changed by the continuous mixing of two different solvents to gradually change the ratio of the two solvents. Alternatively, the concentration of a component in the solvent can be gradually increased. This is most often done by addition of a salt (KCl, NaCl, etc.). Devices are commercially available to prepare predetermined, reproducible gradients.

Collecting the Eluent

The separated components emerging from the column in the eluent are usually collected as discrete fractions. This may be done manually by collecting specified volumes of eluent in Erlenmeyer flasks or test tubes. Alternatively, if many fractions are to be collected, a mechanical fraction collector is convenient and even essential. An automatic fraction collector (see Figure 3.6) directs the eluent into a single tube until a predetermined volume has been collected or until a preselected time period has elapsed; then the collector advances another tube for collection. Specified volumes are collected by a drop counter activated by a photocell, whereas a timer can be set to collect a fraction over a specific period. The fraction size collected is sometimes a critical factor in effective separations. If the solutes to be separated are very similar in structure and properties, it is best to collect relatively small fractions (1 to 3 mL).

Detection of Eluting Components

The completion of a chromatographic experiment calls for a means to detect the presence of solutes in the collected fractions. The detection method used will depend on the nature of the solutes. Smaller molecules such as lipids, amino acids, and carbohydrates can be detected by spotting fractions on a thin-layer plate or a piece of filter paper and treating them with a chemical reagent that produces a color. The same reagents that are used to visualize spots on a thin-layer or paper chromatogram are useful for this. Proteins and nucleic acids are conveniently detected by spectroscopic absorption measurements at 280 and 260 nm, respectively. Enzymes can be detected by measurements of catalytic activity associated with each fraction. Research-grade chromatographic systems are equipped with detectors that continuously monitor some physical property of the eluent (see Figure 3.6). The newest advance in detectors is the diode array (see Chapter 5). Most often the eluent is directed through a flow cell where absorbance or fluorescence characteristics can be measured. The detector is connected to a strip chart recorder for a permanent record of spectroscopic changes. When the location of the various solutes is determined, fractions containing identical components are pooled and stored for later use.

Advantages and Limitations of Adsorption Column Chromatography

Column procedures are relatively straightforward and trouble free; however, several precautions should be taken. Most problems are the result of human error and can be avoided if the experimenter is careful, alert, and knowledgeable.

Proper column conditions in terms of adsorbent preparation, flow rate, and sample loading have already been discussed. A column must never be allowed to go dry during packing or development. This leads to irregular flow or channeling and column cracking. The glass tube must be clamped to a ring stand in the level position before pouring. The top of the column bed must be level and undisturbed during the complete operation.

E. Ion-Exchange Chromatography

Ion-exchange chromatography is a form of adsorbtion chromatography in which ionic solutes display reversible electrostatic interactions with a charged stationary phase. The chromatographic setup is identical to that described in the last section and Figure 3.6. The column is packed with a stationary phase consisting of a synthetic resin that is tagged with ionic functional groups. The steps involved in ion-exchange chromatography are outlined in Figure 3.7. In stage 1, the insoluble resin material (positively charged) in the column is surrounded by buffer counterions. Loading of the column in stage 2 brings solute molecules of different charge into the ion-exchange medium. Solutes entering the column may be negatively charged, positively charged, or neutral under the experimental conditions. Solute molecules that have a charge opposite to that of the resin bind tightly but reversibly to the stationary phase (stage 3). The strength of binding depends on the size of the charge and the charge den-

Figure 3.7
Illustration of the principles of ion-exchange chromatography. See text for explanation.

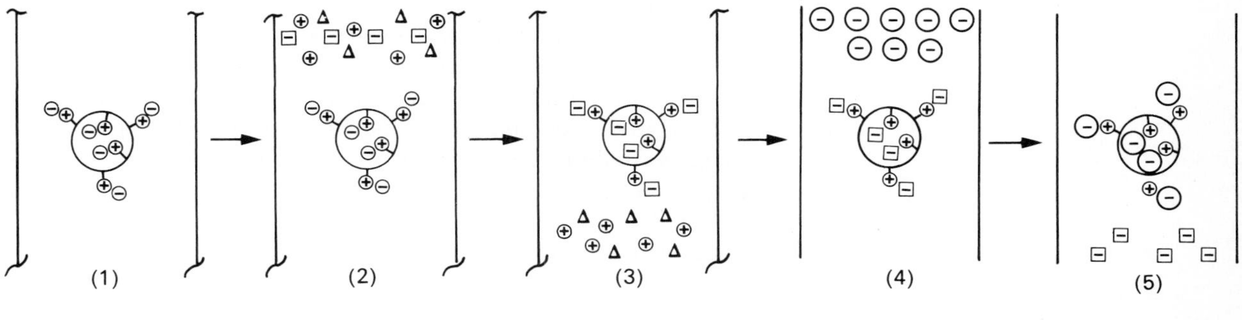

sity (amount of charge per unit volume of molecule) of the solute. The greater the charge or the charge density, the stronger the interaction. Neutral solute molecules (Δ) or those with a charge identical to that of the resin show little or no affinity for the stationary phase and move with the eluting buffer. The bound solute molecules can be released by eluting the column with a buffer of increased ionic strength or pH (stage 4). An increase in buffer ionic strength releases bound solute molecules by displacement. Increasing the buffer pH decreases the strength of the interaction by reducing the charge on the solute or on the resin (stage 5).

Summarizing the action of ion-exchange chromatography, solute molecules are separated on the basis of their affinity for a charged stationary phase. The following sections will focus on the properties of ion-exchange resins, selection of experimental conditions, and applications of ion-exchange chromatography.

Ion-Exchange Resins

Ion exchangers are made up of two parts—an insoluble, three-dimensional matrix and chemically bonded charged groups within and on the surface of the matrix. The resins are prepared from a variety of materials, including polystyrene, acrylic resins, polysaccharides (dextrans), and celluloses. An ion exchanger is classified as **cationic** or **anionic** depending on whether it exchanges cations or anions. A resin that has negatively charged functional groups exchanges positive ions and is a **cation exchanger**. Each type of exchanger is also classified as **strong** or **weak** according to the ionizing strength of the functional group. An exchanger with a quaternary amino group is, therefore, a **strongly basic anion exchanger**, whereas primary or secondary aromatic or aliphatic amino groups would lead to a **weakly basic anion exchanger**. A **strongly acidic cation exchanger** contains the sulfonic acid group. Table 3.3 lists the more common ion exchangers according to each of these classifications.

The ability of an ion exchanger to adsorb counterions is defined quantitatively by *capacity*. The *total capacity* of an ion exchanger is the quantity of charged and potentially charged groups per unit weight of dry exchanger. It is usually expressed as milliequivalents of ionizable groups per milligram of dry weight, and it can be experimentally determined by titration. The capacity of an ion exchanger is a function of the porosity of the resin. The resin matrix contains covalent cross-linking that creates a "molecular sieve." Ionized functional groups within the matrix are not readily accessible to large molecules that cannot fit into the pores. Only surface charges would be available to these molecules for exchange. The purely synthetic resins (polystyrene and acrylic) have cross-linking ranging from 2 to 16%, with 8% being the best for general purposes. The cellulosic ion exchangers have little or no cross-linking, while the dextran-based ion exchangers are available in two cross-link sizes.

Table 3.3
Ion-Exchange Resins

Name	Functional Group	Matrix	Class
Anion Exchangers			
AG 1	Tetramethylammonium	Polystyrene	Strong
AG 3	Tertiary amine	Polystyrene	Weak
DEAE-cellulose	Diethylaminoethyl	Cellulose	Weak
PEI-cellulose	Polyethyleneimine	Cellulose	Weak
DEAE-Sephadex	Diethylaminoethyl	Dextran	Weak
QAE-Sephadex	Diethyl-(2-hydroxyl-propyl)-aminoethyl	Dextran	Strong
Cation Exchangers			
AG 50	Sulfonic acid	Polystyrene	Strong
Bio-Rex 70	Carboxylic acid	Acrylic	Weak
CM-Cellulose	Carboxymethyl	Cellulose	Weak
P-Cellulose	Phosphate	Cellulose	Intermediate
CM-Sephadex	Carboxymethyl	Dextran	Weak
SP-Sephadex	Sulfopropyl	Dextran	Strong

With so many different experimental options and resin properties to consider, it is difficult to select the proper conditions for a particular separation. The next section will outline the choices and offer guidelines for proper experimental design.

Selection of the Ion Exchanger

Before a proper choice of ion exchanger can be made, the nature of the solute molecules to be separated must be considered. For relatively small, stable molecules (amino acids, lipids, nucleotides, carbohydrates, pigments, etc.) the synthetic resins based on polystyrene are most effective. They have relatively high capacity for small molecules because the extensive cross-linking still allows access to the interior of the resin beads. For separations of peptides, proteins, nucleic acids, polysaccharides, and other large biomolecules, one must consider the use of fibrous cellulosic ion exchangers and low-percent cross-linked dextran or acrylic exchangers. The immobilized functional groups in these resins are readily available for exchange even to larger molecules. In addition, most macromolecules are found to be relatively stable on cellulose, dextran, and acrylic-based resins but are sometimes denatured on polystyrene-based exchangers.

The choice of ion exchanger has now been narrowed considerably. The next decision is whether to use a cationic or anionic exchanger. If the solute molecule has only one type of charged group, the choice is simple. A solute that has a positive charge will bind to a cationic exchanger and vice versa. However, many biomolecules have more than one type of ionizing group and may have both negatively and positively charged groups (they are amphoteric). The net charge on such molecules depends on pH. At the isoelectric point, the substance has no net charge and would not bind to any type of ion exchanger.

In principle, amphoteric molecules should bind to both anionic and cationic exchangers. However, when one is dealing with large biomolecules, the pH "range of stability" must also be evaluated. The range of stability refers to the pH range in which the biomolecule is not denatured. Figure 3.8 shows how the net charge of a hypothetical protein changes as a function of pH. Below the isoelectric point, the molecule has a net positive charge and would be bound to a cation exchanger. Above the isoelectric point, the net charge is negative, and the protein would bind to an anion exchanger. Superimposed on this graph is the pH range of stability for the hypothetical protein. Because it is stable in the range of pH 7.0–9.0, the ion exchanger of choice is an anionic exchanger. In most cases, the isoelectric point of the protein is not known. The type of ion exchanger must be chosen by trial and error as follows. Small samples of the solute mixture in buffer are equilibrated for 10 to 15 minutes in separate test tubes, one with each type of ion exchanger. The tubes are then centrifuged or let stand to sediment the ion exchanger. Check each supernatant for the presence of solute (A_{260} for nucleic acids, A_{280} for proteins, catalytic activity for enzymes, etc.). If a supernatant

Figure 3.8
The effect of pH on the net charge of a protein.

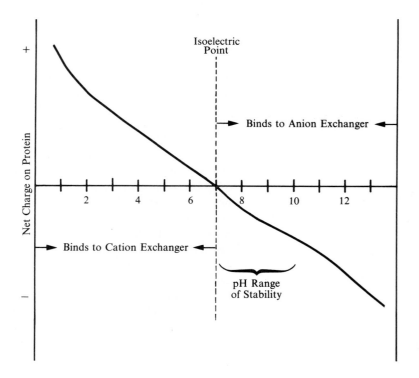

has a relatively low level of added solute, that ion exchanger would be suitable for use. This simple test can also be extended to find conditions for elution of the desired macromolecule from the ion exchanger. The ion exchanger charged with the macromolecule is treated with buffers of increasing ionic strength or pH. The supernatant after each treatment is analyzed as before for release of the macromolecule.

The influence of porosity on ion-exchange function has been mentioned. The general rule to follow is that ion exchangers with a high degree of cross-linking (small pore size) are best for the separation of small molecules. Since cellulose ion exchangers are fibrous and have little or no cross-linking, they are appropriate for large and small molecules. The dextran and acrylic-based ion exchangers are available in various pore sizes. These should be used according to the general guidelines furnished by the supplier.

Choice of Buffer

This decision includes not just the buffer substance but also the pH and the ionic strength. Buffer ions will, of course, interact with ion-exchange resins. Buffer ions with a charge opposite to that on the ion exchanger compete with solute for binding sites and greatly reduce the capacity of the column. Cationic buffers should be used with anionic exchangers; anionic buffers should be used with cationic exchangers.

The pH chosen for the buffer depends first of all on the range of stability of the macromolecule to be separated (see Figure 3.8). Second, the buffer pH should be chosen so that the desired macromolecule will bind to the ion exchanger. In addition, the ionic strength should be relatively low to avoid "damping" of the interaction between solute and ion exchanger. Buffer concentrations in the range 0.05 to 0.1 M are recommended.

Preparation of the Ion Exchanger

The commercial suppliers of ion exchangers provide detailed instructions for the preparation of the adsorbents. Failure to pretreat ion exchangers will greatly reduce the capacity and resolution of a column. Because of the extensive use of the cellulose ion exchangers in biochemical research, a suitable method of preconditioning will be described here. This will also provide an example of the care and planning that must go into the design of a chromatography experiment. Most cellulosic ion exchangers are furnished in a dry powder form and must be swollen in aqueous solution to "uncover" ionic functional groups. The two most common cellulosic resins are diethylaminoethyl (DEAE) and carboxymethyl (CM) cellulose. To pretreat the anion exchanger, DEAE cellulose, 100 g is first suspended in 1 L of 0.5 M HCl and stirred for 30 min. The cation exchanger, CM cellulose, is suspended in 1 L of 0.5 M NaOH for the same period. This initial treatment puts a like charge on all the ionic functional groups, causing them to repel each other; thus, the exchanger swells. The cellulose is then allowed to settle (or is filtered) and washed with water to a neutral pH. DEAE cellulose is then washed for 30 min with 1 L of 0.5 M NaOH (CM cellulose with 0.5 M HCl), followed by water until a neutral pH is attained. This treatment puts the ion exchanger in the free base or free acid form, and it is ready to be converted to the ionic form necessary for the separation. An additional step should be included at this point to remove trace metal ions from the ion exchanger. The presence of heavy metals in the column interferes with the ion-exchange process and also may denature sensitive macromolecules. Treatment of the swollen cellulose with 0.01 M EDTA removes metal ions.

These washing treatments do not remove the small particles that are present in most ion-exchange materials. If left in suspension, these particles, called fines, result in decreased resolution and low column flow rates. The fines are removed from an exchanger by suspending the swollen adsorbent in a large volume of water in a graduate cylinder and allowing at least 90% of the exchanger to settle. The cloudy supernatant containing the fines is decanted. This process is repeated until the supernatant is completely clear. The number of washings necessary to remove most fines is variable, but for a typical cellulose exchanger 8 to 10 times is probably sufficient.

The washed, metal-free, defined cellulose is now ready for equilibration with the starting buffer. The equilibration is accomplished by washing the exchanger with a large volume of the desired buffer, either in a packed column or batchwise.

Using the Ion-Exchange Resin

Ion exchangers are most commonly used in a column form. The column method discussed earlier in this chapter can be directly applied to ion-exchange chromatography.

An alternative method of ion exchange is **batch separation**. This involves mixing and stirring equilibrated exchanger directly with the solute mixture to be separated. After an equilibration time of approximately 1 hour, the slurry is filtered and washed with buffer. The ion exchanger can be chosen so that the desired solute is adsorbed onto the exchanger or remains unbound in solution. If the latter is the case, the desired material is in the filtrate. If the desired solute is bound to the exchanger, it can be removed by suspending the exchanger in a buffer of greater ionic strength or different pH. Batch processes have some advantages over column methods. They are rapid, and the problems of packing, channeling, and dry columns are avoided.

A new development in ion-exchange column chromatography allows the separation of proteins according to their isoelectric points. This technique, **chromatofocusing**, involves the formation of a pH gradient on an ion-exchange column. If a buffer of a specified pH is passed through an ion-exchange column that was equilibrated at a second pH, a pH gradient is formed on the column. Proteins bound to the ion exchanger are eluted in the order of their isoelectric points. In addition, protein band concentration (focusing) takes place during elution. Chromatofocusing is similar to isoelectric focusing, introduced in Chapter 4, in which a column pH gradient is produced by an electric current. The requirements for chromatofocusing include specialized exchangers and buffers (available from Pharmacia-LKB under the trade name Polybuffer). Column equipment is appropriate for chromatofocusing experiments. Further information is available in a booklet, "Chromatofocusing," published by Pharmacia-LKB.

Storage of Resins

Most ion exchangers in the dry form are stable for many years. Aqueous slurried ion exchangers are still useful after several months. One major storage problem with a wet exchanger is microbial growth. This is especially true for the cellulose and dextran exchangers. If it is necessary to store pretreated exchangers, an antimicrobial agent must be added to the slurry. Sodium azide (0.02%) is suitable for cation exchangers and phenylmercuric salts (0.001%) are effective for anion exchangers. Since these preservative reagents are toxic, they must be used with caution.

F. Gel Exclusion Chromatography

The chromatographic methods discussed up to this point allow the separation of molecules according to polarity, volatility, and charge. The method of **gel exclusion chromatography** (also called gel filtration, molecular sieve chromatography, or gel permeation chromatography) exploits the physical property of molecular size to achieve separation. The molecules of nature range in molecular weight from less than 100 to as large as several million. It should be obvious that a technique capable of separating molecules of molecular weight 10,000 from those of 100,000 would be very popular among research biochemists. Gel chromatography has been of major importance in the purification of thousands of proteins, nucleic acids, enzymes, polysaccharides, and other biomolecules. In addition, the technique may be applied to molecular weight determination and quantitative analysis of molecular interactions. In this section the theory and practice of gel filtration will be introduced and applied to several biochemical problems.

Theory of Gel Filtration

The operation of a gel filtration column is illustrated in Figure 3.9. The stationary phase consists of inert particles that contain small pores of a controlled size. Microscopic examination of a particle reveals an interior

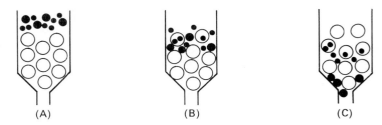

Figure 3.9
Separation of molecules by gel filtration.   (A) Application of sample containing large and small molecules.   (B) Large molecules cannot enter gel matrix, so they move more rapidly through the column.   (C) Elution of the large molecules.

resembling a sponge. A solution containing solutes of various molecular sizes is allowed to pass through the column under the influence of continuous solvent flow. Solute molecules that are larger than the pores cannot enter the interior of the gel beads, so they are limited to the space between the beads. The volume of the column that is accessible to very large molecules is, therefore, greatly reduced. As a result, they are not slowed in their progress through the column and elute rapidly in a single zone. Small molecules that are capable of diffusing in and out of the beads have a much larger volume available to them. Therefore, they are delayed in their journey through the column bed. Molecules of intermediate size migrate through the column at a rate somewhere between those for large and small molecules. Therefore, the order of elution of the various solute molecules is directly related to their molecular dimensions.

Physical Characterization of Gel Chromatography

Several physical properties must be introduced to define the performance of a gel and solute behavior. Some important properties are:

1. Exclusion Limit   This is defined as the molecular mass of the smallest molecule that cannot diffuse into the inner volume of the gel matrix. All molecules above this limit elute rapidly in a single zone. The exclusion limit of a typical gel, Sephadex G-50, is 30,000 daltons. All solute molecules having a molecular size greater than this value would pass directly through the column bed without entering the gel pores.

2. Fractionation Range   Sephadex G-50 has a fractionation range of 1500 to 30,000 daltons. Solute molecules within this range would be separated in a somewhat linear fashion.

3. Water Regain and Bed Volume   Gel chromatography media are most often supplied in dehydrated form and are swollen in a solvent, usually water, before use. The weight of water taken up by 1 g of dry gel is known as the water regain. For G-50, this value is 5.0 ± 0.3 g. This value does not include the water surrounding the gel particles, so it cannot be used as an estimate of the final volume of a packed gel column. Most commercial suppliers of gel materials provide, in addition to water regain, a bed volume value. This is the final volume taken up by 1 g of dry gel when swollen in water. For G-50, bed volume is 9 to 11 mL/g dry gel.

4. Gel Particle Shape and Size   Ideally, gel particles should be spherical to provide a uniform bed with a high density of pores. Particle size is defined either by mesh size or bead diameter (μm). The degree of resolution afforded by a column and the flow rate both depend on particle size. Larger particle sizes (50 to 100 mesh, 100 to 300 μm) offer high flow rates but poor chromatographic separation. The opposite is true for very small particle sizes ("superfine," 400 mesh, 10 to 40 μm). The most useful particle size, which represents a compromise between resolution and flow rate, is 100 to 200 mesh (50 to 150 μm).

5. Void Volume   This is the total space surrounding the gel particles in a packed column. This value is determined by measuring the volume of solvent required to elute a solute that is completely excluded from the gel matrix. Most columns can be calibrated for void volume with a dye, blue dextran, which has an average molecular mass of 2,000,000 daltons.

6. Elution Volume   This is the volume of eluting buffer necessary to remove a particular solute from a packed column.

Chemical Properties of Gels

Four basic types of gels are available: **dextran, polyacrylamide, agarose,** and **combined polyacrylamide-dextran**. The first gels to be developed were those based on a natural polysaccharide, dextran. These are supplied by Pharmacia-LKB under the trade name Sephadex. Table 3.4 gives the physical properties of the various sizes of Sephadex. The number given each gel refers to the water regain multiplied by 10. Sephadex is available in various particle sizes labeled coarse, medium, fine, and superfine. Dextran-based gels cannot be manufactured with an exclusion limit greater than 600,000 daltons because the small extent of cross-linking is not sufficient to prevent collapse of the particles. If the dextran is cross-linked with *N,N'*-methylenebisacrylamide, gels for use in higher fractionation ranges are possible. Table 3.4 lists these gels, which are called Sephacryl and are also supplied by Pharmacia-LKB.

Polyacrylamide gels are produced by the copolymerization of acrylamide and the cross-linking agent *N,N'*-methylenebisacrylamide. These are supplied by Bio-Rad Laboratories (Bio-Gel P). The Bio-Gel media are available in 10 sizes with exclusion limits ranging from 1800 to 400,000 daltons. Table 3.4 lists the acrylamide gels and their physical properties.

The agarose gels have the advantage of very high exclusion limits. Agarose, the neutral polysaccharide component of agar, is composed of alternating galactose and anhydrogalactose units. The gel structure is sta-

Table 3.4
Properties of Gel Filtration Media

Name	Fractionation Range for Proteins (daltons)	Water Regain (mL/g dry gel)	Bed Volume (mL/g dry gel)
Dextran (Sephadex)[1]			
G-10	0–700	1.0 ± 0.1	2–3
G-15	0–1500	1.5 ± 0.2	2.5–3.5
G-25	1000–5000	2.5 ± 0.2	4–6
G-50	1500–30,000	5.0 ± 0.3	9–11
G-75	3000–80,000	7.5 ± 0.5	12–15
G-100	4000–150,000	10 ± 1.0	15–20
G-150	5000–300,000	15 ± 1.5	20–30
G-200	5000–600,000	20 ± 2.0	30–40
Polyacrylamide (Bio-Gels)[2]			
P-2	100–1800	1.5	3.0
P-4	800–4000	2.4	4.8
P-6	1000–6000	3.7	7.4
P-10	1500–20,000	4.5	9.0
P-30	2500–40,000	5.7	11.4
P-60	3000–60,000	7.2	14.4
P-100	5000–100,000	7.5	15.0
P-150	15,000–150,000	9.2	18.4
P-200	30,000–200,000	14.7	29.4
P-300	60,000–400,000	18.0	36.0
Dextran-polyacrylamide (Sephacryl)[1]			
S-100 HR	—	—	—
S-200 HR	5,000–250,000	—	—
S-300 HR	10,000–1,500,000	—	—
S-400 HR	20,000–8,000,000	—	—
Agarose			
Sepharose[1] 6B	10,000–4,000,000	—	—
Sepharose 4B	60,000–20,000,000	—	—
Sepharose 2B	70,000–40,000,000	—	—
Superose[1] 12 HR	1,000–300,000	—	—
Superose 6 HR	5,000–5,000,000	—	—
Bio-Gel[2] A-0.5	10,000–500,000	—	—
Bio-Gel A-1.5	10,000–1,500,000	—	—
Bio-Gel A-5	10,000–5,000,000	—	—
Bio-Gel A-15	40,000–15,000,000	—	—
Bio-Gel A-50	100,000–50,000,000	—	—
Bio-Gel A-150	1,000,000–150,000,000	—	—
Vinyl (Fractogel TSK)[3]			
HW-40	100–10,000	—	—
HW-55	1000–700,000	—	—
HW-65	50,000–5,000,000	—	—
HW-75	500,000–50,000,000	—	—

[1] Pharmacia-LKB Biotechnology.
[2] Bio-Rad Laboratories.
[3] Pierce Chemical Co.

bilized by hydrogen bonds rather than by covalent cross-linking. Agarose gels, supplied by Bio-Rad Laboratories (Bio-Gel A) and by Pharmacia-LKB (Sepharose and Superose), are listed in Table 3.4.

The combined polyacrylamide-agarose gels are commercially available under the trade name Ultragel. These consist of cross-linked polyacrylamide with agarose trapped within the gel network. The polyacrylamide gel allows a high degree of separation and the agarose maintains gel rigidity, so high flow rates may be used.

Selecting a Gel

The selection of the proper gel is a critical stage in successful gel chromatography. Most gel chromatographic experiments can be classified as either **group separations** or **fractionations**. Group separations involve dividing a solute sample into two groups, a fraction of relatively low-molecular-weight solutes and a fraction of relatively high-molecular-weight solutes. Specific examples of this are desalting a protein solution or removing small contaminating molecules from protein or nucleic acid extracts. For group separations, a gel should be chosen that allows complete exclusion of the high-molecular-weight molecules in the void volume. Sephadex G-25, Bio-Gel P-6, and Sephacryl S-100HR are recommended for most group separations. The particle size recommended is 100 to 200 mesh or 50 to 150 μm diameter.

Gel fractionation involves separation of groups of solutes of similar molecular weights in a multicomponent mixture. In this case, the gel should be chosen so that the fractionation range includes the molecular weights of the desired solutes. If the solute mixture contains macromolecules up to 120,000 in molecular weight, then Bio-Gel P-150, Sephacryl S-200HR, or Sephadex G-150 would be most appropriate. If P-100, G-100, or Sephacryl S-100HR were used, some of the higher-molecular-weight proteins in the sample would elute in the void volume. On the other hand, if P-200, P-300, or G-200 were used, there would be a decrease in both resolution and flow rate. If the molecular weight range of the solute mixture is unknown, empirical selection is necessary. The recommended gel grade for most fractionations is 100–200 or 200–400 mesh (20–80 μm or 10–40 μm). The finest grade that allows a suitable flow rate should be selected. For very critical separations, superfine grades offer the best resolution but with very low flow rates.

Gel Preparation and Storage

The dextran and acrylamide gel products are supplied in dehydrated form. They must be allowed to swell in water before use. The swelling time required differs for each gel, but the extremes are 3 to 4 hours at 20°C for highly cross-linked gels and up to 72 hours at 20°C for P-300 or G-200. The swelling time can be shortened if a boiling-water bath is used. Agarose gels and combined polyacrylamide-agarose gels are supplied in a hydrated state, so there is no need for swelling.

Before a gel slurry is packed into the column, it should be defined and deaerated. Defining is necessary to remove very fine particles, which would reduce flow rates. To define, pour the gel slurry into a graduated cylinder and add water equivalent to two times the gel volume. Invert the cylinder several times and allow the gel to settle. After 90 to 95% of the gel has settled, decant the supernatant, add water, and repeat the settling process. Two or three defining operations are usually sufficient to remove most small particles.

Deaerating (removing dissolved gases) should be done on the gel slurry and all eluting buffers. Gel particles that have not been deaerated tend to float and form bubbles in the column bed. Dissolved gases are removed by placing the gel slurry in a side-arm vacuum flask and applying a vacuum from a water aspirator. The degassing process is complete when no more small air bubbles are released from the gel (usually 1 to 2 hours).

Antimicrobial agents must be added to stored, hydrated gels. One of the best agents is sodium azide (0.02%).

Operation of a Gel Column

The procedure for gel column chromatography is very similar to the general description given earlier. The same precautions must be considered in packing, loading, and eluting the column. A brief outline of important considerations follows.

COLUMN SIZE

For fractionation purposes, it is usually not necessary to use columns greater than 100 cm in length. The ratio of bed length to width should be between 25 and 100. For group separations, columns less than 50 cm long are sufficient, and appropriate ratios of bed length to width are between 5 and 10.

ELUTING BUFFER

There are fewer restrictions on buffer choice in gel chromatography than in ion-exchange chromatography. Dextran and polyacrylamide gels are stable in the pH range 1.0 to 10.0, whereas agarose gels are limited to pH 4 to 10. Since there is such a wide range of stability of the gels, the buffer pH should be chosen on the basis of the range of stability of the macromolecules to be separated.

SAMPLE VOLUME

The sample volume is a critical factor in planning a gel chromatography experiment. If too much sample is applied to a column, resolution is decreased; if the sample size is too small, the solutes are greatly diluted.

For group separations, a sample volume of 10 to 25% of the column total volume is suitable. The sample volume for fractionation procedures should be between 1 and 5% of the total volume. Column total volume is determined by measuring the volume of water in the glass column that is equivalent to the height of the packed bed.

COLUMN FLOW RATE

The flow rate of a gel column depends on many factors, including length of column and type and size of the gel. It is generally safe to elute a gel column at a rate slightly less than free flow. A high flow rate reduces sample diffusion or zone broadening but may not allow complete equilibration of solute molecules with the gel matrix.

A specific flow rate cannot be recommended, since each type of gel requires a different range. The average flow rate given in literature references for small-pore-size gels is 8 to 12 mL/cm² of cross-sectional bed area per hour (15 to 25 mL/hr). For large-pore-size gels, a value of 2 to 5 mL/cm² of cross-sectional bed area per hour (5 to 10 mL/hr) is average.

Eluent can be made to flow through a column by either of two methods, gravity or pump elution. Gravity elution is most often used because no special equipment is required. It is quite acceptable for developing a column used for group separations and fractionations when small-pore-sized gels are used. However, if the flow rate must be maintained at a constant value throughout an experiment or if large-pore gels are used, pump elution is recommended.

One variation of gel chromatography is ascending eluent flow. Some investigators report more reproducible results, better resolution, and a more constant flow rate if the eluting buffer is pumped backward through the gel. This type of experiment requires special equipment, including a specialized column and a peristaltic pump.

Applications of Gel Exclusion Chromatography

Several experimental applications of gel chromatography have already been mentioned, but more detail will be given here.

DESALTING

Inorganic salts, organic solvents, and other small molecules are used extensively for the purification of macromolecules. Gel chromatography provides an inexpensive, simple, and rapid method for removal of these small molecules. One especially attractive method for desalting very small samples (0.1 mL or less) of proteins or nucleic acid solutions is to use spin columns. These are prepacked columns of polyacrylamide exclusion gels. They are available in two sizes, one that excludes nucleic acids of more than 5 base pairs or proteins greater than 6000 daltons and another that excludes nucleic acids of more than 20 base pairs or proteins greater than 30,000 daltons. Spin columns are used in a similar fashion to microfiltration centrifuge tubes (Chapter 2, pp. 45–47). The sample is placed on top of the gel column and spun in a centrifuge. Large molecules are eluted from the column and collected in a reservoir. The small molecules to be removed remain in the gel.

PURIFICATION OF BIOMOLECULES

This is probably the most popular use of gel chromatography. Because of the ability of a gel to fractionate molecules on the basis of size, gel filtration complements other purification techniques that separate molecules on the basis of polarity and charge.

ESTIMATION OF MOLECULAR WEIGHT

The elution volume for a particular solute is proportional to the molecular size of that solute. This indicates that it is possible to estimate the molecular weight of a solute on the basis of its elution characteristics on a gel column. An elution curve for several standard proteins on Sephadex G-100 is shown in Figure 3.10. This curve, a plot of protein concentration vs. volume collected, is representative of data obtained from a gel filtration experiment. The elution volume, V_e, for each protein can be estimated as shown in the figure. A plot of log molecular mass vs. elution volume for the proteins is shown in Figure 3.11. Note that the linear portion of the curve in Figure 3.11 covers the molecular mass range of 10,000 to 100,000 daltons. Molecules below 10,000 daltons are not eluted in an elution volume proportional to size. Since they readily diffuse into the gel particles, they are retarded. Molecules larger than

Figure 3.10
Elution curve for a mixture of several proteins.
A = hemoglobin,
B = egg albumin,
C = chymotrypsinogen,
D = myoglobin,
E = cytochrome *c*.

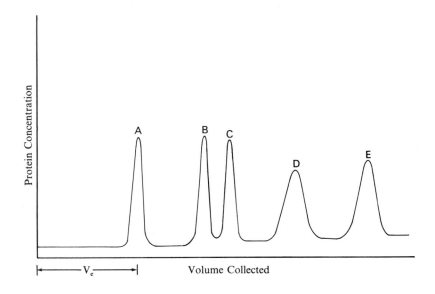

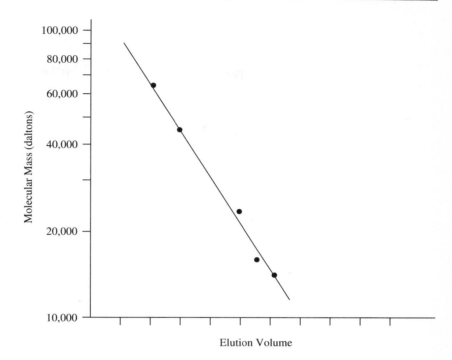

Figure 3.11
A plot of log molecular mass vs. elution volume for the proteins in Figure 3.10.

100,000 daltons are all excluded from the gel in the void volume. A solution of the unknown protein is chromatographed through the calibrated column under conditions identical to those for the standards and the elution volume is measured. The unknown molecular size is then read directly from the graph. This method of molecular weight estimation is widely used because it is simple, accurate, inexpensive, and fast. It can be used with highly purified or impure samples. There are, of course, some limitations to consider. The gel must be chosen so that the molecular weight of the unknown is within the linear section of the curve. The method assumes that only steric and partition effects influence the elution of the standards and unknown. If a protein interacts with the gel by adsorptive or ionic processes, the estimate of molecular weight is lower than the true value. The assumption is also made that the unknown molecules have a general spherical shape, not an elongated or rod shape.

If only small samples of a particular protein are available, the molecular weight can be estimated by **thin-layer gel filtration**. A glass plate is coated with an appropriate hydrated gel (usually superfine grade) and the protein sample is applied along with standards just as in thin-layer chromatography. The loaded plate is set in a special chamber and developed with descending buffer at a rate of 1 to 2 cm/hr. This is particularly useful for molecular weight determination but can also be used for analytical purposes.

GEL CHROMATOGRAPHY IN ORGANIC SOLVENTS

The gels discussed so far in this section are hydrophilic, and the inner matrix retains its integrity only in aqueous solvents. Because there is a need for gel chromatography of nonhydrophilic solutes, gels have been produced that can be used with organic solvents. The trade names of some of these products are Sepharose CL and Sephadex LH (Pharmacia-LKB), Bio Beads S (Bio-Rad), and Styragel (Dow Chemical). Organic solvents that are of value in gel chromatography are ethanol, acetone, dimethylsulfoxide, dimethylformamide, tetrahydrofuran, chlorinated hydrocarbons, and acetonitrile. For specific instructions in the use of these gels, brochures from the individual suppliers should be studied.

G. High-Performance Liquid Chromatography (HPLC)

The previous discussions on the theory and practice of the various chromatographic methods should convince you of the tremendous influence chromatography has had on our biochemical understanding. It is tempting, but unfair, to make comparisons of the methods; however, each serves a specific purpose. There will, for example, always be a need for fast, inexpensive, and qualitative analyses as afforded by TLC. Traditional column chromatography will probably always be preferred in large-scale protein purification.

However, during the past two decades, an analytical method has been developed that currently rivals and may soon surpass the traditional liquid chromatographic techniques in importance for analytical separations. This technique, high-performance liquid chromatography (HPLC), is ideally suited for the separation and identification of amino acids, carbohydrates, lipids, nucleic acids, proteins, pigments, steroids, pharmaceuticals, and many other biologically active molecules.

The future promise of HPLC is indicated by its classification as "modern liquid chromatography" when compared to other forms of column-liquid chromatography, now referred to as "classical" or "traditional." Compared to the classical forms of liquid chromatography (paper, TLC, column), HPLC has several advantages:

1. Resolution and speed of analysis far exceed the classical methods.
2. HPLC columns can be reused without repacking or regeneration.
3. Reproducibility is greatly improved because the parameters affecting the efficiency of the separation can be closely controlled.
4. Instrument operation and data analysis are easily automated.
5. HPLC is adaptable to large-scale, preparative procedures.

Gas chromatography has had and will continue to have a major impact in biochemical research. The major disadvantage of GC is that many biochemical samples either are not volatile or are thermally unstable.

HPLC, which can be operated at ambient temperatures, does not have this limitation.

The advantages of HPLC are the result of two major advances: (1) the development of stationary supports with very small particle sizes and large surface areas, and (2) the improvement of elution rates by applying high pressure to the solvent flow.

The great versatility of HPLC is evidenced by the fact that all chromatographic modes, including partition, adsorption, ion exchange, chromatofocusing, and gel exclusion, are possible. In a sense, HPLC can be considered as automated liquid chromatography. The theory of each of these chromatographic modes has been discussed and needs no modification for application to HPLC. However, there are unique theoretical and practical characteristics of HPLC that should be introduced.

The **retention time** of a solute in HPLC (t_R) is defined as the time necessary for maximum elution of the particular solute. This is analogous to retention time measurements in GC. **Retention volume** (V_R) of a solute is the solvent volume required to elute the solute and is defined by Equation 3.4, where F is the flow rate of the solvent.

▶ $$V_R = Ft_R$$ (Equation 3.4)

In all forms of chromatography, a measure of column efficiency is **resolution**, R. Resolution indicates how well solutes are separated; it is defined by Equation 3.5, where t_R and t'_R are the retention times of two solutes and w and w' are the base peak widths of the same two solutes.

▶ $$R = 2\frac{t_R - t'_R}{w + w'}$$ (Equation 3.5)

Instrumentation

The increased resolution achieved in HPLC compared to classical column chromatography is primarily the result of adsorbents of very small particle sizes (less than 20 μm) and large surface areas. The smallest gel beads used in gel exclusion chromatography are "superfine" grade with diameters of 20 to 50 μm. Recall that the smaller the particle size, the lower the flow rate; therefore, it is not feasible to use very small gel beads in liquid column chromatography because low flow rates lead to solute diffusion and the time necessary for completion of an analysis would be impractical. In HPLC, increased flow rates are obtained by applying a pressure differential across the column. A combination of high pressure and adsorbents of small particle size leads to the high resolving power and short analysis times characteristic of HPLC.

A schematic diagram of a typical high-pressure liquid chromatograph is shown in Figure 3.12. The basic components are a solvent reservoir, high-pressure pump, packed column, detector, and recorder. The simi-

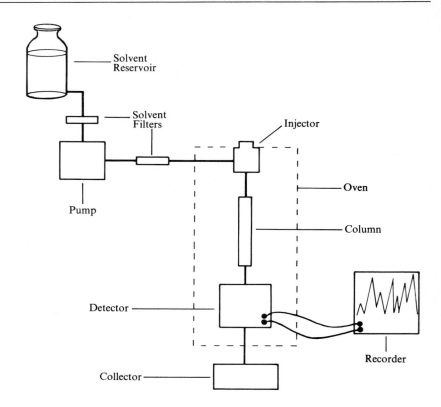

Figure 3.12
A schematic diagram of a high-performance liquid chromatograph.

larities between a gas chromatograph and an HPLC are obvious. The tank of carrier gas in GC is replaced by the solvent reservoir and high-pressure pump in HPLC.

SOLVENT RESERVOIR

The solvent chamber should have a capacity of at least 500 mL for analytical applications, but larger reservoirs are required for preparative work. In order to avoid bubbles in the column and detector, the solvent must be degassed. Several methods may be used to remove unwanted gases, including refluxing, filtration through a vacuum filter, ultrasonic vibration, and purging with an inert gas. The solvent should also be filtered to remove particulate matter that would be drawn into the pump and column.

PUMPING SYSTEMS

The purpose of the pump is to provide a constant, reproducible flow of solvent through the column. Two types of pumps are available—constant pressure and constant volume. Typical requirements for a pump are:

1. It must be capable of pressure outputs of at least 500 psi and preferably up to 5000 psi.

2. It should have a controlled, reproducible flow delivery of at least 3 mL/min for analytical applications and up to 100 mL/min for preparative applications.

3. It should yield pulse-free solvent flow.

4. It should have a small holdup volume.

Although neither type of pump meets all these criteria, constant-volume pumps maintain a more accurate flow rate and a more precise analysis is obtained.

INJECTION PORT

A sample must be introduced onto the column in an efficient and reproducible manner. One of the most popular injectors is the syringe injector. The sample, in a microliter syringe, is injected through a neoprene/Teflon septum. This type of injection can be used at pressures up to 3000 psi.

COLUMNS

HPLC columns are prepared from stainless steel or glass-Teflon tubing. Typical column inside diameters are 2.1, 3.2, or 4.5 mm for analytical separations and up to 30 mm for preparative applications. The length of the column can range from 5 to 100 cm, but 10 to 20 cm columns are common.

DETECTOR

Liquid chromatographs are equipped with a means to continuously monitor the column effluent and recognize the presence of solute. Only small sample sizes are used with most HPLC columns, so a detector must have high sensitivity. The type of detector that has the most universal application is the **differential refractometer**. This device continuously monitors the refractive index difference between the mobile phase (pure solvent) and the mobile phase containing sample (column effluent). The sensitivity of this detector is on the order of 0.1 μg, which, compared to other detectors, is only moderately sensitive. The major advantage of the refractometer detector is its versatility; its main limitation is that there must be at least 10^{-7} refractive index units between the mobile phase and sample. However, the refractometer cannot be used with gradient elution (see below); this presents a major limitation to its use.

The most widely used HPLC detectors are the **photometric detectors**. These detectors measure the extent of absorption of ultraviolet or visible radiation by a sample. Since few compounds are colored, visible detectors are of limited value. Ultraviolet detectors are the most widely used in HPLC. The typical UV detector functions by focusing radiation from a low-pressure mercury lamp on a flow cell that contains column effluent. The mercury lamp provides a primary radiation at 254 nm. The use of filters or other lamps provides radiation at 220, 280, 313, 334, and

365 nm. Many compounds absorb strongly in this wavelength range, and sensitivities on the order of 1 ng are possible. The extent of absorption is measured with a photocell. The ultraviolet detector is suitable for compounds containing π-bonding and nonbonded electrons. Most biochemicals are detected, including proteins, nucleic acids, pigments, vitamins, some steroids, and aromatic amino acids. Aliphatic amino acids, carbohydrates, lipids, and other biochemicals that do not absorb UV can be detected by chemical derivatization with UV-absorbing functional groups. UV detectors have many positive characteristics, including high sensitivity, small sample volumes, linearity over wide concentration ranges, nondestructiveness to sample, and suitability for gradient elution.

A third type of detector that has only limited use is the **fluorescence detector**. This type of detector is extremely sensitive; its use is limited to samples containing trace quantities of biological materials. Its response is not linear over a wide range of concentrations, but it may be up to 100 times more sensitive than the UV detector.

COLLECTION OF ELUENT

All of the detectors described here are nondestructive to the samples, so column effluent can be collected for further chemical and physical analysis.

ANALYSIS OF HPLC DATA

The response from the detector is directed to a strip chart recorder for a trace of solute elution vs. time. These data are useful for qualitative or quantitative analysis. As in GC, sample components may be identified by **retention time, peak enhancement,** or **physical/chemical characterization** of collected samples.

Most HPLC instruments are on line with an integrator and a microprocessor for data handling. For quantitative analysis of HPLC data, operating parameters such as rate of solvent flow must be controlled. In modern instruments, the whole system (including the pump, injector, detector, and data system) is under the control of a central microprocessor.

Figure 3.13 illustrates the separation by HPLC of several phenylhydantoin derivatives of amino acids.

Stationary Phases in HPLC

The adsorbents in HPLC are typically small-diameter, porous materials. Two types of stationary phases are available. **Porous layer beads** (Figure 3.14A) have an inert solid core with a thin porous outer shell of silica, alumina, or ion-exchange resin. The average diameter of the beads ranges from 20 to 45 μm. They are especially useful for analytical applications, but, because of their short pores, their capacities are too low for preparative applications.

Column: Microsorb Cyano, 5 μm,
    4.6 mm ID × 25 cm L
Mobile phase: 15.5% THF and 17.1% acetonitrile
    in 6 mM phosphate, pH 3.2
Flow: 1 mL/min
Temperature: 35°C
Detection: UV, 254 nm

PEAK IDENTIFICATION

1. Cys-A	12. Cys
2. Asn	13. Tyr
3. Gln	14. Pro
4. Ser	15. Val
5. Thr	16. Met
6. Asp	17. Ile
7. Gly	18. Phe
8. His	19. Leu
9. Glu	20. nor-Leu
10. Ala	21.Trp
11. Arg	22. Lys

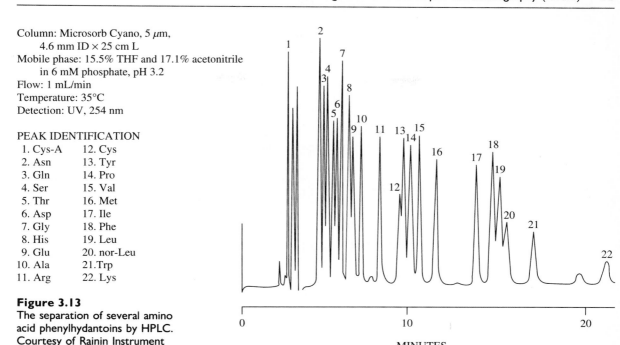

Figure 3.13
The separation of several amino acid phenylhydantoins by HPLC. Courtesy of Rainin Instrument Co., Woburn, MA.

Figure 3.14
Adsorbents used in HPLC. (A) Porous layer with short pores.   (B) Microporous particle with longer pores.

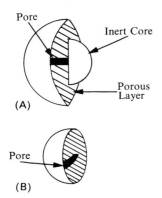

Microporous particles are available in two sizes: 20 to 40 μm diameter with longer pores and 5 to 10 μm with short pores (see Figure 3.14B). These are now more widely used than the porous layer beads because they offer greater resolution and faster separations with lower pressures. The microporous beads are prepared from alumina, silica, ion-exchanger resins, and chemically bonded phases (see next section).

The HPLC can function in several chromatographic modes. Each type of chromatography will be discussed below, with information about stationary phases.

LIQUID-SOLID (ADSORPTION) CHROMATOGRAPHY

HPLC in the adsorption mode can be carried out with silica or alumina porous-layer-bead columns. Small glass beads are often used for the inert core. Some of the more widely used packings are μPorasil (Waters Associates), BioSilA (Bio-Rad Laboratories), LiChrosorb Si-100 Partisil, Vydac, ALOX 60D (several suppliers), and Supelcosil (Supelco).

In high-pressure adsorption chromatography, solutes adsorb with different affinities to binding sites in the solid stationary phase. Separation of solutes in a sample mixture occurs because polar solutes adsorb more strongly than nonpolar solutes. Therefore, the various components in a sample are eluted with different retention times from the column. This form of HPLC is usually called **normal phase** (polar stationary phase and a nonpolar mobile phase).

LIQUID-LIQUID (PARTITION) CHROMATOGRAPHY

In the early days of HPLC (1970–78), solid supports were coated with a liquid stationary phase as in gas chromatography. Columns with these packings had short lifetimes and a gradual decrease in resolution because there was continuous loss of the liquid stationary phase with use of the column.

This problem was remedied by the discovery of methods for chemically bonding the active stationary phase to the inert support. Most chemically bonded stationary phases are produced by covalent modification of the surface silica. Three modification processes are shown in Equations 3.6–3.8.

▶ Silicate esters

$$—Si—OH + ROH \longrightarrow —Si—O—R \qquad \text{(Equation 3.6)}$$

▶ Silica-carbon

$$—Si—Cl + \begin{matrix} RMgBr \\ or \\ RLi \end{matrix} \longrightarrow —Si—R \qquad \text{(Equation 3.7)}$$

▶ Siloxanes

$$—Si—OH + \begin{matrix} ClSiR_3 \\ or \\ ROSiR_3 \end{matrix} \longrightarrow —Si—O—SiR_3 \qquad \text{(Equation 3.8)}$$

The major advantage of a bonded stationary phase is stability. Since it is chemically bonded, there is very little loss of stationary phase with column use. The siloxanes are the most widely used silica supports. Functional groups that can be attached as siloxanes are alkylnitriles ($—Si—CH_2CH_2—CN$), phenyl ($—Si—C_6H_5$), alkylamines ($—Si(CH_2)_n—NH_2$), and alkyl side chains ($—Si—C_8H_{17}$; $—Si—C_{18}H_{37}$).

The use of nonpolar chemically bonded stationary phases with a polar mobile phase is referred to as **reverse-phase HPLC**. This technique separates sample components according to hydrophobicity. It is widely used for the separation of all types of biomolecules, including peptides, nucleotides, carbohydrates, and derivatives of amino acids. Typical solvent systems are water-methanol, water-acetonitrile, and water-tetrahydrofuran mixtures. Figure 3.15 shows the results of protein separation on a silica-based reverse-phase column.

ION-EXCHANGE CHROMATOGRAPHY

Ion-exchange HPLC uses column packings with charged functional groups. Structures of typical ion exchangers are shown in Figure 3.16. They are prepared by chemically bonding the ionic groups to the support

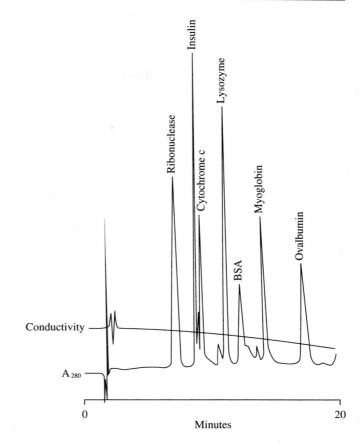

Conditions

Column:	Hi-Pore RP-304 column, 250 × 4.6 mm
Sample:	Protein standard
Buffers:	A. H₂O, 0.1% B
	B. 95% Acetonitrile, 5% H₂O, 0.1% TFA
Gradient:	25% – 100% B in 30 minutes
Flow rate:	1.5 mL/min
Temperature:	Ambient
Detection:	UV @ 220 nm; conductivity

Figure 3.15
Reverse-phase chromatography of a mixture of standard proteins. Separation courtesy of Bio-Rad Laboratories, Chemical Division, Richmond, CA. Used by permission.

via silicon atoms or by using polystyrene-divinylbenzene resins. These stationary phases may be used for the separation of proteins, peptides, and other charged biomolecules. Two widely used phases are DEAE-silica and CM-silica. Both of these columns can be used with aqueous buffers in the pH range 2.5 to 7.5. Chromatofocusing columns are available from commercial suppliers.

GEL EXCLUSION CHROMATOGRAPHY

The combination of HPLC and gel exclusion chromatography is used extensively for the separation of large biomolecules, especially proteins and nucleic acids.

The exclusion gels discussed in Section F are not appropriate for HPLC use because they are soft and, except for small-pore beads

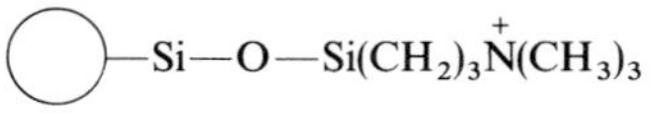

Anion Exchanger

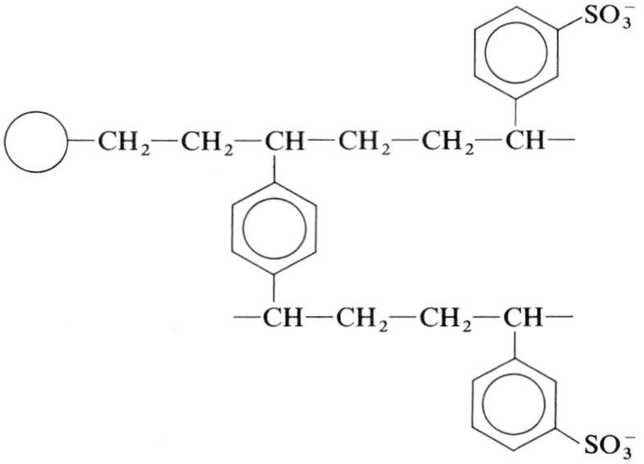

Cation Exchanger

Figure 3.16
Ion exchangers used in HPLC.

(G-25 and less), collapse under high-pressure conditions. Semirigid gels based on cross-linked styrene-divinylbenzene, polyacrylamide, and vinyl-acetate copolymer are available with various fractionation ranges useful for the separation of molecules up to 10,000,000 daltons. The semirigid gels are useful only with aqueous-based solvents and a few organic solvents.

Rigid packings for HPLC gel exclusion are prepared from porous glass or silica. They have several advantages over the semirigid gels, including several fractionation ranges, ease of packing, and compatibility with water and organic solvents.

The Mobile Phase

Selection of a column packing that is appropriate for a given analysis does not ensure a successful HPLC separation. A suitable solvent system must also be chosen. Several critical solvent properties will be considered here.

Purity Very-high-purity solvents with no particulate matter are required. Many laboratory workers do not purchase expensive prepurified solvents, but rather they purify lesser grade solvents by microfiltration through a Millipore system or distillation in glass.

Reactivity The mobile phase must not react with the analytical sample or column packing. This does not present a major limitation since many relatively unreactive hydrocarbons, alkyl halides, and alcohols are suitable.

Detector Compatibility   A solvent must be carefully chosen to avoid interference with the detector. Most UV detectors monitor the column effluent at 254 nm. Any UV-absorbing solvent, such as benzene or olefins, would be unacceptable because of high background. Since refractometer detectors monitor the difference in refractive index between solvent and column effluent, a greater difference leads to greater ability to detect the solute.

Solvents for HPLC Operation

It has long been recognized that the eluting power of a solvent is related to its polarity. Chromatographic solvents have been organized into a list according to their ability to displace adsorbed solutes (eluting power, $\varepsilon°$). This list of solvents, called an **eluotropic series**, is shown in Table 3.5. It should be noted that $\varepsilon°$ increases with an increase in polarity. Using the eluotropic series makes solvent choice less a matter of trial and error. Quite often, trial experiments can be performed on a silica gel thin-layer plate spotted with the analytical sample. A solvent, ethyl acetate, for example, is used to develop the plate. If ethyl acetate is too nonpolar to cause suitable movement and separation of the sample components, a solvent of higher eluent strength (ethanol, for example) is tried. If the solutes move too far, a solvent of lower $\varepsilon°$ is used (toluene, for example).

Table 3.5
Eluotropic Series of HPLC Solvents

Solvent	$\varepsilon°^{1}$	n_0^{2}	UV Cutoff (nm)
n-Pentane	0.00	1.358	210
n-Hexane	0.01	1.375	210
Cyclohexane	0.04	1.427	210
Carbon tetrachloride	0.18	1.466	265
2-Chloropropane	0.29	1.378	225
Toluene	0.29	1.496	285
Ethyl ether	0.38	1.353	220
Chloroform	0.40	1.443	245
Tetrahydrofuran	0.45	1.408	230
Acetone	0.56	1.359	330
Ethyl acetate	0.58	1.370	260
Dimethylsulfoxide	0.62	1.478	270
Acetonitrile	0.65	1.344	210
2-Propanol	0.82	1.380	210
Ethanol	0.88	1.361	210
Methanol	0.95	1.329	210
Ethylene glycol	1.11	1.427	210

[1] Eluent strength.
[2] Refractive index.

Occasionally, a single solvent does not provide suitable resolution of solutes. Solvent **binary mixtures** can be prepared with eluent strengths intermediate between the $\varepsilon°$ values for the individual solvents.

Gradient Elution in HPLC

Figure 3.17A illustrates the separation of a multicomponent sample using a single solvent and a silica column. Note that some earlier components

Figure 3.17
The general elution problem.   (A) Normal elution.   (B) Gradient elution. From L. R. Snyder, *J. Chromatogr. Sci.* **8**, 692 (1970) Reproduced by permission of Preston Publications.

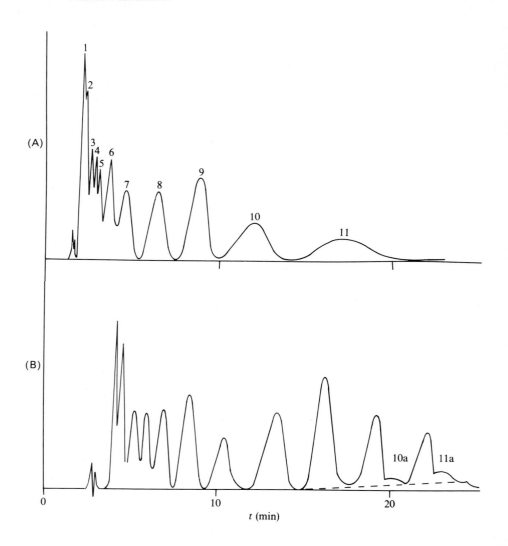

t (min)

(1 to 5) are poorly resolved and later peaks (9 to 11) display extensive broadening. This common difficulty encountered in the analysis of multi-component samples is referred to as the **general elution problem**. It is due to the fact that the components have a wide range of K_D values and no single solvent system is equally effective in displacing all components from the column.

The general elution problem is solved by the use of **gradient elution**. This is achieved by varying the composition of the mobile phase during elution. The similarity between gradient elution and temperature pro-gramming in gas chromatography should be readily apparent. In practice, gradient elution is performed by beginning with a weakly eluting solvent (low $\varepsilon°$) and gradually increasing the concentration of a more strongly eluting solvent (higher $\varepsilon°$). The weaker solvent is able to improve resolu-tion of components having low K_D values (peaks 1 to 5, Figure 3.17B). In addition, the gradual increase in $\varepsilon°$ of the solvent mixture decreases line broadening of components 9 to 11 and provides a more effective separation.

The choice of solvents for gradient elution is still somewhat empirical; however, using the data from Table 3.5 narrows the choices. Modern HPLC instruments are equipped with **solvent programming units** that control gradient elution in a stepwise or continuous manner.

Gradient elution cannot be used with the differential refractometer de-tector because it is difficult to match the refractive indices of the column effluent and the reference (a steady baseline is difficult to achieve). De-spite this limitation, gradient elution provides the conditions for maxi-mum resolution for most multicomponent samples.

Sample Preparation and Selection of HPLC Operating Conditions

During the initial stages of biochemical sample preparation, the sample is often quite crude; it may contain hundreds of components in addition to the desired biomolecules. Most sample preparations must be pretreated before optimum HPLC results can be expected. The following proce-dures may be needed in order to convert a crude sample into a clean one: desalting, removal of anions and cations, removal of metal ions, concen-tration of the desired macromolecules, removal of detergent, and particu-late removal. Sample preparation techniques used to achieve these results are gel exclusion chromatography, ion-exchange chromatography, micro-filtration, and metal affinity chromatography. These procedures may be completed by commercially available prefilters or precolumns.

Each type of HPLC instrument has its own characteristics and operat-ing directions. It is not feasible to describe those here. However, it is ap-propriate to outline the general approach taken when an HPLC analysis is desired. The following items must be considered:

1. Chemical nature and proper preparation of the sample.
2. Selection of type of chromatography (partition, adsorption, ion exchange, gel exclusion).
3. Choice of solvent system and mode of elution.
4. Selection of column packing.
5. Choice of equipment (type of detector).

FPLC—A Modification of HPLC

In 1982 Pharmacia-LKB introduced a new and innovative chromatographic method that provides a link between classical column chromatography and HPLC. This new technique, called **fast protein liquid chromatography (FPLC)**, uses experimental conditions intermediate between those of column chromatography and HPLC. The typical FPLC system requires a pump that will deliver solvent to the column in the flow rate range 1–499 mL/hr with operating pressures of 0–40 bar. (HPLC pumps deliver solvent in a flow rate range of 0.010–10 mL/min with operating pressures of 1–400 bar. Classical chromatography columns are operated at atmosphere pressure.) Also required for FPLC are a controller, detector, and fraction collector. Since lower pressures are used in FPLC than in HPLC, a wider range of column supports is possible. Chromatographic techniques that may be incorporated in an FPLC system are gel filtration, ion exchange, affinity (see next section), hydrophobic interaction, reversed phase, and chromatofocusing. Gels and prepacked columns to accomplish these procedures are commercially available from many sources including Pharmacia-LKB.

H. Affinity Chromatography and Immunoadsorption

The more conventional chromatographic procedures that we have studied up to this point rely on rather nonspecific physicochemical interactions between a stationary support and solute. The molecular characteristics of net charge, size, and polarity do not provide a basis for high selectivity in the separation and isolation of biomolecules. The desire for more specificity in chromatographic separations led to the discovery of **affinity chromatography**. This technique offers the ultimate in specificity—separation on the basis of biological interactions. The biological function displayed by most macromolecules (antibodies, transport proteins, enzymes, nucleic acids, polysaccharides, receptor proteins, etc.) is a result of recognition of and interaction with specific molecules called **ligands**. This is illustrated by Equation 3.9 where A represents a macromolecule and B is a smaller molecule or ligand. The two molecules interact in a specific manner to form a complex, A:B.

$$\blacktriangleright \quad A + B \; \rightleftharpoons \; A{:}B \; \longrightarrow \; \text{biological response} \qquad \text{(Equation 3.9)}$$

In a biological system, the formation of the complex often triggers some response such as immunological action, control of a metabolic process, hormone action, catalytic breakdown of a substrate, or membrane transport. The biological response depends on proper molecular recognition and binding as shown in the reaction. The most common example of Equation 3.9 is the interaction that occurs between an enzyme molecule, E, and a substrate, S, with reversible formation of an ES complex. The biological event resulting from this interaction is the transformation of S to a metabolic product, P. Only the first step in Equation 3.9, formation of the complex, is of concern in affinity chromatography.

In practice, affinity chromatography requires the preparation of an insoluble stationary phase, to which appropriate ligand molecules (B) are covalently affixed. Thus, ligand molecules are immobilized on the stationary support. The affinity support is packed into a column through which a mixture containing the desired macromolecule, A, is allowed to percolate. There are many types of molecules in the mixture, especially if it is a crude cell extract, but only macromolecules that recognize and bind to immobilized B are retarded in their movement through the column. After the nonbinding molecules have washed through the column, the desired macromolecules, B, are eluted by gentle disruption of the A:B complex. Study Figure 3.18 for an illustration of affinity chromatography.

Affinity chromatography can be applied to the isolation and purification of virtually all biological macromolecules. It has been used to purify nucleic acids, enzymes, transport proteins, antibodies, hormone receptor proteins, drug-binding proteins, neurotransmitter proteins, and many others.

Successful application of affinity chromatography requires careful design of experimental conditions. The essential components, which are

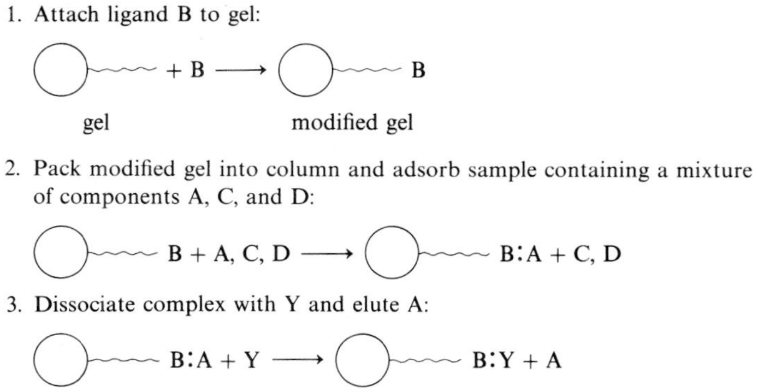

Figure 3.18
The steps of affinity chromatography.

1. Attach ligand B to gel:

   gel   modified gel

2. Pack modified gel into column and adsorb sample containing a mixture of components A, C, and D:

   B + A, C, D $\longrightarrow$ B:A + C, D

3. Dissociate complex with Y and elute A:

   B:A + Y $\longrightarrow$ B:Y + A

outlined below, are (1) creation and preparation of a stationary matrix with immobilized ligand and (2) design of column development and eluting conditions.

Chromatographic Media

Selection of the matrix used to immobilize a ligand requires consideration of several properties. The stationary supports used in gel exclusion chromatography are found to be quite suitable for affinity chromatography because (1) they are physically and chemically stable under most experimental conditions, (2) they are relatively free of nonspecific adsorption effects, (3) they have satisfactory flow characteristics, (4) they are available with very large pore sizes, and (5) they have reactive functional groups to which an appropriate ligand may be attached.

Four types of media possess most of these desirable characteristics: agarose, polyvinyl, polyacrylamide, and controlled-porosity glass (CPG) beads. Highly porous agarose beads such as Sepharose 4B (Pharmacia-LKB) and Bio-Gel A-150 m (Bio-Rad Laboratories) have virtually all of these characteristics and are the most widely used matrices. Polyacrylamide gels such as Bio-Gel P-300 (Bio-Rad) display many of the recommended features; however, the porosity is not especially high. Fractogel HW-65F, a vinyl polymer supplied by Pierce Chemical Company, has many of the above characteristics and a protein fractionation range of 50,000 to 5,000,000 daltons. CPG beads have not yet found wide acceptance, primarily because of extensive nonspecific protein adsorption and small numbers of reactive functional groups for ligand attachment. Future developments in the manufacture of these beads should minimize these difficulties. The advantages of the CPG beads are mechanical strength and chemical stability.

The Immobilized Ligand

The ligand (B in Equation 3.9 and Figure 3.18) can be selected only after the nature of the macromolecule to be isolated is known. When a hormone receptor protein is to be purified by affinity chromatography, the hormone itself is an ideal candidate for the ligand. For antibody isolation, an antigen or hapten may be used as ligand. If an enzyme is to be purified, a substrate analog, inhibitor, cofactor, or effector may be used as the immobilized ligand. The actual substrate molecule may be used as a ligand, but only if column conditions can be modified to avoid catalytic transformation of the bound substrate.

In addition to the foregoing requirements, the ligand must display a strong, specific, but reversible interaction with the desired macromolecule and it must have a reactive functional group for attachment to the matrix. It should be recognized that several types of ligand may be used for affinity purification of a particular macromolecule. Of course, some ligands will work better than others, and empirical binding studies can be performed to select an effective ligand.

Attachment of Ligand to Matrix

Several procedures have been developed for the covalent attachment of the ligand to the stationary support. All procedures for gel modification proceed in two separate chemical steps: (1) activation of the functional groups on the matrix and (2) joining of the ligand to the functional group on the matrix.

A wide variety of activated gels is now commercially available. The most widely used are described in the following.

CYANOGEN BROMIDE-ACTIVATED AGAROSE

This gel is especially versatile because all ligands containing primary amino groups are easily attached to the agarose. It is available under the trade name CNBr-activated Sepahrose 4B (Pharmacia-LKB). Since the gel is extremely reactive, very gentle conditions may be used to couple the ligand. One disadvantage of CNBr activation is that small ligands are coupled very closely to the matrix surface; macromolecules, because of steric repulsion, may not be able to interact fully with the ligand. The procedure for CNBr activation and ligand coupling is outlined in Figure 3.19A.

6-AMINOHEXANOIC ACID (CH)-AGAROSE AND 1,6-DIAMINOHEXANE (AH)-AGAROSE

These activated gels overcome the steric interference problems stated above by positioning a six-carbon spacer arm between the ligand and the matrix. Ligands with free primary amino groups can be covalently attached to CH-agarose, whereas ligands with free carboxyl groups can be coupled to AH-agarose. CH-agarose and AH-agarose are marketed by several companies including Pharmacia-LKB, Pierce Chemical Co., and Bio-Rad Laboratories. The attachment of ligands to AH and CH gels is outlined in Figure 3.19B,C.

CARBONYLDIIMIDAZOLE (CDI)-ACTIVATED SUPPORTS

Reaction with CDI produces gels that contain uncharged N-alkylcarbamate groups (see Figure 3.19D). CDI-activated agarose, dextran, and polyvinyl acetate are sold by Pierce Chemical Co. under the trade name Reacti-Gel.

EPOXY-ACTIVATED AGAROSE

The structure of this gel is shown in Figure 3.19E. It provides for the attachment of ligands containing hydroxyl, thiol, or amino groups. The

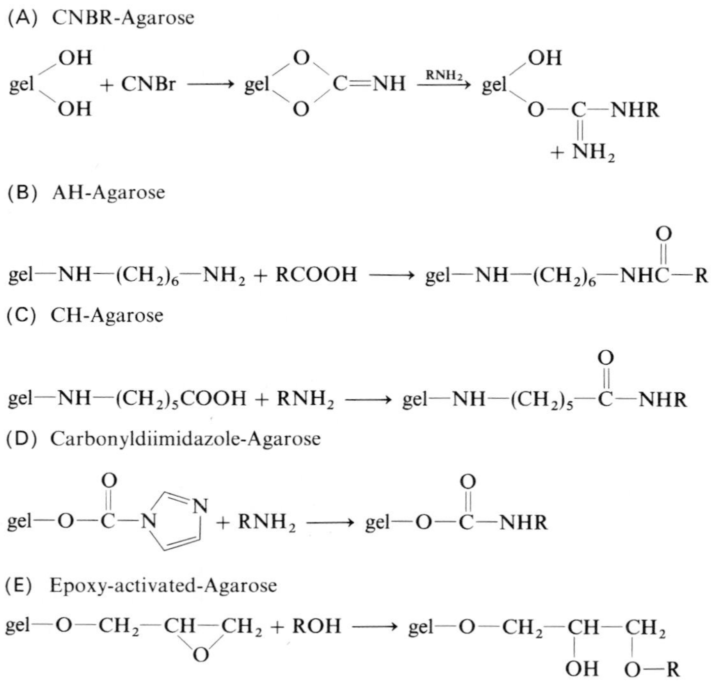

(A) CNBR-Agarose

$$\text{gel}\diagup_{OH}^{OH} + CNBr \longrightarrow \text{gel}\diagup_{O}^{O}\!\!\diagdown C=NH \xrightarrow{\ RNH_2\ } \text{gel}\diagup_{O-C-NHR}^{OH} \atop \quad\quad \overset{\|}{\underset{+\,NH_2}{}}$$

(B) AH-Agarose

$$\text{gel}-NH-(CH_2)_6-NH_2 + RCOOH \longrightarrow \text{gel}-NH-(CH_2)_6-NH\overset{\displaystyle O}{\overset{\|}{C}}-R$$

(C) CH-Agarose

$$\text{gel}-NH-(CH_2)_5COOH + RNH_2 \longrightarrow \text{gel}-NH-(CH_2)_5-\overset{\displaystyle O}{\overset{\|}{C}}-NHR$$

(D) Carbonyldiimidazole-Agarose

$$\text{gel}-O-\overset{\displaystyle O}{\overset{\|}{C}}-N\!\diagdown\!\!\diagup^{N} + RNH_2 \longrightarrow \text{gel}-O-\overset{\displaystyle O}{\overset{\|}{C}}-NHR$$

(E) Epoxy-activated-Agarose

$$\text{gel}-O-CH_2-CH\underset{O}{-}CH_2 + ROH \longrightarrow \text{gel}-O-CH_2-\underset{\underset{OH}{|}}{CH}-\underset{\underset{O-R}{|}}{CH_2}$$

Figure 3.19
Attachment of specific ligands to activated gels. R = ligand.

hydroxyl groups of mono-, oligo-, and polysaccharides can readily be attached to the gel. Epoxy-activated Sepharose 6B is available from Pharmacia-LKB.

GROUP-SPECIFIC ADSORBENTS

The affinity materials described up to this point are modified with a ligand having specificity for a particular macromolecule. Therefore, each time a biomolecule is to be isolated by affinity chromatography, a new adsorbent must be designed and prepared. Ligands of this type are called substance specific. In contrast, group-specific adsorbents contain ligands that have affinity for a class of biochemically related substances. Table 3.6 shows several commercially available group-specific adsorbents and their specificities. The principles behind binding of nucleic acids and proteins to group-specific adsorbents depend on the actual affinity adsorbent. In most cases, the immobilized ligand and macromolecule (protein or nucleic acid) interact through one or more of the following forces: hydrogen bonding, hydrophobic interactions, and/or covalent interactions. Some group-specific adsorbents deserve special attention. Phenyl- and octyl-Sepharose are gels used for **hydrophobic interaction chromatography.** These adsorbent materials separate proteins on the basis of their hydrophobic character. Because most proteins contain hydrophobic amino acid side chains, this method is widely used. Octyl-Sepharose is

Table 3.6
Group-Specific Adsorbents Useful in Biochemical Applications

Group-Specific Adsorbent	Group Specificity
5'-AMP-agarose	Enzymes that have NAD$^+$ cofactor; ATP-dependent kinases
Benzamidine-Sepharose	Serine proteases
Boronic acid–agarose	Compounds with *cis*-diol groups; sugars, catecholamines, ribonucleotides, glycoproteins
Cibracron blue–agarose	Enzymes with nucleotide cofactors (dehydrogenases, kinases, DNA polymerases); serum albumin
Concanavalin A–agarose	Glycoproteins and glycolipids
Heparin-Sepharose	Nucleic acid–binding proteins, restriction endonucleases, lipoproteins
Iminodiacetate-agarose	Proteins with affinity for metal ions, serum proteins, interferons
Lentil lectin–Sepharose	Detergent–soluble membrane proteins
Lysine-Sepharose	Nucleic acids
Octyl-Sepharose	Weakly hydrophobic proteins, membrane proteins
Phenyl-Sepharose	Strongly hydrophobic proteins
Poly(A)-agarose	Nucleic acids containing poly (U) sequences, mRNA-binding proteins
Poly(U)-agarose	Nucleic acids containing poly (A) sequences, poly(U)-binding proteins
Protein A–agarose	IgG-type antibodies
Thiopropyl-Sepharose	–SH containing proteins

strongly hydrophobic; hence it binds nonpolar proteins very strongly. Phenyl-Sepharose is more weakly hydrophobic; therefore, it is more likely to reversibly bind strongly hydrophobic proteins.

The use of thiopropyl-Sepharose and boronic acid–agarose is an example of **covalent chromatography,** since relatively strong but reversible covalent bonds are formed between the affinity gel and specific macromolecules. In Experiment 10, hemoglobin A$_{1c}$, a glycoprotein, is isolated by covalent chromatography using boronic acid–agarose.

Metal affinity chromatography is a relatively new method that separates proteins on the basis of metal binding. This technique is used in Experiment 3 to isolate α-lactalbumin from milk.

The availability of a great variety of group-specific adsorbents in prepacked columns makes possible the combination of FPLC and affinity chromatography for the separation and purification of proteins.

One of the most specific modifications of affinity chromatography is **immunoaffinity**. The unique high specificity of antibodies for their antigens is valuable for the purification of antigens. In practice, the antibody is immobilized on a column support. When a mixture containing several other proteins along with the protein antigen for the antibody is passed through the column, only the antigen binds; the other proteins, which have no affinity for the antibody, wash off the column. Protein A–agarose in Table 3.6 is an example of immunoaffinity; however, this adsorbent does not recognize specific antibodies but, rather, the general family of immunoglobulin G antibodies.

Prepacked columns for group-specific affinity chromatography are now commercially available. These columns save much time since the affinity gel is provided and need not be prepared and characterized.

Experimental Procedure for Affinity Chromatography

Although the procedure is different for each type of substance isolated, a general experimental plan is outlined here. Figure 3.20 provides a step-by-step plan in flowchart form. Many types of matrix-ligand systems are commercially available and the costs are reasonable, so it is not always necessary to spend valuable laboratory time for affinity gel preparation. Even if a specific gel is not available, time can be saved by purchasing preactivated gels for direct attachment of the desired ligand. Once the gel is prepared, the procedure is similar to that described earlier. The major difference is the use of shorter columns. Most affinity gels have high capacities, and column beds less than 10 cm in length are commonly used. A second difference is the mode of elution. Ligand-macromolecule complexes immobilized on the column are held together by hydrogen bonding, ionic interactions, and hydrophobic effects. Any agent that diminishes these forces causes the release and elution of the macromolecule from the column. The common methods of elution are change of buffer pH, increase of buffer ionic strength, affinity elution, and chaotropic agents. The choice of elution method depends on many factors, including the types of forces responsible for complex formation and the stability of the ligand matrix and isolated macromolecule.

BUFFER pH OR IONIC STRENGTH

If ionic interactions are important for complex formation, a change in pH or ionic strength weakens the interaction by altering the extent of ionization of ligand and macromolecule. In practice, either a decrease in pH or a gradual increase in ionic strength (continual or stepwise gradient) is used.

AFFINITY ELUTION

In this method of elution, a selective substance added to the eluting buffer competes for binding to the ligand or for binding to the adsorbed macromolecule. An example of the first case is the elution of glycoproteins from a boronic acid-agarose gel by addition of a free sugar, sorbitol, to the eluting buffer (Experiment 10).

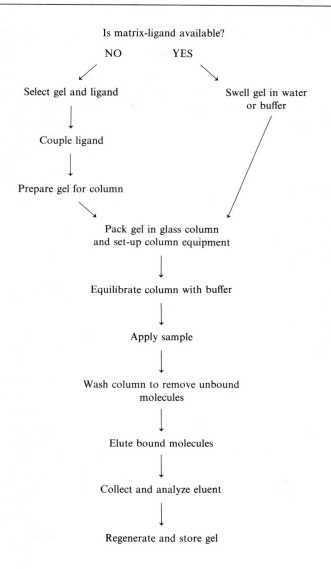

Figure 3.20
Experimental procedure for affinity chromatography.

CHAOTROPIC AGENTS

If gentle and selective elution methods do not release the bound macromolecule, then mild denaturing agents can be added to the buffer. These substances deform protein and nucleic acid structure and decrease the stability of the complex formed on the affinity gel. The most useful agents are urea, guanidine · HCl, CNS^-, ClO_4^-, and CCl_3COO^-. These substances should be used with care, because they may cause irreversible structural changes in the isolated macromolecule.

The application of affinity chromatography is limited only by the imagination of the investigator. Every year literally hundreds of research

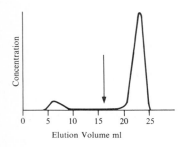

Figure 3.21
Purification of α-chymotrypsin by affinity chromatography on immobilized D-tryptophan methyl ester. From *Affinity Chromatography: Principles and Methods,* Pharmacia-LKB, Uppsala, Sweden.

papers appear with new and creative applications of affinity chromatography. Figure 3.21 illustrates the purification of α-chymotrypsin by affinity chromatography on immobilized D-tryptophan methyl ester. α-Chymotrypsin can recognize and bind, but not chemically transform, D-tryptophan methyl ester. The enzyme catalyzes the hydrolysis of L-tryptophan methyl ester. The impure α-chymotrypsin mixture was applied to the gel, D-tryptophan methyl ester coupled to CH-Sepharose 4B, and the column washed with Tris buffer. At the point shown by the arrow, the eluent was changed to 0.1 M acetic acid. The decrease in pH caused release of α-chymotrypsin from the column.

The best sources of further information and more applications are the many excellent brochures available from commercial suppliers of affinity gels.

I. Perfusion Chromatography

A new separation method that improves resolution and decreases the time required for analysis of biomolecules has been introduced. This method, called **perfusion chromatography,** relies on a new type of particle support called POROS, which may be used in low-pressure and high-pressure liquid chromatography applications. In conventional chromatographic separations, some biomolecules in the sample move rapidly around and past the media particles while other molecules diffuse slowly through the particles (Figure 3.22A). The result is loss of resolution because some biomolecules exit the column before others. To improve resolution, the researcher with conventional media found it necessary to reduce the flow rate to allow for diffusion processes, increasing the time required for analysis. In other words, before the development of perfusion chromatography, the researcher had to choose between high speed–low resolution and low speed–high resolution. POROS particles have two types of pores—**through pores** (6000–8000 Å in diameter), which provide channels through the particles, and connected **diffusion pores** (800–1500 Å in diameter), which line the through pores and have very short diffusion path lengths (Figure 3.22B). This combination pore system increases the porosity and the effective surface area of the particles and results in improved resolution and shorter analysis times (30 seconds to 3 minutes for POROS versus 30 minutes to several hours for conventional media).

POROS media, made by copolymerization of styrene and divinylbenzene, have high mechanical strength and are resistant to many solvents and chemicals. The functional surface chemistry of the particles can be modified to provide supports for many types of chromatography, including ion exchange, hydrophobic interaction, immobilized metal affinity, reversed phase, group-selective affinity, and conventional bioaffinity. Perfusion chromatography has been applied with success to

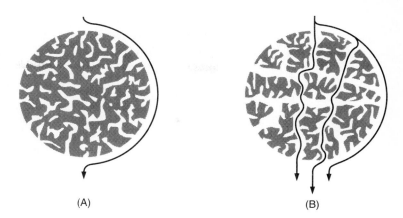

(A)   (B)

Figure 3.22
Transport of biomolecules through chromatographic media.   (A) Conventional
support particles.   (B) POROS particles for perfusion chromatography. Courtesy
of PerSeptive Biosystems, Cambridge, MA.

the separation of peptides, proteins, and polynucleotides on both preparative and analytical scales.

In addition to high resolution and short analysis times, perfusion chromatography has the advantag⌐ of improved recovery of biological activity because active biomolecules spend less time on the column, where
denaturing conditions may exist. This state-of-the-art separation method
is patented by PerSeptive Biosystems, Inc.

References

N. Afeyan, S. Fulton, N. Gordon, I. Mazsaroff, L. Varady and F. Regnier, *Bio/Technology* **8**(3), 203–206 (1990). "Perfusion Chromatography: An Approach to Purifying Biomolecules."

H. Barth and B. Boyes, *Anal. Chem.* **62**, 382R–394R (1990). "Size Exclusion Chromatography."

P. R. Brown, Editor, *HPLC In Nucleic Acid Research: Methods and Applications* (1984), Marcel Dekker (New York). Analysis of nucleotides
and nucleic acids.

I. Chaiken, *J. Chromatogr.* **488**, 145–160 (1989). "Bioaffinity Chromatography."

S. C. Churms, *J. Chromatogr.* **500**, 555–583 (1990). "Recent Developments in the Chromatographic Analysis of Carbohydrates."

R. E. Clement, Editor, *Gas Chromatography: Biochemical, Biomedical
and Clinical Applications* (1990), John Wiley & Sons (New York).
Over 14 chapters of modern GC techniques.

M. Colpan and D. Riesner, in *Modern Physical Methods in Biochemistry,* Part B, A. Neuberger and L. van Deenen, Editors (1988), Elsevier (Amsterdam), pp. 85–105. "HPLC of Nucleic Acids."

T. Cooper, *The Tools of Biochemistry* (1977), John Wiley & Sons (New York). Chapters on ion-exchange, gel permeation, and affinity chromatography.

A. Fallon, R. Booth, and L. Bell, in *Laboratory Techniques in Biochemistry and Molecular Biology,* R. Burdon and P. Knippenberg, Editors (1987), Vol. 17, Elsevier (Amsterdam). "Applications of HPLC in Biochemistry."

D. Freifelder, *Physical Biochemistry,* 2nd ed. (1982). W. H. Freeman (San Francisco), pp. 216–275. Theory and applications of chromatography.

M. T. Gilbert, *High Performance Liquid Chromatography* (1987), Wright Publishers (Bristol). Introduction to basic theory and principles.

N. Grinberg, Editor, *Modern Thin-Layer Chromatography* (1990), Marcel Dekker (New York). Up-to-date developments in TLC.

M. Hearn and M. Aguilar, in *Modern Physical Methods in Biochemistry* (1988), A. Neuberger and L. van Deenen, Editors, Elsevier (Amsterdam), Vol. 11B, pp. 107–142. "Reversed Phase HPLC of Peptides and Proteins."

K. Hicks, in *Advances in Carbohydrate Chemistry and Biochemistry,* Vol. 46, R. Tipson and D. Horton, Editors (1988), Academic Press (San Diego), pp. 17–72. "HPLC of Carbohydrates."

W. Jakoby, Editor, *Methods in Enzymology,* Vol. XXII (1971), Academic Press (New York), pp. 273–378. Chapters on ion-exchange, gel filtration, and hydroxyapatite chromatography.

W. Jakoby, Editor, *Methods in Enzymology,* Vol. 34B (1974), Academic Press (New York). Entire volume is devoted to affinity chromatography.

J. Janson and T. Kristiansen, *Chromatogr. Sci.* **47**, 747–81 (1990). "Packings in Affinity Chromatography."

W. Jennings, *Gas Chromatography with Glass Capillary Columns,* 2nd ed. (1980), Academic Press (New York). Detailed coverage of GC.

J. A. Jonsson, Editor, *Chromatographic Theory and Basic Principles* (1987), Marcel Dekker (New York). Good modern introduction to chromatography.

A. Krstulovic' and P. Brown, *Reversed Phase High Performance Liquid Chromatography* (1982), Wiley-Interscience (New York). Theory, practice, and biomedical applications.

C. Mathews and K. van Holde, *Biochemistry* (1990), Benjamin/Cummings (Redwood City, CA) pp. 156–161. A brief introduction to macromolecule purification by chromatography.

P. Mohr and K. Pommerening, *Affinity Chromatography* (1985), Marcel Dekker (New York). Practical and theoretical aspects.

T. Ngo, Editor, *Nonisotopic Immunoassay* (1988), Plenum Publishing (New York). Detailed discussion of several analytical techniques using immunoassays.

A. Niederwieser, in *Methods in Enzymology,* Vol. XXV, C. H. W. Hirs and S. N. Timasheff, Editors (1972), Academic Press (New York), pp. 60–99. Thin-layer chromatography of amino acids.

I. Papadoyannis, *HPLC in Clinical Chemistry* (1990), Marcel Dekker (New York). A comprehensive reference book.

Pharmacia-LKB Biotechnology, *Gel Filtration: Principles and Methods,* 5th ed. (1991), Pharmacia-LKB Biotechnology, S-751 82 Uppsala, Sweden, or 800 Centennial Avenue, Piscataway, NJ 08854. An excellent guide on theory, practice, and applications of gel filtration.

R. Plattner, in *Methods in Enzymology,* Vol. 72, J. M. Lowenstein, Editor (1981), Academic Press (New York), pp. 21–33. HPLC of triglycerides.

N. Porter, *ibid.* pp. 34–40. HPLC of complex lipids.

J. D. Rawn, *Biochemistry* (1989), Neil Patterson (Burlington, NC) pp. 61–65. Chromatography.

F. Regnier, *Nature* **350**, 634 (1991). "Perfusion Chromatography."

J. F. Robyt and B. J. White, *Biochemical Techniques: Theory and Practice* (1987). Brooks/Cole (Monterey, CA), pp. 73–128. An introduction to chromatographic techniques.

E. F. Rossomando, *HPLC in Enzymatic Analysis* (1987), John Wiley & Sons (New York). Purification of proteins by HPLC.

W. H. Scouten, *Affinity Chromatography* (1981), Wiley-Interscience (New York). Theory, practice, and application of affinity chromatography.

J. Sherma, *Anal. Chem.* **62,** 371R–381R (1990). "Planar Chromatography."

L. R. Snyder, J. L. Glajch, and J. J. Kirkland, *Practical HPLC Method Development* (1988), John Wiley & Sons (New York). Theory and applications of HPLC.

L. Stryer, *Biochemistry* (1988), Freeman (New York) pp. 47–48. Affinity, ion-exchange, and gel-filtration chromatography.

R. Synovec, E. Johnson, L. Moore, and C. Renn, *Anal. Chem.* **62**, 357R–370R (1990). "Column Liquid Chromatography: Equipment and Instrumentation."

J. C. Touchstone, Editor, *Planar Chromatography in the Life Sciences* (1990), John Wiley & Sons (New York). Modern aspects of TLC.

K. K. Unger, Editor, *Packings and Stationary Phases in Chromatographic Techniques* (1990), Marcel Dekker (New York). Extensive review of HPLC stationary phases.

D. Voet and J. Voet, *Biochemistry* (1990), John Wiley & Sons (New York), pp. 81–94. Use of chromatography in protein purification.

P. A. Williams and M. J. Hudson, Editors, *Recent Developments in Ion Exchange* (1990), Elsevier Applied Science (Amsterdam). Modern discussion of ion-exchange chromatography.

W. Worthy, *Chem. Eng. News,* November 18, pp. 25–26 (1991). Applications of perfusion chromatography.

Sources of Chromatography Information and Materials

Bio-Rad Laboratories
2200 Wright Avenue
Richmond, CA 94804

Pharmacia-LKB Biotechnology, Inc.
800 Centennial Avenue
Piscataway, NJ 08854

Pierce Chemical Company
P.O. Box 117
Rockford, IL 61105

PerSeptive Biosystems, Inc.
38 Sidney Street
Cambridge, MA 02139

CHAPTER 4

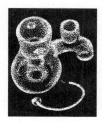

Characterization of Proteins and Nucleic Acids by Electrophoresis

■ Electrophoresis is the study of the movement of charged molecules in an electric field. The technique, which has been used by biochemists for over 50 years, is especially applicable to the characterization and analysis of charged biological polymers. Molecular migration in an electric field is influenced by the size, shape, charge, and chemical composition of the molecule. Electrophoresis, which is a relatively inexpensive and convenient technique, is capable of analyzing and purifying several different types of biomolecules, but especially proteins and nucleic acids.

Modern electrophoretic techniques use a buffer-saturated gel-type matrix as a support medium. The sample to be analyzed is applied to the medium as a spot or thin band, hence the term "zonal" electrophoresis. The migration of molecules is influenced by the applied electric field and the rigid, mazelike matrix of the gel support. Although it is difficult to provide an exact theoretical description of the electrophoretic movement of molecules in a gel support, zonal electrophoresis has gained widespread use in purifying and analyzing biochemicals and in determining their molecular size.

A. Theory of Electrophoresis

The movement of a charged molecule subjected to an electric field is represented by Equation 4.1.

$$v = \frac{Eq}{f}$$

(Equation 4.1)

where

E = the electric field in volts/cm

q = the net charge on the molecule

f = frictional coefficient, which depends on the mass and shape of the molecule

v = the velocity of the molecule

The charged particle moves at a velocity that depends directly on the electrical field (E) and charge (q), but inversely on a counteracting force generated by the viscous drag (f). The applied voltage represented by E in Equation 4.1 is usually held constant during electrophoresis, although some experiments are run under conditions of constant current (where the voltage changes with resistance) or constant power (the product of voltage and current). Under constant-voltage conditions, Equation 4.1 shows that the movement of a charged molecule depends only on the ratio q/f. For molecules of similar conformation (for example, a collection of linear DNA fragments or spherical proteins), f varies with size but not shape; therefore, the only remaining variables in Equation 4.1 are the charge (q) and mass dependence of f, meaning that under such conditions molecules migrate in an electric field at a rate proportional to their charge-to-mass ratio.

The movement of a charged particle in an electric field is often defined in terms of mobility, μ, the velocity per unit of electric field (Equation 4.2).

▶ $$\mu = \frac{v}{E}$$ (Equation 4.2)

This equation can be modified using Equation 4.1.

▶ $$\mu = \frac{Eq}{Ef} = \frac{q}{f}$$ (Equation 4.3)

In theory, if the net charge, q, on a molecule is known, it should be possible to measure f and obtain information about the hydrodynamic size and shape of that molecule by investigating its mobility in an electric field. Attempts to define f by electrophoresis have not been successful, primarily because Equation 4.3 does not adequately describe the electrophoretic process. Important factors that are not accounted for in the equation are interaction of migrating molecules with the support medium and shielding of the molecules by buffer ions. This means that electrophoresis is not useful for describing specific details of the size and shape of a molecule. Instead, it has been applied to the analysis of purity and identification of macromolecules. Each molecule in a mixture is expected to have a unique charge and size, and its mobility in an electric field will therefore be unique. This expectation forms the basis for analysis and separation by all electrophoretic methods. The technique is especially useful for the analysis of amino acids, peptides, proteins, nucleotides, and nucleic acids.

B. Methods of Electrophoresis

All types of electrophoresis are based on the principles just outlined. The major difference between methods is the type of support medium, which can be either cellulose or thin gels. Cellulose is used as a support medium for low-molecular-weight biochemicals such as amino acids and carbohydrates, and polyacrylamide and agarose gels are widely used as support media for larger molecules. Geometries (vertical and horizontal), buffers, and electrophoretic conditions for these two types of gels provide several different experimental arrangements, as described below.

Polyacrylamide Gel Electrophoresis (PAGE)

Gels formed by polymerization of acrylamide have several positive features in electrophoresis: (1) high resolving power for small and moderately sized proteins and nucleic acids (up to approximately 1×10^6 daltons), (2) acceptance of relatively large sample sizes, (3) minimal interactions of the migrating molecules with the matrix, and (4) physical stability of the matrix. Recall from the earlier discussion of gel filtration (Chapter 3) that gels can be prepared with different pore sizes by changing the concentration of cross-linking agents. Electrophoresis through polyacrylamide gels leads to enhanced resolution of sample components because the separation is based on both molecular sieving and electrophoretic mobility. The order of molecular movement in gel filtration and PAGE is very different, however. In gel filtration (Chapter 3), large molecules migrate through the matrix faster than small molecules. The opposite is the case for gel electrophoresis, where there is no void volume in the matrix, only a continuous network of pores throughout the gel. The electrophoresis gel is comparable to a single bead in gel filtration. Therefore, large molecules do not move easily through the medium, and the rate of movement is small molecules > large molecules.

Polyacrylamide gels are prepared by the free radical polymerization of acrylamide and the cross-linking agent N,N'-methylene-bis-acrylamide (Figure 4.1). Chemical polymerization is controlled by an initiator-catalyst system, ammonium persulfate–N,N,N',N'-tetramethylethylenediamine (TEMED). Photochemical polymerization may be initiated by riboflavin in the presence of ultraviolet (UV) radiation. A standard gel for protein separation is 7.5% polyacrylamide. It can be used over the molecular size range of 10,000 to 1,000,000 daltons; however, the best resolution is obtained in the range of 30,000 to 300,000 daltons. The re-

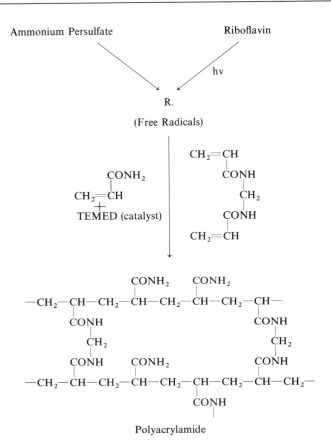

Figure 4.1
Chemical reactions illustrating the copolymerization of acrylamide and *N,N'*-methylene-bis-acrylamide. See text for details.

solving power and molecular size range of a gel depend on the concentrations of acrylamide and bis-acrylamide. Lower concentrations give gels with larger pores, allowing analysis of higher-molecular-weight biomolecules. In contrast, higher concentrations of acrylamide give gels with smaller pores, allowing analysis of lower-molecular-weight biomolecules (Table 4.1).

A new electrophoresis gel that eliminates the need for the highly toxic monomer acrylamide has recently been introduced by International Biotechnologies, Inc., a subsidiary of Eastman Kodak Company. This new water-soluble gel system is called Instacryl. It comes as a 20% (weight basis) stock solution that quickly and safely polymerizes with dithiothreitol. There is no need to mix acrylamide, bis-acrylamide, ammonium persulfate, and TEMED. One type of gel, Instacryl L Copolymer, can be used to separate proteins in the molecular size range 14,000 to 330,000 daltons when used in gel concentrations of 8–15%.

Polyacrylamide electrophoresis can be done using either of two arrangements, column or slab. Figure 4.2 shows the typical arrangement for a column gel. Glass tubes (10 cm × 6 mm i.d.) are filled with a mix-

Table 4.1
Effective Range of Separation of DNA by PAGE

Acrylamide[1] (% w/v)	Range of Separation (bp)	Bromphenol Blue[2]	Xylene Cyanol[2]
3.5	1000–2000	100	450
5.0	80–500	65	250
8.0	60–400	50	150
12.0	40–200	20	75
20.0	5–100	10	50

[1] Ratio of acrylamide to bis-acrylamide, 20:1.
[2] The numbers (in bp) represent the size of DNA fragment with the same mobility as the dye.

ture of acrylamide, *N,N'*-methylene-bis-acrylamide, buffer, and free radical initiator-catalyst. Polymerization occurs in 30 to 40 minutes. The gel column is inserted between two separate buffer reservoirs. The upper reservoir usually contains the cathode and the lower the anode. Gel electrophoresis is usually carried out at basic pH, where most biological polymers are anionic; hence, they move down toward the anode. The sample to be analyzed is layered on top of the gel and voltage is applied to the

Figure 4.2
A column gel for polyacrylamide electrophoresis.

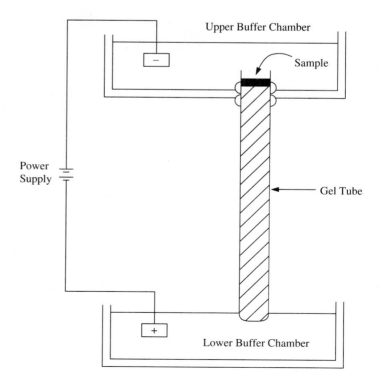

system. A "tracking dye" is also applied, which moves more rapidly through the gel than the sample components. When the dye band has moved to the opposite end of the column, the voltage is turned off and the gel is removed from the column and stained with a dye. Chambers for gel electrophoresis are commercially available or can be constructed from inexpensive materials.

A typical vertical slab gel apparatus is shown in Figure 4.3. The polyacrylamide slab is prepared between two glass plates that are separated by spacers (Figure 4.4). The spacers allow a uniform slab thickness of 0.5 to 2.0 mm, which is appropriate for analytical procedures. Slab gels may be as large as 20 × 40 cm and can accommodate multiple samples. A plastic "comb" inserted into the top of the gel during polymerization forms indentations in the gel that serve as sample wells. Up to 20 sample wells may be formed. After polymerization, the tape and comb are carefully removed and the wells are rinsed thoroughly with buffer to remove salts and any unpolymerized acrylamide. The gel plate is clamped into place between two buffer reservoirs, a sample is loaded into each well, and voltage is applied. For visualization, the slab is removed and stained with an appropriate dye.

Several modifications of PAGE have greatly increased its versatility and usefulness as an analytical tool.

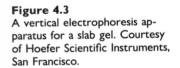

Figure 4.3
A vertical electrophoresis apparatus for a slab gel. Courtesy of Hoefer Scientific Instruments, San Francisco.

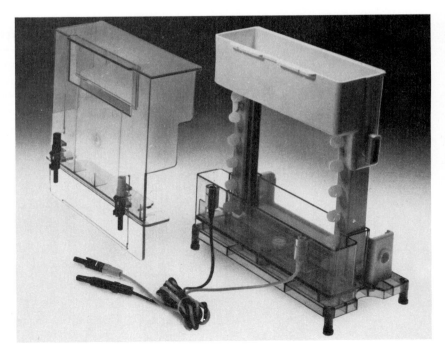

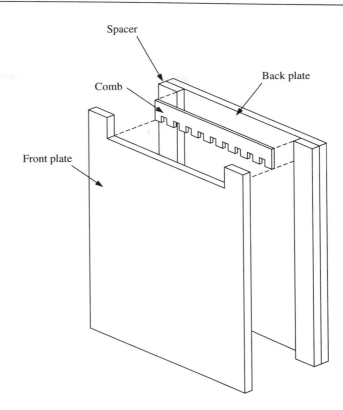

Spacer

Comb

Back plate

Front plate

Figure 4.4
Arrangement of two glass plates
with spacers to form a slab gel.
The comb is used to prepare
wells for placement of samples.

DISCONTINUOUS GEL ELECTROPHORESIS

The experimental arrangement for "disc" gel electrophoresis is shown
in Figure 4.5. The three significant differences in this method are
(1) there are now two gel layers, a lower or **resolving gel** and an upper
or **stacking gel**; (2) the buffers used to prepare the two gel layers are of
different ionic strengths and pH; and (3) the stacking gel has a lower
acrylamide concentration, so its pore sizes are larger. These three
changes in the experimental conditions cause the formation of highly
concentrated bands of sample in the stacking gel and greater resolution of
the sample components in the lower gel. Sample concentration in the upper gel occurs in the following manner. The sample is usually dissolved
in glycine-chloride buffer, pH 8 to 9, before loading on the gel. Glycine
exists primarily in two forms at this pH, a zwitterion and an anion (Equation 4.4).

▶ $H_3\overset{+}{N}CH_2COO^- \rightleftharpoons H_2NCH_2COO^- + H^+$ (Equation 4.4)

The average charge on glycine anions at pH 8.5 is about -0.2. When the
voltage is turned on, buffer ions (glycinate and chloride) and protein or

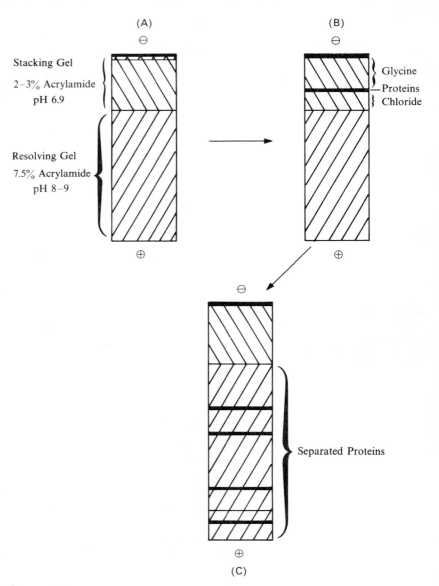

Figure 4.5
The process of disc gel electrophoresis.   (A) Before electrophoresis.   (B) Movement of chloride, glycinate, and protein through the stacking gel.   (C) Separation of
protein samples by the resolving gel.

nucleic acid sample move into the stacking gel, which has a pH of 6.9.
Upon entry into the upper gel, the equilibrium of Equation 4.4 shifts toward the left, increasing the concentration of glycine zwitterion, which
has no net charge and hence no electrophoretic mobility. In order to
maintain a constant current in the electrophoresis system, a flow of anions must be maintained. Since most proteins and nucleic acid samples

are still anionic at pH 6.9, they replace glycinate as mobile ions. Therefore, the relative ion mobilities in the stacking gel are chloride > protein or nucleic acid sample > glycinate. The sample will tend to accumulate and form a thin, concentrated band sandwiched between the chloride and glycinate as they move through the upper gel. Since the acrylamide concentration in the stacking gel is low (2 to 3%), there is little impediment to the mobility of the large sample molecules.

Now, when the ionic front reaches the lower gel with pH 8 to 9 buffer, the glycinate concentration increases and anionic glycine and chloride carry most of the current. The protein or nucleic acid sample molecules, now in a narrow band, encounter both an increase in pH and a decrease in pore size. The increase in pH would, of course, tend to increase electrophoretic mobility, but the smaller pores decrease mobility. The relative rate of movement of anions in the lower gel is chloride > glycinate > protein or nucleic acid sample. The separation of sample components in the resolving gel occurs as described in an earlier section on gel electrophoresis. Each component has a unique charge/mass ratio and a discrete size and shape, which directly influence its mobility.

Disc gel electrophoresis yields excellent resolution and is the method of choice for analysis of proteins and nucleic acid fragments. Protein or nucleic acid bands containing as little as 1 or 2 μg can be detected by staining the gels after electrophoresis. Disc gel is not limited to the column arrangement, but is also adaptable to slab gel electrophoresis.

SODIUM DODECYL SULFATE–POLYACRYLAMIDE GEL ELECTROPHORESIS (SDS-PAGE)

The electrophoretic techniques previously discussed are not applicable to the measurement of the molecular weights of biological molecules because mobility is influenced by both charge and size. If protein samples are treated so that they have a uniform charge, electrophoretic mobility then depends primarily on size (see Equation 4.3). The molecular weights of proteins may be estimated if they are subjected to electrophoresis in the presence of a detergent, sodium dodecyl sulfate (SDS), and a disulfide reducing agent, mercaptoethanol.

When protein molecules are treated with SDS, the detergent disrupts the secondary, tertiary, and quaternary structure to produce linear polypeptide chains coated with negatively charged SDS molecules. The presence of mercaptoethanol assists in protein denaturation by reducing all disulfide bonds. The detergent binds to hydrophobic regions of the denatured protein chain in a constant ratio of about 1.4 g of SDS per gram of protein. The bound detergent molecules carrying negative charges mask the native charge of the protein. In essence, polypeptide chains of a constant charge/mass ratio and uniform shape are produced. The electrophoretic mobility of the SDS-protein complexes is influenced primarily by molecular size: the larger molecules are retarded by the molecular sieving effect of the gel, and the smaller molecules have greater mobility. Empirical measurements have shown a linear relationship between the log molecular weight and the electrophoretic mobility (Figure 4.6).

In practice, a protein of unknown molecular weight and subunit structure is treated with 1% SDS and 0.1M mercaptoethanol in electrophoresis buffer. A standard mixture of proteins with known molecular weights must also be subjected to electrophoresis under the same conditions. Two sets of standards are commercially available, one for low-molecular-weight proteins (molecular weight range 14,000 to 100,000) and one for high-molecular-weight proteins (45,000 to 200,000). Figure 4.7 shows a stained gel after electrophoresis of a standard protein mixture. After electrophoresis and dye staining, mobilities are measured and molecular weights determined graphically.

SDS-PAGE is valuable for estimating the molecular weight of protein subunits. This modification of gel electrophoresis finds its greatest use in characterizing the sizes and different types of subunits in oligomeric proteins. SDS-PAGE is limited to a molecular weight range of 10,000 to 200,000. Gels of less than 2.5% acrylamide must be used for determining molecular weights above 200,000, but these gels do not set well and

Figure 4.6
Graph illustrating the linear relationship between electrophoretic mobility of a protein and its molecular weight. Thirty-seven different polypeptide chains with a molecular weight range of 11,000 to 70,000 are shown. From K. Weber and M. Osborn, *J. Biol. Chem* **244**, 4406 (1969). By permission of the copyright owner, the American Society for Biochemistry and Molecular Biology, Inc.

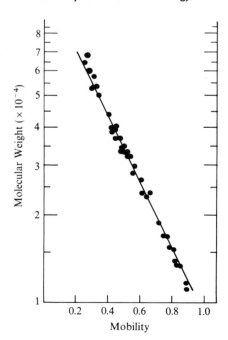

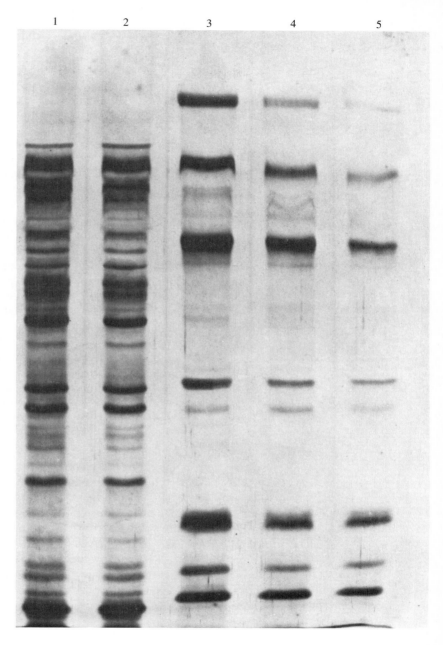

Figure 4.7
A silver-stained gel obtained by electrophoresis of a standard protein mixture. Samples were run on 12% polyacrylamide gels, 0.75 mm. Lane 1: Bovine brain homogenate soluble fraction, 20 μL. Lane 2: Bovine brain homogenate soluble fraction, 10 μL. Lanes 3, 4, 5: Bio-Rad SDS-PAGE low-molecular-weight standards, three different dilutions. Courtesy of Bio-Rad Laboratories, Richmond, CA.

are very fragile because of minimal cross-linking. A modification using gels of agarose-acrylamide mixtures allows the measurement of molecular weights above 200,000.

Nucleic Acid Sequencing Gels

The amino acid sequence of a protein is determined by identifying amino acid residues as they are sequentially cleaved from the intact protein (see Experiment 4). Sequence analysis of nucleic acids is based on the generation of sets of DNA or RNA fragments with common ends and the separation of these oligonucleotide fragments by polyacrylamide electrophoresis. Two methods have been developed for sequencing nucleic acids: (1) the partial chemical degradation method of Maxam and Gilbert, which uses four specific chemical reactions to modify bases and cleave phosphodiester bonds, and (2) the chain termination method developed by Sanger, which requires a single-stranded DNA template and chain extension processes, followed by chain termination caused by the presence of dideoxynucleoside triphosphates. Both sequencing methods result in nested sets of DNA or RNA fragments that have one common end and chains varying in length. The smallest possible size difference of nucleic acid fragments is one nucleotide. Separation of the nucleic acid fragments by polyacrylamide electrophoresis allows one to "read" the sequence of nucleotides from the gel. The experimental arrangement is the same as that previously described for PAGE; however, the gel is prepared with many sample wells to accommodate a large number of samples. Sequence gels of 6, 8, 12, and 20% polyacrylamide are routinely used. Gels of 20% may be used to sequence the first 50 to 100 nucleotides of a nucleic acid, and lower percentage gels allow sequencing out to 250 nucleotides. Sequencing gels are large (up to 40×100 cm) and power supplies must provide more power than for conventional methods. Denaturants such as urea and formamide are required to prevent renaturing of the nucleic acid fragments during electrophoresis. For detection, nucleic acid chains for sequencing must be end labeled with ^{32}P or with a fluorescent tag. ^{32}P-labeled nucleic acids on gels are detected by autoradiography (see later). Nucleic acids end labeled with fluorescent molecules are detected by fluorimeter scanning of the gels. Many researchers working on the large and expensive human genome project[1] are generating huge amounts of DNA sequence data. Much of this information is stored in computer data banks for use by researchers around the world.

Agarose Gel Electrophoresis

The electrophoretic techniques discussed up to this point are useful for analyzing proteins and small fragments of nucleic acids up to 350,000

---

[1] The human genome project is a federal government-sponsored program to sequence all DNA in human chromosomes.

daltons (500 bp) in molecular size; however, the small pore sizes in the gel are not appropriate for analysis of large nucleic acid fragments or intact DNA molecules. The standard method used to characterize RNA and DNA in the range 200 to 50,000 base pairs (50 kilobase) is electrophoresis with agarose as the support medium.

Agarose, a product extracted from seaweed, is a linear polymer of galactopyranose derivatives. Gels are prepared by dissolving agarose in warm electrophoresis buffer. After cooling the gel mixture to 50°C, the agarose solution is poured between glass plates as described for polyacrylamide. Gels with less than 0.5% agarose are rather fragile and must be used in a horizontal arrangement (Figure 4.8). The sample to be separated is placed in a sample well made with a comb, and voltage is applied until separation is complete.

The new electrophoresis gel system Instacryl, introduced by Eastman Kodak, is also available in a form to replace agarose. Instacryl H Copolymer, which polymerizes in the presence of dithiothreitol, is available as a 10% stock solution. Gel concentrations of 3–6% may be used to separate double-stranded DNA in the range 10 to 3500 base pairs (bp).

Nucleic acids can be visualized on the slab gel after separation by soaking in a solution of ethidium bromide, a dye that displays enhanced fluorescence when intercalated between stacked nucleic acid bases. Ethidium bromide may be added directly to the agarose solution before gel formation. This method allows monitoring of nucleic acids during electrophoresis. Irradiation of ethidium bromide–treated gels by UV light results in orange-red bands where nucleic acids are present.

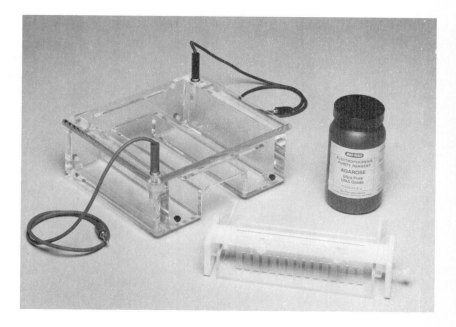

Figure 4.8
An apparatus for horizontal slab gel electrophoresis. Courtesy of Bio-Rad Laboratories, Richmond, CA.

The mobility of nucleic acids in agarose gels is influenced by the agarose concentration and the molecular size and molecular conformation of the nucleic acid. Agarose concentrations of 0.3 to 2.0% are most effective for nucleic acid separation (Table 4.2). Experiment 21 illustrates the separation of DNA fragments on agarose gels. Like proteins, nucleic acids migrate at a rate that is inversely proportional to the logarithm of their molecular weights; hence, molecular weights can be estimated from electrophoresis results using standard nucleic acids or DNA fragments of known molecular weight. The DNA conformations most frequently encountered are superhelical circular (form I), nicked circular (form II), and linear (form III). The small, compact, supercoiled form I molecules usually have the greatest mobility, followed by the rodlike, linear form III molecules. The extended, circular form II molecules migrate more slowly. The relative electrophoretic mobility of the three forms of DNA, however, depends on experimental conditions such as agarose concentration and ionic strength.

The versatility of agarose gels is obvious when one reviews their many applications in nucleic acid analysis. The rapid advances in our understanding of nucleic acid structure and function in recent years are due primarily to the development of agarose gel electrophoresis as an analytical tool. Two of the many applications of agarose gel electrophoresis will be described here.

ANALYSIS OF DNA FRAGMENTS AFTER DIGESTION BY
RESTRICTION ENDONUCLEASES

As described in Experiment 21, restriction endonucleases recognize a specific base sequence in double-stranded DNA and cause cleavage (hydrolysis of phosphodiester bonds) in or near that specific region. Many viral, bacterial, or animal DNA molecules are substrates for the enzymes. When each type of DNA is treated with a restriction endonuclease, a specific number of DNA fragments is produced. The base sequence recognized by the enzyme occurs only a few times in any particular DNA molecule; therefore, the smaller the DNA molecule, the fewer specific sites there are. Viral or phage DNA, for example, is cleaved into up to 50

Table 4.2
Effective Range of Separation
of DNA by Agarose

Agarose (% w/v)	Effective Range (kb)
0.3	5–50
0.5	2–25
0.7	0.8–10
1.2	0.4–5
1.5	0.2–3
2.0	0.1–2

fragments depending on the enzyme used, whereas larger bacterial or animal DNA may be cleaved into hundreds or thousands of fragments. Smaller DNA molecules, upon cleavage with a particular enzyme, will produce a limited set of fragments. It is unlikely that this set of fragments will be the same for any two different DNA molecules, so the fragmentation pattern can be considered a "fingerprint" of the DNA substrate. The **restriction pattern** is produced by electrophoresis of the cleavage reaction mixture through agarose gels, followed by staining with ethidium bromide (Figure 4.9). The separation of the fragments is based on molecular size, with large fragments remaining near the origin and smaller fragments migrating farther down the gel. In addition to characterization of DNA structure, endonuclease digestion coupled with agarose gel electrophoresis is a valuable tool for plasmid mapping (Experiment 21) and DNA recombination experiments.

CHARACTERIZATION OF SUPERHELICAL STRUCTURE OF DNA

The structure of plasmid, viral, and bacterial DNA is often closed circular with negative superhelical turns. It is possible under various experimental conditions to induce reversible changes in the conformation of DNA. The intercalating dye ethidium bromide causes an unwinding of supercoiled DNA that affects its centrifugal sedimentation rate and electrophoretic mobility. Electrophoresis of DNA on agarose in the presence of increasing concentrations of ethidium bromide provides an unambiguous method for distinguishing between closed circular and other DNA conformations.

Figure 4.9
Restriction patterns produced by agarose electrophoresis of DNA fragments after restriction endonuclease action. Courtesy of Bio-Rad Laboratories, Richmond, CA.

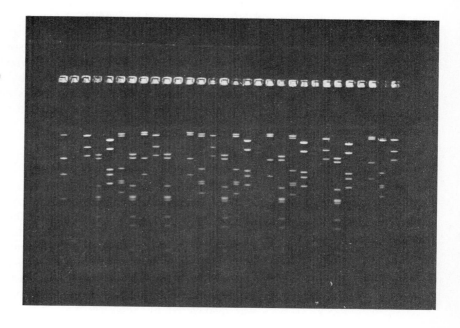

Closed circular, negatively supercoiled DNA (form I) usually has the greatest electrophoretic mobility of all DNA forms because supercoiled DNA molecules tend to be compact. If ethidium bromide is added to form I DNA, the dye intercalates between the stacked DNA bases, causing unwinding of some of the negative supercoils. As the concentration of ethidium bromide is increased, more and more of the negative supercoils are removed until no more are present in the DNA. The conformational change of the DNA supercoil can be monitored by electrophoresis because the mobility decreases with each unwinding step. With increasing concentration of ethidium bromide, the negative supercoils are progressively unwound and the electrophoretic mobility decreases to a minimum. This minimum represents the free dye concentration necessary to remove all negative supercoils. (The free dye concentration at this minimum has been shown to be related to the superhelix density, which is a measure of the extent of supercoiling in a DNA molecule.) The circular DNA at this point is equivalent to the "relaxed" form. If more ethidium bromide is added to the relaxed DNA, positive superhelical turns are induced in the structure and the electrophoretic mobility increases. Forms II and III DNA, under the same conditions of increasing ethidium bromide concentration, show a gradual decrease in electrophoretic mobility throughout the entire concentration range.

Agarose gel electrophoresis is able to resolve topoisomers of native, covalently closed, circular DNA that differ only in their degree of supercoiling. This technique has proved useful in the analysis and characterization of a new class of enzymes that catalyze changes in the conformation or topology of native DNA. These enzymes, called topoisomerases, have been isolated from bacterial and mammalian cells. They change DNA conformations by catalyzing nicking and closing of phosphodiester bonds in circular duplex DNA. Agarose gel electrophoresis is an ideal method for identifying and assaying topoisomerases because the intermediate DNA molecules can be resolved on the basis of the extent of supercoiling. Topoisomerases may be assayed by incubating native DNA with an enzyme preparation, removing aliquots after various periods of time, and subjecting them to electrophoresis on an agarose gel with standard supercoiled and relaxed DNA.

Pulsed Field Gel Electrophoresis (PFGE)

Conventional agarose gel electrophoresis is limited in use for the separation of nucleic acid fragments smaller than 50,000 bp (50 kb). In practice, that limit is closer to 20,000 to 30,000 bp if high resolution is desired. DNA molecules larger than 50 kb are so large that they get stuck in the pores of agarose gel and remain essentially at the origin of the gel. Since chromosomal DNA from most organisms contains thousands and even millions of base pairs, the DNA must be cleaved by restriction enzymes before analysis by standard electrophoresis. In the early 1980s it was discovered by Schwartz and Cantor at Columbia University that large molecules of DNA (yeast chromosomes, 200–3000 kb) could be separated by **pulsed field gel electrophoresis (PFGE)**. There is one major distinction between standard gel electrophoresis and PFGE. In PFGE, the electric field is not constant as in the standard method but is changed repeatedly (pulsed) in direction and strength during the separation (Figure 4.10). The physical mechanism for separation of the large DNA molecules as they move through the gel under these conditions is not yet well understood. An early explanation was that the electrical pulses abruptly perturbed the conformation of the DNA molecules. They would be oriented by the influence of the electric field coming from one direction and then reoriented as a new electric field at a different angle to the first was turned on. According to this explanation, it takes longer for larger molecules to reorient, so smaller fragments respond faster to the new pulse and move faster. More recent experiments on dyed DNA moving in gels have shown that conformational changes of the DNA are not abrupt but more gradual in response to the electrical pulse, and DNA molecules tend to "slither" through the gel matrix. In addition, it has been discovered that the gel becomes more fluid during electrical pulsing.

Even though our theoretical understanding of PFGE is lacking, practical applications and experimental advances are expanding rapidly. The availability of PFGE has sparked changes in DNA research. New methods for isolating intact DNA molecules have been developed. Because of mechanical breakage, the average size of DNA isolated from cells in the presence of lysozyme, detergent, and EDTA (Experiment 17) is about 400–500 kb. Intact chromosomal DNA can be isolated by embedding

Figure 4.10
Pulsed field gel electrophoresis. Generalized PFGE separation of four DNA fragments of different sizes in one lane.   (A) DNA molecules of various shapes and configurations move toward the positive field.   (B) The new field orientation pulls the DNA in a different direction, realigning the molecules.   (C) The field returns to the original configuration.   (D) The bands show the final position of a large collection of the molecules. Reprinted with permission from the *Journal of NIH Research* **1**, 115 (Nov.-Dec. 1989). Illustration by Terese Winslow.

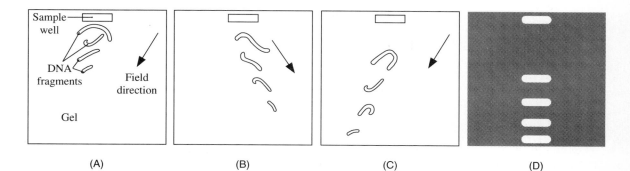

(A)   (B)   (C)   (D)

cells in an agarose matrix and disrupting the cells with detergents and enzymes. Slices or "plugs" of the agarose with intact DNA are then placed on the gel for PFGE analysis. Newly discovered restriction endonucleases that cut DNA only rarely can now be used to subdivide chromosome-sized DNA. Two important endonucleases with eight-base recognition sites are *Not* I and *Sfi* I.

There are also many instrumental advances that allow changes in the experimental design of PFGE. Some of the variables that can be changed for each experiment are voltage, pulse length, number of electrodes, relative angle of electrodes, gel box design, temperature, agarose concentration, buffer pH, and time of electrophoresis.

Like all laboratory techniques, PFGE has its disadvantages and problems. Long periods of electrophoresis (several days) are often required for good resolution, and migration of fragments is extremely dependent on experimental conditions. Therefore it is difficult to compare gels even when they are run under similar conditions. In spite of these shortcomings, PFGE will continue to advance as a significant tool for the characterization of very large molecules. The technique, which is being widely used in the human genome project, will greatly increase our understanding of chromosome structure and function.

Isoelectric Focusing of Proteins

Another important and effective use of electrophoresis for the analysis of proteins is **isoelectric focusing (IEF),** which examines electrophoretic mobility as a function of pH. The net charge on a protein is pH dependent. Proteins below their isoelectric pH (pH_I, or the pH at which they have zero net charge) are positively charged and migrate in a medium of fixed pH toward the negatively charged cathode. At a pH above its isoelectric point, a protein is deprotonated and negatively charged and migrates toward the anode. If the pH of the electrophoretic medium is identical to the pH_I of a protein, the protein has a net charge of zero and does not migrate toward either electrode. Theoretically, it should be possible to separate protein molecules and to estimate the pH_I of a protein by investigating the electrophoretic mobility in a series of separate experiments in which the pH of the medium is changed. The pH at which there is no protein migration should coincide with the pH_I of the protein. Because such a repetitive series of electrophoresis runs is a rather tedious and time-consuming way to determine the pH_I, IEF has evolved as an alternative method for performing a single electrophoresis run in a medium of gradually changing pH (i.e., a pH gradient).

Figure 4.11 illustrates the construction and operation of an IEF pH gradient. Acid, usually phosphoric, is placed at the cathode; a base, such as triethanolamine, is placed at the anode. Between the electrodes is a medium in which the pH gradually increases from 2 to 10. The pH gradient can be formed before electrophoresis is conducted or formed during the course of electrophoresis. It can be either broad (pH 2–10) for sepa-

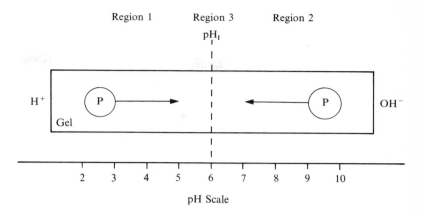

Figure 4.11

Illustration of isoelectric focusing. See text for details.

rating several proteins of widely ranging pH_I values or narrow (pH 7–8) for precise determination of the pH_I of a single protein. P in Figure 4.11 represents different molecules of the same protein in two different regions of the pH gradient. Assuming that the pH in region 1 is less than the pH_I of the protein and the pH in region 2 is greater than the pH_I of the protein, molecules of P near region 1 will be positively charged and will migrate in an applied electric field toward the cathode. As P migrates, it will encounter an increasing pH, which will influence its net charge. As it migrates *up* the pH gradient, P will become increasingly deprotonated and its net charge will decrease toward zero. When P reaches a region where its net charge is zero (region 3), it will stop migrating. The pH in this region of the electrophoretic medium will coincide with the pH_I of the protein and can be measured with a surface microelectrode, or the position of the protein can be compared to that of a calibration set of proteins of known pH_I values. P molecules in region 2 will be negatively charged and will migrate toward the anode. In this case, the net charge on P molecules will gradually decrease to zero as P moves *down* the pH gradient, and P molecules originally in region 2 will approach region 3 and come to rest. The P molecules move in opposite directions, but the final outcome of IEF is that P molecules located anywhere in the gradient will migrate toward the region corresponding to their isoelectric point and will eventually come to rest in a sharp band; that is, they will "focus" at a point corresponding to their pH_I.

Since different protein molecules in mixtures have different pH_I values, it is possible to use IEF to separate proteins. In addition, the pH_I of each protein in the mixture can be determined by measuring the pH of the region where the protein is focused.

The pH gradient is prepared in a horizontal glass tube or slab. Special precautions must be taken so that the pH gradient remains stable and is not disrupted by diffusion or convective mixing during the electrophoresis experiment. The most common stabilizing technique is to form the gradient in a polyacrylamide, agarose, or dextran gel. The pH gradient is formed in the gel by electrophoresis of synthetic polyelectrolytes, called **ampholytes,** which migrate to the region of their pH_I values just as proteins do and establish a pH gradient that is stable for the duration of the IEF run. Ampholytes are low-molecular-weight polymers that have a wide range of isoelectric points because of their numerous amino and carboxyl or sulfonic acid groups. The polymer mixtures are available in specific pH ranges (pH 5–7, 6–8, 3.5–10, etc.) under the trade names Ampholine, Pharmalyte, Immobiline, and Bio-Rad-Lyte. It is critical to select the appropriate pH range for the ampholyte so that the proteins to be studied have pH_I values in that range. The best resolution is, of course, achieved with an ampholyte mixture over a small pH range (about two units) encompassing the pH_I of the sample proteins. If the pH_I values for the proteins under study are unknown, an ampholyte of wide pH range (pH 3–10) should be used first and then a narrower pH range selected for use.

The gel medium is prepared as previously described except that the appropriate ampholyte is mixed prior to polymerization. The gel mixture is poured into the desired form (column tubes, horizontal slabs, etc.) and allowed to set. Immediately after casting of the gel, the pH is constant throughout the medium, but application of voltage will induce migration of ampholyte molecules to form the pH gradient. The standard gel for proteins with molecular sizes up to 100,000 daltons is 7.5% polyacrylamide; however, if larger proteins are of interest, gels with larger pore sizes must be prepared. Such gels can be prepared with a lower concentration of acrylamide (about 2%) and 0.5 to 1% agarose to add strength.

The protein sample can be loaded on the gel in either of two ways. A concentrated, salt-free sample can be layered on top of the gel as previously described for ordinary gel electrophoresis. Alternatively, the protein can be added directly to the gel preparation, resulting in an even distribution of protein throughout the medium. The protein molecules move more slowly than the low-molecular-weight ampholyte molecules, so the pH gradient is established before significant migration of the proteins occurs.

Very small protein samples can be separated by IEF. For analytical purposes, 10 to 50 μg is a typical sample size. Larger sample sizes (up to 20 mg) can be used for preparative purposes. Electrofocusing is now established as a standard method for determining the purity of proteins.

Electrofocusing is conducted using an applied current of 2 mA per tube and usually takes 30 to 120 minutes. The time period for electrofocusing is not as critical as for standard electrophoresis. Trial and error can be used to find a period that results in an effective separation; however, longer periods do not move the protein samples out of the medium as in other types of gel electrophoresis.

IEF gels cannot be stained directly for protein detection because the commonly used stains also bind ampholytes. Carrier ampholytes must first be removed from gels by soaking in 5% trichloroacetic acid, after which the protein zones are stained with Coomassie Blue dye.

Two-Dimensional Electrophoresis of Proteins

The separation of proteins by IEF is based on charge, whereas SDS-PAGE separates molecules based on molecular size. A combination of the two methods leads to enhanced resolution of complex protein mixtures. Such an experiment was first reported by O'Farrell (1975), and the combined method has since become a routine and powerful separatory technique. Figure 4.12 shows the results of O'Farrell's analysis of total *Escherichia coli* protein. The sample was first separated in one dimension by IEF. The tube gel was then transferred to an SDS-PAGE slab and electrophoresis was continued in the second dimension. At least 1000 discrete protein spots are visible. This technique is becoming increasingly valuable in developmental biochemistry, where the increase or decrease in intensity of a spot representing a specific protein can be monitored as a function of cell growth. In addition, two-dimensional electrophoresis is a standard method for judging protein purity.

Capillary Electrophoresis (CE)

Capillary electrophoresis is a new technique that combines the high resolving power of electrophoresis with the speed, versatility, and automation of high-performance liquid chromatography (HPLC). It offers the ability to analyze very small samples (5–10 nL) utilizing up to 1 million theoretical plates to achieve high resolution and sensitivity to the attomole level (10^{-18} mole). It will undoubtedly become one of the most widely used techniques in the analysis of amino acids, peptides, proteins, nucleic acids, and pharmaceuticals.

IEF $\rightarrow$ SDS
                                                                      $\downarrow$

Figure 4.12
SDS–isoelectric focusing gel electrophoresis of total *E. coli* protein. Photo courtesy of Dr. P. O'Farrell.

A general experimental design is diagrammed in Figure 4.13. The equipment consists of a power supply, two buffer reservoirs, a buffer-filled capillary tube, and an on-line detector. Platinum electrodes connected to the power supply are immersed in each buffer reservoir. A high voltage is applied along the capillary and a small plug of sample solution is injected into one end of the capillary. Components in the solution migrate along the length of the capillary under the influence of the electric field. Molecules are detected as they exit from the opposite end of the capillary. The detection method used depends on the type of molecules separated, but the most common are UV-VIS fixed-wavelength detectors and diode-array detectors (see Chapter 5). The capillaries used are flexible, fused, silica tubes of 50–100 μm i.d. and 25–100 cm length that may or may not be filled with chromatographic matrix.

A major advantage of capillary electrophoresis is that many analytical experimental designs are possible, just as in the case of HPLC. In HPLC, a wide range of molecules can be separated by changing the column support (see Chapter 3). In CE, the capillary tube may be coated or filled with a variety of materials. For separation of small, charged molecules, bare silica or polyimide-coated capillaries are often used. If separation by molecular sieving is desired, the tube is filled with polyacrylamide or SDS-polyacrylamide. If the capillary is filled with electrolyte and an ampholyte pH gradient, isoelectric focusing experiments on proteins may be done. We can expect to see numerous applications of CE in all aspects of biochemistry and molecular biology. New applications will include DNA sequencing, analysis of single cells and separations of neutral molecules.

Immunoelectrophoresis (IE)

In immunoelectrophoresis two sequential procedures are applied to the analysis of complex protein mixtures: (1) separation of the protein mixture by agarose gel electrophoresis, followed by (2) interaction with

Figure 4.13
Experimental setup for capillary electrophoresis. Courtesy of Bio-Rad Laboratories, Life Science Group, Hercules, CA.

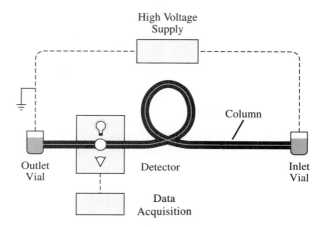

specific antibodies to examine the antigenic properties of the separated proteins.

The technique of IE was first reported by Grabar and Williams in 1953 for the separation and immunoanalysis of serum proteins, but it can be applied to the analysis of any purified protein or complex mixture of proteins. In practice (Figure 4.14), a protein mixture is separated by standard electrophoresis in an agarose gel prepared on a small glass plate. This is followed by exposing the separated proteins to a specific antibody preparation. The antibody is added to a trough cut into the gel, as shown in Figure 4.14, and is allowed to diffuse through the gel toward the separated proteins. If the antibody has a specific affinity for one of the proteins, a visible precipitin arc forms. This is an insoluble complex formed at the boundary of antibody and antigen protein. The technique is most useful for the analysis of protein purity, composition, and antigenic properties. The basic IE technique described here allows only qualitative examination of antigenic proteins. If quantitative results in the form of protein antigen concentration are required, the advanced modifications, rocket immunoelectrophoresis and two-dimensional (crossed) immunoelectrophoresis, may be used.

Figure 4.14
Immunoelectrophoresis.   (A) Antigen is placed in sample wells.   (B) Electrophoresis.   (C) Antiserum containing antibody is placed in trough.   (D) Insoluble antigen-antibody complexes form precipitin arcs.

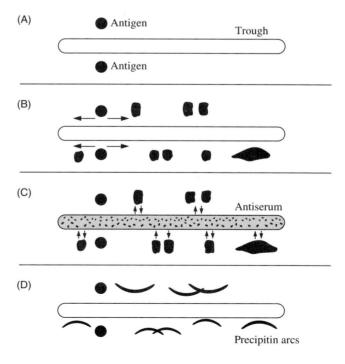

C. Practical Aspects of Electrophoresis

Instrumentation

The basic components required for electrophoresis are a power supply and an electrophoresis chamber (gel box). A power supply that provides a constant current is suitable for most conventional electrophoresis experiments. Power supplies that generate both constant voltage (up to 3 kV) and constant current (up to 200 mA) are commercially available. In order to implement the many modifications of electrophoresis, such versatile power supplies are essential in a research laboratory. If isoelectric focusing experiments are planned, a power supply that furnishes a constant voltage is necessary. DNA sequencing experiments require power supplies capable of generating 2–3 kV. Because of the high power requirements of these experiments, the glass plates sandwiching the gel must be covered with conductive aluminum plates to dissipate heat and prevent gel melting and glass plate breakage. CE also requires the use of high voltage, which generates heat as well. However, the heat is quickly dissipated through the thin walls of the capillary tubing. Pulsed field electrophoresis experiments have special requirements, including a high-voltage power supply, electrical switching devices for control of the field, equipment for temperature control, and a specially designed gel box with an array of electrodes. All types of electrophoresis chambers for both horizontal and vertical placements of gels are available from Bio-Rad Laboratories, Pharmacia-LKB, Hoefer, and Beckman.

Reagents

High-quality, electrophoresis grade chemicals must be used, since impurities may influence both the gel polymerization process and electrophoretic mobility. The reagents used for gel formation should be stored in a refrigerator.

⚠ Caution

> Acrylamide, N,N,N',N'-tetramethylethylenediamine, N,N'-methylene-bis-acrylamide, and ammonium persulfate are toxic and must be used with care. Acrylamide is a neurotoxin and a potent skin irritant, so gloves and a mask must be worn while handling it in the unpolymerized form.

Buffers appropriate for electrophoresis gels include Tris-glycine, Tris-acetate, Tris-phosphate, and Tris-borate at concentrations of about 0.05 M. The availability of the new Instacryl copolymers for protein and nucleic acid separations eliminates the need for acrylamide.

Staining and Detecting Electrophoresis Bands

During the electrophoretic process, it is important to know when to stop applying the voltage or current. If the process is run for too long, the desired components may pass entirely through the medium and into the buffer; if too short a period is used, the components may not be completely resolved. It is common practice to add a "tracking dye," usually bromphenol blue and/or xylene cyanol, to the sample mixture. These dyes, which are small and anionic, move rapidly through the gel ahead of most proteins or nucleic acids. After electrophoresis, bands have a tendency to widen by diffusion. Because this broadening may decrease resolution, gels should be analyzed as soon as possible after the power supply has been turned off. In the case of proteins, gels can be treated with an agent that "fixes" the proteins in their final positions, a process that is often combined with staining.

Reagent dyes that are suitable for visualization of biomolecules after electrophoresis were discussed earlier. The most commonly used stain for proteins is **Coomassie Brilliant Blue.** This dye can be used as a 0.25% aqueous solution. It is followed by destaining (removing excess background dye) by repeated washing of the paper or gel with 7% acetic acid. Alternatively, gels may be stained by soaking in 0.25% dye in H_2O,methanol,acetic acid (5:5:1) followed by repeated washings with the same solvent.

The most time-consuming procedure in the visualization process is destaining, which often requires days of washing. Rapid destaining of gels may be brought about electrophoretically. The gel, after soaking in the stain, is replaced in the glass tube and subjected to electrophoresis again, using a buffer or higher concentration to remove excess stain.

The search for more rapid and sensitive methods of protein detection after electrophoresis led to the development of fluorescent staining techniques. Two commonly used fluorescent reagents are fluorescamine and anilinonaphthalene sulfonate. New dyes based on silver salts (silver diamine or silver–tungstosilicic acid complex) have been developed for protein staining. They are 10 to 100 times more sensitive than Coomassie Blue.

There is often a need for a visualization procedure that is specific for a certain biomolecule, for example, an enzyme. If the enzyme remains in an active form while in the gel, any substrate that produces a colored product could be used to locate the enzyme on the gel. Although it is less desirable for detection, the electrophoresis support medium may be cut into small segments and each part extracted with buffer and analyzed for the presence of the desired component.

Nucleic acids are visualized in agarose and polyacrylamide gels using the fluorescent dye ethidium bromide. The gel is soaked in a solution of the dye and washed to remove excess dye. Illumination of the rinsed slab with UV light reveals red-orange stains where nucleic acids are located.

Although ethidium bromide stains both single- and double-stranded nucleic acids, the fluorescence is much greater for double-stranded molecules. The electrophoresis may be performed with the dye incorporated in the gel and buffer. This has the advantage that the gel can be illuminated with UV light during electrophoresis to view the extent of separation. The mobility of double-stranded DNA may be reduced 10 to 15% in the presence of ethidium bromide. Destaining of the gel is not necessary because ethidium bromide–DNA complexes have a much greater fluorescent yield than free ethidium bromide, so relatively small amounts of DNA can be detected in the presence of free ethidium bromide. The detection limit for DNA is 10 ng.

⚠   **Caution**

Ethidium bromide must be used with great care as it is a potent mutagen. Gloves should be worn at all times while using dye solutions or handling gels.

Because single-stranded nucleic acids do not stain deeply with ethidium bromide, other techniques must be used for detection, especially when only small amounts of biological material are available for analysis. One of the most sensitive techniques is to use radiolabeled molecules. For nucleic acids, this usually means labeling the 5′ or 3′ end with ^{32}P, a strong β emitter (see Chapter 6). Bands of labeled nucleic acids on an electrophoresis gel can easily be located by **autoradiography**. For this technique, the electrophoresed slab gel is transferred to heavy chromatography paper. After covering the gel and paper with plastic wrap, they are placed on X-ray film and wrapped in a folder to avoid external light exposure. This procedure must be done in a darkroom. The gel-film combination is stored at $-70°C$ for exposure. The low temperature maintains the gel in a rigid form and prevents diffusion of gel bands. The exposure time depends on the amount of radioactivity but can range from a few minutes to several days. Figure 4.15 shows the autoradiogram of a typical DNA sequencing gel. The autoradiogram also provides a permanent record of the gel for storage and future analysis. The actual gel is very fragile and difficult to store. Also, because of the short half-life of ^{32}P (14 days), an autoradiogram of the gel cannot be obtained in the distant future. Proteins labeled with ^{32}P or ^{125}I can be dealt with in a manner similar to nucleic acids.

Because of concerns about the safety of radioisotope use, researchers are developing fluorescent and chemiluminescent methods for detection of small amounts of biomolecules on gels. One attractive approach is to label biomolecules before analysis with the coenzyme biotin. Biotin forms a strong complex with enzyme-linked streptavidin. Some dynamic property of the enzyme is then measured to locate the biotin-labeled

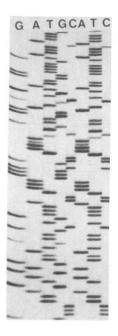

Figure 4.15
Autoradiogram of a DNA sequencing gel. From Zyskind and Bernstein, *Recombinant DNA Laboratory Manual* (1989), Academic Press (San Diego, CA), Figure 7.4.

biomolecule on the gel. These new methods are not yet as sensitive as methods involving radiolabeled molecules, but rapid advances are being made.

If an autoradiogram of a gel can be prepared, a permanent record of the experimental data is available. For biomolecules on a gel stained with Coomassie Blue (proteins) or ethidium bromide (nucleic acids), the best method for permanent storage is conventional photography.

Protein and Nucleic Acid Blotting

Only a minute amount of protein or nucleic acid is present in bands on electropherograms. In spite of this, there is often a need to extract the desired biomolecule from the gel for further investigation. This sometimes involves the tedious and cumbersome process of crushing slices of the gel in a buffer to release the trapped proteins or nucleic acids. Techniques are now available for removing specific nucleic acids and proteins from gels and characterizing them using probes to detect certain structural features or functions. After electrophoresis, the desired biomolecules are transferred or "blotted" out of the gel onto a nitrocellulose filter or nylon membrane. The biomolecule is now accessible on the filter for further analysis. The first blotting technique was reported by E. Southern in 1975. Using labeled complementary DNA probes, he searched for certain nucleotide sequences among DNA molecules blotted from the gel. This technique of detecting DNA-DNA hybridization is called Southern blotting. The general blotting technique has now been extended to the transfer and detection of specific RNA with labeled complementary DNA probes (Northern blotting) and the transfer and detection of proteins that react with specific antibodies (Western blotting). In practice, the electropherogram is alkali treated, neutralized, and placed in contact with the filter or nylon membrane. A buffer is used to facilitate the transfer. Figure 4.16 shows the setup for a blotting experiment. The transferred molecules are fixed to the filter by heating or UV cross-linking. The location of the desired nucleic acid or protein is then detected either by incubation of the membrane with a radiolabeled probe and autoradiography or by use of a biotinylated probe.

Blotting techniques have many applications, including mapping the genes responsible for inherited diseases by using restriction fragment length polymorphisms (RFLPs), screening collections of cloned DNA fragments (DNA libraries), and "DNA fingerprinting" for analysis of biological material remaining at the scene of a crime.

Analysis of Electrophoresis Results

By separating biochemicals on the basis of charge, size, and conformation, electrophoresis can provide valuable information, such as purity,

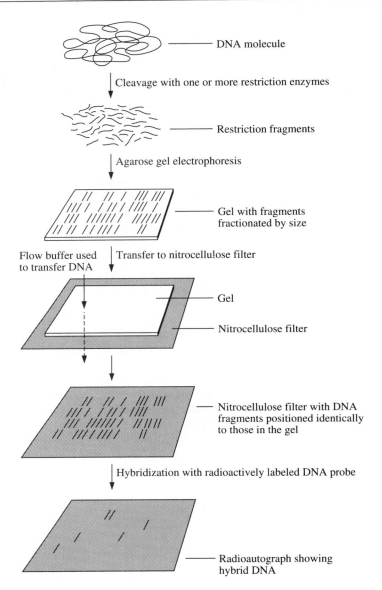

DNA molecule

Cleavage with one or more restriction enzymes

Restriction fragments

Agarose gel electrophoresis

Gel with fragments
fractionated by size

Flow buffer used   Transfer to nitrocellulose filter
to transfer DNA

Gel

Nitrocellulose filter

Nitrocellulose filter with DNA
fragments positioned identically
to those in the gel

Hybridization with radioactively labeled DNA probe

Radioautograph showing
hybrid DNA

Figure 4.16
Diagram of a blotting experiment. From C. Mathews and K. van Holde, *Biochemistry*
(1990), Benjamin/Cummings Publishing Company (Redwood City, CA), Figure T21.1,
p. 906.

identity, and molecular weight. Purity is indicated by the number of
stained bands in the electropherogram. One band usually means that only
one detectable component is present; that is, the sample is homogeneous
or "electrophoretically pure." Two or more bands usually indicate that
the sample contains two or more components, contaminants, or impurities and is therefore heterogeneous or impure. There are, of course, ex-

ceptions to this description. Other proteins or nucleic acids may be present in what appears to be a homogeneous sample, but they may be below the limit of detection of the staining method. Occasionally a homogeneous sample may result in two or more bands because of degradation during the electrophoresis process. Information on purity may be obtained by all of the electrophoresis methods discussed.

The identity of unknown biomolecules can be confirmed by electrophoresis on the same gel, the unknown alongside known standards. This is similar to the identification of unknowns by gas chromatography and HPLC as discussed in Chapter 3.

As previously discussed in this chapter, the molecular size of protein or nucleic acid samples may be determined by electrophoresis. This requires the preparation of standard curves of log molecular weight versus μ (mobility) using standard proteins or nucleic acids.

References

R. Anands, *Trends Genet*. **2**, 278–283 (1986). "Pulsed Field Gel Electrophoresis: A Technique for Fractionating Large DNA Molecules."

W. Bauer and J. Vinograd, *J. Mol. Biol*. **33**, 141–171 (1968). "The Interaction of Closed Circular DNA with Intercalative Dyes."

Bio-Rad Life Science Group, An Introduction to Capillary Electrophoresis, Bulletin No. 1701 (1991), 3300 Regatta Blvd., Richmond, CA 94804.

C. Cantor and P. Schimmell, *Biophysical Chemistry,* Part II (1980), W. H. Freeman (San Francisco), pp. 676–682. A detailed theoretical discussion of electrophoresis.

G. Carle, M. Frank, and M. Olson, *Science* **232**, 65–68 (1986). "Electrophoretic Separations of Large DNA Molecules by Periodic Inversion of the Electric Field."

Chem. Eng. News (March 18, 1991), 28–40. "Capillary Electrophoresis."

A. Chrambach, M. Dunn, and B. Radola, Editors, *Advances in Electrophoresis,* Vol. I (1987), VCH (Weinheim). Eight chapters on applications to nucleic acids and proteins.

A. Ewing, R. Wallingford, and T. Olefirowicz, *Anal. Chem*. **61**, 292A–303A (1989). "Capillary Electrophoresis."

J. Garrels, in *Methods in Enzymology,* R. Wu, L. Grossman, and K. Moldave, Editors, Vol. 100B (1983), Academic Press (New York), pp. 411–423. "Quantitative Two-Dimensional Gel Electrophoresis of Proteins."

P. Grabar and C. Williams, *Biochim. Biophys. Acta* **10**, 193–194 (1953). Application of immunoelectrophoresis to the separation of proteins.

B. Hames and D. Rickwood, Editors, *Gel Electrophoresis of Proteins: A Practical Approach,* 2nd ed. (1990), IRL Press (Oxford). An excellent book for starters.

P. Hayes, C. Wolf, and J. Hayes, *Br. Med. J.* **299**, 965–968 (1989). "Blotting Techniques for the Study of DNA, RNA and Proteins."

J. Jorgenson and M. Phillips, Editors, *New Directions in Electrophoretic Methods* (1987), American Chemical Society (Washington, DC). References to nucleic acids and proteins.

B. Karger, A. Cohen, and A. Guttman, *J. Chromatogr.* **492**, 585–614 (1989). "High Performance Capillary Electrophoresis in the Biological Sciences."

J. Knox and K. McCormack, *J. Liq. Chromatogr.* **12**, 2435–2470 (1989). "Capillary Electroseparation Methods."

W. Kuhr, *Anal. Chem.* **62**, 403R–414R (1990). "Capillary Electrophoresis."

E. Lai, B. Birren, S. Clark, M. Simon, and L. Hood, *Biotechniques* **7**, 34–41 (1989). "Pulsed Field Gel Electrophoresis."

T. Maniatis, E. Fritsch, and J. Sambrook, *Molecular Cloning, a Laboratory Manual,* 2nd ed. (1989). Cold Spring Harbor Press (Cold Spring Harbor, NY). An excellent, detailed manual for techniques and experimental protocols in molecular biology.

C. Mathews and K. van Holde, *Biochemistry* (1990), Benjamin/Cummings (Redwood City, CA), pp. 54–57, 905–908. An introduction to electrophoresis and blotting.

A. Maxam and W. Gilbert, *Proc. Natl. Acad. Sci. U.S.A.* **74**, 560–564 (1977). "A New Method for Sequencing DNA."

P. O'Farrell, *J. Biol. Chem.* **250**, 4007–4021 (1975). "High Resolution Two-Dimensional Electrophoresis of Proteins."

R. Ogden and D. Adams, in *Methods in Enzymology,* S. Berger and A. Kimmel, Editors, Vol. 152 (1987), Academic Press (Orlando, FL). "Electrophoresis in Agarose and Acrylamide Gels."

M. Olson, in *Genetic Engineering,* J. Setlow, Editor, Vol. 11 (1989), Plenum Publishing (New York), pp. 183–227. "Pulsed Field Gel Electrophoresis."

D. Rickwood and B. Hames, Editors, *Gel Electrophoresis of Nucleic Acids: A Practical Approach,* 2nd ed. (1990), IRL Press (Oxford). An excellent reference book.

P. Righetti, in *Laboratory Techniques in Biochemistry and Molecular Biology,* R. Burdon and P. Knippenberg, Editors, Vol. 20 (1990), Elsevier (Amsterdam). "Immobilized pH Gradients: Theory and Methodology."

P. Righetti and E. Gianazza, in *Methods of Biochemical Analysis,* D. Glick, Editor, Vol. 32 (1987) pp. 215–278, John Wiley & Sons (New

York). "Isoelectric Focusing in Immobilized pH Gradients: Theory and Newer Methodology."

F. Sanger, S. Nicklen, and A. Coulson, *Proc. Natl. Acad. Sci. U.S.A.* **74**, 5463–5467 (1977). "DNA Sequencing with Chain-Terminating Inhibitors."

D. Schwartz and C. Cantor, *Cell* **37**, 67–75 (1984). "Separation of Yeast Chromosome-Sized DNAs by Pulsed Field Gradient Gel Electrophoresis."

E. Southern, in *Methods in Enzymology,* R. Wu, Editor, Vol. 68 (1979), Academic Press (New York), pp. 152–176. "Gel Electrophoresis of Restriction Fragments."

K. van Holde, *Physical Biochemistry,* 2nd ed. (1985), Prentice-Hall (Englewood Cliffs, NJ), pp. 137–163. Theoretical description and applications of electrophoresis.

O. Vesterberg, in *Methods in Enzymology,* W. B. Jakoby, Editor, Vol. XXII (1971), Academic Press (New York), pp. 389–412. "Isoelectric Focusing of Proteins."

D. Voet and J. Voet, *Biochemistry* (1990), John Wiley & Sons (New York), pp. 94–100. An introduction to electrophoresis.

R. Wallingford and A. Ewing, *Adv. Chromatogr. (N.Y.)* **29**, 1–76 (1989). "Capillary Electrophoresis."

D. Weller and P. Gariepy, *J. Chem. Educ.* **68**, 81–82 (1991). "Field Inversion Agarose Gel Electrophoresis of DNA."

CHAPTER **5**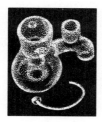

Spectroscopic Analysis of Biomolecules

■ Some of the earliest experimental measurements of biomolecules involved studies of their interactions with electromagnetic radiation. It was observed that when light impinges on solutions of molecules, at least two distinct processes occur: **light scattering** and **light absorption**. Both processes have become the basis of useful techniques for characterizing and analyzing biomolecules. With some molecules, the process of absorption is followed by emission of light of a different wavelength. This process, called **fluorescence,** depends on molecular structure and environmental factors and serves as a valuable tool for the characterization and analysis of biologically significant molecules and dynamic processes occurring between molecules. Using X-rays and waves from other regions of the electromagnetic spectrum, such as radio-frequency waves for NMR, it has been possible to elucidate very specific structural characteristics of biomolecules.

A. Ultraviolet-Visible Absorption Spectrophotometry

Principles

The electromagnetic spectrum, as shown in Figure 5.1, is composed of a continuum of waves with different properties. The regions of primary importance in biochemistry are the ultraviolet (UV, 180–350 nm) and visible (VIS, 350–800 nm). Light in these regions has sufficient energy to excite the valence electrons of molecules. Figure 5.2 shows that the propagation of light is due to an electrical field component, E, and a magnetic field component, H, that are perpendicular to each other. The **wavelength** of light, defined by Equation 5.1, is the distance between adjacent wave peaks as shown in Figure 5.2.

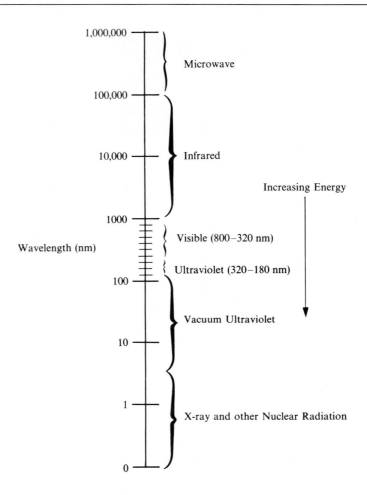

Figure 5.1
The electromagnetic spectrum.

$$\lambda = \frac{c}{\nu}$$
(Equation 5.1)

where
λ = wavelength
c = speed of light
ν = frequency, the number of waves passing a certain point per unit time

Light also behaves as though it were composed of energetic particles. The amount of energy, E, associated with these particles (or photons) is given by Equation 5.2.

$$E = h\nu$$
(Equation 5.2)

where h is Planck's constant

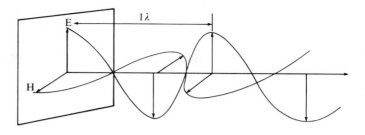

Figure 5.2
An electromagnetic wave, showing the E and H components.

When a photon of specified energy interacts with a molecule, one of two processes may occur. The photon may be **scattered**, or it may transfer its energy to the molecule, producing an **excited state** of the molecule. The former process, called **Rayleigh scattering**, occurs when a photon collides with a molecule and is diffracted or scattered with unchanged frequency. Light scattering is the physical basis of several experimental methods used to characterize macromolecules. Before the development of electrophoresis, light scattering techniques were used to measure the molecular weights of macromolecules. The widely used techniques of X-ray diffraction (crystal and solution), electron microscopy, laser light scattering, and neutron scattering all rely in some way on the light scattering process.

The other process mentioned above, the transfer of energy from a photon to a molecule, is **absorption**. For a photon to be absorbed, its energy must match the energy difference between two energy levels of the molecule.

Molecules possess a set of quantized energy levels, as shown in Figure 5.3. Although several states are possible, only two electronic states are shown, a **ground state**, G, and the **first excited state**, S_1. These two states differ in the distribution of valence electrons. When electrons are promoted from a ground state orbital in G to an orbital of higher energy in S_1, an **electronic transition** is said to occur. The energy associated with ultraviolet and visible light is sufficient to promote molecules from

Figure 5.3
Energy-level diagram showing the ground state, G, and the first excited state, S_1.

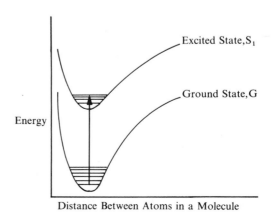

Distance Between Atoms in a Molecule

one electronic state to another, that is, to move electrons from one orbital to another.

Within each electronic energy level is a set of **vibrational levels**. These represent changes in the stretching and bending of covalent bonds. The importance of these energy levels will not be discussed here, but transitions between these levels are the basis of infrared spectroscopy.

The electronic transition for a molecule from G to S_1, represented by the vertical arrow in Figure 5.3, has a high probability of occurring if the energy of the photon corresponds to the energy necessary to promote an electron from energy level E_1 to energy level E_2:

$$\blacktriangleright \quad E_2 - E_1 = \Delta E = \frac{hc}{\lambda} \qquad \text{(Equation 5.3)}$$

A transition may occur from any vibrational level in G to some other vibrational level in S_1, for example, $v = 3$; however, not all transitions have equal probability. The probability of absorption is described by quantum mechanics and will not be discussed here.

A UV-VIS spectrum is obtained by measuring the light absorbed by a sample as a function of wavelength. Since only discrete packets of energy (specific wavelengths) are absorbed by molecules in the sample, the spectrum theoretically should consist of sharp discrete lines. However, the many vibrational levels of each electronic energy level increase the number of possible transitions. This results in several spectral lines, which together make up the familiar spectrum of broad peaks as shown in Figure 5.4.

An absorption spectrum can aid in the identification of a molecule because the wavelength of absorption depends on the functional groups or arrangement of atoms in the sample. The spectrum of oxyhemoglobin in

Figure 5.4
The visible absorbance spectrum of hemoglobin.

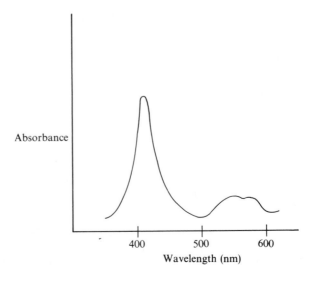

Figure 5.4 is due to the presence of the iron porphyrin moiety and is useful for the characterization of heme derivatives or hemoproteins. Note that the spectrum consists of several peaks at wavelengths where absorption reaches a maximum (415, 542, and 577 nm). These points, called λ_{max}, are of great significance in the identification of unknown molecules and will be discussed later in the chapter.

Quantitative measurements in spectrophotometry are evaluated using the Beer-Lambert law:

▶ $A = Elc$ (Equation 5.4)

where

A = absorbance or $-\log (I/I_0)$

I_0 = intensity of light irradiating the sample

I = intensity of light transmitted through the sample

E = absorption coefficient or absorptivity

l = path length of light through the sample, or thickness of the cell

c = concentration of absorbing material in the sample

Since the absorbance, A, is derived from a ratio ($-\log I/I_0$), it is unitless. The term E, which is a proportionality constant, defines the efficiency or extent of absorption. If this is defined for a particular chromophore at a specific wavelength, the term **absorption coefficient** or **absorptivity** is used. However, students should be aware that in the older biochemical literature, the term extinction coefficient is often used. The units of E depend on the units of l (usually cm) and c (usually molar) in Equation 5.4. For biomolecules, E is often used in the form **molar absorption coefficient**, ϵ, which is defined as the absorbance of a 1 M solution of pure absorbing material in a 1-cm cell under specified conditions of wavelength and solvent. The units of ϵ are M^{-1} cm^{-1}. To illustrate the use of Equation 5.4, consider the following calculation.

Example I   The absorbance, A, of a 5×10^{-4} M solution of the amino acid tyrosine, at a wavelength of 280 nm, is 0.75. The path length of the cuvette is 1 cm. What is the molar absorption coefficient, ϵ?

$A = \epsilon lc = 0.75$

$l = 1$ cm

$c = 5 \times 10^{-4}$ M

$\epsilon = \dfrac{0.75}{(1 \text{ cm})(5 \times 10^{-4} \text{ mole/liter})}$

$= 1500 \dfrac{\text{liter}}{\text{mole} \times \text{cm}} = 1500 \ M^{-1} \text{ cm}^{-1}$

Notice that the units of ϵ are defined by the concentration units of the tyrosine solution (M) and the dimension units of the cuvette (cm). Although E is most often expressed as a molar absorption coefficient, you may encounter other units such as $E_\lambda^\%$, which is the absorbance of a 1% (w/v) solution of pure absorbing material in a 1-cm cuvette at a specified wavelength, λ.

Instrumentation

The **spectrophotometer** is used to measure absorbance experimentally. This instrument produces light of a preselected wavelength, directs it through the sample (usually dissolved in a solvent and placed in a cuvette), and measures the intensity of light transmitted by the sample. The major components are shown in Figure 5.5. These consist of a light source, a monochromator (including various filters, slits, and mirrors), a sample chamber, a detector, and a meter or recorder.

LIGHT SOURCE

For absorption measurements in the ultraviolet region, a high-pressure hydrogen or deuterium lamp is used. These lamps produce radiation in the 200 to 340 nm range. The light source for the visible region is the tungsten-halogen lamp, with a wavelength range of 340 to 800 nm. Instruments with both lamps have greater flexibility and can be used for the study of most biologically significant molecules.

MONOCHROMATOR

Both lamps discussed above produce continuous emissions of all wavelengths within their range. Therefore, a spectrophotometer must have an optical system to select monochromatic light (light of a specific wavelength). Modern instruments use a prism or, more often, a diffraction grating to produce the desired wavelengths. It should be noted that light emitted from the monochromator is not entirely of a single wavelength but is enhanced in that wavelength. That is, most of the light is of a single wavelength, but shorter and longer wavelengths are present.

Before the monochromatic light impinges on the sample, it passes through a series of slits, lenses, filters, and mirrors. This optical system

Figure 5.5
A diagram of a conventional, single-beam spectrophotometer.

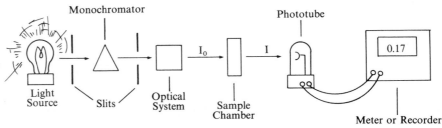

concentrates the light, increases the spectral purity, and focuses it toward the sample. The operator of a spectrophotometer has little control over the optical manipulation of the light beam, except for adjustment of slit width. Light passing from the monochromator to the sample encounters a "gate" or slit. On some instruments this is fixed to allow a light beam of a certain width to pass. More expensive instruments have variable slits, allowing the operator to adjust the slit width. The slit width determines both the intensity of light impinging on the sample and the spectral purity of that light. Decreasing the slit width increases the spectral purity of the light, but the amount of light directed toward the sample decreases. The efficiency or sensitivity of the detector then becomes a limiting factor. Setting the proper slit width requires finding a balance between spectral purity and detector sensitivity.

In instruments equipped with photodiode array detectors (see Figure 5.6 and the following section on detectors), polychromatic light from a source passes through the sample and is focused on the entrance slit of a polychromator (a holographic grating that disperses the light into its wavelengths). The wavelength-resolved light beam is then focused onto the photodiode array detector. The relative positions of the sample and grating are reversed compared to conventional spectrometry; hence the new configuration is often called reversed optics. (Compare Figures 5.5 and 5.6.)

SAMPLE CHAMBER

The processed monochromatic light is then directed into a sample chamber, which can accommodate a wide variety of sample holders. Most UV-VIS measurements on biomolecules are taken on solutions of

Figure 5.6
Optical diagram for a diode array spectrometer.

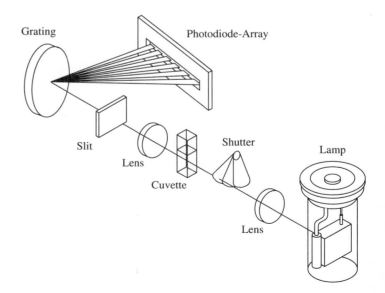

the molecules. The sample is placed in a tube or cuvette made of glass, quartz, or other transparent material. Figure 5.7 shows the design of the most common sample holders and the transmission properties of several transparent materials used in cuvette construction.

Glass cuvettes are the least expensive but, because they absorb UV light, they can be used only above 340 nm. Quartz or fused silica cuvettes may be used throughout the UV and visible regions (~200–800 nm). Disposable cuvettes are now commercially available in polymethacrylate (280–800 nm) and polystyrene (350–800 nm).

Figure 5.7
(A) An assortment of cuvettes. Courtesy of VWR Scientific, Division of Univar.   (B) A standard 3-mL cuvette. Courtesy of Beckman Instruments, Inc.   (C) The transmission properties of several materials used in cuvettes. A = silica (quartz); B = NIR silica; C = polymethacrylate; D = polystyrene; E = glass.

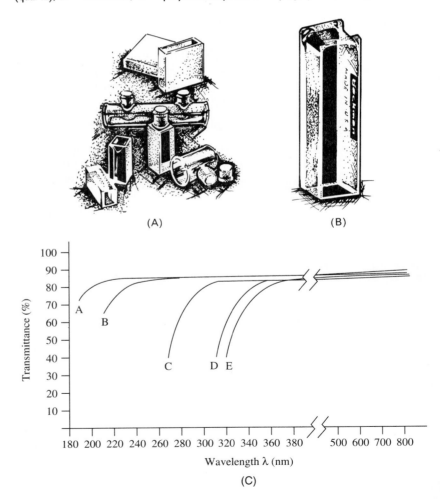

(A)   (B)

(C)

The quality and condition of cuvettes are critical factors in spectrophotometric use. High-quality cuvettes are expensive and must be carefully maintained. The care and use of cuvettes were discussed in Chapter 1.

Less expensive spectrophotometers are "single-beam" instruments that allow insertion of one cuvette at a time in the sample holder. More sophisticated instruments are "double beam" and accept two cuvettes, one containing sample dissolved in a solvent (sample position in Figure 5.5) and the other containing pure solvent. The use of both types of instruments will be outlined in the applications section.

Spectrophotometers are often used to monitor the rates of chemical reactions. For these measurements, the sample cuvette must be held at a constant temperature. Many sample chambers allow for circulating a constant-temperature liquid around the sample holder.

DETECTOR

The intensity of the light that passes through the sample under study depends on the amount of light absorbed by the sample. Intensity is measured by a light-sensitive detector, usually a photomultiplier tube (PMT). The PMT detects a small amount of light energy, amplifies this by a cascade of electrons accelerated by dynodes, and converts it into an electrical signal that can be fed into a meter or recorder.

A new technology has been introduced during the past few years that greatly increases the speed of spectrophotometric measurements. New detectors called photodiode arrays are being used in modern spectrometers. Photodiodes are composed of silicon crystals that are sensitive to light in the wavelength range 170–1100 nm. Upon photon absorption by the diode, a current is generated in the photodiode that is proportional to the number of photons. Linear arrays of photodiodes are self-scanning and have response times on the order of 100 milliseconds; hence, an entire UV-VIS spectrum can be obtained with an extremely brief exposure of the sample to polychromatic light. New spectrometers designed by Hewlett-Packard and Perkin-Elmer use this technology and can produce a full spectrum from 190 to 820 nm in one-tenth of a second.

METERS OR RECORDERS

Less expensive instruments give a direct readout of absorbance and/or transmittance in analog or digital form. These instruments are suitable for single-wavelength measurements; however, if a scan of absorbance vs. wavelength (Figure 5.4) is desired, a recorder must be used.

Modern, research-grade spectrometers are available that offer the latest in technology. All of the components discussed above are integrated into a single package and are completely under the control of a computer. By simply pushing a button, one can obtain the UV-VIS spectrum of a sample displayed on a computer screen in less than 1 second. In addition, these modern instruments with computers can be programmed to carry out several functions, such as spectral overlay, storage, difference spectra, derivative spectra, and calculation of concentrations and rate constants.

Applications

Now that you are familiar with the theory and instrumentation of absorption spectrophotometry, you will more easily understand the actual operation and typical applications of a spectrophotometer. Since virtually all UV-VIS measurements are made on samples dissolved in solvents, only those applications will be described here. Although many different types of operations can be carried out on a spectrophotometer, all applications fall in one of two categories: (1) measurement of absorbance at a fixed wavelength and (2) measurement of absorbance as a function of wavelength. Measurements at a fixed wavelength are most often used to obtain quantitative information such as the concentration of a solute in solution or the absorption coefficient of a chromophore. Absorbance measurements as a function of wavelength provide qualitative information that assists in solving the identity and structure of a pure substance by detecting characteristic groupings of atoms in a molecule.

For fixed-wavelength measurements with a single-beam instrument, a cuvette containing solvent only is placed in the sample beam and the instrument is adjusted to read "zero" absorbance. A matched cuvette containing sample plus solvent is then placed in the sample chamber and the absorbance is read directly from the meter or recorder. The adjustment to zero absorbance with only solvent in the sample chamber allows the operator to obtain a direct reading of absorbance for the sample.

Fixed-wavelength measurements using a double-beam spectrophotometer are made by first zeroing the instrument with no cuvette in either the sample or reference holder. Alternatively, the spectrophotometer can be balanced by placing matched cuvettes containing water or solvent in both sample chambers. Then, a cuvette containing pure solvent is placed in the reference position and a matched cuvette containing solvent plus sample is set in the sample position. The absorbance reading given by the instrument is that of the sample; that is, the absorbance due to solvent is subtracted by the instrument.

An **absorbance spectrum** of a compound is obtained by scanning a range of wavelengths and plotting the absorbance at each wavelength. Most double-beam spectrophotometers automatically scan the desired wavelength range and record the absorbance as a function of wavelength. If solvent is placed in the reference chamber and solvent plus sample in the sample position, the instrument will continuously and automatically subtract the solvent absorbance from the total absorbance (solvent plus sample) at each wavelength; hence, the recorder output is really a difference spectrum (absorbance of sample plus solvent, minus absorbance of solvent).

Both types of measurements (fixed wavelength and absorbance spectrum) are common in biochemistry, and you should be able to interpret results from each. The following examples are typical of the kinds of problems readily solved by spectrophotometry.

MEASUREMENT OF THE CONCENTRATION OF A SOLUTE IN SOLUTION

According to the Beer-Lambert law, the absorbance of a material in solution is directly dependent on the concentration of that material. Two methods are commonly used to measure concentration. If the absorption coefficient is known for the absorbing species, the concentration can be calculated after experimental measurement of the absorbance of the solution.

Example 2   A solution of the nucleotide base uracil, in a 1-cm cuvette, has an absorbance at λ_{max} (260 nm) of 0.65. Pure solvent in a matched quartz cuvette has an absorbance of 0.07. What is the molar concentration of the uracil solution? Assume the molar absorption coefficient, ϵ, is $8.2 \times 10^3 \ M^{-1} \ cm^{-1}$.

$$A = \epsilon l c$$

$$A = (\text{absorbance of solvent} + \text{sample}) - (\text{absorbance of solvent})$$

$$A = 0.65 - 0.07 = 0.58$$

$$\epsilon = 8.2 \times 10^3 \ M^{-1} \ cm^{-1}$$

$$l = 1 \ cm$$

$$c = \frac{A}{\epsilon l} = \frac{0.58}{(8.2 \times 10^3 M^{-1} \ cm^{-1})(1 \ cm)}$$

$$c = 7.1 \times 10^{-5} M$$

If the absorption coefficient for an absorbing species is known, the concentration of that species in solution can be calculated as outlined. However, there are limitations to this application. Most spectrophotometers are useful for measuring absorbances up to 1, although more sophisticated instruments can measure absorbances as high as 2. Also, some substances do not obey the Beer-Lambert law; that is, absorbance may not increase in a linear fashion with concentration. Reasons for deviation from the Beer-Lambert law are many; however, the majority are instrumental, chemical, or physical. Spectrophotometers often display a nonlinear response at high absorption levels because of stray light. Physical reasons for nonlinearity include hydrogen bonding of the absorbing species with the solvent and intermolecular interactions at high concentrations. Chemical reasons may include reaction of the solvent with the absorbing species and the presence of impurities. Linearity is readily tested by preparing a series of concentrations of the absorbing species and measuring the absorbance of each. A plot of A vs. concentration should be linear if the Beer-Lambert law is valid. If the absorption coefficient for a species is unknown, its concentration in solution can be measured if the absorbance of a standard solution of the compound is known.

Example 3 The absorbance of a 1% (w/v) solution of the enzyme tyrosinase, in a 1-cm cell at 280 nm, is 24.9. What is the concentration of a tyrosinase solution that has an A_{280} of 0.25?

Since the absorption coefficient, $E\%$, is the same for both solutions, the concentration can be calculated by a direct ratio:

$$\frac{A_{std}}{C_{std}} = \frac{A_x}{C_x}$$

A_{std} = absorbance of the 1% standard solution = 24.9

C_{std} = concentration of the standard solution = 1% (1g/dL)

A_x = absorbance of the unknown solution = 0.25

C_x = concentration of the unknown solution in %

$$\frac{24.9}{1\%} = \frac{0.25}{C_x}$$

$$C_x = 0.01\% = 0.01 \text{ g/dL} = 0.1 \text{ mg/mL}$$

Alternatively, the concentration of a species in solution can be determined by preparing a standard curve of absorbance vs. concentration.

Example 4 The Bradford protein assay is one of the most used spectrophotometric assays in biochemistry. (For a discussion of the Bradford assay, see Chapter 2.) Solutions of varying amounts of a standard protein are mixed with reagents that cause the development of a color. The amount of color produced depends on the amount of protein present. The absorbance at 595 nm of each reaction mixture is plotted against the known protein concentration. A protein sample of unknown concentration is treated with the Bradford reagents and the color is allowed to develop.

The following absorbance measurements are typical for the standard curve of a protein:

Protein (μg per assay)	A_{595}
15	0.07
25	0.15
50	0.28
100	0.55
150	0.90
0.1 mL unknown protein solution	0.10
0.2 mL unknown protein solution	0.22

A standard curve for the data is plotted in Figure 2.14. Note the linearity, indicating that the Beer-Lambert law is obeyed over this concentration range of standard protein. Two different volumes of unknown protein were tested. This was to ensure that one volume would be in the concentration range of the standard curve. Since the accuracy of the assay is dependent on identical times for color development, the unknowns must be assayed at the same time as the standards. From graphical analysis, an A_{595} of 0.22 corresponds to 40 μg of protein per 0.20 mL. This indicates that the original protein solution concentration was approximately 200 μg/mL or 0.20 mg/mL. Using computer graphics, students can now quickly visualize experimental data and determine the need for further analysis. Most modern computer programs use the method of least squares to calculate automatically the slope, intercepts, and correlation coefficients.

IDENTIFICATION OF UNKNOWN BIOMOLECULES
BY SPECTROPHOTOMETRY

The UV-VIS spectrum of a biomolecule reveals much about its molecular structure. Therefore, a spectral analysis is one of the first experimental measurements made on an unknown biomolecule. Natural molecules often contain chromophoric (color-producing) functional groups that have characteristic spectra. Figure 5.8 displays spectra of the well-characterized biomolecules DNA, FMN, $FMNH_2$, NAD, NADH, and nucleotides. Spectral analysis in the visible region is used in Experiment 11 to identify pigments isolated from plants.

The procedure for obtaining a UV-VIS spectrum begins with the preparation of a solution of the species under study. A standard solution should be prepared in an appropriate solvent. An aliquot of the solution is transferred to a cuvette and placed in the sample chamber of a spectrophotometer. A cuvette containing solvent is placed in the reference holder. The spectrum is scanned over the desired wavelength range and an absorption coefficient is calculated for each major λ_{max}.

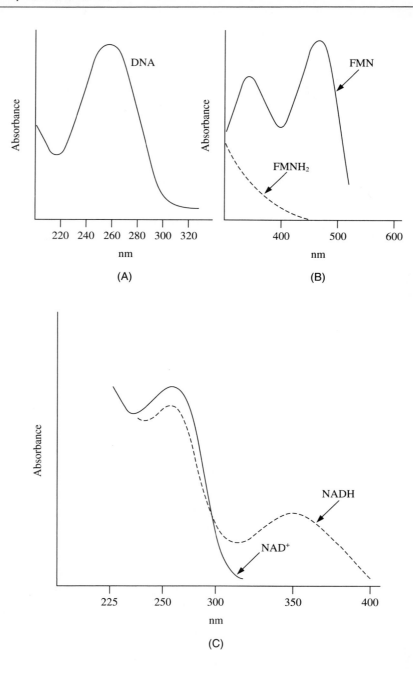

Figure 5.8
Absorbance spectra of significant biomolecules. (A) DNA; (B) FMN and
FMNH₂; (C) NAD⁺ and NADH; (D) d-GMP; (E) thymine.

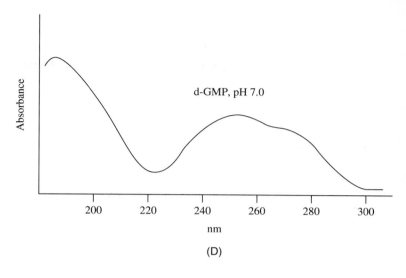

(D)

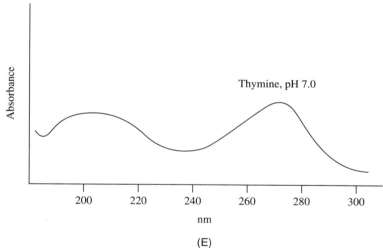

(E)

Figure 5.8 *(continued)*

KINETICS OF CHEMICAL REACTIONS

Spectrophotometry is one of the best methods available for measuring the rates of chemical reactions. Consider a general chemical reaction as shown in Equation 5.5.

$$\blacktriangleright \quad A + B \longrightarrow C + D \qquad \text{(Equation 5.5)}$$

If reactants A or B absorb in the UV-VIS region of the spectrum at some wavelength λ_1, the rate of the reaction can be measured by monitoring the decrease of absorbance at λ_1 due to loss of A or B. Alternatively, if products C or D absorb at a specific wavelength λ_2, the kinetics of the reaction can be evaluated by monitoring the absorbance increase at λ_2. According to the Beer-Lambert law, the absorbance change of a reactant or product is proportional to the concentration change of that species occurring during the reaction. This method is widely used to assay enzyme-catalyzed processes. Experiment 6 utilizes spectrophotometry to characterize the kinetics of the tyrosinase-catalyzed oxidation of 3,4-dihydroxyphenylalanine.

CHARACTERIZATION OF MACROMOLECULE-LIGAND INTERACTIONS BY DIFFERENCE SPECTROSCOPY

Many biological processes depend on a specific interaction between molecules. The interaction often involves a macromolecule (protein or nucleic acid) and a smaller molecule, a **ligand**. Specific examples include enzyme-substrate interactions and receptor protein–hormone interactions. One of the most effective and convenient methods for detecting and characterizing such interactions is **difference spectroscopy**. The interaction of small molecules with the transport protein hemoglobin is a classic example of the utility of difference spectroscopy. If a small molecule or ligand, such as inositol hexaphosphate, binds to hemoglobin, there is a change in the heme spectral properties. The spectral change is small and would be difficult to detect if the experimenter recorded the spectrum in the usual fashion. Normally, one would obtain a spectrum of a heme protein by placing a solution of the protein in a cuvette in the sample compartment of a spectrophotometer and the neat solvent in the reference compartment. Any absorption due to solvent is subtracted because the solvent is present in both light beams, so the spectrum is that due to the heme protein. Then the ligand to be tested would be added to the heme protein, and the spectrum would be obtained for this mixture vs. solvent. If the free or bound ligand molecule did not absorb light in the wavelength range studied, there would be no need to have ligand in the reference cell. The two spectra (heme protein in solvent vs. solvent and heme protein and ligand in solvent vs. solvent) can then be compared and differences noted.

A difference spectrum is faster than the preceding method because only one spectral recording is necessary. Two cuvettes are prepared in the following manner. The reference cuvette contains heme protein and solvent, whereas the sample cuvette contains heme protein, solvent, and ligand. There must be equal concentrations of the heme protein in the two cuvettes. (Why?) The two cuvettes are placed in a double-beam spectrophotometer and the spectrum is recorded. If the spectrum of the heme protein is not influenced by the ligand, the result would be a zero difference spectrum, that is, a straight line (Figure 5.9A). Both samples have identical spectral properties, indicating that there is probably little or no interaction between heme protein and ligand. However, such data should be treated with caution because it is possible that the heme group

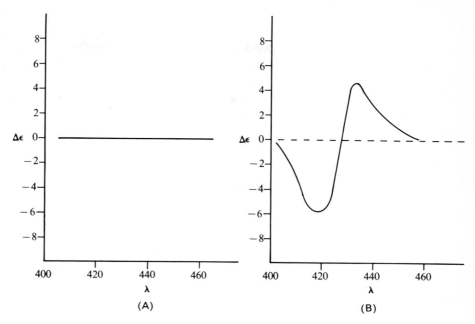

Figure 5.9
Difference spectroscopy.   (A) Hemoglobin vs. hemoglobin.   (B) Hemoglobin vs. hemoglobin + inositol hexaphosphate.

is not affected by ligand binding. A nonzero difference spectrum indicates that the ligand interacts with the heme protein and induces a change in the environment of the heme group (Figure 5.9B).

A difference spectrum can be analyzed and used in several ways. This is a useful technique for demonstrating qualitatively whether an interaction occurs between a macromolecule and a ligand. Quantitative analysis of difference spectra requires measurement of λ_{max} and ΔA at λ_{max}. This method can be used to quantify the strength of ligand-protein interaction. Note that every time you record a spectrum in a double beam spectrometer, you are obtaining a difference spectrum between sample and reference. Although a heme protein was used in this example, this does not imply that only heme interactions can be characterized by difference spectroscopy. Any protein that contains a chromophoric group, whether it be an aromatic amino acid, cofactor, prosthetic group, or metal ion, can be studied by difference spectroscopy (Herskovits, 1967; Robyt and White, 1987).

LIMITATIONS AND PRECAUTIONS IN SPECTROPHOTOMETRY

The use of a spectrophotometer is relatively straightforward and can be mastered in a short period of time. There are, however, difficulties that must be considered. A common problem encountered with biochemical measurements is turbidity or cloudiness of biological samples. This can lead to great error in absorbance measurements because much of the light entering the cuvette is not absorbed but is scattered. This causes artificially high absorbance readings. Occasionally, absorbance readings on turbid solutions are desirable (as in measuring the rate of bacterial growth in a culture), but in most cases turbid solutions must be avoided or clarified by filtration or centrifugation.

A difficulty encountered in measuring the concentration of an unknown absorbing species in solution is deviation from the Beer-Lambert law. For reasons stated earlier in this chapter, some absorbing species do not demonstrate an increase in absorbance that is proportional to an increase in concentration. (In reality, most compounds follow the Beer-Lambert relationship over a relatively small concentration range.) When measuring solution concentration, adherence to the Beer-Lambert law must always be tested in the concentration range under study.

Finally, care in sample preparation and cleanliness are especially important in performing spectrophotometric measurements. The cause of many errors in absorbance measurements is improper preparation and dilution of solutions used to test the Beer-Lambert law. Cuvettes used for UV-VIS measurements must be clean and lint-free. Always handle the cuvettes on the sides not in the light beam; better yet, use tissue paper to handle them.

B. Fluorescence Spectrophotometry

Principles

In our discussion of absorption spectroscopy, we noted that the interaction of photons with molecules resulted in the promotion of valence electrons from ground state orbitals to higher energy level orbitals. The molecules were said to be in an excited state.

Molecules in the excited state do not remain there long, but spontaneously relax to the more stable ground state. With most molecules, the relaxation process is brought about by collisional energy transfer to solvent or other molecules in the solution. Some excited molecules, however, return to the ground state by emitting the excess energy as light. This process, called **fluorescence**, is illustrated in Figure 5.10. The solid vertical arrow in the figure indicates the photon absorption process in which the molecule is excited from G to some vibrational level in S. The excited molecule loses vibrational energy by collision with solvent and ground state molecules. This relaxation process, which is very rapid, leaves the molecule in the lowest vibrational level of S, as indicated by the wavy arrow. The molecule may release its energy in the form of light (fluorescence, dashed arrow) to return to some vibrational level of G.

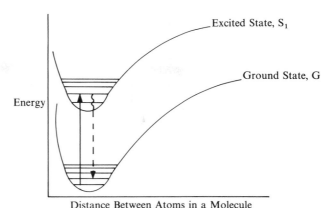

Energy

Excited State, S_1

Ground State, G

Distance Between Atoms in a Molecule

Figure 5.10
Energy-level diagram describing the fluorescence process.

Two important characteristics of the emitted light should be noted: (1) It is of longer wavelength (lower energy) than the excitation light. This is because part of the energy initially associated with the S state is lost as heat energy, and the energy lost by emission may be sufficient only to return the excited molecule to a higher vibrational level in G. (2) The emitted light is composed of many wavelengths, which results in a fluorescence spectrum as shown in Figure 5.11. This is due to the fact that fluorescence from any particular excited molecule may return the molecule to one of many vibrational levels in the ground state. Just as in the case of an absorption spectrum, a wavelength of maximum fluorescence is observed, and the spectrum is composed of a wavelength distribution centered at this emission maximum. Note that all emitted wavelengths are longer and of lower energy than the initial excitation wavelength.

In our discussion above, it was pointed out that a molecule in the excited state can return to lower energy levels by collisional transfer or by light emission. Since these two processes are competitive, the **fluorescence intensity** of a fluorescing system depends on the relative importance of each process. The fluorescence intensity is often defined in terms of **quantum yield**, represented by Q. This describes the efficiency or probability of the fluorescence process. By definition, Q is the ratio of the number of photons emitted to the number of photons absorbed (Equation 5.6).

▶ $$Q = \frac{\text{number of photons emitted}}{\text{number of photons absorbed}}$$ (Equation 5.6)

Measurement of quantum yield is often the goal in fluorescence spectroscopy experiments. Q is of interest because it may reveal important characteristics of the fluorescing system. Two types of factors affect the intensity of fluorescence, internal and external (environmental)

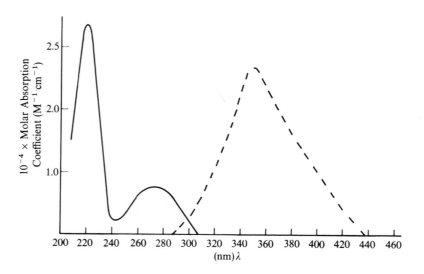

Figure 5.11
Absorption (——) and fluorescence (- - - -) spectra of tryptophan. From D. M. Freifelder, *Physical Biochemistry*, W. H. Freeman (San Francisco). Copyright © 1982. All rights reserved.

influences. Internal factors, such as the number of vibrational levels available for transition and the rigidity of the molecules, are associated with properties of the fluorescent molecules themselves. Internal factors will not be discussed in detail here because they are of more interest in theoretical studies. The external factors that affect Q are of great interest to biochemists because information can be obtained about macromolecule conformation and molecular interactions between small molecules (ligands) and larger biomolecules (proteins, nucleic acids). Of special value is the study of experimental conditions that result in **quenching** or **enhancement** of the quantum yield. Quenching in biochemical systems can be caused by chemical reactions of the fluorescent species with added molecules, transfer of energy to other molecules by collision (actual contact between molecules), and transfer of energy over a distance (no contact, resonance energy transfer). The reverse of quenching, enhancement of fluorescent intensity, is also observed in some situations. Several fluorescent dye molecules are quenched in aqueous solution, but their fluorescence is greatly enhanced in a nonpolar or rigidly bound environment (the interior of a protein, for example). This is a convenient method for characterizing ligand binding.

Both fluorescence quenching and fluorescence enhancement studies can yield important information about biomolecular structure and function. Several applications will be described in a later section of this chapter and in Experiment 14.

Instrumentation

The basic instrument for measuring fluorescence is the **spectro-fluorometer**. It contains a light source, two monochromators, a sample holder, and a detector. A typical experimental arrangement for fluorescence measurement is shown in Figure 5.12. The setup is similar to that for absorption measurements, with two significant exceptions. First, there are two monochromators, one for selection of the excitation wavelength, another for wavelength analysis of the emitted light. Second, the detector is at an angle (usually 90°) to the excitation beam. This is to eliminate interference by the light that is transmitted through the sample. Upon excitation of the sample molecules, the fluorescence is emitted in all directions and is detected by a photocell at right angles to the excitation light beam.

The lamp source used in most instruments is a xenon arc lamp that emits radiation in the ultraviolet, visible, and near-infrared regions (200 to 1400 nm). The light is directed by an optical system to the excitation monochromator, which allows either preselection of a wavelength or scanning of a certain wavelength range. The exciting light then passes into the sample chamber, which contains a fluorescence cuvette with dissolved sample. Because of the geometry of the optical system, a typical fused absorption cuvette with two opaque sides cannot be used; instead, special fluorescence cuvettes with four translucent quartz or glass sides must be used. When the excitation light beam impinges on the sample cell, molecules in the solution are excited and some will emit light.

Light emitted at right angles to the incoming beam is analyzed by the emission monochromator. In most cases, the wavelength analysis of emit-

Figure 5.12
A diagram of a typical fluorometer.

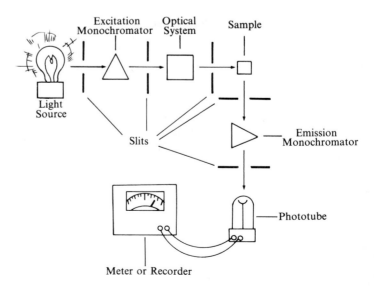

ted light is carried out by measuring the intensity of fluorescence at a preselected wavelength (usually the wavelength of emission maximum). The analyzer monochromator directs emitted light of only the preselected wavelength toward the detector. A photomultiplier tube serves as a detector to measure the intensity of the light. The output current from the photomultiplier is fed to some measuring device that indicates the extent of fluorescence. The final readout is not in terms of Q but in units of the photomultiplier tube current (microamperes) or in relative units of percent of full scale. Therefore, the scale must be standardized with a known. Some newer instruments provide, as output, the ratio of emitted light to incident light intensity. This type of information is advantageous because the xenon arc lamp is not a particularly stable light source, and its output may vary with time.

Applications

Two types of measurements are most common in fluorescence experiments, measurements of **relative fluorescence intensities** and measurements of the **quantum yield**. Experiments introduced in this book will require only relative fluorescence intensity measurements, and they proceed as follows. The fluorometer is set to "zero" or "full scale" fluorescence intensity (microamps or %) with the desired biochemical system under standard conditions. Some perturbation is then made in the system (pH change, addition of a chemical agent in varying concentrations, change of ionic strength, etc.) and the fluorescence intensity is determined relative to the standard conditions. This is a straightforward type of experiment because it consists of replacing one solution with another in the fluorometer and reading the detector output for each. For these experiments, the excitation wavelength and the emission wavelength are preselected and set for each monochromator.

The measurement of quantum yield is a more complicated process. Before these measurements can be made, the instrument must be calibrated. A thermopile or chemical actinometer may be used to measure the absolute intensity of incident light on the sample. Alternatively, quantum yields may be measured relative to some accepted standard. Two commonly used fluorescence standards are quinine sulfate in 1 N H_2SO_4 (Q = 0.70) and fluorescein in 0.1 N NaOH (Q = 0.93). The quantum yield of the unknown, Q_x, is then calculated by Equation 5.7.

$$\blacktriangleright \quad \frac{Q_x}{Q_{std}} = \frac{F_x}{F_{std}} \qquad \text{(Equation 5.7)}$$

where

Q_x = quantum yield of unknown

Q_{std} = quantum yield of standard

F_x = experimental fluorescence intensity of unknown

F_{std} = experimental fluorescence intensity of standard

Some biomolecules are **intrinsic fluors**; that is, they are fluorescent themselves. The amino acids with phenyl rings (phenylalanine, tyrosine, and tryptophan) are fluorescent; hence, proteins containing these amino acids have intrinsic fluorescence. The purine and pyrimidine bases in nucleic acids (adenine, guanine, cytosine, uracil) and some coenzymes (NAD, FAD) are also intrinsic fluors. Intrinsic fluorescence is most often used to study protein conformational changes and to probe the location of active sites and coenzymes in enzymes.

Valuable information can also be obtained by the use of **extrinsic fluors**. These are fluorescent molecules that are added to the biochemical system under study. Many fluorescent dyes have enhanced fluorescence when they are in a nonpolar solution or bound in a rigid hydrophobic environment. Some of these dyes bind to specific sites on proteins or nucleic acid molecules, and the resulting fluorescence intensity depends on the environmental conditions at the binding site. Extrinsic fluorescence is of value in characterizing the binding of natural ligands to biochemically significant macromolecules. This is because many of the extrinsic fluors bind in the same sites as natural ligands. Extrinsic fluorescence has been used to study the binding of fatty acids to serum albumin, to characterize the binding sites for cofactors and substrates in enzyme molecules, to characterize the heme binding site in various hemoproteins, and to study the intercalation of small molecules into the DNA double helix (Experiment 18).

Figure 5.13 shows the structures of extrinsic fluors that have been of value in studying biochemical systems. ANS, dansyl chloride, and fluorescein are used for protein studies, whereas ethidium, proflavine, and various acridines are useful for nucleic acid characterization. Ethidium bromide has the unique characteristic of enhanced fluorescence when bound to double-stranded DNA but not to single-stranded DNA. Aminomethyl coumarin (AMC) is of value as a fluorogenic leaving group in measuring peptidase activity.

The potential applications of fluorescence spectroscopy are too numerous to mention here. The use of a fluorometer is becoming routine in biochemical research and can be applied, as illustrated here, to many aspects of biochemistry.

Difficulties in Fluorescence Measurements

Fluorescence measurements have much greater sensitivity than absorption measurements. Therefore, the experimenter must take special precautions in making fluorescence measurements because any contaminant or impurity in the system can lead to poor results. The following factors must be considered when preparing for a fluorescence experiment.

XENON ARC LAMP

As noted earlier, this lamp is not particularly stable. It is best to make several measurements on a sample and average the readings. Instruments

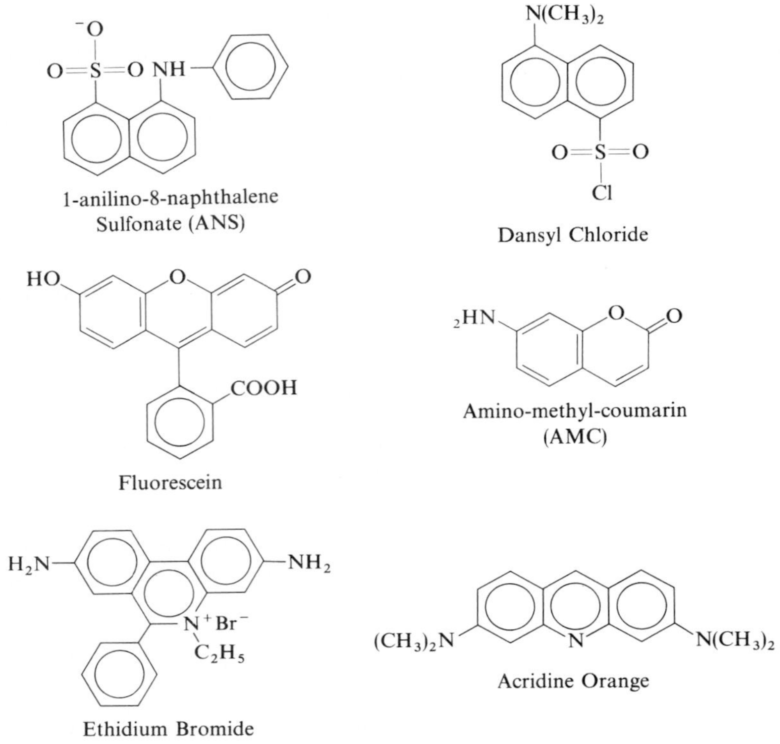

Figure 5.13
Fluorescent molecules useful in
biochemical studies.

that provide data in terms of the ratio of light emitted to incident light intensity have a special advantage.

PREPARATION OF REAGENTS AND SOLUTIONS

Since fluorescence measurements are very sensitive, dilute solutions
of biomolecules and other reagents are appropriate. Special precautions
must be taken to maintain the integrity of these solutions. All solvents
and reagents must be checked for the presence of fluorescent impurities,
which can lead to large errors in measurement. "Blank" readings should
be taken on all solvents and solutions, and any background fluorescence
must be subtracted from the fluorescence of the complete system under
study. Solutions should be stored in the dark, in clean glass-stoppered
containers, in order to avoid photochemical breakdown of the reagents
and contamination by corks and rubber stoppers. Some biomolecules, especially proteins, tend to adsorb to glass surfaces, which can lead to loss
of fluorescent material or to contamination of fluorescence cuvettes. All
glassware must be scrupulously cleaned with a mild detergent. Solutions
for use in fluorometry should not be stored for extended periods of time;
ideally, they should be prepared just before use. All turbid solutions must
be clarified by centrifugation or filtration.

CONTROL OF TEMPERATURE

Fluorescence measurements, unlike absorption, are temperature dependent. All solutions, especially if relative fluorescent measurements are taken, must be thermostated at the same temperature.

References

J. Bell, Editor, *Spectroscopy in Biochemistry,* Vols. I and II (1981), CRC Press (Boca Raton, FL). Applications of spectroscopic techniques to biochemistry.

I. Campbell and R. Dwek, *Biological Spectroscopy* (1984), Benjamin/Cummings (Menlo Park, CA). Application of spectroscopy to biological systems.

C. Cantor and P. Schimmel, *Biophysical Chemistry*, Part II (1980), W. H. Freeman (San Francisco), pp. 349–480, 687–846. An excellent theoretical discussion of spectroscopy and scattering phenomena.

T. Dewey, Editor, *Biophysical and Biochemical Aspects of Fluorescence Spectroscopy,* (1991), Plenum Press (New York). Theory and applications.

D. Freifelder, *Physical Biochemistry*, 2nd ed (1982), W. H. Freeman (San Francisco), pp. 494–572, 573–579, 691–698. Theory and applications of absorption, fluorescence spectroscopy, and light scattering.

R. Gendreau, Editor, *Spectroscopy in the Biomedical Sciences* (1986), CRC Press (Boca Raton, FL). Analytical techniques for biological scientists.

B. Giardina, F. Ascoli, and M. Brunori, *Nature* **256**, 761 (1975). Spectroscopic study of the binding of inositol hexaphosphate to hemoglobin.

G. Guilbault, Editor, *Practical Fluorescence,* 2nd ed. (1990), Marcel Dekker (New York). Introduction to fluorescence.

T. Herskovits, in *Methods in Enzymology*, Vol. XI, C. H. W. Hirs, Editor (1967), Academic Press (New York), pp. 748–775. "Difference Spectroscopy."

D. Jameson and G. Reinhart, Editors, *Fluorescent Biomolecules: Methodologies and Applications* (1989), Plenum Publishing (New York). Recent developments in biochemical fluorescence.

A. Knowles and C. Burgess, Editors, *Practical Absorption Spectrometry* (1984), Chapman & Hall (London). Useful applications.

J. Lakowicz, in *Modern Physical Methods in Biochemistry*, Part B, A. Neuberger and L. van Deenen, Editors (1988), Elsevier (Amsterdam), pp. 1–26. "Fluorescence Spectroscopy: Principles and Applications to Biological Macromolecules."

C. Mathews and K. van Holde, *Biochemistry* (1990), Benjamin/Cummings (Redwood City, CA), pp. 205–212. Spectroscopic methods.

A. Owen, *The Diode-Array Advantage in UV/VIS Spectroscopy* (1988), Hewlett-Packard Co. Theory and practice of diode array spectroscopy.

J. Rawn, *Biochemistry*, 2nd ed (1989), Neil Patterson (Burlington, NC), p. 64. Introduction to spectroscopy.

J. Robyt and B. White, *Biochemical Techniques: Theory and Practice* (1987), Brooks/Cole (Monterey, CA), pp. 40–72. Spectroscopic methods.

H. Strobel and W. Heineman, *Chemical Instrumentation: A Systematic Approach* (1989), John Wiley & Sons (New York), pp. 540–592, 593–632. Photodiodes and principles and instrumentation for UV-VIS absorption.

K. van Holde, *Physical Biochemistry*, 2nd ed. (1985), Prentice-Hall (Englewood Cliffs, NJ), pp. 182–208. Theory of spectroscopy.

D. Voet and J. Voet, *Biochemistry* (1990), John Wiley & Sons (New York), pp. 590–591. Spectral analysis of biomolecules.

CHAPTER **6**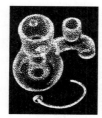

Radioisotopes in Biochemical Research

■ The use of radioactive isotopes in experimental biochemistry has provided us with a wealth of information about biological processes. In the earlier days of biochemistry, radioisotopes were primarily used to elucidate metabolic pathways. In modern biochemistry, there is probably no experimental technique that offers such a diverse range of applications as that provided by radioactive isotopes. The measurement of radioactivity is now a tool in all areas of experimental biochemistry, including enzyme assays, biochemical pathways of synthesis and degradation, analysis of biomolecules, measurements of antibodies, binding and transport studies, and many others. Radioisotope use has two advantages over other analytical techniques: (1) sensitive instrumentation allows the detection of minute quantities of radioactive material (some radioactive substances can be measured in the picomole or 10^{-12} M range), and (2) radioisotope techniques offer the ability to differentiate physically between substances that are chemically indistinguishable.

This chapter will explore the origin, properties, detection, and biochemical applications of radioisotopes.

A. Origin and Properties of Radioactivity

Introduction

Radioactivity results from the spontaneous nuclear disintegration of unstable isotopes. The hydrogen nucleus, consisting of a proton, is represented as ^1_1H. Two additional forms of the hydrogen nucleus contain one and two neutrons; they are represented by ^2_1H and ^3_1H. These forms of hydrogen, which are commonly called deuterium and tritium, respectively, are **isotopes** of hydrogen. They have an identical number of protons (constant charge, $+1$) but they differ in the number of neutrons. Only tritium is radioactive.

The stability of a nucleus depends on the ratio of neutrons to protons. Some nuclei are unstable and undergo spontaneous nuclear disintegration accompanied by emission of particles. Unstable isotopes of this type are called **radioisotopes**. Three main types of radiation are emitted during nuclear decay: α particles, β particles, and γ rays. The α particle, a helium nucleus, is emitted only by elements of mass number greater than 140. These elements are seldom used in biochemical research.

Most of the radioisotopes commonly used in biochemistry are β emitters. The β particles, which exist in two forms (β^+, positrons, and β^-, electrons), are emitted from a given radioisotope with a continuous range of energies. However, an average energy (called the *mean* energy) of β particles from an element can be determined. The mean energy is a characteristic of that isotope and can be used to identify one β emitter in the presence of a second (see Figure 6.1). Equations 6.1 and 6.2 illustrate the disintegration of two β emitters. Here $\overline{v}_e$ represents the antineutrino and v_e the neutrino.

▶ $$^{32}_{15}P \longrightarrow {}^{32}_{16}S + \beta^- + \overline{v}_e \qquad \text{(Equation 6.1)}$$

▶ $$^{65}_{30}Zn \longrightarrow {}^{65}_{29}Cu + \beta^+ + v_e \qquad \text{(Equation 6.2)}$$

A few radioisotopes of biochemical significance are γ emitters. Emission of a γ ray (a photon of electromagnetic radiation) is often a secondary process occurring after initial decay by β emission. The disintegration of the isotope ^{131}I is an example of this multistep process.

▶ $$^{131}_{53}I \longrightarrow \beta^- + {}^{131}_{54}Xe \longrightarrow {}^{131}_{54}Xe + \gamma \qquad \text{(Equation 6.3)}$$

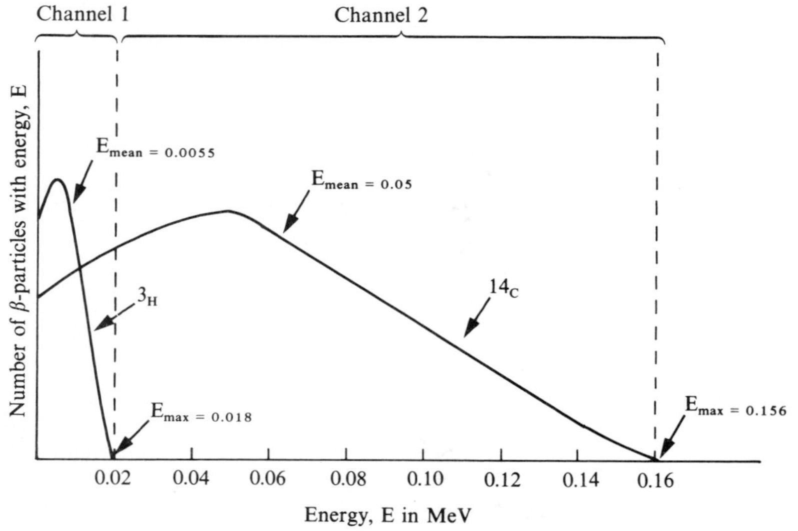

Figure 6.1
The energy spectra for the β emitters ^3H and ^{14}C. The dashed lines indicate the upper and lower limits for discrimination counting.

Each radioisotope emits γ rays of a distinct energy, which can be measured for identification of the isotope.

The spontaneous disintegration of a nucleus is, kinetically, a first-order process. That is, the rate of radioactive decay of N atoms ($-dN/dt$, the change of N with time, t) is proportional to the number of radioactive atoms present (Equation 6.4).

$$\blacktriangleright \quad -\frac{dN}{dt} = \lambda N \qquad \text{(Equation 6.4)}$$

The proportionality constant, λ, is called the **disintegration** or **decay constant**. Equation 6.4 can be transformed to a more useful equation by integration within the limits of $t = 0$ and t. The result is shown in Equation 6.5.

$$\blacktriangleright \quad N = N_0 e^{-\lambda t} \qquad \text{(Equation 6.5)}$$

where

$N_0 =$ the number of radioactive atoms present at $t = 0$

$N =$ the number of radioactive atoms present at time $= t$

Equation 6.5 can be expressed in the natural logarithmic form as in Equation 6.6.

$$\blacktriangleright \quad \ln \frac{N}{N_0} = -\lambda t \qquad \text{(Equation 6.6)}$$

Combining Equations 6.5 and 6.6 leads to a relationship that defines the **half-life**. Half-life, $t_{1/2}$, is a term used to describe the time necessary for one-half of the radioactive atoms initially present in a sample to decay. At the end of the first half-life, N in Equation 6.6 becomes $\frac{1}{2}N_0$, and the result is shown in Equation 6.7 and 6.8.

$$\blacktriangleright \quad \ln \frac{\frac{1}{2}N_0}{N_0} = -\lambda t_{1/2} \qquad \text{(Equation 6.7)}$$

$$\ln \tfrac{1}{2} = -\lambda t_{1/2}$$

$$\blacktriangleright \quad t_{1/2} = \frac{0.693}{\lambda} \qquad \text{(Equation 6.8)}$$

Equations 6.6 and 6.8 allow calculation of the ratio N/N_0 at any time during an experiment. This calculation is especially critical when radioisotopes with short half-lives are used.

Isotopes in Biochemistry

The properties of several radioisotopes that are important in biochemical research are listed in Table 6.1. Note that many of the isotopes are β emitters; however, a few are γ emitters.

The half-life, defined in the previous section and listed for each isotope in Table 6.1, is an important property when designing experiments using radioisotopes. Using an isotope with a short half-life (for example, ^{24}Na with $t_{1/2} = 15$ hr) is difficult because the radioactivity lost during the course of the experiment is significant. Quantitative measurements made before and after the experiment must be corrected for this loss of activity. Radioactive phosphorus, ^{32}P, an isotope of significant value in biochemical research, has a relatively short half-life (14 days), so if quantitative measurements are made they must be corrected as described in Equations 6.7 and 6.8. More information about the choice of a radioisotope in an experiment, the detection and measurement of that radioisotope, and safety rules for handling radioisotopes will be given later in the chapter.

Table 6.1
Properties of Biochemically Important Radioisotopes

Isotope	Particle Emitted	Energy of Particle (MeV)	Half-Life, $t_{1/2}$
^{3}H	β^-	0.018	12.3 yr
^{14}C	β^-	0.155	5570 yr
^{22}Na	β^+	0.55	2.6 yr
	γ	1.28	
^{24}Na	β^-	1.39	15 hr
	γ	1.7, 2.75	
^{32}P	β^-	1.71	14.2 days
^{35}S	β^-	0.167	87.1 days
^{36}Cl	β^-	0.714	3×10^5 days
^{40}K	β^-	1.33	1.3×10^9 yr
^{45}Ca	β^-	0.25	165 days
^{59}Fe	β^-	0.46	45 days
	γ	1.1	
^{60}Co	β^-	0.318	5.3 yr
	γ	1.3	
^{65}Zn	β^+	0.33	245 days
	γ	1.14	
^{90}Sr	β^-	0.54	29 yr
^{125}I	γ	0.035	60 days
	γ	0.027	
^{131}I	β^-	0.61, 0.33	8.1 days
	γ	0.64, 0.36, 0.28	
^{137}Cs	β^-	0.51	30 yr
	γ	0.66	
^{226}Ra	α	4.78	1600 yr
	γ	0.19	

Mention should also be made of short-lived isotopes that are becoming important in biotechnology and medical biochemistry. The isotopes ^{11}C, ^{13}N, ^{15}O, and ^{18}F, which are positron emitters, are crucial for use in positron emission tomography (PET).

Units of Radioactivity

The basic unit of radioactivity is the **curie,** Ci. One curie is the amount of radioactive material that emits particles at a rate of 3.7×10^{10} disintegrations per second (dps), or 2.2×10^{12} min^{-1} (dpm). Amounts that large are seldom used in experimentation, so subdivisions are convenient. The **millicurie** (mCi, 2.2×10^9 min^{-1}) and **microcurie** (μCi, 2.2×10^6 min^{-1}) are standard units for radioactive measurements (see Table 6.2). The radioactivity unit of the meter-kilogram-seconds (MKS) system is the becquerel (Bq). A becquerel, named in honor of Antoine Becquerel, who studied uranium radiation, represents one disintegration per second. The two systems of measurement are related by the definition 1 curie = 3.70×10^{10} becquerels. Since the becquerel is such a small unit, radioactive units are sometimes reported in MBq (mega, 10^6) or TBq (tera, 10^{12}). Both unit systems are in common use today, and radioisotopes received through commercial sources are labeled in curies and bequerels.

The number of disintegrations emitted by a radioactive sample depends on the purity of the sample (number of radioactive atoms present) and the decay constant, λ. Therefore, radioactive decay is also expressed in terms of **specific activity,** the disintegration rate per unit mass of radioactive atoms. Typical units for specific activity are mCi/mmole and μCi/μmole.

Although radioactivity is defined in terms of nuclear disintegrations per unit of time, rarely does one measure this absolute number in the laboratory. Instruments that detect and count emitted particles respond to only a small fraction of the particles. Data from a radiation counter are in counts per minute (cpm), which can be converted to actual disintegra-

Table 6.2
Units of Radioactivity

Units Name	Multiplication Factor (relative to curie)	Activity (dps)
Curie (Ci)	1.0	3.70×10^{10}
Millicurie (mCi)	10^{-3}	3.70×10^7
Microcurie (μCi)	10^{-6}	3.70×10^4
Nanocurie (nCi)	10^{-9}	3.70×10
Becquerel (Bq)	2.7×10^{-11}	1.0
Mega becquerel (MBq)	2.7×10^{-5}	1.0×10^6

tions per minute if the counting efficiency of the instrument is known. The percent efficiency of an instrument is determined by counting a standard compound of known radioactivity and using the ratio of detected activity (observed cpm) to actual activity (disintegrations per minute). Equation 6.9 illustrates the calculation using a standard of 1 μCi.

▶ $$\% \text{ efficiency} = \frac{\text{observed cpm of standard}}{\text{dpm of 1 } \mu\text{Ci of standard}} \times 100$$

$$= \frac{\text{observed cpm}}{2.2 \times 10^6 \text{ dpm}} \times 100 \qquad \text{(Equation 6.9)}$$

B. Detection and Measurement of Radioactivity

Since most of the radioisotopes used in biochemical research are β emitters, only methods that detect and measure β particles will be emphasized. Two counting techniques are in current use, scintillation counting and Geiger-Müller counting.

Liquid Scintillation Counting

Samples for liquid scintillation counting consist of three components: (1) the radioactive material; (2) a solvent, usually aromatic, in which the radioactive substance is dissolved or suspended; and (3) one or more organic fluorescent substances. Components 2 and 3 make up the **liquid scintillation system**. β particles emitted from the radioactive sample interact with the scintillation system, producing small flashes of light or scintillations. The light flashes are detected by a photomultiplier tube (PMT). Electronic pulses from the PMT are amplified and registered by a counting device called a scaler. A schematic diagram of a typical scintillation counter is shown in Figure 6.2.

The scintillation process, in detail, begins with the collision of emitted β particles with solvent molecules, S (Equation 6.10).

▶ $\beta + S \longrightarrow S^* + \beta$ (less energetic) $\qquad$ (Equation 6.10)

Contact between the energetic β particles and S in the ground state results in transfer of energy and conversion of an S molecule into an excited state, S^*. Aromatic solvents are most often used because their electrons are easily promoted to an excited state orbital (see discussion of fluorescence, Chapter 5). The β particle after one collision still has sufficient energy to excite several more solvent molecules. The excited solvent molecules normally return to the ground state by emission of a photon, $S^* \longrightarrow S + h\nu$. Photons from the typical aromatic solvent are of short wavelength and are not efficiently detected by photocells. A

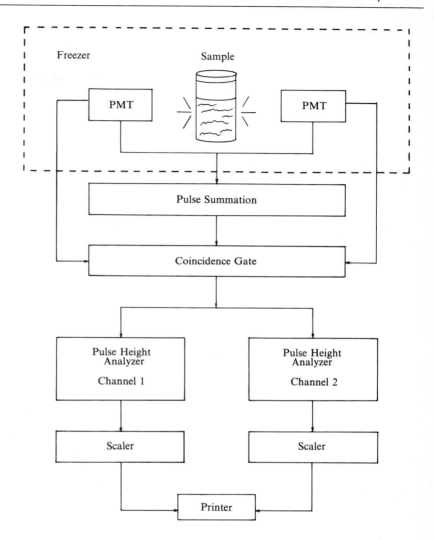

Figure 6.2
Diagram of a typical scintillation counter, showing coincidence circuitry.

convenient way to resolve this technical problem is to add one or more fluorescent substances (fluors) to the scintillation mixture. Excited solvent molecules interact with a **primary fluor,** F_1, as shown in Equation 6.11.

▶ $$S^* + F_1 \longrightarrow S + F_1^*$$ (Equation 6.11)

Energy is transferred from S^* to F_1, resulting in ground state S molecules and excited F_1 molecules, F_1^*. F_1^* molecules are fluorescent and emit light of a longer wavelength than S^* (Equation 6.12).

▶ $$F_1^* \longrightarrow F_1 + h\nu_1$$ (Equation 6.12)

If the light emitted during the decay of F_1^* is still of a wavelength too short for efficient measurement by a PMT, a **secondary fluor,** F_2, that accepts energy from F_1^* may be added to the scintillation system. Equations 6.13 and 6.14 outline the continued energy transfer process and fluorescence of F_2^*.

▶ $F_1^* + F_2 \longrightarrow F_1 + F_2^*$ (Equation 6.13)

▶ $F_2^* \longrightarrow F_2 + h\nu_2$ (Equation 6.14)

The light, $h\nu_2$, from F_2^* is of longer wavelength than $h\nu_1$ from F_1^* and is more efficiently detected by a PMT. Two widely used primary and secondary fluors are 2,5-diphenyloxazole (PPO) with an emission maximum of 380 nm and 1,4-bis-2-(5-phenyloxazolyl)benzene (POPOP) with an emission maximum of 420 nm.

The most basic elements in a liquid scintillation counter are the PMT, a pulse amplifier, and a counter, called a **scaler**. This simple assembly may be used for counting; however, there are many problems and disadvantages with this setup. Many of the difficulties can be alleviated by more sophisticated instrumental features. Some of the problems and practical solutions are outlined below.

THERMAL NOISE IN PHOTOMULTIPLIER TUBES

The energies of the β particles from most β emitters are very low. This, of course, leads to low-energy photons emitted from the fluors and relatively low-energy electrical pulses in the PMT. In addition, photomultiplier tubes produce thermal background noise with 25 to 30% of the energy associated with the fluorescence-emitted photons. This difficulty cannot be completely eliminated, but its effects can be lessened by placing the samples and the PMT in a freezer at -5 to $-8°C$ in order to decrease thermal noise.

A second way to help resolve the thermal noise problem is to use two photomultiplier tubes for detection of scintillations. Each flash of light that is detected by the photomultiplier tubes is fed into a **coincidence circuit**. A coincidence circuit counts only the flashes that arrive simultaneously at the two photodetectors. Electrical pulses that are the result of simultaneous random emission (thermal noise) in the individual tubes are very unlikely. A schematic diagram of a typical scintillation counter with coincidence circuitry is shown in Figure 6.2. Coincidence circuitry has the disadvantage of inefficient counting of very low energy β particles emitted from ^3H and ^{14}C.

COUNTING MORE THAN ONE ISOTOPE IN A SAMPLE

The basic liquid scintillation counter with coincidence circuitry can be used only to count samples containing one type of isotope. Many experiments in biochemistry require the counting of just one isotope; however, more valuable experiments can be performed if two radioisotopes can be simultaneously counted in a single sample (double-labeling experiments). The basic scintillation counter just described has no means of discriminating between electrical pulses of different energies.

The size of the current generated in a photocell is nearly proportional to the energy of the β particle initiating the pulse. Recall that β particles from different isotopes have characteristic energy spectra with an average energy (see Figure 6.1). Modern scintillation counters are equipped with **pulse height analyzers** that measure the size of the electrical pulse and count only the pulses within preselected energy limits set by **discriminators**. The circuitry required for pulse height analysis and energy discrimination of β particles is shown in Figure 6.2. Discriminators are electronic "windows" that can be adjusted to count β particles within certain energy or voltage ranges called **channels**. The channels are set to a **lower limit** and an **upper limit**, and all voltages within those limits are counted. Figure 6.1 illustrates the function of discriminators for the counting of ^3H and ^{14}C in a single sample. Discriminator channel 1 is adjusted to accept typical β particles emitted from ^3H and channel 2 is adjusted to receive β particles of the energy characteristic of ^{14}C.

QUENCHING

Any chemical agent or experimental condition that reduces the efficiency of the scintillation and detection process leads to a reduced level of counting, or **quenching**. Quenching may be caused by a decrease in the amount of fluorescence from the primary and secondary scintillators (fluors) or a decrease in light activating the PMT. There are four common origins of quenching.

Color quenching is a problem if chemical substances that absorb photons from the secondary fluors are present in the scintillation mixture. Since the secondary fluors emit light in the visible region between 410 and 420 nm, colored substances may absorb the emitted light before it is detected by the photocells. Radioactive samples may be treated to remove colored impurities before mixing with the scintillation solvent.

Chemical quenching occurs when chemical substances in the scintillation solution interact with excited solvent and fluor molecules and decrease the efficiency of the scintillation process. To avoid this type of quenching the sample can be purified or the fluors can be increased in concentration. Modern scintillation counters have computer programs to correct for color and chemical quenching.

Point quenching occurs if the radioactive sample is not completely dissolved in the solvent. The emitted β particles may be absorbed before they have a chance to interact with solvent molecules. The addition of solubilizing agents such as Cab-O-Sil or Thixin decreases point quenching by converting the liquid scintillator to a gel.

Dilution quenching results when a large volume of liquid radioactive sample is added to the scintillation solution. In most cases, this type of quenching cannot be eliminated, but it can be corrected by one of the techniques discussed on the following page.

Since quenching can occur during all experimental counting of radioisotopes, it is important to be able to determine the extent of the reduced counting efficiency. Two methods for quench correction are in common use.

In the **internal standard ratio** method, the sample, X, under study is first counted; then a known amount of a standard radioactive solution is added to the sample, and it is counted again. The absolute activity of the original sample A_x, is represented by Equation 6.15.

$$\blacktriangleright \quad A_x = A_s \frac{C_x}{C_s} \qquad \text{(Equation 6.15)}$$

where

A_x = activity of the sample, X

C_x = radioactive counts from sample

A_s = activity of the standard

C_s = radioactive counts from standard

The absolute value of C_s is determined from Equation 6.16.

$$\blacktriangleright \quad C_T = C_x + C_s \qquad \text{(Equation 6.16)}$$

where

C_T = total radioactive counts from sample plus standard

Equation 6.15 can then be modified to Equation 6.17.

$$\blacktriangleright \quad A_x = \frac{A_s C_x}{C_T - C_x} \qquad \text{(Equation 6.17)}$$

The internal standard ratio method for quench correction is tedious and time-consuming and it destroys the sample, so it is not an ideal method. Modern scintillation counters are equipped with a standard radiation source inside the instrument but outside the scintillation solution. The radiation source, usually a gamma emitter, is mechanically moved into a position next to the vial containing the sample, and the combined system of standard and sample is counted. Gamma rays from the standard excite solvent molecules in the sample, and the scintillation process occurs as previously described. However, the instrument is adjusted to register only scintillations due to γ particle collisions with solvent molecules. This method for quench correction, called the **external standard method,** is fast and precise.

SCINTILLATION COCKTAILS AND SAMPLE PREPARATION

Many of the quenching problems discussed earlier can be lessened if an experiment is carefully planned. Many sample preparation methods and scintillation liquids are available for use. Only a few of the more common techniques will be mentioned here. Most liquid or solid radioactive samples can be counted by mixing with a **scintillation cocktail,** a mixture of solvent and fluor(s). The radioactive sample and scintillation solution are placed in a glass, polyethylene, polypropylene, polyester, or polycarbonate vial for counting. Traditional counting vials include a 20-mL-standard vial, 6-mL miniature vial, 1-mL-Eppendorf tube, or 200-μL microfuge tube. If glass is used, it must contain a low potassium content because naturally occurring ^{40}K is a β emitter. Two major types of solvent systems are available: (1) those that are immiscible with water, in which only organic samples may be used, and (2) those that dissolve aqueous samples. During the early years of liquid scintillation counting the most common solvents used were toluene and xylene for organic samples and dioxane for aqueous samples. These solvents provide high counting efficiency. However, on the downside, they are flammable (flash points between 4 and 25°C), highly toxic, and present major disposal costs and problems. In addition, they can penetrate plastic counting vials, releasing vapors and liquids into the laboratory. A new environmental awareness has prompted the development of biosafe liquid scintillation cocktails. The use of highly alkylated aromatic solvents has produced cocktails that are fully biodegradable (converted to CO_2 and H_2O), have higher flash points (above 120°C), and offer high counting efficiency. The newer solvents may be disposed of after use by incineration, which greatly reduces costs compared to road transport to a disposal site. Occupational risks are also lessened because the solutions are defined by the Environmental Protection Agency as nontoxic. These new products are available from several manufacturers, including ICN Biomedicals, Pharmacia-LKB, National Diagnostics, and Research Products International. Most suppliers offer a range of products for aqueous and organic samples.

Solid radioactive samples or those that are insoluble in either type of solvent may be quantified by collection onto small pieces of a solid support (filter paper or cellulose membrane) and added directly to the cocktail for counting. The efficiency of counting these samples depends on the support but is usually less than that of counting a homogeneous sample.

Liquid scintillation counting as described above is especially useful for counting weak β emitters such as ^{3}H, ^{14}C, and ^{35}S. Counting high-energy β emission, such as that of ^{32}P, does not require a fluor because the β particle can be detected directly by the PMT. Samples containing ^{32}P may be counted directly with relatively high efficiency in water solution. This mode of direct measurement is called **Cerenkov counting**.

New and innovative products will continue to become available to reduce the problems of radioisotope use in the laboratory. One example of a new product is the Redi-Cap from Beckman. Here a small volume of radioactive sample is evaporated onto a solid scintillant, which is held in a clear plastic cup. The cup is placed in a vial and counted in a normal scintillation counter. A similar product, the Betaplate, designed by LKB-Wallac/Pharmacia, allows solid samples on filters to be counted in a plastic sample bag that contains a small volume of liquid scintillation cocktail.

Geiger-Müller Counting of Radioactivity

Another method of radiation detection that has great value in biochemistry is the use of gas ionization chambers. The most common device that uses this technique is the **Geiger-Müller tube** (G-M tube). When β particles pass through a gas, they collide with atoms and may cause ejection of an electron from a gas atom. This results in the formation of an ion pair made up of the negatively charged electron and the positively charged atom. If this ionization occurs between two charged electrodes (an anode and a cathode), the electron will be attracted to the anode and the positive ion to the cathode. This results in a small current in the electrode system. If only a low voltage difference exists between the anode and cathode, the ion pairs will move slowly and will, most likely, recombine to form neutral atoms. Clearly, this will result in no pulse in the electric circuit because the individual ions do not reach the respective electrodes. At higher voltages, the charged particles are greatly accelerated toward the electrodes and collide many times with un-ionized gas atoms. This leads to extensive ionization and a cascade or avalanche of ions. If the voltage is high enough (1000 volts for most G-M tubes), all ions are collected at the electrodes. A Geiger-Müller counting system uses this voltage region for ion acceleration and detection. A typical G-M tube is diagramed in Figure 6.3. It consists of a mica window for entry of β particles from the radiation source, an anode down the center of the tube, and a cathode surface inside the walls. A high voltage is applied between the electrodes. Current generated from electron movement toward the anode is amplified, measured, and converted to counts per minute. The cylinder contains an inert gas that readily ionizes (argon, helium, or neon) plus a quenching gas (Q gas, usually butane) to reduce continuous ionization of the inert gas. Beta particles of high energy emitted from atoms such as ^{24}Na, ^{32}P, and ^{40}K have little difficulty entering the cylinder by penetrating the mica window. Particles from weak β emitters (especially ^{14}C and ^{3}H) cannot efficiently pass through the window to induce ionization inside the chamber. Modified G-M tubes with thin Mylar windows, called **flow window tubes,** may be used to count weak β emitters.

G-M counting has several disadvantages compared to liquid scintillation counting. The counting efficiency of a G-M system is not as high. The response time for G-M tubes is longer than for photomultiplier tubes; therefore, samples of high radioactivity are not efficiently counted by the G-M tube. G-M meters are seldom used when careful, accurate measurements are required. They are most useful in the biochemical laboratory as survey meters to monitor and detect radioactive contamination on lab benches, glassware, equipment, and laboratory personnel.

Figure 6.3
A Geiger-Müller tube.

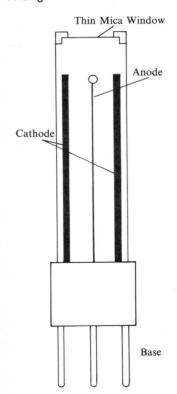

Thin Mica Window

Anode

Cathode

Base

Scintillation Counting of γ Rays

[24]Na and ^{131}I are both β and γ emitters and are often used in biochemical research. Gamma rays are more energetic than β particles, and denser materials are necessary for absorption. A gamma counter consists of a sample well, a sodium iodide crystal as fluor, and a photomultiplier tube (see Figure 6.4). The high-energy γ rays are not absorbed by the scintillation solution or vial, but they interact with a crystal fluor, producing scintillations. The scintillations are detected by photomultiplier tubes and electronically counted.

Background Radiation

All radiation counting devices register counts even if there is no specific or direct source of radiation near the detectors. This is due to **background radiation**. It has many sources, including natural radioactivity, cosmic rays, radioisotopes in the construction materials of the counter, radioactive chemicals stored near the counter, and contamination of sample vials or counting equipment. Background count in a scintillation counter will depend on the scintillation solution used, but is usually 30 to 50 cpm. Background counts that are the result of natural sources cannot be eliminated; however, those due to contamination can be greatly reduced if the laboratory and counting instrument are kept clean and free of radioactive contamination. Since it is difficult to regulate background radiation, it is usually necessary to monitor its level. The typical procedure is to count a radioactive sample and then count a background sample containing the same scintillation system but no radioisotope. The actual or corrected activity is obtained by subtracting the background counts from the results for the radioactive sample.

Applications of Radioisotopes

The applications of radioisotopes in biochemical measurements are too numerous to outline here. Excellent reference books, some of which are listed at the end of this chapter, provide experimental procedures for ra-

Figure 6.4
Diagram of a γ counter.

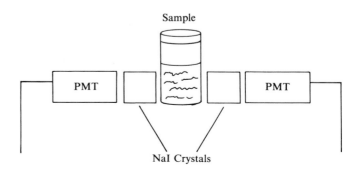

dioisotope utilization. The technique of **autoradiography**, however, should be mentioned. This procedure allows the detection and localization of a radioactive substance (molecule or atom) in a tissue, cell, or cell organelle. Briefly, the technique involves placing a radioactive material directly onto or close to a photographic emulsion. Radioisotopes in the material emit radiation that impinges on the photographic plate, activating silver halide crystals in the emulsion. Upon development of the plate, a pattern or image is displayed that yields information about the location and amount of radioactive material in the sample. A densitometer may be used to scan the autoradiograph and quantify the amount of radioactivity in each region. One of the most common uses of autoradiography in biochemistry and molecular biology is in the detection and quantification of ^{32}P-labeled nucleic acids on polyacrylamide or agarose gels after electrophoresis (Figure 6.5). Autoradiography has also been useful in concentration and localization studies of biomolecules in cells and cell organelles.

Statistical Analysis of Radioactivity Measurements

Radioactive decay is a random process. It is impossible to predict when a radioactive event will occur. Therefore, counting measurements provide only average rates of decay. However, counting measurements may be

Figure 6.5
Autoradiograph showing the transfer of ^{32}P from γ-^{32}P–labeled ATP to the 5' end of a polynucleotide, (dT)$_8$, catalyzed by polynucleotide kinase. Reagents were incubated for 30 min and electrophoresed on a 20% polyacrylamide, 7 M urea gel and autoradiographed. Note that as more units of kinase are added, more ^{32}P-labeled polynucleotide is made. The dark regions indicate the presence of ^{32}P. Source: Copyright © 1992 *New English Biolabs Catalog*. Reprinted with permission.

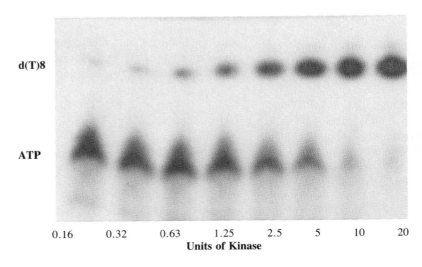

treated by Gaussian distribution analysis to determine an average counting rate, standard deviation, percent confidence level, and other statistical parameters. An introduction to statistical analysis of radioactivity counting data and other experimental measurements is given in Chapter 1.

C. Safety Rules for Handling Radioactive Materials

Safety should be of major concern in performing any chemistry experiment, but when radioisotopes are involved, special precautions should be taken. Many chemistry and biochemistry departments have specially equipped laboratories for radioisotope work. Alternatively, your instructor may set aside a specific area of your laboratory for using radioactive materials. In either situation, specific guidelines should be followed.

Preparation for the Experiment

⚠ Caution

> If you are pregnant or may become pregnant during this course, special precautions in the use of radioisotopes must be considered. Since each institution has its own rules and regulations, consult your instructor. Some institutions may require the use of a personnel badge in order to monitor the exposure for each individual. Your instructor will provide this if necessary.

1. The first responsibility of the student is to become knowledgeable about the properties and hazards of the radioactive substances to be used. You must know which radioisotopes are to be handled and the form, liquid or solid, of the material. It is also important to know whether the isotope is a β and/or γ emitter and whether it is "weak" or "strong." ^3H and ^{14}C are considered weak β emitters and ^{32}P is a strong β emitter.

2. Confine your work with radioisotopes to a small area in the laboratory. A convenient plan is to use a stainless steel tray lined with absorbent blotter paper coated on the bottom side with polyethylene. The paper must be replaced every day. If the radioactive materials are volatile, the work should be done in a fume hood. If spills occur, a small work area such as a tray is much easier to clean than a large lab bench. If ^{32}P or other strong β emitter is used, it is necessary to work at all times with shielding between yourself and the radioactive samples. The most cost-effective and convenient shielding material is Plexiglas. The thicknesses of shielding required for various materials are given in Table 6.3.

Table 6.3
Maximum Range of β Particles from
^{32}P Through Various Materials

Material	Maximum Range
Air	6 m
Water	0.85 cm
Plexiglas	0.64 cm
Glass	0.38 cm
Lead	0.045 cm

3. Label all glassware and equipment that will be used with special adhesive tape labeled "Radioactive." Plan the experiment so that a minimal number of transfers of radioactive materials is required. This will reduce the amout of contaminated glassware.

4. Two types of containers should be available for disposal purposes. One should be labeled "Liquid Radioactive Waste" and used for all waste solutions; the other, "Solid Radioactive Waste," for blotter paper, broken glassware, etc. **Liquid wastes must not be poured down any drain, nor solid wastes deposited in normal trash cans.**

5. Take all precautions against contaminating yourself or fellow workers. Always wear gloves and a lab coat. No smoking, eating, or drinking is allowed in the laboratory. It is especially hazardous to ingest radioactive materials.

6. The laboratory should be equipped with a portable G-M survey meter in order to check spills or possible self-contamination.

7. A specific area with a sink should be set aside for washing weakly contaminated glassware and equipment.

8. All radioactive materials should be stored in well-labeled, glass containers. The label must include your name, the type of isotope, the radioactive compound, the total amount of radioactivity, the specific activity, and the date of measurement.

Performing the Experiment

1. **Do not pipet any solutions by mouth; always use a mechanical pipet filler.**

2. Contaminated glassware should be kept separated from uncontaminated. Contaminated beakers and flasks are placed in the special sink or other container for washing. Clean and wash all equipment with soap and water immediately after the experiment has been completed. If water-insoluble materials are being used, the first washing should be done with an organic solvent such as acetone. Soak contaminated pipets in a container filled with water. All broken glassware is disposed of in the "Solid Radioactive Waste" container.

3. Minor spills must be cleaned first with absorbent blotter paper and then *thoroughly* rinsed with water. Always wear gloves during clean-up. Dispose of the blotter paper in the "Solid Radioactive Waste" container. The spill area should then be checked with a portable G-M counter.

4. All skin cuts, accidental ingestion of radioactive materials, and major spills must be reported immediately to the instructor.

5. After final clean-up of all contaminated glassware and the work area, remove your gloves and thoroughly wash your hands with soap and warm water. Use a portable G-M counter to check for contamination on your hands, clothes, and work area in the laboratory. Report any areas of unusually high count (above 200 cpm) to your instructor.

References

H. Ahern, *The Scientist,* June 25, p. 23 (1990). "Environmentalists Toast New Liquid Scintillation Cocktails."

H. Andrews, *Radiation Biophysics,* 2nd ed. (1974), Prentice-Hall (Englewood Cliffs, NJ). The entire book is excellent for an introduction to radiochemistry theory and techniques.

D. Billington, G. Jayson and P. Maltby, *Radioisotopes* (1992), Bios Scientific Publishers (Oxford). Excellent reference on principles, methods, techniques and applications.

W. Bonner, in *Methods in Enzymology,* Vol. 152, S. Berger and A. Kimmel, Editors (1987), Academic Press (Orlando, FL), pp. 55–61. "Autoradiograms: ^{35}S and ^{32}P."

J. Chapman and G. Ayrey, *The Uses of Radioactive Isotopes in the Life Sciences,* (1981), George Allen and Unwin (London). A useful discussion of applications.

G. Chase and J. Rabinowitz, *Principles of Radioisotope Methodology,* 3rd ed. (1967), Burgess (Minneapolis), pp. 75–108. Although somewhat outdated, this is still a useful book.

T. Cooper, *The Tools of Biochemistry* (1977), John Wiley & Sons (New York), pp. 65–135. One chapter on the theory and application of radioisotopes.

E. Evans and K. Oldham, Editors, *Radiochemicals in Biomedical Research* (1988), Wiley (Chichester). Advances in the use of radioisotopes.

D. Freifelder, *Physical Biochemistry,* 2nd ed. (1982), W. H. Freeman (San Francisco), pp. 129–192. Theory and applications of radioisotopes.

E. Hahn (1983), *Am. Lab.* **15,** 64–71 (1983), "Autoradiography—A Review of Basic Principles."

C. Mathews and K. van Holde, *Biochemistry* (1990), Benjamin/Cummings (Redwood City, CA), pp. 426–430. An introduction to radioisotopes and the liquid scintillation counter.

New England Nuclear, ^{32}P Handling and Hazards (1985), Boston, MA.

J. Robyt and B. White, *Biochemical Techniques: Theory and Practice* (1987), Brooks/Cole (Monterey, CA), pp. 73–128. Theory, measurement, and use of radioisotopes.

R. Slater, Editor, *Radioisotopes in Biology—A Practical Approach* (1990), IRL Press (Oxford). Theory and techniques for the use of isotopes.

H. Strobel and W. Heineman, *Chemical Instrumentation: A Systematic Approach,* 3rd ed. (1989), John Wiley & Sons (New York), pp. 836–859. Methods using radioisotopes.

D. Voet and J. Voet, *Biochemistry* (1990), John Wiley & Sons (New York), pp. 405–408. Isotopes in biochemistry.

Y. Wang, Editor, *CRC Handbook of Radioactive Nuclides* (1969), Chemical Rubber Co. (Cleveland). An excellent reference for properties and use of radioisotopes.

P. Warwick, *Anal. Proc. (London)* **27**, 109–110 (1990). "Sample Preparation and Presentation in Liquid Scintillation Counting."

R. Zoon, in *Methods in Enzymology,* Vol. 152, S. Berger and A. Kimmel, Editors (1987), Academic Press (Orlando, FL), pp. 25–29. "Safety with ^{32}P- and ^{35}S- Labeled Compounds."

CHAPTER **7**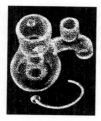

Centrifugation of Biomolecules

■ A centrifuge of some kind is found in nearly every biochemistry laboratory. Centrifuges have many uses, but they are used primarily for the preparation of biological samples and the analytical measurement of the hydrodynamic properties (shape, size, density, etc.) of purified macromolecules or cellular organelles. This is done by subjecting a biological sample to an intense force by spinning the sample at high speed. These experimental conditions cause sedimentation of particles, cell organelles, or macromolecules at a rate that depends on their masses, sizes, and densities.

In this chapter we will explore the underlying principles of centrifugation and discuss the application of this technique to the isolation and characterization of biological molecules and cellular components.

A. Basic Principles of Centrifugation

A particle, whether it is a precipitate, a macromolecule, or a cell organelle, is subjected to a centrifugal force when it is rotated at a high rate of speed. The **centrifugal force,** *F,* is defined by Equation 7.1.

▶ $F = m\omega^2 r$ (Equation 7.1)

where

F = intensity of the centrifugal force

m = effective mass of the sedimenting particle

ω = angular velocity of rotation in rad/sec

r = distance of the migrating particles from the central axis of rotation

The force on a sedimenting particle increases with the velocity of the rotation and the distance of the particle from the axis of rotation. A more common measurement of F, in terms of the earth's gravitation force, g, is **relative centrifugal force**, RCF, defined by Equation 7.2.

▶ $$RCF = (1.119 \times 10^{-5})(rpm)^2(r) \qquad \text{(Equation 7.2)}$$

This equation relates RCF to revolutions per minute of the sample. Equation 7.2 dictates that the RCF on a sample will vary with r, the distance of the sedimenting particles from the axis of rotation (see Figure 7.1). It is convenient to determine RCF by use of the nomogram in Figure 7.2. It should be clear from Figures 7.1 and 7.2 that, since RCF varies with r, it is important to define r for an experimental run. Often an average RCF is determined using a value for r midway between the top and bottom of the sample container. The RCF value is reported as "a number times gravity, g." To illustrate this, consider a sample with an average r of 7 cm being rotated at 20,000 rpm. From Figure 7.2, RCF is 32,000 × g.

Although this simple introduction outlines the basic principles of centrifugation, it does not take into account other factors that influence the rate of particle sedimentation. Centrifuged particles migrate at a rate that depends on the mass, shape, and density of the particle and the density of the medium. The centrifugal force felt by the particle is defined by Equation 7.1. The term m is the **effective mass** of the particle, that is, the actual mass, m_0, minus a correction factor for the weight of water displaced (buoyancy factor) (Equation 7.3).

Figure 7.1
A diagram illustrating the variation of RCF with r, the distance of the sedimenting particles from the axis of rotation. Courtesy of Beckman Instruments, Inc.

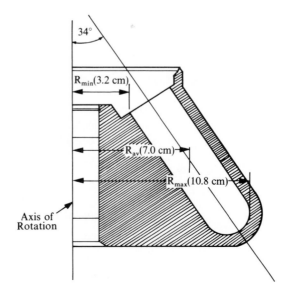

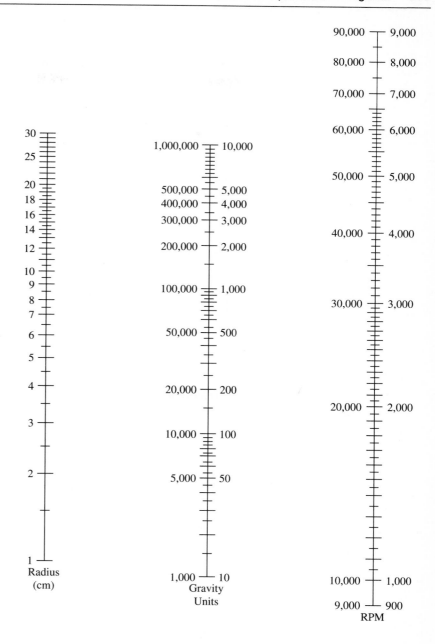

Figure 7.2
A nomogram for estimating RCF. To use, place a straight edge connecting the instrumental revolutions (right column) with the distance from the center of rotation (left column). The RCF is read where the straight edge crosses the middle column. Courtesy of Beckman Instruments, Inc.

▶ $m = m_0 - m_0\overline{v}\rho$ (Equation 7.3)

where

m = effective mass of the sedimenting particle

m_0 = actual mass of the particle

$\overline{v}$ = partial specific volume, the volume change occurring when a particle is placed in a large excess of solvent

ρ = density of the solvent or medium

Equations 7.1 and 7.3 may be combined to describe the centrifugal force on a particle (Equation 7.4).

▶ $F = m_0(1 - \overline{v}\rho)\omega^2 r$ (Equation 7.4)

As particles sediment under the influence of the centrifugal field, their movement is countered by a resistance force, the **frictional force**. The frictional force is defined by Equation 7.5.

▶   Frictional force = fv (Equation 7.5)

where

f = frictional coefficient

v = velocity of the sedimenting particle (sedimentation velocity)

The **frictional coefficient,** f, depends on the size and shape of the particle, as well as the viscosity of the solvent. The frictional force increases with the velocity of the particle until a constant velocity is reached. At this point, the two forces are balanced (Equation 7.6).

▶ $m_0(1 - \overline{v}\rho)\omega^2 r = fv = f\left(\dfrac{dr}{dt}\right)$ (Equation 7.6)

The rate of sedimentation, sometimes called **sedimentation velocity,** v, is defined by Equation 7.7.

▶ $v = \dfrac{dr}{dt} = \dfrac{m_0(1 - \overline{v}\rho)\omega^2 r}{f}$ (Equation 7.7)

It is cumbersome and sometimes impractical to express sedimentation velocity in terms of ρ, $\overline{v}$, and f, since these factors are difficult to measure. A new term, **sedimentation coefficient,** s (the ratio of sedimentation velocity to centrifugal force) is introduced by rearranging Equation 7.7 to Equation 7.8.

▶ $s = \dfrac{v}{\omega^2 r} = \dfrac{m_0(1 - \overline{v}\rho)}{f}$ (Equation 7.8)

The term s is most often defined under standard conditions, 20°C and water as the medium, and denoted by $s_{20,w}$. The s value is a physical characteristic used to classify biological macromolecules and cell organelles. Sedimentation coefficients are in the range 1×10^{-13} to $10,000 \times 10^{-13}$ second. For numerical convenience, sedimentation coefficients are expressed in Svedberg units, S, where $1\ S = 1 \times 10^{-13}$ second. Human hemoglobin has an s value of 4.5×10^{-13} second or 4.5 S. The value of S for several biomolecules, bacterial cells, and cell organelles is shown in Figure 7.3. Note in the figure that there appears to be a direct relationship between the S value and the molecular weight or particle size. This, however, is not always true, as in the case of nonspherical molecules.

The goal of many centrifugation experiments is the measurement of s and S. This value is important because it can be used to calculate the size (molecular weight, kilo base pairs, etc.) of a molecule or cell organelle. The units of s are not obvious from Equation 7.8. Dimensional analysis shows the following: v in cm/sec, ω in radians/sec, r in cm, m_0 in grams, $\overline{v}\ in\ cm^3/g$, ρ in g/cm^3, and f in g/sec. Therefore, the unit for s is second.

B. Instrumentation for Centrifugation

The basic centrifuge consists of two components, an electric motor with drive shaft to spin the sample and a **rotor** to hold tubes or other containers of the sample. A wide variety of centrifuges is available, ranging from a low-speed centrifuge used for routine pelleting of relatively heavy particles to sophisticated instruments that include accessories for making analytical measurements during centrifugation. Here we will describe three types, the benchtop or clinical centrifuge, the high-speed centrifuge, and the ultracentrifuge. Major characteristics and applications of each type are compared in Table 7.1.

Benchtop Centrifuges

Most laboratories have a standard low-speed centrifuge used for routine sedimentation of relatively heavy particles. The common centrifuge has a maximum speed in the range of 4000 to 5000 rpm, with RCF values up to $3000 \times g$. These instruments usually operate at room temperature with no means of temperature control of the samples. Two types of rotors, **fixed angle** and **swinging bucket,** may be used in the instrument. Centrifuge tubes or bottles that contain 12 or 50 mL of sample are commonly used. Low-speed centrifuges are especially useful for the rapid sedimentation of coarse precipitates or red blood cells. The sample is centrifuged until the particles are tightly packed into a **pellet** at the bottom of the tube. The liquid portion, the **supernatant,** is then separated by decantation.

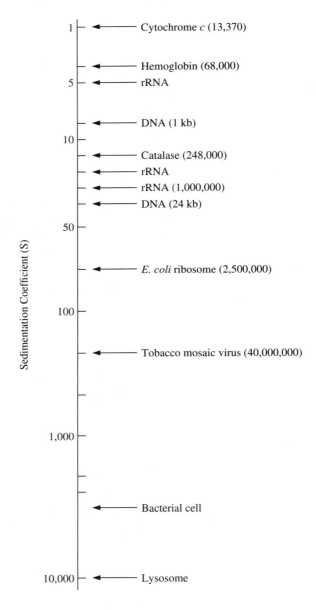

Figure 7.3
Range of S values for biomolecules, cell organelles, and cells. kb = kilo base pairs (1000 base pairs). Molecular weight or kb is shown in parentheses.

Table 7.1

Types of Centrifuges and Applications

Characteristic	Type of Centrifuge		
	Low-speed	High-speed	Ultracentrifuge
Range of speed (rpm)	1–6000	1000–25,000	20–80,000
Maximum RCF (g)	6000	50,000	600,000
Refrigeration	Some	Yes	Yes
Applications			
Pelleting of cells	Yes	Yes	Yes
Pelleting of nuclei	Yes	Yes	Yes
Pelleting of organelles	No	Yes	Yes
Pelleting of ribosomes	No	No	Yes
Pelleting of macromolecules	No	No	Yes

High-Speed Centrifuges

For more sensitive biochemical applications, higher speeds and temperature control of the rotor chamber are essential. A typical high-speed centrifuge is shown in Figure 7.4. The operator of this instrument can carefully control speed and temperature, which is especially important for carrying out reproducible centrifugations of temperature-sensitive biological samples. Rotor chambers in most instruments are maintained at or near 4°C. Three types of rotors are available for high-speed centrifugation, the fixed-angle, the swinging-bucket, and the vertical rotor (Figure 7.5A–C). Fixed-angle rotors are especially useful for differential pelleting of particles (Figure 7.6A). In swinging-bucket rotors (Figure 7.5B), the sample tubes move to a position perpendicular to the axis of rotation during centrifugation, as shown in Figure 7.7. These are used most often for density gradient centrifugation (see below). In the vertical rotor (Figure 7.5C), the sample tubes remain in an upright position (Figure 7.8). These rotors are used often for gradient centrifugation. Modern instruments are equipped with a brake to slow the rotor rapidly after centrifugation.

A recent addition to the category of medium-speed centrifuges is the "microfuge" (Figure 7.9). These instruments, which are designed for the benchtop, are used for rapid pelleting of small samples. Fixed-angle rotors are available to hold up to eighteen 1.5- or 0.5-mL tubes. The maximum speed of most commercial microfuges is between 12,000 and 13,000 rpm, which delivers a force of 11,000–12,000 × g. Some instruments can accelerate to full speed in 6 seconds and decelerate within 18 seconds. Most instruments have a variable speed control and a momentary button for minispins.

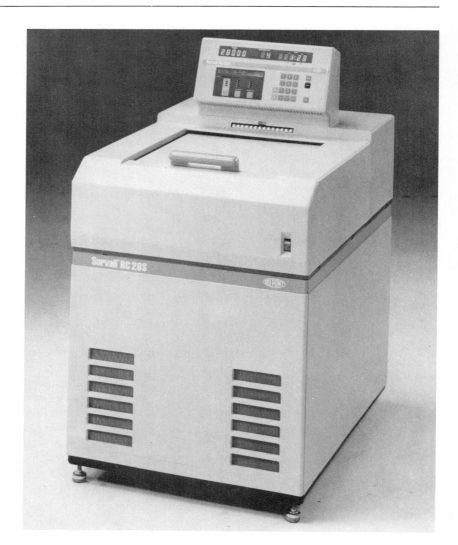

Figure 7.4
A typical high-speed, refrigerated centrifuge. Courtesy of Du Pont Company.

The preparation of biological samples almost always requires the use of a high-speed centrifuge. Specific examples will be described later, but high-speed centrifuges may be used to sediment (1) cell debris after cell homogenization, (2) ammonium sulfate precipitates of proteins, (3) microorganisms, and (4) cellular organelles such as chloroplasts, mitochondria, and nuclei.

Ultracentrifuges

The most sophisticated of the centrifuges are the **ultracentrifuges**. Because of the high speeds attainable (see Table 7.1), intense heat is gener-

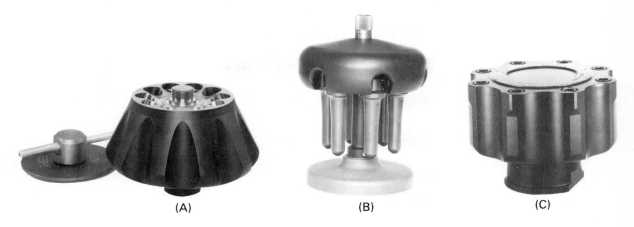

(A)   (B)   (C)

Figure 7.5
Rotors for a high-speed centrifuge.   (A) Fixed angle;   (B) Swinging bucket;   (C) Vertical. Courtesy of Beckman Instruments, Inc.

ated in the rotor, so the spin chamber must be refrigerated and placed under a high vacuum to reduce friction. The sample in a cell or tube is placed in an aluminum or titanium rotor, which is then driven by an electric motor. Although it is relatively uncommon, these rotors, when placed under high stress, sometimes break into fragments. The rotor chamber on all ultracentrifuges is covered with protective steel armor plate. The drive shaft of the ultracentrifuge is constructed of a flexible material to accommodate any "wobble" of the rotor due to imbalance of the samples. It is still important to counterbalance samples as carefully as possible.

The two previously discussed centrifuges—the benchtop and the high speed—are of value only for preparative work, that is, for the isolation and separation of precipitates and biological samples. Ultracentrifuges can be used both for preparative work and for analytical measurements. Thus, two types of ultracentrifuges are available, **preparative models,** primarily used for separation and purification of samples for further analysis, and **analytical models,** which are designed for performing physical measurements on the sample during sedimentation. One of the most versatile models is the Beckman TLX, a microprocessor-controlled tabletop ultracentrifuge (Figure 7.10). With a typical fixed-angle rotor, which holds six, 0.2- to 2.2-mL samples, the instrument can generate 100,000 rpm and a RCF of 540,000 $\times$ g.

Analytical ultracentrifuges have the same basic design as preparative models except that they are equipped with optical systems to monitor directly the sedimentation of the sample during centrifugation. The first commercial instrument was the Beckman Model E, introduced in 1947.

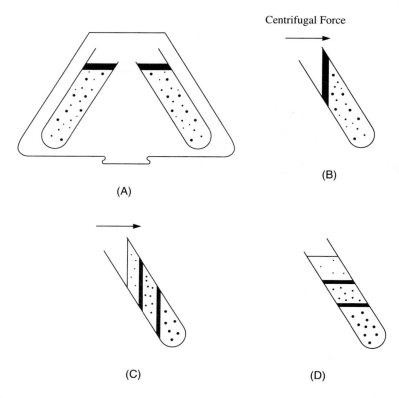

Figure 7.6
Operation of a fixed angle rotor. (A) Loading of sample. (B) Sample at start of centrifugation. (C) Bands form as molecules sediment. (D) Rotor at rest showing separation of two components.

Because of a renewed interest in the use of ultracentrifugation, Beckman in 1990 began marketing the Optima XL-A, which spins samples up to 60,000 rpm.

For analysis, a sample of nucleic acid or protein (0.1 to 1.0 mL) is sealed in a special analytical cell and rotated. Light is directed through the sample parallel to the axis of rotation, and measurements of absorbance by sample molecules are made. (The new Beckman instrument can scan the sample over the wavelength range 190 to 800 nm.) If sample molecules have no significant absorption bands in the wavelength range, then optical systems that measure changes in the refractive index may be used. Optical systems aided by computers are capable of relating absorbance changes or index of refraction changes to the rate of movement of particles in the sample. The optical system actually detects and measures the front edge or moving boundary of the sedimenting molecules.

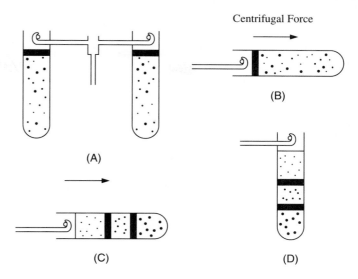

Figure 7.7
Operation of a swinging-bucket rotor.   (A) Loading of sample.   (B) Sample at start
of centrifugation.   (C) Sample during centrifugation separates into two components.   (D) Rotor at rest.

These measurements can lead to an analysis of concentration distribu-
tions within the centrifuge cell. Applications of these measurements will
be discussed in the next section.

C. Applications of Centrifugation

Preparative Techniques

Centrifuges in undergraduate biochemistry laboratories are used most often for preparative-scale separation procedures. This technique is quite
straightforward, consisting of placing the sample in a tube or similar container, inserting the tube in the rotor, and spinning the sample for a fixed
period. The sample is removed and the two phases, pellet and supernatant (which should be readily apparent in the tube), may be separated by
careful decantation. Further characterization or analysis is usually carried
out on the individual phases. This technique, called **velocity sedimentation centrifugation,** separates particles ranging in size from coarse precipitates to cellular organelles. Relatively heavy precipitates are sedimented in low-speed, benchtop centrifuges, whereas lighter organelles
such as ribosomes require the high centrifugal forces of an ultracentrifuge.

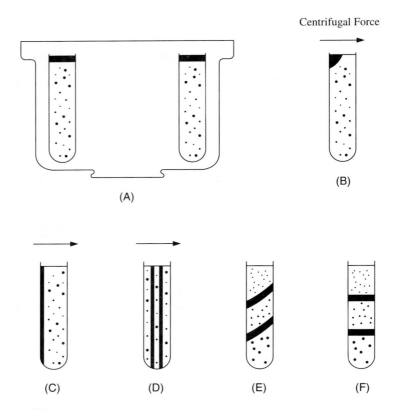

Figure 7.8
Operation of a vertical rotor.   (A) Loading of sample.   (B) Beginning of centrifugation.   (C, D) During centrifugation.   (E) Deceleration of sample.   (F) Rotor at rest.

Much of our current understanding of cell structure and function depends on separation of subcellular components by centrifugation. The specific method of separation, called **differential centrifugation,** consists of successive centrifugations at increasing rotor speeds. Figure 7.11 illustrates the differential centrifugation of a cell homogenate, leading to the separation and isolation of the common cell organelles. For most biochemical applications, the rotor chamber must be kept at low temperatures to maintain the native structure and function of each cellular organelle and its component biomolecules. A high-speed centrifuge equipped with a fixed-angle rotor is most appropriate for the first two centrifugations at $600 \times g$ and $20,000 \times g$. After each centrifuge run, the supernatant is poured into another centrifuge tube, which is then rotated at the next higher speed. The final centrifugation at $100,000 \times g$ to sediment microsomes and ribosomes must be done in an ultracentrifuge.

Figure 7.9
Variable speed micro centrifuges. Photo of Eppendorf micro centrifuges courtesy of Brinkmann Instruments, Inc.

The $100,000 \times g$ supernatant, the **cytosol,** is the soluble portion of the cell and consists of soluble proteins and smaller molecules. Differential centrifugation is used to isolate beef heart mitochondria for enzymatic characterization in Experiment 13.

Analytical Measurements

A variety of analytical measurements can be made on biological samples during and after a centrifuge run. Most often, the measurements are taken to determine molecular weight, density, and purity of biological samples. All analytical techniques require the use of an ultracentrifuge and can be classified as **differential** or **density gradient**.

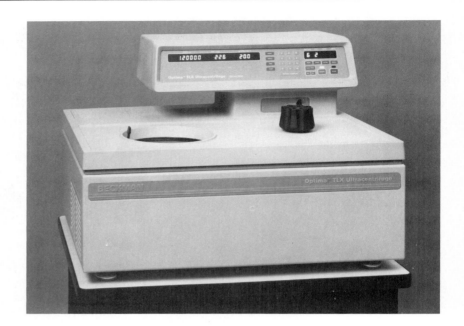

Figure 7.10
Outside (A) and inside (B) views of the Beckman Optima TLX ultracentrifuge. Courtesy of Beckman Instruments, Inc.

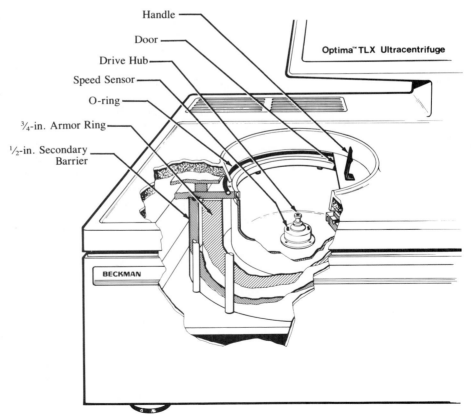

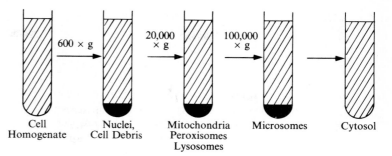

Figure 7.11
Differential centrifugation of a cell homogenate. See text for description.

DIFFERENTIAL CENTRIFUGATION

Differential methods involve sedimentation of particles in a medium of homogeneous density. Although the technique is similar to preparative differential centrifugation as previously discussed, the goal of an experiment is to measure the sedimentation coefficient of a particle. The underlying principles of this technique are illustrated in Figure 7.12A. During centrifugation, a moving boundary is generated between pure solvent and sedimenting particles. An analytical ultracentrifuge is capable of detecting and measuring the rate of movement of the boundary. Hence, the sedimentation velocity, v, can be experimentally determined. By using Equation 7.8 the sedimentation coefficient, s, can be calculated. The value of s for a sedimenting particle is related to the molecular weight of that particle by the Svedberg equation. The Svedberg equation is derived from Equation 7.8 by recognizing that the frictional force, f, may be defined by Equation 7.9.

▶ $$f = \frac{RT}{ND}$$ (Equation 7.9)

where

R = the gas constant, 8.3×10^7 g cm^2/sec/deg/mole

T = the absolute temperature

D = the diffusion coefficient of the solute in units of cm^2/sec

N = Avogadro's number

Thus, Equation 7.8 may be transformed into Equation 7.10.

▶ $$s = \frac{m_0(1 - \overline{v}\rho)}{RT/ND}$$ (Equation 7.10)

$$RTs = m_0ND(1 - \overline{v}\rho)$$

Since molecular weight, MW, is equal to m_0N, Equation 7.10 is converted to Equation 7.11, the Svedberg equation.

▶ $$\text{MW} = \frac{RTs}{D(1 - \overline{v}\rho)}$$ (Equation 7.11)

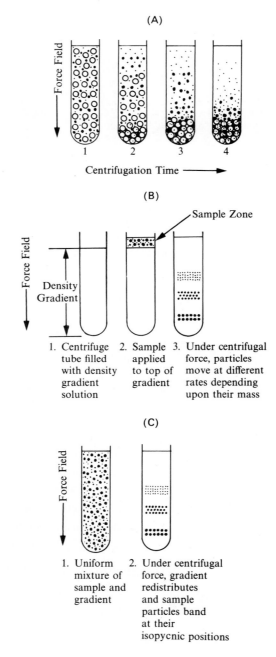

Figure 7.12
A comparison of differential and density gradient measurements. (A) Differential centrifugation in a medium of unchanging density. (B) Zonal centrifugation in a prepared density gradient. (C) Isopycnic centrifugation; the density gradient forms during centrifugation. Illustration courtesy of Beckman Instruments, Inc.

This equation provides an accurate calculation of molecular weight and is applicable to macromolecules such as proteins and nucleic acids. However, its usefulness is limited because diffusion coefficients are difficult to measure and are not readily available in the literature.

An alternative method sometimes used to determine molecular weights of macromolecules is **sedimentation equilibrium**. In the previous example, using the Svedberg equation, the sample is rotated at a rate sufficient to sediment the particles. Here, the sample is rotated at a lower rate, and the particles sediment until they reach an equilibrium position at the point where the centrifugal force is equal to the frictional component opposing their movement (see Equation 7.6). The molecular weight is then calculated using Equation 7.12.

$$\blacktriangleright \quad MW = \frac{RT}{(1 - \overline{v}\rho)\omega^2}\left(\frac{1}{rc}\right)\left(\frac{dc}{dr}\right) \qquad \text{(Equation 7.12)}$$

where

$\dfrac{dc}{dr}$ = change in particle concentration as a function of distance from the rotation center, as measured in the ultracentrifuge

and MW, R, T, $\overline{v}$, ρ, ω, and r have the definitions previously given. This technique is time consuming, as the system may require 1 to 2 days to reach equilibrium. Also, the use of Equation 7.12 is complicated by the difficulty of making concentration measurements in the ultracentrifuge.

Differential ultracentrifugation methods may also be applied to analysis of the purity of macromolecular samples. If one sharp moving boundary is observed in a rotating centrifuge cell, it indicates that the sample has one component and therefore is pure. In an impure sample, each component would be expected to form a separate moving boundary upon sedimentation.

DENSITY GRADIENT CENTRIFUGATION

In differential procedures, the sample is uniformly distributed in a cell before centrifugation, and the initial concentration of the sample is the same throughout the length of the centrifuge cell. Although useful analytical measurements can be made with this technique, it has disadvantages when applied to impure samples or samples with more than one component. Large particles that sediment faster pass through a medium consisting of solvent and particles of smaller size. Therefore, clear-cut separations of macromolecules are seldom obtained. This can be avoided if the sample is centrifuged in a fluid medium that gradually increases in density from top to bottom. This technique, called **density gradient centrifugation,** permits the separation of multicomponent mixtures of macromolecules and the measurement of sedimentation coefficients. Two methods are used, **zonal centrifugation,** in which the sample is centrifuged in a preformed gradient, and **isopycnic centrifugation,** in which a self-generating gradient forms during centrifugation.

Zonal Centrifugation

Figure 7.12B outlines the procedure for zonal centrifugation of a mixture of macromolecules. A density gradient is prepared in a tube prior to centrifugation. This is accomplished with the use of an automatic gradient mixer. Solutions of low-molecular-weight solutes such as sucrose or glycerol are allowed to flow into the centrifuge cell. The sample under study is layered on top of the gradient and placed in a swinging-bucket rotor. Sedimentation in an ultracentrifuge results in movement of the sample particles at a rate dependent on their individual *s* values. As shown in Figure 7.12B, the various types of particles sediment as *zones* and remain separated from the other components. The centrifuge run is terminated before any particles reach the bottom of the gradient. The various zones in the centrifuge tubes are then isolated by collecting fractions from the bottom of the tube and analyzing them for the presence of macromolecules. The zones of separated macromolecules are relatively stable in the gradient because it slows diffusion and convection. The gradient conditions can be varied by using different ranges of sucrose concentration. Sucrose concentrations up to 60% can be used, with a density limit of 1.28 g/cm^3. However, solutions above 20% become sticky and viscous. Ficoll, a polysaccharide composed of sucrose monomers, may also be used as a gradient medium. A 20% Ficoll solution provides a density of 1.07 g/cm^3.

The zonal method can be applied to the separation and isolation of macromolecules (preparative ultracentrifuge) and to the determination of *s* (analytical ultracentrifuge).

Isopycnic Centrifugation

In the isopycnic technique, the density gradient is formed during the centrifugation. Figure 7.12C outlines the operation of isopycnic centrifugation. The sample under study is dissolved in a solution of a dense salt such as cesium chloride or cesium sulfate. The cesium salts may be used to establish gradients to an upper density limit of 1.8 g/cm^3. The solution of biological sample and cesium salt is uniformly distributed in a centrifuge tube and rotated in an ultracentrifuge. Under the influence of the centrifugal force, the cesium salt redistributes to form a continuously increasing density gradient from the top to the bottom. The macromolecules of the biological sample seek an area in the tube where the density is equal to their respective densities. That is, the macromolecules move to a region where the sum of the forces (centrifugal and frictional) is zero (Equation 7.6). The macromolecules either sediment or float to this region of equal density. Stable zones or bands of the individual components are formed in the gradient (see Figure 7.12C). These bands can be isolated as previously described. Up to 2 days of centrifugation may be required to completely establish a gradient of cesium salt. Cesium salt gradients are especially valuable for separation of nucleic acids. Other media that are appropriate for isopycnic gradients are iodinated compounds (metrizamide and Nycodenz) and colloidal silica (Ludox and Percoll).

Density gradients are widely used in separating and purifying biological samples. In addition to this preparative application, measurements of s can be made. Gradient techniques have been used to isolate and purify the subcellular components, microsomes, ribosomes, lysosomes, mitochondria, peroxisomes, chloroplasts, and others. After isolation, they have been biochemically characterized as to their protein, lipid, and enzyme contents.

Nucleic acids, in particular, have been extensively studied by density gradient techniques. Both RNA and DNA are routinely classified according to their s values. The different structural forms of DNA discussed in Chapter 4 can be determined by density gradient centrifugation. In Experiment 19, plasmid DNA is purified by centrifugation in a cesium salt gradient. The DNA band in the gradient is detected with ethidium bromide, an intercalating dye.

Space does not allow an exhaustive review of centrifuge applications. Interested students should consult the references at the end of the chapter for recent developments.

Care of Centrifuges and Rotors

Centrifuge equipment represents a sizable investment for a laboratory, so proper maintenance is essential. In addition, poorly maintained equipment is unsafe. Since many instruments are now available, specific instructions will not be given here, but general guidelines are outlined.

1. Carefully read the operating manual or receive proper instructions before you use any centrifuge.
2. Select the proper operating conditions on the instrument. If refrigeration is necessary, set the temperature to the appropriate level and allow 1 to 2 hours for temperature equilibration.
3. Check the rotor chamber for cleanliness and for damage. Clean with soap and warm water.
4. Select the proper rotor. Many sizes and types are available. Follow guidelines already stated in this chapter or consult your instructor.
5. Be sure the rotor is clean and undamaged. Observe any nicks, scratches, or other damage that may cause imbalance. If dirty, the rotor should be cleaned with warm water and a mild, nonbasic detergent. A soft brush can be used inside the cavities. Rinse well with distilled water and dry. Scratches should not be made on the surface coating, as corrosion may result.

6. Filled centrifuge tubes or bottles should be weighed carefully and balanced before centrifugation.

7. Rotor manufacturers provide a maximum allowable speed limit for each rotor. Do not exceed that limit.

8. Keep an accurate record of centrifuge and rotor use. Just as your automobile needs service after a certain number of miles, the centrifuge should be serviced after certain intervals of use. Centrifuge maintenance is usually determined by hours of use and total revolutions of the rotor. It is also essential to maintain a record of the use for each rotor. Rotors weaken with use, and the maximum allowable speed limit decreases. Rotor manufacturers usually provide guidelines for decreasing the allowable speed for a rotor.

9. If an unusual noise or vibration develops during centrifugation, immediately turn the centrifuge off.

10. Carefully clean the rotor chamber and rotor after centrifugation.

References

K. Aune, in *Methods in Enzymology,* Vol. 48, C. H. W. Hirs and S. N. Timasheff, Editors (1978), Academic Press (New York), pp. 163–185. "Molecular Weight Measurements by Sedimentation Equilibrium: Some Common Pitfalls and How to Avoid Them."

C. Cantor and P. Schimmel, *Biophysical Chemistry,* Part II (1980), W. H. Freeman (San Francisco), pp. 591–641. Excellent but advanced treatment of theory and applications of ultracentrifugation.

T. Ford and J. Graham, *An Introduction to Centrifugation* (1991), Bios Scientific Publishers (Oxford). Practical aspects of centrifugation.

D. Freifelder, *Physical Biochemistry,* 2nd ed. (1982), W. H. Freeman (San Francisco), pp. 362–454. Theoretical background with many applications, especially to the study of nucleic acids.

G. Howlett, in *Methods in Enzymology,* Vol. 150, G. DiSabato, Editor (1987), Academic Press (San Diego, CA), pp. 447–463. "Air-Driven Ultracentrifuge for Sedimentation Equilibrium and Binding Studies."

C. Mathews and K. van Holde, *Biochemistry* (1990), Benjamin/Cummings (Redwood City, CA), pp. 156–158. Centrifugation in biochemistry.

A. Minton, *Anal. Biochem.* **176,** 209–216 (1989). "Analytical Centrifugation with Preparative Ultracentrifuges."

R. Pollet, in *Methods in Enzymology,* Vol. 117, C. H. W. Hirs and S. N. Timasheff, Editors (1985), Academic Press (Orlando, FL), pp. 3–27. "Characterization of Macromolecules by Sedimentation Equilibrium in the Air-Turbine Ultracentrifuge."

D. Rickwood, Editor, *Centrifugation—A Practical Approach,* 2nd ed. (1984), IRL Press (Oxford). Theory and practice of centrifugation with many applications.

J. Robyt and B. White, *Biochemical Techniques: Theory and Practice* (1987), Brooks/Cole (Monterey, CA), pp. 256–260. Centrifugation in the preparation of biological materials.

H. Schachman, *Nature* **341**, 259–260 (1989). "Analytical ultracentrifugation reborn".

P. Sharpe, in *Laboratory Techniques in Biochemistry and Molecular Biology,* Vol. 18, R. Burdon and P. Knippenberg, Editors (1988), pp. 18–68, Elsevier (Amsterdam). "Centrifugation."

L. Stryer, *Biochemistry,* 3rd ed. (1988), Freeman (New York), pp. 49–50. Ultracentrifugation for separation and characterization of biomolecules.

K. van Holde, *Physical Biochemistry,* 2nd ed. (1985), Prentice Hall (Englewood Cliffs, NJ), pp. 110–136. An introduction to sedimentation.

D. Voet and J. Voet, *Biochemistry* (1990), John Wiley & Sons (New York), pp. 100–105. Use of centrifugation in the purification of proteins.

PART II Experiments

PART II Experiments

EXPERIMENT I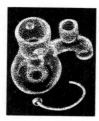

Using the Biochemical Literature

■ **Synopsis**

The objective of this exercise is to introduce students to the biochemistry books and journals found in most libraries. In addition, the applications of computers and bibliographic search services will be described. Several problems typically encountered in biochemical research are posed, and students are asked to solve the problem or answer questions.

I. Introduction and Theory

The Biochemical Literature

Experimental biochemists do not spend all their working time in the laboratory. An important component of a biochemistry research project is a search of the biochemical literature. The library should be considered a tool for experimental biochemistry in the same way as any scientific instrument. The use of the biochemical literature by the student in biochemistry laboratory is not as extensive as that of a full-time researcher, but you must be aware of what is available in the library and how to use it.

The library is used in all stages of research. Before an investigator can begin experimentation, a research idea must be generated. This idea develops only after extensive reading and study of the literature. A research project usually begins in the form of a question to be answered or problem to be solved. For ease of solution, a major project is subdivided into questions that may be answered by experimentation. Before laboratory work can begin, the researcher must have a knowledge of the past and current literature dealing with the research area. In a sense, this can be reduced to two questions: What is the current state of knowledge in the area? What are the significant unknowns? These questions can be answered only by developing a familiarity with the biochemical literature. The researcher will find that this knowledge of the literature is also invaluable for the design of experiments. The development of experiments requires knowledge of techniques and laboratory procedures. Excellent methods books and journals are available that provide experimental details. Finally, while performing experiments, the researcher often needs physical and chemical constants and miscellaneous information. Various handbooks and encyclopedias are excellent for this purpose.

The beginning student in biochemistry laboratory will not be expected to proceed through all these stages in the design of an experiment. However, a familiarity with the literature will increase your understanding of the experiment and may aid in the development of more effective methods. When you do begin a research program, you will be able to use the library to fullest advantage.

The biochemical literature is massive and expanding rapidly. It is almost a full-time job just to maintain a current awareness of a specialized research area. There are no discipline boundaries in the study of biochemistry. The biochemical literature overlaps into the biological sciences, the physical sciences, and the basic medical sciences. The intent of this discussion is to bring some order to the many textbooks, reference books, research journals, computer information retrieval services, and handbooks that are available. Each component of the biochemical literature will be dealt with individually.

TEXTBOOKS

The student's first exposure to biochemistry is probably a lecture course accompanied by the reading of a general textbook of biochemistry. By providing an in-depth survey of biochemistry, textbooks allow students to build a strong foundation of important principles and concepts. By the time most books are in print, the information in them is 1 to 2 years old, but textbooks still should be considered the starting point for mastery of biochemistry. The textbooks published since 1988 that have been most widely adopted for the 1-year undergraduate biochemistry course are:

B. Alberts, D. Bray, J. Lewis, M. Raff, K. Roberts, and J. Watson, *Molecular Biology of the Cell* (1989), Garland (New York).

R. C. Bohinski, *Modern Concepts in Biochemistry*, 5th ed. (1988), Allyn & Bacon (Boston).

J. Darnell, H. Lodish, and D. Baltimore, *Molecular Cell Biology*, 2nd ed. (1990), Scientific American Books (New York).

C. Mathews and K. van Holde, *Biochemistry* (1990), Benjamin/Cummings (Redwood City, CA).

J. D. Rawn, *Biochemistry,* 2nd ed. (1989), Neil Patterson Publishers (Burlington, NC).

L. Stryer, *Biochemistry,* 3rd ed. (1988), W. H. Freeman (New York).

D. Voet and J. Voet, *Biochemistry* (1990), John Wiley & Sons (New York).

G. Zubay, *Biochemistry,* 2nd ed. (1988), Macmillan (New York).

REFERENCE BOOKS AND REVIEW PUBLICATIONS

For more specialized and detailed biochemical information that is not offered by textbooks, reference books must be used. Reference works range from general surveys to specialized series. The best works are multivolume sets that continue publication of volumes on a periodic basis. Each volume usually covers a specialized area with articles written by recognized authorities in the field. Some of the more familiar reference and review publications are listed in Table E1.1. It should be noted that reference articles of interest to biochemists are often found in publications that are not strictly biochemical. The best known and most widely used review publication is *Annual Review of Biochemistry*. Each volume in this series, which was introduced in 1932, contains several detailed and extensive articles written by experts in the field. For shorter reviews emphasizing current topics, *Trends in the Biochemical Sciences* (TIBS) is widely read.

RESEARCH JOURNALS

The core of the biochemical literature consists of research journals. It is essential for a practicing biochemist to maintain a knowledge of biochemical advances in his or her field of research and related areas. Scores of research journals are published with the intent of keeping scientists up-to-date. With the expansion of scientific information has come the need for efficient storage and use of research journals. Many publishers are now providing journals in forms such as microcards, microfilm, microfiche, and more recently CD-ROM disks. Some of the major research journals in biochemistry and related fields are listed in Table E1.2. The concept of core journals in biochemistry is important. Some research journals have achieved a reputation as the most significant, and articles therein are considered to be of the highest quality. A 1979 ranking of the biochemical journals, based on the number of citations received, produced the following order for the top six: *Journal of Biological Chemistry, Biochimica et Biophysica Acta, Biochemistry, Proceedings of the National Academy of Sciences (USA), Biochemical Journal,* and *Biochemical and Biophysical Research Communications.* The core journals used by an individual depend on the area of speciality and are best determined from experience. When journals that contain articles relevant to the speciality are identified, these can be scanned regularly by the individual.

Table EI.I
Reference Books in Biochemistry and Related Areas

General Reference/Survey Books

P. Boyer, *The Enzymes,* 3rd ed., Academic Press, more than 16 volumes.
C. Cantor and P. Schimmel, *Biophysical Chemistry,* Freeman, three volumes.
  Volume I: *The Conformation of Biological Macromolecules*
  Volume II: *Techniques for the Study of Biological Structure and Function*
  Volume III: *The Behavior of Biological Macromolecules*
M. Florkin and E. Stolz, *Comprehensive Biochemistry,* Elsevier, more than 35 volumes.

Reviews

Advances in Cancer Research
Advances in Carbohydrate Chemistry
Advances in Cell Biology
Advances in Comparative Physiology and Biochemistry
Advances in Enzyme Regulation
Advances in Enzymology and Related Areas of
  Molecular Biology
Advances in Immunology
Advances in Lipid Research
Advances in Pharmacology and Chemotherapy
Advances in Protein Chemistry
Annals of the New York Academy of Science
Annual Reports in Medicinal Chemistry
Annual Reports of the Chemical Society
Annual Review of Biochemistry
Annual Review of Biophysics and Bioengineering
Annual Review of Genetics
Annual Review of Microbiology
Annual Reports of Pharmacology
Annual Review of Physiology
Annual Review of Plant Physiology
Bacteriological Reviews

Biochemical Society Symposia
Biological Reviews
Chemical Reviews
Cold Spring Harbor Symposia in Quantitative
  Biology
Essays in Biochemistry
Harvey Lectures
International Review of Cytology
Physiological Reviews
Progress in Biophysics and Biophysical Chemistry
Progress in the Chemistry of Fats and Other Lipids
Progress in Nucleic Acid Research and Molecular
  Biology
Quarterly Review of Biology
Quarterly Reviews of Biophysics
Quarterly Review of the Chemical Society
Trends in Biochemical Sciences
Trends in Biotechnology
Trends in Neurosciences
Trends in Pharmacological Sciences
Trends in Genetics
Vitamins and Hormones

Table EI.2
Research Journals in Biochemistry and Related Areas

Accounts of Chemical Research
Agricultural and Biological Chemistry
American Journal of Human Genetics
American Journal of Physiology
Analytical Biochemistry
Angewandte Chemie (English International Edition)
Archives of Biochemistry and Biophysics
Biochemical and Biophysical Research Communications
Biochemical Education
Biochemical Genetics
Biochemical Journal
Biochemical Medicine and Metabolic Biology
Biochemical Pharmacology
Biochemistry
Biochemistry International

Table E1.2 (continued)

Biochimica et Biophysica Acta
Biochemie
Biokhimiya (Russian, translated into English)
Bioelectrochemistry and Bioenergetics
Bioelectrochemistry and Bioengineering
Biology of Membranes
Biology of Metals
Biology of Trace Elements Research
Biomedical Biochemical Acta
Bioorganic Chemistry
Biophysical Chemistry
Biophysical Journal
Biopolymers
Biotechniques
Biotechnology and Bioengineering
Biotechnology Education
Biotechnology and Applied Biochemistry
Biopolymers
Brain Research
Bulletin de la Societe Chimie Biologique
Canadian Journal of Biochemistry
Canadian Journal of Microbiology
Carbohydrate Research
Cell
Cell and Biochemical Function
Cell and Molecular Biology
Cell and Tissue Kinetics
Cell Biochemistry and Function
Cell Biophysics
Chemico-Biological Interactions
Chemistry and Physics of Lipids
Comparative Biochemistry and Physiology
Comptes Rendus de l'Academie des Sciences (Paris), Parts C and D
Comptes Rendus des Seances de la Societe de Biologie
Developmental Biology
Electrophoresis
EMBO Journal
Environmental and Molecular Mutagenesis
Enzyme
Enzyme Microbiological Technology
European Journal of Biochemistry
Experimental Cell Research
Experientia
FASEB Journal
Federation Proceedings
Federation of European Biochemical Societies (FEBS Letters)
Food Chemistry
Free Radical Biology and Medicine
General Cytochemical Methods
Genetic Pharmacology
Genetical Research
Genetics
Genome (formerly Canadian Journal of Genetics and Cytology)
Hoppe-Seyler's Zeitschrift für Physiologische Chemie
Immunology

(continued on next page)

Table E1.2 *(continued)*

Indian Journal of Biochemistry
Inorganica Chemica Acta
International Journal of Biochemistry
International Journal of Biological Macromolecules
International Journal of Vitamin and Nutrition Research
Journal of Agricultural and Food Chemical
Journal of the American Chemical Society
Journal of the American Oil Chemists' Society
Journal of Applied Biochemistry
Journal of Bacteriology
Journal of Biochemical Toxicology
Journal of Biochemical and Biophysical Methods
Journal of Biochemistry (Tokyo)
Journal of Bioenergetics
Journal of Bioenergetics and Biomembranes
Journal of Biological Chemistry
Journal of Biomolecular and Structural Dynamics
Journal of Biotechnology
Journal of Cell Biology
Journal of Cell Biochemistry
Journal of Cell Science
Journal of Cellular Physiology
Journal of Chemical Education
Journal of Chromatography
Journal of Clinical Investigation
Journal of Electron Microscopy
Journal of Experimental Biology
Journal of Food Chemistry
Journal of General Microbiology
Journal of General Virology
Journal of Histochemistry
Journal of Immunology
Journal of Inorganic Biochemistry
Journal of Lipid Research
Journal of Membrane Biology
Journal of Molecular Biology
Journal of Neurochemistry
Journal of Nutrition
Journal of Nutrition Science and Vitaminology
Journal of Organic Chemistry
Journal of Pharmacology and Experimental Therapeutics
Journal of Photochemistry and Photobiology
Journal of Physical Chemistry
Journal of Physiology
Journal of Plant Physiology
Journal of Protein Chemistry
Journal of Steroid Biochemistry
Journal of the Chemical Society (London)
Journal of Theoretical Biology
Journal of Ultrastructure Research
Journal of Virology
Lipids
Macromolecules
Membrane Biochemistry
Metabolism—Clinical and Experimental

Table E1.2 *(continued)*

Microchemical Journal
Molecular and Cellular Biochemistry
Molecular and General Genetics
Molecular Immunology
Molecular Pharmacology
Molecular Physiology
Mutation Research
Nature
Naturwissenschaften
New England Journal of Medicine
Nucleic Acids Research
Nucleosides and Nucleotides
Peptides
Pharmacological Toxicology
Pharmacology
Pharmacology Research Communications
Photobiochemistry and Photobiophysics
Photosynthesis Research
Physiological Chemistry and Physics
Phytochemistry
Planta
Plant Molecular Biology
Plant Physiology
Preparative Biochemistry
Proceedings of the National Academy of Sciences (USA)
Proceedings of the Royal Society (London), Part B
Proceedings of the Society for Experimental Biology and Medicine
Prostaglandins
Protoplasma
Recombinant DNA
Science
Steroids
Tetrahedron Letters
The Journal of Trace Elements in Experimental Medicine
The Plant Cell
Toxicology and Applied Pharmacology
Toxicology Letters
Xenobiotics
Zeitschrift für Naturforschung

METHODOLOGY REFERENCES

The active researcher has a continuing need for new methods and techniques. Several publications specialize in providing details of research methods. Some of the major biochemical methodology publications are:

Analytical Biochemistry, a monthly journal.

Analytical Chemistry, a monthly journal.

Basic Cloning Techniques, R. Pritchard and I. Holland, Editors. A manual of experimental procedures.

Biochemical Preparations, an annual volume.

Current Protocols in Molecular Biology, P. Ausabel et al. Editors. A manual of techniques in two volumes that are updated quarterly.

Laboratory Techniques in Biochemistry and Molecular Biology, T. S. Work and R. G. Burdon, Editors (formerly T. S. Work and E. Work). Each volume in the series is concentrated in an area of biochemistry and written by recognized authorities.

Methods of Biochemical Analysis, D. Glick, Editor. Contains techniques for the analysis and preparation of biochemicals.

Methods of Enzymatic Analysis, 10th ed., H. Bergmeyer, Editor. Contains methods for enzyme purification and assay, in several volumes.

Methods in Enzymology, various editors. The most valuable methods series available. Each volume contains numerous articles describing biochemical techniques. The series is well indexed and easy to use. Over 200 volumes.

Molecular Cloning: A Laboratory Manual, 2nd ed., J. Sambrook, E. Fritsch, and T. Maniatis. A standard manual in three volumes for laboratory protocols in molecular biology.

A Practical Guide to Molecular Cloning, 2nd ed., B. Perbal. Useful for setting up research projects in molecular cloning.

HANDBOOKS OF CHEMICAL AND BIOCHEMICAL DATA

Students in introductory biochemistry laboratory may use books in this category more than any other type. While doing biochemical experiments, you will need physical, chemical and biochemical information such as R_f values, molecular weights, amino acid sequences of peptides and proteins, pK values, and sequences of nucleotides in DNA and RNA. This information is most easily found in the many handbooks and collections of biochemical data. You should be introduced to the use of these references early in your training. Some of the most useful handbooks are listed below with a brief description of contents.

Concise Encyclopedia of Biochemistry, T. Scott and M. Brewer, Editors (1983), Walter De Gruyter (Hawthorne, NY). Contains 4200 entries of biochemical terms.

Data for Biochemical Research, 3rd ed., R. Dawson, D. Elliott, W. Elliott, and K. Jones (1986), Clarendon Press (Oxford). Information needed by biochemists in the laboratory.

Dictionary of Biochemistry and Molecular Biology, 2nd ed. J. Stenesh (1989), Wiley Interscience (New York). Contains definitions of 16,000 terms.

Dictionary of Microbiology and Molecular Biology, 2nd ed. P. Singleton and D. Sainsbury (1987), Wiley (Chichester). Definitions of terms and phrases.

Glossary of Biochemistry and Molecular Biology, D. Glick (1990), Raven Press (New York). Emphasis on words and phrases that are unique to biochemistry and molecular biology.

Merck Index, 11th ed., S. Budavari, Editor (1989), Merck and Co. (Rahway, NJ). An encyclopedia of chemicals, drugs, and biological substances.

Practical Handbook of Biochemistry and Molecular Biology, G. Fasman, Editor (1989), CRC Press (Boca Raton, FL). Excellent source of chemical and physical data for nucleic acids, proteins, lipids, and carbohydrates.

Worthington Enzyme Manual, 2nd ed., C. Worthington, Editor (1988), Worthington Biochemical Corp. (Freehold, NJ). Contains detailed assay information and references for over 75 enzymes.

Brochures of Commercial Suppliers. Some of the most up-to-date information on biochemical techniques, methods, and instrumentation is available from the many companies that manufacture and supply equipment and reagents. There is usually no charge for these brochures.

COMPUTER-BASED SEARCHES AND OTHER AIDS TO THE LITERATURE

As you study and work in biochemistry, you will often need to complete a thorough literature search on some specialized area or topic. It is not practical to survey the hundreds of books, journals, and reports that may contain information related to the topic. Two publications that provide brief summaries of published articles, reviews, and patents are *Chemical Abstracts* and *Biological Abstracts.* If you are not familiar with the use of these abstracts, refer to the next section or ask your instructor or librarian for assistance.

Research articles of interest to biochemists may appear in many types of research journals. Research libraries do not have the funds necessary to subscribe to every journal, nor do scientists have the time to survey every current journal copy for articles of interest. Two publications that help scientists to keep up with published articles are *Chemical Titles* (published every 2 weeks by the American Chemical Society) and the weekly *Current Contents* available in hard copy and computer disks (published by the Institute of Science Information). The Life Science edition of *Current Contents* is the most useful for biochemists. The computer revolution has reached into the chemical and biochemical literature, and many college and university libraries now subscribe to computer bibliographic search services. One such service is STN International, the scientific and technical information network. This on-line system allows direct access to some of the world's largest scientific databases. The STN databases of most value to life scientists include BIOSIS Previews/RN (produced by Bio Sciences Information Service; covers original research reports, reviews, and U.S. patents in biology and biomedicine), CA (produced by Chemical Abstracts service; covers research reports in all areas of chemistry), MEDLINE and MEDLARS (produced by the U.S. National Library of Medicine and *Index Medicus,* respectively; cover all areas of biomedicine), and PHAR (produced by PJB Publications; covers articles on pharmaceutical development). These networks provide on-line service so the databases can be accessed from personal computers in the office, laboratory, or library. The equipment needed for on-line searching consists of a microcomputer, a modem (or a communicating word processor), knowledge of the required password(s), and general instructions.

II. Procedure—Library Projects

The first step toward the solution of a problem in the biochemistry laboratory is often a visit to the library. During almost every experiment you will need information that can be found only in the literature. You may require the pK for an amino acid, the molecular weight of a protein, or information on separating the fragments produced by cleavage of DNA by a restriction endonuclease. Therefore, you should have some experience in the library early in the term so that the use of available resources is familiar.

Outlined here are several problems or questions that might be encountered by research workers in biochemistry. Different types of references will be necessary to solve each of these problems. The references that will be of most value are research journals, biochemistry textbooks, biochemistry dictionaries, biochemistry handbooks, *Chemical Abstracts,* and methods books.

The exercise will begin with some examples with hints to help initiate the library search.

1. *Science* magazine declared the polymerase chain reaction (PCR) the major scientific development of 1989 and chose the DNA polymerase necessary for the reaction as "Molecule of the Year." What is the PCR?

   Solution: For basic definitions, a biochemistry dictionary or your biochemistry textbook is your best source. For example, PCR is defined as follows in the *Dictionary of Biochemistry and Molecular Biology,* 2nd ed., J. Stenesh (1989): "A technique for the synthesis of large quantities of specific DNA segments; consists of a series of repetitive cycles, one step of which involves a high temperature. The latter inactivates the DNA polymerase used originally, thus requiring the addition of fresh enzyme at each cycle." More details of this procedure may be found in your textbook or research articles.

2. What is the molecular weight of the protein hemoglobin A (HbA)?
   Solution: Physical constants such as molecular weight, pK values, $[\alpha]_D$, and absorption coefficients are found in biochemistry textbooks or handbooks such as the *CRC Practical Handbook of Biochemistry and Molecular Biology*, G. Fasman (1989). From *Biochemistry*, 3rd ed., L. Stryer (1988), the molecular weight of normal adult hemoglobin is 68,000.

3. List one research article published in the past 2 years on the topic of ribosomal RNA formation by self-splicing RNA.
   Solution: When research articles are required, *Chemical Abstracts* should be the starting point. Since this form of RNA is a chemical substance, the *Chemical Abstracts Chemical Substance Index* is used. RNA formation is listed in alphabetic order. Under the category of RNA formation are several subtopics including messenger, transfer, and ribosomal. For example, in Volume 113 (1990) under the main heading *RNA formation* and subheading *ribosomal* is a listing *self-splicing of Group I introns*. The abstract number is R110038c. The R indicates that this reference is a review article. Find the number 110038c in the *Abstracts,* Volume 113. This abstract lists the title of the article (Self-splicing of Group I introns), the author (Cech, Thomas, R.), and the address of the author. The article is published in *Annual Review of Biochemistry,* Vol. 59, 1990, pages 543–568. The abstract also contains a brief description of the article.

   To illustrate another use of *Chemical Abstracts,* find the author's name in the *Author Index* (Volume 113). Under the listing "Cech, Thomas, R." is the same abstract, R110038c.

4. Briefly define each of the following biochemical terms. Give the reference used for each in the following order: Name of book, author, publisher, year of publication, and page number.
   (a) Coomassie Blue (a chemical used in Experiment 5).
   (b) Chloroplasts (isolated in Experiment 12).
   (c) Chemiosmotic coupling hypothesis (see Experiment 12).
   (d) Restriction endonucleases (see Experiment 21).
   (e) Codons.
   (f) Metal ion affinity chromatography (used in Experiment 3).
   (g) Sedimentation coefficient.
   (h) Cytochrome P-450.
   (i) Galactosemia.
   (j) Catalytic antibodies.
   (k) Plasmids.

5. What is the molecular weight of each of the following biomolecules?

   (a) Sucrose

   (b) Plasmid pBR322

   (c) Egg albumin

   (d) Histidine

   (e) ATP

6. What are the pK_a values for the buffer Tris and the amino acids aspartic acid and histidine?

7. What is the isoelectric pH (pH_I) of human ceruloplasmin?

8. What is the specific rotation, $[\alpha]_D$, for the carbohydrate D-ribose?

9. Draw the structure of ethidium bromide, a dye that binds to DNA.

10. List five journal articles published in the past 2 years that describe the use of the polymerase chain reaction (PCR). List by title of article, author(s), journal name, volume, pages, and year.

11. Professor Peter Dervan of California Institute of Technology has written several articles on specific cleavage of DNA by metal complexes. Write a brief summary of one of the articles.

12. Find a reference that gives complete details of the purification of the calcium-binding protein calmodulin from animal tissue.

13. List a reference that gives complete details of the performance of sodium dodecyl sulfate–polyacrylamide gel electrophoresis (SDS-PAGE).

14. Find a reference that gives a theoretical background and explanation for the technique of pulsed-field electrophoresis.

15. Choose a research journal article related to biochemistry published within the last month. Write a brief summary (8–10 sentences). Emphasize the analytical techniques used in the article.

III. References

H. Collier, Editor, *Chemical Information* (1989), Springer-Verlag (Berlin). The role of literature searching in research.

E. Davis, *Using the Biological Literature: A Practical Guide* (1981), Marcel Dekker (New York). An excellent introduction to the literature; however, somewhat outdated.

R. Maizell, *How to Find Chemical Information—A Guide for Practicing Chemists, Teachers and Students,* 2nd ed. (1987), John Wiley & Sons (New York). Emphases on chemistry.

R. Smith, W. Reid, and A. Luchsinger, *Smith's Guide to the Literature of the Life Sciences,* 9th ed. (1980), Burgess (Minneapolis). Somewhat outdated but good introductory background.

J. Turner, in *Information Sources in the Life Sciences,* H. Wyatt, Editor, (1987) Butterworths (London), pp. 57–79. "Biochemical Sciences."

Y. Wolman, *Chemical Information—A Practical Guide to Utilization,* 2nd ed. (1988), Wiley (Chichester). A useful introduction.

EXPERIMENT **2**

Analytical Methods for Amino Acid Separation and Identification

■ **Recommended Reading**

Chapter 2, Section A; Chapter 3, Sections A, B, E; Chapter 4, Sections A, B.

■ **Synopsis**

Separation and identification of amino acids are operations that must be performed frequently by biochemists. The 20 amino acids present in proteins have similar structures but are unique in polarity and ionic characteristics. In this experiment, the student will use a combination of ion-exchange chromatography and paper chromatography to separate and identify the components of an unknown amino acid mixture.

I. Introduction and Theory

Twenty common amino acids are the fundamental building blocks of proteins. All proteins found in nature are constructed by amide bond linkages between α-amino acids. The amino acids isolated from protein material all have common structural characteristics. They have at least one carboxyl group and at least one amino group (Figure E2.1). The distinctive physical, chemical, and biological properties associated with an amino acid are the result of the R group, which is unique for each amino acid.

The structure and biological function of a protein depend on its amino acid content. It is a matter of basic importance to understand practical methods used for the separation and identification of the 20 common amino acids.

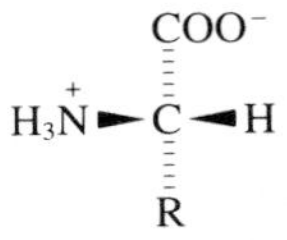

Figure E2.1
The general structure of the zwitterionic form of an amino acid. R represents the side chain.

Acid-Base Chemistry of the Amino Acids

When proteins are heated in aqueous acid or base, the amide bonds are hydrolyzed, releasing the constituent amino acids in a free form. Separating and identifying the amino acids present in this complex mixture, which may contain up to 20 different molecules, require a thorough knowledge of the physical and chemical properties of amino acid molecules. Recall from your biochemistry lectures that amino acids are amphoteric; that is, they can act as Brønsted acids (proton donating) or Brønsted bases (proton accepting) (Equations E2.1 and E2.2). The zwitterionic form (II) is produced by proton dissociation of I (Equation 1) and it can also participate in proton dissociation (Equation 2), which forms anion III.

$$\underset{\text{I}}{H_3\overset{\oplus}{N}CH(R)COOH} \rightleftharpoons \underset{\text{II}}{H_3\overset{\oplus}{N}CH(R)COO^{\ominus}} + H^{\oplus} \qquad \text{(Equation E2.1)}$$

$$\underset{\text{II}}{H_3\overset{\oplus}{N}CH(R)COO^{\ominus}} \rightleftharpoons \underset{\text{III}}{H_2NCH(R)COO^{\ominus}} + H^{\oplus} \qquad \text{(Equation E2.2)}$$

Each ionic reaction is defined by an acid ionization constant, K_a, and a pK_a ($-\log K_a$). All of the α-carboxyl groups of the 20 amino acids have similar but different pK_a values ($pK_a \approx 2$ to 3), as do the α-amino groups ($pK_a \approx 9$ to 10). (See Appendix III.) The pK_a value of an amino acid represents the pH at which two ionic forms are present in equal concentrations. Some amino acids also have ionizable groups on the R side chain. The extra acidic or basic groups, of course, add to the complexity of acid-base reactions of amino acids; but, as we shall see, this allows greater resolution in analyzing amino acid mixtures. If you are not completely familiar with the acid-base properties of amino acids, refer to your biochemistry textbook.

Separation of Amino Acid Mixtures

Let us consider a typical problem that is often encountered in protein chemistry. When a protein is isolated in purified form, it is first characterized as to its amino acid composition. The protein is hydrolyzed, usually under acidic conditions, and the resulting mixture of free amino acids is subjected to qualitative and quantitative analysis. This determines what amino acids are present and how many molecules of each amino acid are present in the protein.

At first glance, this seems to be an insurmountable problem. The mixture of amino acids may contain up to 20 different, but similar, molecular structures. How does one resolve such a mixture of similar molecules? Several techniques ranging from chemical methods to chromatographic methods have been developed. Chapters 3, 4, and 7 describe three general methods useful for separation of biomolecules: chromatography, electrophoresis, and centrifugation. Centrifugation, which is used to separate molecules on the basis of size and specific gravity, would not be helpful since amino acids show little difference in these properties. Analysis of mixtures by electrophoresis is primarily dependent on the sign and size of the charge on the individual molecules. Since amino acids are charged molecules at most pH values, this technique should have promise. Electrophoresis of the 20 amino acids at pH 1.0 might lead to separation into only two groups; one group of 3 amino acids nearer the cathode (Lys, His, Arg) and one group of 17 amino acids between the origin and the group of three. In theory, this would not lead to satisfactory separation of the amino acids. In practice, however, electrophoresis is a valuable technique because, in addition to charge, molecules are separated on the basis of polarity. This method is most appropriate when a solution of only a few amino acids or peptides is to be analyzed. In addition, two-dimensional electrophoresis and use of more effective stationary media are improvements that make electrophoresis a viable technique in amino acid and protein analysis.

We should also examine the third option mentioned above—chromatography. Does it have any potential in separating amino acid mixtures? The major differences among the amino acid structures are polarity and charge, and chromatographic methods that can exploit these differences are paper, thin-layer, ion-exchange, gas, and high-performance liquid chromatography. Differences in molecular weight of the amino acids are too small for effective separation by gel filtration, and affinity chromatography does not lend itself well to separation of small molecules. As discussed in Chapter 3, paper, thin-layer, ion-exchange, high-performance liquid, and gas chromatographies are especially suitable for the separation of molecules that have primarily polarity differences. High-performance liquid and gas chromatographies are excellent choices for the separation of amino acids, but they are limited for your use because derivatives of the amino acids must be prepared for GC analysis and expensive instrumentation is required for both. However, it should be noted that these two methods have certain advantages and are widely used. A technician who is familiar with the recent advances in HPLC and GC and who has the necessary instrumentation can analyze complex amino acid mixtures more effectively and more rapidly than with any other technique. A distinct advantage is the small sample size required for analysis. Students interested in recent advances in amino acid analysis should review the references at the end of the experiment.

A widely used and simple method for separating and identifying amino acids in a mixture is ion-exchange chromatography. As a first approximation, separation by ion-exchange chromatography is based on charge differences between the amino acid molecules. At a specific pH value, all amino acids may have similar charge signs. For example, most amino acids are cationic below pH 4 and most are anionic above pH 8.

However, since each has a different series of ionization constants, the size of the charge at a specific pH is different. The exact size of the charge can be calculated using the Henderson-Hasselbalch equation. The charge differences are small for most amino acids, and complete separation by ion-exchange chromatography is not predicted. This means that under certain conditions, the three groups of acidic, neutral, and basic amino acids can be separated as described for electrophoresis. However, in the actual practice of cation-exchange chromatography, much better resolution than expected is attained.

Figure E2.2 illustrates the separation of amino acids by ion-exchange chromatography. Careful study of the order of elution of 17 amino acids reveals that within a charge group they are separated on the basis of polarity. For example, glutamic acid and aspartic acid have the same ap-

Figure E2.2
Separation of amino acids by cation-exchange chromatography. Three buffers, listed at the top, were used for elution. From *Biochemistry, A Problems Approach,* 2nd ed., by William B. Wood, John H. Wilson, Robert M. Benbow, and Leroy E. Hood. © 1981. The Benjamin/Cummings Publishing Company, Inc.

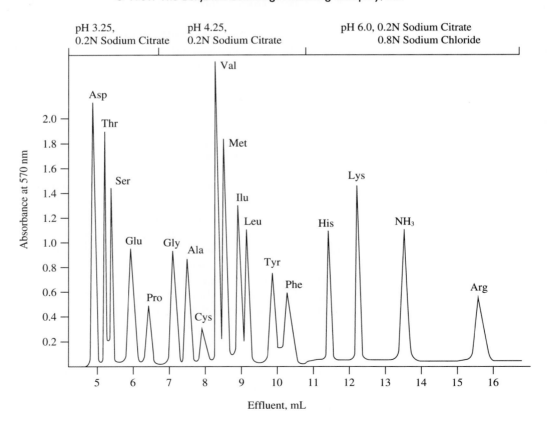

proximate net charge at pH 3.25, yet they are well resolved by cation-exchange chromatography. The pH_I, the pH at which the amino acid has no net charge, is $(2.09 + 3.86)/2 = 2.98$ for aspartic acid and $(2.19 + 4.25)/2 = 3.22$ for glutamic acid. This means that at pH 3.25 aspartic acid has a net charge of approximately 0 and glutamic acid has a net charge between 0 and +1. With net charge of approximately 0, aspartic acid would experience little attraction to the anionic sites on the resin, and the partial positive charge on glutamic acid would delay elution but probably not enough to explain the wide separation. Perhaps some additional factor is affecting the rate of elution of the amino acid.

To examine this possibility, let's study another pair of amino acids, alanine and serine. At pH 3.0 both are slightly cationic, yet serine elutes much earlier than alanine from a cation-exchange column. In fact, alanine does not elute until after a buffer change from pH 3.25 to pH 4.25. In both cases, glutamic acid vs. aspartic acid and alanine vs. serine, note that the more polar amino acid elutes first. Aspartic acid, with one less methylene group than glutamic acid, would be expected to be more polar ($—CH_2—$ vs. $—CH_2—CH_2—$). The hydroxyl group of serine makes it more polar than alanine ($—CH_2OH$ vs. $—CH_3$). We believe that the slower elution of glutamic acid and alanine is due to more favorable hydrophobic interactions between the side chain of the amino acid and the nonpolar regions of the ion-exchange resin. Further confirmation that polarity aids greatly in resolving the amino acids comes from the position of phenylalanine on the elution chart. Phenylalanine, which has a net charge similar to alanine and serine, is exceedingly late in column elution. Why? In summary, the separation of amino acids by cation-exchange chromatography is based on two primary factors: (1) *charge*—the greater the positive charge on the amino acid at a specific pH, the slower the elution time, and (2) *polarity*—the more polar the amino acid side chain, the greater the rate of elution.

Before we discuss the experiment to be performed, some peculiarities of Figure E2.2 should be considered. Why does the chromatographic analysis shown in the figure record 17 amino acids and not 20? Which amino acids are missing and why? The amino acid mixture represented by the chromatogram is the product of acid-catalyzed hydrolysis of a typical protein. Under these conditions, tryptophan is destroyed and the amide side chains of asparagine and glutamine are hydrolyzed to free carboxyl groups (Asn → Asp; Gln → Glu). It should also be noted that serine and threonine are damaged by acid hydrolysis, so their quantitative recovery from protein hydrolysis is not possible.

The last four peaks in Figure E2.2, which elute with pH 6 buffer, represent basic molecules—histidine, lysine, ammonia, and arginine. What is the source of the ammonia? Note that the three amino acids elute with the pH 6 buffer even though they are cationic at this pH. (Draw the structures at pH 6 to prove this.) The ionic interactions between the amino acids (+ charged) and the resin (− charged) are disrupted by the addition of a salt (in this case, 0.8 N NaCl) to the buffer. This results in an increase in ionic strength of the eluting buffer. The large excess of Na^+ ions causes the displacement of the cationic amino acids, which are present in lower concentration. Consequently, we now recognize two methods for elution of bound amino acids from an ion-exchange resin. First, the pH of the eluting buffer can be changed, which alters the net charge on the bound ions. Second, the ionic strength of the buffer may be increased, thus disfavoring the interaction between the amino acid and the charged resin.

Practical Aspects of Ion-Exchange Chromatography

In practice, ion-exchange chromatography begins with the preparation of a buffered ion-exchange column as described in Chapter 3. Application of an amino acid mixture to the top of the resin and elution with buffers result in differential migration of the amino acids. Fractions that are collected from the column may be analyzed qualitatively and quantitatively. Quantitative analysis of the individual amino acids is best done by reaction of each fraction with ninhydrin, followed by absorbance or fluorescence measurement of the intensity of the blue-purple color. Qualitative identification of eluted amino acids is accomplished by comparison with a profile of standard amino acids, as in Figure E2.2. Amino acid analyzers automatically elute the ion-exchange column, treat the eluent with ninhydrin, and provide a recorded profile of elution as shown in Figure E2.2.

Successful completion of this experiment requires careful preparation of the ion-exchange medium. This process is time-consuming, so it has been completed for you. Your instructor obtained a cation-exchange resin such as AG50W (Bio-Rad) or Dowex 50 (Dow) in the H^+ form (see Figure E2.3). The resin was first washed with 4 N HCl to remove contaminating metal ions and then washed with 2 N NaOH to convert the H^+ form to the Na^+ form. Now the resin can function as a cation exchanger.

Now that the ionic state of the resin has been defined, how does this form of the resin function in the separation of amino acids? Assume that your amino acid mixture contains three components, lysine, alanine, and aspartic acid. They are dissolved in a pH 3.0 buffer. The predominant ionic form of each is shown below:

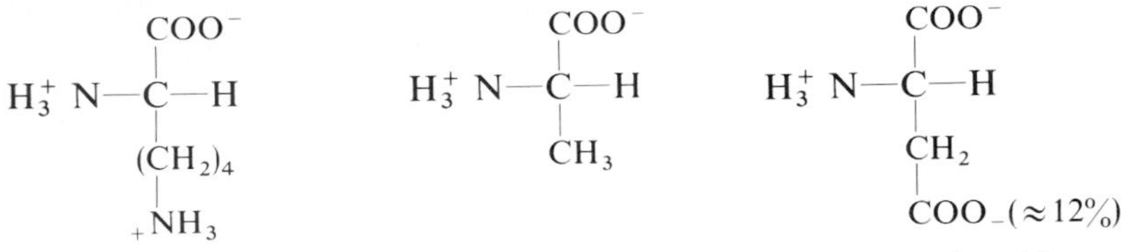

lysine
at
pH 3.0

net charge = +

alanine
at
pH 3.0

net charge $\approx$ 0

aspartic acid
at
pH 3.0

net charge = slightly −

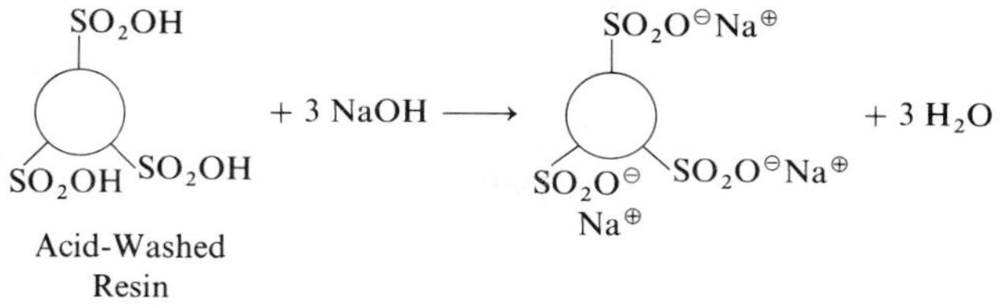

Figure E2.3
Conversion of the protonated
ion-exchange resin to the
cation-exchange form.

$$SO_2OH \bigcirc \begin{matrix} SO_2OH \\ SO_2OH \end{matrix} \quad + 3\ NaOH \longrightarrow \quad SO_2O^{\ominus}Na^{\oplus} \bigcirc \begin{matrix} SO_2O^{\ominus}Na^{\oplus} \\ SO_2O^{\ominus} \\ Na^{\oplus} \end{matrix} \quad + 3\ H_2O$$

Acid-Washed
Resin

Since the stationary resin is negatively charged, only cations interact favorably. Aspartic acid, with a negative charge, percolates through the column with buffer solution and elutes first. Both lysine and alanine are cationic and interact with the resin by exchanging with bound sodium ions. The rate of elution of lysine and alanine from the column depends on the strength of their interaction with resin. Lysine has a greater positive charge and binds more tightly to the anionic resin at pH 3. Alanine, with a slightly positive charge, also remains bound to the resin. At this point, one amino acid (aspartic acid) has eluted and, therefore, separated from the mixture. How is it possible to achieve selective elution of each of the two amino acids remaining on the column? Increasing the pH of the buffer should decrease the positive charge on the amino acids and weaken the ionic interactions between the bound amino acids and the resin. What pH should be chosen? At pH values greater than 6.0, alanine becomes negative and hence would repel the resin and be eluted. At pH values greater than the pH_I of lysine (9.74) it, too, becomes negative. Since stepwise elution of the amino acids is preferred, small pH changes are more effective than large pH changes. Changing from a buffer of pH 3.0 to pH 6.0 would elute the alanine, and then elution with a buffer of pH 9 would elute lysine. By a stepwise increase in pH, the separation of the three amino acids can be achieved. What would happen if you skipped the pH 6.0 buffer and eluted the column with pH 3 buffer and then pH 9 buffer? If you wished to maintain a constant buffer pH throughout the experiment, how would you elute the more basic amino acids?

How do we know that the amino acids have been eluted from the column? They are not colored, so their progress through the column cannot be directly monitored. Ninhydrin reacts with all of the amino acids except proline to yield a purple pigment. Proline reacts with ninhydrin to form a yellow reaction mixture. The intensity of the color is proportional to the amount of amino acid present and can be used to estimate the concentration of the amino acid. You should refer to your textbook for a further discussion of the ninhydrin reaction.

It is more difficult to confirm the identity of each amino acid isolated from the ion-exchange column. Standard amino acids can be applied to the column, and the volume of buffer needed to elute each amino acid can be measured, but it is better to confirm the identity of the amino acid by some other physical or chemical method such as paper or thin-layer chromatography. Paper chromatography will be used in this experiment because of its versatility, convenience, and low cost.

Overview of the Experiment

In this experiment you will analyze a solution containing one to three unknown amino acids. The individual amino acids in the mixture will first be separated by cation-exchange chromatography. Each amino acid will then be identified by paper chromatographic analysis using known amino acids as standards. Study Figure E2.4 for an outline of the experiment.

The experiment can be completed in 3 hours. It is recommended that students work in pairs. While one student begins the preparation of the ion-exchange column (part A), the other should begin spotting amino acid standards on the paper chromatogram (part B).

Figure E2.4
Flowchart for analysis of an amino acid mixture.

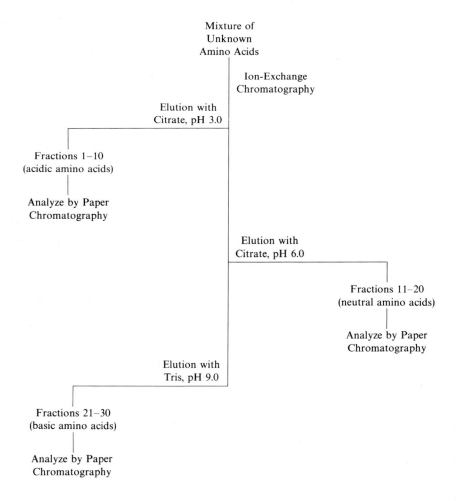

II. Materials and Supplies

Dowex 50-x8 or Amberlite IR-120 cation-exchange resin in 0.05 M citrate buffer, pH 3.0

Citrate buffer, 0.05 M, pH 3.0 and 6.0

Glass column, 1 × 25 cm, or glass buret

Tris buffer, 0.05 M, pH 9.0

Amino acid unknown mixture in citrate buffer, 2% w/v, pH 3.0

Ninhydrin solution in aerosol bottle

Small plug of glass wool

Chromatography paper, Whatman 3MM, 3 × 10 cm and 20 × 20 cm

Small glass funnel

Test tubes (30, 10 × 100 mm)

Solvent system for chromatography, acetonitrile and 0.1 M ammonium acetate (60:40), pH 4.0

Chromatography chamber

Capillary tubes or microcapillary pipets, 10 μL

Standard amino acid solutions, 2% w/v in H_2O

Oven at 110°C

III. Experimental Procedure

A. Separation of Amino Acid Mixture by Ion-Exchange Column Chromatography

PREPARING THE COLUMN

Clamp a glass column or buret to a ring stand with two clamps, making sure the column is exactly vertical. With the stopcock closed, pour 2 to 3 mL of 0.05 M citrate buffer, pH 3.0, into the column. With a long glass rod, put a small plug of glass wool into the bottom of the column. Press the air bubbles out of the glass wool. This glass wool barrier prevents the ion-exchange resin from running out of the column. Use just enough glass wool to fill the hole in the column; otherwise the flow rate will be greatly diminished. Obtain about 15 to 20 mL of cation-exchange resin in a small graduated beaker. Transferring the resin from the beaker to the column is not an easy task. Place a small glass funnel into the top of the column, mix the resin well by swirling or stirring with a glass rod, and pour the slurry into the column. Allow the resin to settle into a column 15 to 20 cm in height. When all the resin beads have settled, open the stopcock and allow the buffer to drain from the column until the buffer meniscus is just above the top of the resin. **Do not allow the column to run dry.**

APPLICATION OF SAMPLE AND COLLECTION OF FRACTIONS

Prepare 30 test tubes for fraction collection by numbering the tubes and marking the 1 mL level on each tube with a wax pencil. To do this, pour 1 mL of water into a test tube and mark the top level of water with the pencil. Use the height of the water to mark the other test tubes.

Using a graduated pipet, carefully apply 0.5 mL of an unknown amino acid mixture to the top of the resin. Apply the mixture dropwise so the resin is not disturbed. Open the stopcock and begin collecting fractions in test tube #1. When the top level of the buffer has just entered the resin, turn off the stopcock and carefully add 1 mL of citrate buffer (pH 3.0) to the column to wash any amino acid from the inside wall of the column. Allow the buffer to enter the resin and continue to collect fractions in tube #1 until 1 mL is collected; then begin collecting in tube #2.

When the buffer has all entered the resin, turn off the stopcock and carefully add 10 mL of buffer to the top of the column. Continue to collect 1-mL fractions. If an acidic amino acid is present in the unknown mixture, it will be eluted during this procedure. To test fractions for the presence of an amino acid, place 2 drops from each fraction onto a strip of chromatography paper (3 × 10 cm) as shown in Figure E2.5. Allow the spots to air-dry, and spray them in a hood with ninhydrin. Place in an oven at 110°C for 10 minutes. Red-purple spots indicate the presence of an amino acid.

If an amino acid has been completely removed from the column (no colored spot for fraction #10), change the eluting buffer by draining all the citrate buffer from the column (continue to collect fractions) and add 10 mL of 0.05 M citrate buffer, pH 6.0, to the column. Collect fractions until the top of the buffer has reached the top of the resin. Test each fraction as before for amino acid or check for pH change with pH paper. Neutral amino acids in the unknown mixture should have been eluted during this wash. Now wash the column with 10 mL of Tris, pH 9.0. Collect fractions until all the buffer has entered the resin. Spot 2 drops of each fraction onto chromatography paper and neutralize each spot by applying two drops of glacial acetic acid. Allow the paper to dry, spray with ninhydrin, and place in the oven as before. Determine which fractions contain amino acids, and save them for part B.

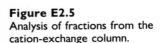

Figure E2.5
Analysis of fractions from the cation-exchange column.

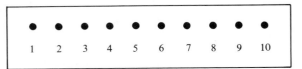

B. Separation and Identification of Amino Acids by Paper Chromatography

Obtain a sheet of 20×20 cm chromatography paper. Handle the paper only with gloves. The chromatogram is prepared by applying spots of each amino acid solution along one edge of the paper. The spots should be no closer than 2 cm from the bottom or side edges. It is best to lay the sheet on a piece of paper and mark the 2-cm line on the underlying paper on both sides of the plate. On the same sheet of paper, mark 2-cm intervals to use as guides for spot application. You should have room for nine different spots—five standard amino acids, the original unknown mixture, and up to three unknown fractions from the chromatography column. Be sure to prepare a "key" for the identity of each application (see Figure E2.6). Using small capillary tubes drawn to a fine point or microcapillary pipets, apply each standard amino acid provided. Each spot on the chromatogram should be prepared with four to six applications of an amino acid solution. This will ensure that there is enough amino acid to be detected after solvent development of the paper. The diameter of each application should be kept as small as possible (0.2 to 0.3 cm) so that the developed spots do not overlap. When amino acid fractions are available from the column, spot these on the paper in the same manner. You may have one, two, or three unknown fractions, depending on how many amino acids were present in the unknown mixture. Use the unknown fractions that contain the most amino acid, as judged by the deepest red-purple color after ninhydrin spray (part A).

After all of the spots have dried, roll the paper into a cylinder and staple along the edges. Do not overlap the paper at the stapled edges. Place the sheet in the chromatography jar containing acetonitrile and 0.1 M ammonium acetate (60:40), pH 4.0. Place the edge with the applied spots in the solvent, but be sure the solvent level is below the row of spots. Allow the solvent to rise to within 1 cm of the top of the paper (about 1 hour), remove it from the chromatography chamber, and dry it in a hood. Spray lightly with ninhydrin (**Hood**!) and develop the color in a 110°C oven for 10 minutes.

Figure E2.6
Preparation of a sheet for paper chromatography.

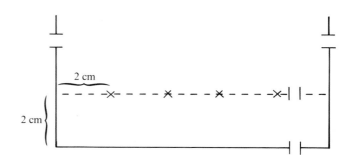

IV. Analysis of Results

A. Ion-Exchange Chromatography

Determine which fractions contain amino acids by studying the papers sprayed with ninhydrin. How many different amino acids are present in your unknown, and to what class (acidic, basic, neutral) does each belong? Explain the order of elution of each amino acid. From this information, is it possible to identify positively each amino acid that was eluted from your column?

B. Paper Chromatography

Prepare a table of R_f values of the standard amino acids. Review Chapter 3 concerning the calculation of R_f. Describe in your notebook the color of each ninhydrin spot. Since the color depends on the amino acid, this helps in the identification of an unknown amino acid. Calculate the R_f values for the unknown amino acids. You should now be able to identify the amino acids present in your unknown mixture. Can you identify the components of your unknown mixture using only the paper chromatography data?

V. Questions and Problems

► 1. Write the predominant ionic structure of lysine that would exist in aqueous solutions at the following pH values: pH 2, pH 5, and pH 11.

► 2. Draw all of the possible ionization states of glutamic acid.

► 3. Predict the relative order of R_f values for the amino acids in the following mixture: Ser, Lys, Leu, Val, and Ala. Assume that the separation was carried out on paper with the solvent system described in this experiment.

► 4. Calculate the net charge on each of the following amino acids at the given pH.

> alanine, pH 3.0
> aspartic acid, pH 4.0
> lysine, pH 8.0
> serine, pH 6.0
> histidine, pH 9.0

► 5. A mixture of amino acids is subjected to paper electrophoresis at pH 6.0. The amino acids present in the mixture are Val, Glu, Phe, Ser, and Lys. Which amino acids will migrate toward the cathode? Toward the anode?

6. Compare the effectiveness of the two techniques you used to separate the unknown amino acid mixture. Which one leads to better resolu-

tion of the mixture? Which technique is better for identification of the individual amino acids?

▶ 7. Predict the order of elution of the following mixture of amino acids from a cation-exchange column: Glu, Ala, Ser, Pro, and Arg. Assume the procedure was the same as in this experiment. Explain your answer.

▶ 8. For each pair of amino acids listed below, predict which one would be first to elute from a cation-exchange column. Assume the buffer pH begins at 3.25 and is increased in a stepwise fashion to pH 6.

(a) Asp, Glu   (d) Lys, Phe

(b) Ser, Val   (e) Leu, Val

(c) Phe, Ala   (f) Gly, Ala

VI. References

S. Black and M. Coon, *Anal. Biochem.* **121,** 281–285 (1982). "HPLC of Phenylthiohydantoin-Amino Acids."

Y. Cheng and N. Dovichi, *Science* **242,** 562–564 (1988). "Subattomole Amino Acid Analysis by Capillary Zone Electrophoresis and Laser-Induced Fluorescence."

T. Cooper, *The Tools of Biochemistry* (1977), John Wiley & Sons (New York), pp. 136–168. An entire chapter on the theory and practice of ion-exchange chromatography.

W. Hancock, Editor, *CRC Handbook of HPLC for the Separation of Amino Acids, Peptides and Proteins* (1984), CRC Press (Boca Raton, FL). Contains many references and methods for analysis.

F. Lottspeich and A. Henschen, in *HPLC in Biochemistry,* A. Henschen, K. Hupe, F. Lottspeich, and W. Voelter, Editors (1985), VCH (Weinheim), pp. 139–216. "Amino Acids, Peptides and Proteins."

C. Mathews and K. van Holde, *Biochemistry* (1990), Benjamin/Cummings (Redwood City, CA), pp. 133–170. An introduction to amino acids and analysis.

J. Rawn, *Biochemistry,* 2nd ed. (1989), Neil Patterson Publishers (Burlington, NC), pp. 61–69. Analysis of amino acids.

R. Somack, *Anal. Biochem.* **104,** 464–468 (1980). "Complete Phenylthiohydantoin Amino Acid Analysis by High-Performance Liquid Chromatography on ULTRASPHERE-Octadecyltrimethyloxysilane."

L. Stryer, *Biochemistry,* 3rd ed. (1988), W. H. Freeman (San Francisco), pp. 15–42. A brief discussion of ion-exchange chromatography.

D. Voet and J. Voet, *Biochemistry* (1990), John Wiley & Sons (New York), pp. 81–87. Amino acid structure and analysis.

EXPERIMENT 3

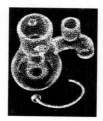

Purification and Characterization of α-Lactalbumin, a Milk Protein

■ **Recommended Reading**

Chapter 2; Chapter 3, Sections A, D, E, F, G, H; Chapter 4, Sections A, B; Chapter 5, Section A; Chapter 7, Sections A, B.

■ **Synopsis**

α-Lactalbumin is one of the major proteins found in milk. Its function is to regulate the synthesis of the milk sugar lactose. It can be isolated readily from milk and purified by Sephadex gel filtration or affinity chromatography. This experiment introduces the student to a series of protein purification steps that yield α-lactalbumin in sufficient amounts and purity for further physical characterization by UV spectroscopy and polyacrylamide gel electrophoresis.

I. Introduction and Theory

Protein purification is an activity that has occupied the time of biochemists throughout the history of biochemistry. In fact, a large percentage of the biochemical literature is a description of how specific proteins have been separated from the thousands of other proteins and biomolecules in tissues, cells, and biological fluids. Biochemical investigations of all biological processes require, at some time, the isolation, purification, and characterization of a protein. Of course, there is no single technique or sequence of techniques that can be followed to purify all proteins. Most investigators approach the problem by trial and error. Fortunately, the experiences and discoveries of hundreds of biochemists have been combined so that, today, a general and somewhat systematic approach is used for protein purification. The best purification procedure is one that yields a maximum amount of the purified protein in a minimum amount of time. The following discussion outlines the basic steps that must be considered in order to develop a protein purification scheme (Table E3.1).

Development of Protein Assay

First and foremost in any protein purification scheme is the development of an assay for the protein. This procedure, which may have a physical, chemical, or biological basis, is necessary in order to determine quantitatively and/or qualitatively the presence of the specific protein. During the early stages of purification, the particular protein desired must be distinguished from thousands of other proteins present in crude cell homogenates. The desired protein may be less than 0.1% of the total protein content of the crude extract. If the desired protein is an enzyme, the obvious assay will be based on biological function, that is, a measurement of the enzymatic activity after each isolation-purification step. The development of enzymatic assays is considered in greater detail in Experiment 6. If the protein to be isolated is not an enzyme or if the biological activity of the protein is unknown, physical or chemical methods must be used. One of the most useful analytical methods is electrophoresis (Chapter 4).

Source of the Protein

The selection of a source from which the desired protein is to be isolated should be considered. If the objective is simply to obtain a certain quantity of a protein for further study, you would choose a source that contains large amounts of the protein. The best choice is an organ from a

Table E3.1
Typical Sequence for the Purification of a Protein

1. Develop an assay for the desired protein.
2. Select the biological source of the protein.
3. Release the protein from the source and solubilize it in an aqueous buffer system.
4. Fractionate the cell components by physical methods (centrifugation).
5. Fractionate the cell components by differential solubility.
6. Ion-exchange chromatography.
7. Adsorption chromatography.
8. Gel filtration.
9. Affinity chromatography.
10. Isoelectric focusing.

large animal that can be obtained from a local slaughterhouse. Microorganisms are also good sources because they can be harvested in large quantities. If, however, you desire a specific protein from a specific type of cell, tissue, cell organelle, or biological fluid, the source is limited. If the desired protein is known to be located in an organelle or subcompartment of the cell, partial purification of the protein is achieved by isolating the organelle. (See Chapter 7, Figure 7.11.) If the protein is, instead, in the soluble cytoplasm of the cell, it will remain dissolved in the final supernatant obtained after centrifugation at $100,000 \times g$.

With the introduction of new technology in DNA recombinant research, it is possible to transfer via a plasmid the specific gene for a protein into another organism (usually *Escherichia coli* cells) and allow that organism to synthesize the desired protein (see Experiments 19, 20, 21).

Preparation of Crude Extract

Once the protein source has been selected, the next step is to release the desired protein from its natural cellular environment and solubilize it in aqueous solution. This calls for disruption of the cell membrane without damage to the cell contents. Proteins are relatively fragile molecules and only gentle procedures are allowed at this stage. The gentlest methods for cell breakage are osmotic lysis, gentle grinding in hand-operated glass homogenizers, and disruption by ultrasonic waves. These methods are useful for "soft tissue" as found in green plants and animals. When dealing with bacterial cells, where rigid cell walls are present, the most effective methods are grinding in a mortar with an inert abrasive such as sand or alumina, treatment with lysozyme, or both. Lysozyme is an enzyme that catalyzes the hydrolysis of polysaccharide moieties present in cell walls. When very resistant cell walls are encountered (yeast, for example), the French press must be used. Here the cells are disrupted by passage, under high pressure, through a small hole.

Osmotic lysis consists of suspending cells in a solution of relatively high ionic strength. This causes water inside the cell to diffuse out through the membrane. The cells are then isolated by centrifugation and transferred to pure water. Water rapidly diffuses into the cell, bursting the membrane.

Another alternative, gentle grinding, is best accomplished with a glass or Teflon homogenizer. This consists of a glass tube with a close-fitting piston. Several varieties are shown in Figure E3.1A, B. The cells are forced against the glass walls under the pressure of the piston, and the cell components are released into an aqueous solution.

Ultrasonic waves, produced by a sonicator, are transmitted into a suspension of cells by a metal probe (Figure E3.1C). The vibration set up by the ultrasonic waves disrupts the cell membrane, releasing the cell components into the surrounding aqueous solution.

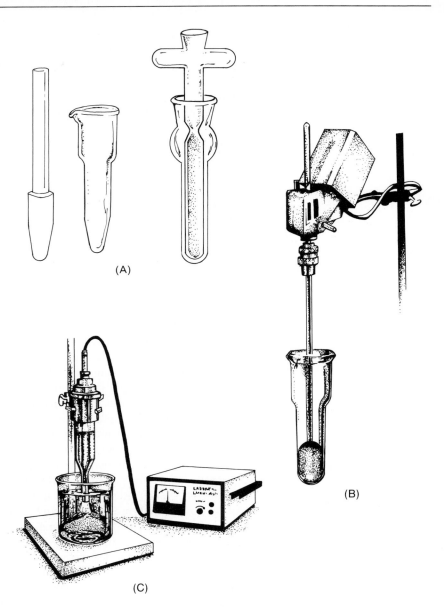

Figure E3.1
Tools for the preparation of a crude cell extract.   (A) Hand-operated homogenizers, courtesy of Ace Glass, Inc., Vineland, NJ.   (B) Homogenizer with electric motor, courtesy of VWR Scientific, Division of Univar.   (C) Sonicator, courtesy of
Curtin Matheson Scientific, Inc.

A less gentle, but widely used, solubilization device is the common electric blender. This method may be used for plant or animal tissue, but it is not effective for disruption of bacterial cell walls.

Since so many cell disruption methods are available, the experimenter must, by trial and error, find a convenient method that yields the maximum quantity of the protein with minimal damage to the molecules.

Stabilization of Proteins in a Crude Extract

Continued stabilization of the protein must always be considered in protein purification. While the protein is inside the cell, it is in a highly regulated environment. Cell components in these surroundings are protected against sudden changes in pH, temperature, or ionic strength and against oxidation and enzymatic degradation. Once the cell wall barrier is destroyed, the protective processes are no longer functional and degradation of the desired protein is likely to begin. An artificial environment that mimics the natural one must be maintained so that the protein retains its chemical integrity and biological function throughout the purification procedure. What factors are important in maintaining an environment in which proteins are stable? Although numerous factors must be considered, the most critical are (1) ionic strength and polarity, (2) pH, (3) the presence or absence of metal ions, (4) oxidation, (5) endogenous proteases, and (6) temperature.

The standard cellular environment is, of course, aqueous; because of the presence of inorganic salts, though, its ionic strength is relatively high. Addition of KCl, NaCl, or $MgCl_2$ to the cell extract may be necessary to maintain this condition. Proteins that are normally found in the hydrophobic regions of cells (i.e., membranes) are generally more stable in aqueous environments in which the polarity has been reduced by addition of 1 to 10% glycerol or sucrose.

The pH of a biological cell is controlled by the presence of natural buffers. Since protein structure is often irreversibly altered by extremes in pH, a buffer system must be maintained for protein stabilization. The importance of proper selection of a buffer system cannot be overemphasized. The criteria that must be considered in selecting a buffer have been discussed in Chapter 2. Obviously, a "perfect" buffer system is not available, but some are much better (more natural) than others. Even at this advanced stage in our understanding of solution buffering, trial and error are still essential in selecting an effective, noninterfering buffer system. For most cell homogenates at physiological pH values, Tris and phosphate buffers are widely used.

The presence of metal ions in mixtures of biomolecules can be both beneficial and harmful. Metal ions such as Na^+, K^+, Ca^{2+}, Mg^{2+}, and Fe^{3+} may actually increase the stability of dissolved proteins. Heavy metal ions such as Ag^+, Cu^{2+}, Pb^{2+}, and Hg^{2+} are deleterious, particularly to proteins that depend on sulfhydryl groups for structural and functional integrity. The main sources of contaminating metals are buffer salts, water used to make buffer solutions, and metal containers and equipment. To avoid heavy metal contamination, you should use high-purity buffers, glass-distilled water, and glassware specifically cleaned to remove extraneous metal ions (Chapter 1). If metal contamination still persists, a chelating agent such as ethylenediaminetetraacetic acid (EDTA, $1 \times 10^{-4}M$) may be added to the buffer. This, of course, should not be used if the desired protein is an enzyme that is inhibited by EDTA.

Many proteins are susceptible to oxidation. This is especially a problem with proteins having free sulfhydryl groups, which are easily oxidized and converted to disulfide bonds. A reducing environment can be maintained by adding mercaptoethanol, cysteine, or dithiothreitol ($1 \times 10^{-3}M$) to the buffer system.

Many biological cells contain degradative enzymes (proteases) that catalyze the hydrolysis of peptide linkages. In the intact cell, functional proteins are protected from these destructive enzymes because the enzymes are stored in cell organelles (lysosomes, etc.) and released only when needed. The proteases are freed upon cell disruption and immediately begin to catalyze the degradation of protein material. This detrimental action can be slowed by the addition of specific protease inhibitors such as phenylmethylsulfonyl fluoride or certain bioactive peptides. These inhibitors are to be used with extreme caution because they are potentially toxic.

Many of the above conditions that affect the stability of proteins in solution are dependent on chemical reactions. In particular, metal ions, oxidative processes, and proteases bring about chemical changes in proteins. It is a well-accepted tenet in chemistry that lower temperatures slow down chemical processes. We generally assume that proteins are more stable at low temperatures. Although there are a few exceptions to this, it is fairly common practice to carry out all procedures of protein isolation under reduced-temperature conditions (0 to 4°C).

After cell disruption, gross fractionation of the properly stabilized, crude cell homogenate may be achieved by physical methods, specifically centrifugation. Figure 7.11, Chapter 7, outlines the stepwise procedure commonly used to separate subcellular organelles such as nuclei, mitochondria, lysosomes, and microsomes.

Separation of Proteins Based on Solubility Differences

Proteins are soluble in aqueous solutions primarily because their charged and polar amino acid residues are solvated by water. Any agent that disrupts these protein-water interactions decreases protein solubility because the protein-protein interactions become more important. Protein-protein aggregates are no longer sufficiently solvated, and they precipitate from solution. Because each specific type of protein has a unique amino acid composition and sequence, the degree and importance of water solvation vary from protein to protein. Therefore, different proteins precipitate at different concentrations of precipitating agent. The agents most often used for protein precipitation are (1) inorganic salts, (2) organic solvents, (3) polyethylene glycol (PEG), (4) pH, and (5) temperature. The most commonly used inorganic salt, ammonium sulfate, is highly solvated in water and actually reduces the water available for interaction with protein. As ammonium sulfate is added to a protein solution, a concentration of salt is reached at which there is no longer sufficient water present to maintain a particular type of protein in solution. The protein precipitates or is "salted out" of solution. The concentration of ammonium sulfate at which the desired protein precipitates from solution cannot be calculated but must be established by trial and error. In practice, ammonium sulfate precipitation is carried out in stepwise intervals. For example, a crude cell extract is treated by slow addition of dry, solid, high-purity ammonium sulfate in order to achieve a change in salt concentration from 0 to 25% in ammonium sulfate, is gently stirred for up to 60 minutes, and is subjected to centrifugation at $20,000 \times g$. The precipitate that separates upon centrifugation and the supernatant are analyzed for the desired protein. If the protein is still predominantly present in the supernatant, the salt concentration is increased from 25 to 35%. This process of ammonium sulfate addition and centrifugation is continued until the desired protein is salted out. Although other inorganic salts are used, ammonium sulfate has many advantages, including high solubility in water and less chance of causing protein denaturation.

Organic solvents also decrease protein solubility, but they are not as widely used as ammonium sulfate. They are thought to function as precipitating agents in two ways: (1) by dehydrating proteins, much as ammonium sulfate does, and (2) by decreasing the dielectric constant of the solution. The organic solvents used (which, of course, must be miscible with water) include methanol, ethanol, and acetone. Organic solvents must be used with care because they are more likely than ammonium sulfate to cause protein denaturation.

A relatively new method of selective protein precipitation involves the use of nonionic polymers. The most widely used agent in this category is polyethylene glycol. The polymer is available in a variety of molecular weights ranging from 400 to 7500. The biochemical literature reports successful use of different sizes, but lower-molecular-weight polymer has been shown to be the most specific. The principles behind the action of PEG as a protein precipitating agent are not completely understood. Possible modes of action include (1) complex formation between protein and polymer and (2) exclusion of the protein from part of the solvent (dehydration) followed by protein aggregation and precipitation. This method of protein precipitation should receive considerable attention in the future because it has several advantages over other fractionation techniques. Possibly the most attractive advantage of PEG is its ability to fractionate proteins on the basis of size and shape as in gel filtration. Although it is too early to state a general rule, most fractionation experiments with PEG show that the larger the protein, the less soluble it is in highly concentrated polymer solution.

Finally, changes in pH and temperature have been used effectively to promote selective protein precipitation. A change in the pH of the solution alters the ionic state of a protein and may even bring some proteins to a state of charge neutrality. Charged protein molecules tend to repel each other and remain in solution; however, neutral protein molecules do not repel each other, so they tend to aggregate and precipitate from solution. A protein is least soluble in aqueous solution when it has no net charge, that is, when it is isoelectric. This characteristic can be used in protein purification, since different proteins usually have different isoelectric pH values.

An increase in temperature generally causes an increase in the solubility of solutes. This general rule is followed by most proteins up to about 40°C. Above this temperature, however, many proteins aggregate and precipitate from solution. If the protein of interest is heat stable and still water soluble above 40°C, a major step in protein purification can be achieved because most other proteins precipitate below 40°C and can be removed by centrifugation.

With so many different protein-fractionating methods available, it becomes difficult to choose the correct one. As a guideline, it is fair to state that all precipitation methods discussed will provide some purification for most proteins. Their effectiveness is very nearly the same, so the selection should be based on the convenience of the method and the stability of the desired protein under the experimental conditions.

Selective Techniques in Protein Purification

After gross fractionation of proteins, as discussed above, more refined methods with greater resolution can be attempted. These methods, in order of increasing resolution, are ion-exchange chromatography, adsorption chromatography, gel filtration, affinity chromatography, and isoelectric focusing. Since the basis of protein separation is different for each of these techniques, it is often most effective and appropriate to use all of the techniques in the order given.

Chromatographic methods for protein purification have been discussed in Chapter 3. However, we must consider another important topic, preparation of protein solutions for chromatography. Fractionation of heterogeneous protein mixtures by inorganic salts, organic solvents, or PEG usually precedes ion-exchange chromatography. The presence of the precipitating agents will interfere with the later chromatographic steps. In particular, the presence of ammonium sulfate increases the ionic strength of the protein solution and damps the ionic protein–ion exchange resin interactions. Procedures that are in current use to remove undesirable small molecules from protein solutions include ultrafiltration (Chapter 2), dialysis (Chapter 2), and gel filtration (Chapter 3). All three methods effectively remove small molecules without serious damage to or loss of proteins.

Chromatography is now and will continue to be the most effective method for selective protein purification. The more conventional methods (ion exchange and gel filtration) rely on rather nonspecific physicochemical interactions between a stationary support and protein molecule. These techniques, which separate proteins on the basis of net charge, size, and polarity, do not have a high degree of specificity. All of these methods must be attempted during the purification of a new protein.

The highest level of selectivity in protein purification is offered by affinity chromatography—the separation of proteins on the basis of specific biological interactions (Chapter 3). A modified form of affinity chromatography, immobilized metal-ion affinity chromatography (IMAC), was introduced by Porath in 1975. In this method, the bioactive support consists of metal ions chelated to an insoluble matrix. A typical affinity support for IMAC, iminodiacetic acid (IDA) covalently linked to agarose, is shown in Figure E3.2. Proteins are separated according to their individual abilities to bind to the immobilized metal ions. Electron-donating amino acid side chains on the surfaces of protein molecules interact with the metal ions. Important electron-rich groups on the protein include the imidazole ring of histidine, the indole ring of tryptophan, and the sulfhydryl group of cysteine. These amino acid side chains are able to displace weakly bound ligands such as water. A typical immobilized metal ion as shown in Figure E3.2 may have up to three adsorption sites for protein binding. Proteins bound to affinity supports may be displaced by free ligands, such as imidazole, ammonia, pyrophosphate, or amino acids, that bind to the metal ion.

Figure E3.2
Structure of metal ion–IDA-agarose affinity support. Me = metal ion such as Cu(II), Zn(II), Ni(II), or Fe(III).

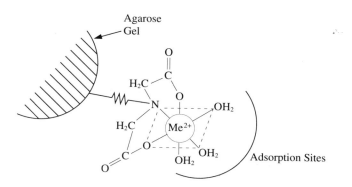

To some individuals, especially those who recall their experiences in organic chemistry laboratory, the ultimate step in purification of a molecule is crystallization. The desire to obtain crystalline protein has long been powerful and many proteins have been crystallized. However, there is a common misconception that the ability to form crystals of a protein ensures that the protein is homogeneous. For many reasons (entrapment of contaminants within crystals, aggregation of protein molecules, etc.), the ability to crystallize a protein should not be used as a criterion of purity. However, protein crystallization is a worthwhile endeavor in the final stages of protein purification because it would be expected to remove minor impurities. The interest in protein crystals today has its origin in the demand for X-ray crystallographic analysis of protein structure.

Purification of α-Lactalbumin

Milk is composed of several proteins, including the caseins (phosphoproteins), β-lactoglobulin, α-lactalbumin, albumin, and immunoglobulins. A crude preparation of α-lactalbumin was first obtained from milk in 1899. Since that time, several investigators have developed procedures for isolating and purifying the protein. α-Lactalbumin has a molecular weight of 15,500 and contains 129 amino acid residues. It is present in milk at concentrations averaging 1 mg/mL and was at first thought to serve a nutritional function. We now know that it is an essential component of the lactose synthase system. Thorough discussions of the history and biological role of α-lactalbumin are available in the literature (Ebner, 1970; Hall and Campbell, 1988).

The primary proteins in milk are the caseins. These may be removed from milk by acid and heat-induced precipitation at pH 4.6 followed by centrifugation. The remaining supernatant (called whey) contains primarily α-lactalbumin and β-lactoglobulin. These two proteins may be separated by gel filtration or with a Cu(II)–IDA-agarose affinity column.

The specific amino acid side chains on α-lactalbumin responsible for binding to the metal support are not known; however, α-lactalbumin is a metalloprotein. Under physiological conditions, it carries one Ca(II) per molecule; hence, there are metal binding sites on the protein. Column-bound α-lactalbumin is eluted by a solution of the free ligand imidazole. A flowchart outlining these procedures is shown in Figure E3.3.

Analysis and Characterization of α-Lactalbumin

Recall from the previous discussion of protein purification that an important step is the development of a specific assay for the desired protein. The assay must be specific so that the desired protein can be detected both quantitatively and qualitatively in a solution containing thousands of other proteins. This is difficult for α-lactalbumin because it has no unique

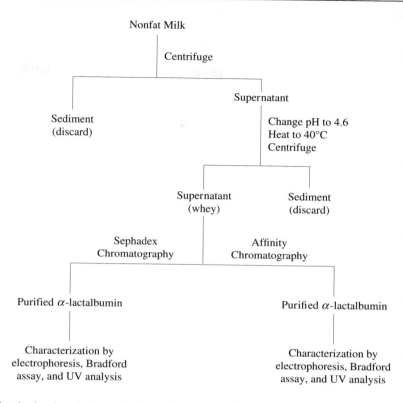

Figure E3.3
Flowchart for isolation and purification of α-lactalbumin from milk.

physical, chemical or biological property that can be readily measured. Measuring its biological activity as a modifier protein in the lactose synthase system is the most specific way to analyze α-lactalbumin; however, the assay, which requires a coupled enzyme system and preparation of numerous solutions, is too time-consuming to complete in this laboratory period.

Most proteins have a broad characteristic absorption spectrum centered at about 280 nm. The major absorption is due to the presence of aromatic moieties in the amino acids phenylalanine, tyrosine, and tryptophan. During the α-lactalbumin purification described in this experiment, you will monitor the process by measuring the absorption at 280 nm (A_{280}) of column fractions to be sure the experiment is proceeding correctly. You must recognize that you are measuring not the concentration or presence of α-lactalbumin specifically but the total amount of all protein present.

The procedure outlined here will provide a final product that has an ultraviolet absorption at 280 nm, indicating that protein material is likely present in solution. However several questions remain unanswered:

1. How much protein was isolated?

2. What is the purity of the protein sample?

3. What is the identity of the protein?

Several useful methods for the quantitative determination of protein solutions were discussed in Chapter 2. Two of those methods, the Bradford protein assay and the spectrophotometric assay, will be applied to the α-lactalbumin solutions. Neither of these assays is specific for a certain type of protein; rather they both estimate total protein content.

It is unlikely that the protein fractions from this experiment contain a single type of protein. How many different proteins are present? What is the relative abundance of each protein? Is α-lactalbumin the predominant protein in the isolated fractions? These questions may be answered by analysis of the isolated fractions by electrophoretic methods. The technique of polyacrylamide gel electrophoresis will be introduced and applied to the column-purified fractions, crude whey fraction, and standard α-lactalbumin.

The UV-VIS absorption spectrum of a protein is a characteristic property of that specific protein and can aid in the identification of an unknown. However, it is helpful to have available standard spectra of known proteins for comparison. The α-lactalbumin fraction will first be characterized by measurement of the spectrum from 240 to 340 nm. From this, the absorption coefficient is calculated using the Beer-Lambert relationship. Finally, the A_{280}/A_{290} ratio is calculated; this may be used as an indicator of the purity of the isolated α-lactalbumin.

Overview of the Experiment

The complete isolation, purification, and characterization of α-lactalbumin as described here require about 9 hours. The preparation of whey is completed in approximately 3 hours and the chromatographic step (Sephadex or affinity) requires another 3 hours. The analysis procedures require 3 hours. The whey fraction may be stored in a freezer for several weeks if desired. As an alternative to isolation, students may be provided commercial α-lactalbumin, which is then further purified by chromatography. The time for gel electrophoresis can be greatly reduced if commercially prepared gel plates are used.

II. Materials and Supplies

A. Preparation of Milk Whey

Nonfat milk, raw or pasteurized, 100 mL

HCl, 12 M

HCl, 0.5 M

Refrigerated centrifuge and tubes, capable of 16,000 × g

pH meter

Heater/stirrer

Syringe cartridge filter, 0.45 μm

B. Chromatography

SEPHADEX CHROMATOGRAPHY

Sephadex G-50, fine mesh, in 0.02 M Tris, pH 7.0 (100 mL slurry)

Glass column, 2.5 × 40 cm

UV monitor and fraction collector

AFFINITY CHROMATOGRAPHY

IDA-agarose, prepacked in column or provided in slurry form

Buffer A: 0.020 M Tris, 0.5 M NaCl, pH 7.0

Buffer B: 0.020 M Tris, 0.5 M NaCl, 0.020 M imidazole, pH 7.0

$CuSO_4$ in H_2O, 0.1 M

UV monitor and fraction collector

Quartz cuvettes

UV-VIS spectrometer

C. Electrophoresis of α-Lactalbumin

Glass plates (10 × 20 cm), spacers, comb, tape (Scotch electrical, 2 inches in diameter), and clamps

Power supply

α-Lactalbumin samples (standards and fractions from parts A and B)

Stock solutions:

  A. Tris-glycine buffer, pH 8.3

  B. Tris buffer containing TEMED, pH 8.9

  C. 28% acrylamide plus 0.74% bis-acrylamide

  D. Tris buffer containing TEMED, pH 6.9

  E. 10% acrylamide plus 2.5% bis-acrylamide

  F. Riboflavin, 4 mg/dL H_2O

  G. Ammonium persulfate, 0.14% in H_2O; prepare just before use

  H. Sucrose, 40 g/dL H_2O

  I. Bromphenol blue tracking dye, 0.02% in H_2O

Fluorescent lamp

Staining with Coomassie Blue:

  Dye solution, 0.25% Coomassie Blue in methanol–acetic acid–water (5:1:5)

  Destaining solution, acetic acid–methanol–water (7:7:86)

Staining with silver:

  Prefixing solution 1, methanol–water–acetic acid (50:40:10)

  Prefixing solution 2, water–acetic acid–methanol (88:7:5)

Fixing solution, 10% glutaraldehyde in H_2O

Silver nitrate, 0.1% in H_2O

Developing solution, 50 μL of 37% formaldehyde in 1 dL of 3% sodium carbonate

Dithiothreitol solution, 4 μg/mL H_2O

Citric acid, 2.3 M

D. Bradford Protein Assay

α-Lactalbumin from parts A and B

Bovine gamma globulin standard, 0.1 mg/mL in H_2O

Bradford dye reagent. This is a commercially available mixture of Coomassie Brilliant Blue G-250 dye, phosphoric acid, and methanol. One volume of dye reagent should be diluted with four volumes of distilled water before use. Filter and store in a glass container at room temperature. The dilute solution is stable for 2 weeks.

Spectrophotometer for reading A_{595} with glass cuvettes (1 or 3 mL)

Test tubes, 10 × 100 mm

III. Experimental Procedure

A. Preparation of Milk Whey

Obtain 10 mL of nonfat milk and centrifuge at 16,000 × g for 45 minutes in a refrigerated centrifuge. (During this step, proceed directly to the chromatography step.) Decant the supernatant into a small beaker. The sediment and floating lipid layer should remain in the centrifuge tube and be discarded. Adjust the pH of the supernatant to 4.6 in two steps:

1. Adjust to approximately pH 5 with dropwise addition of 12 M HCl.
2. Adjust to the desired pH (4.5) with dropwise addition of 0.5 M HCl.

Heat the coagulated solution with constant stirring at 40°C for 30 minutes. Centrifuge at 16,000 × g for 30 minutes. Collect the whey (supernatant) and clarify by filtering through a 0.45-μm syringe cartridge filter. The whey is stored for short periods of time (hours) in crushed ice or for longer periods in a freezer.

B. Chromatography

SEPHADEX CHROMATOGRAPHY

Packing the Sephadex Column

Sephadex G-50 has been prepared for your use. The dry beads have been preswollen in water for several hours, fine particles that may slow column flow have been removed, and the gel has been equilibrated in Tris buffer. If chromatography columns are not available, one can easily be constructed from a glass tube (2.5 × 40 cm), cork stoppers, Tygon tubing, a screw clamp, and small glass tubing as shown in Figure E3.4. With the screw clamp closed, add 10 to 15 mL of Tris buffer and insert a small piece of cotton into the bottom of the column. Force air bubbles out of the cotton with a long glass rod. Pour a well-mixed slurry of Sephadex G-50 into the column. Open the screw clamp to allow a slow column flow. As the gel settles into a bed, add more slurried G-50 until the settled bed reaches a height of about 30 cm. During this packing pro-

Figure E3.4
A column for chromatography.

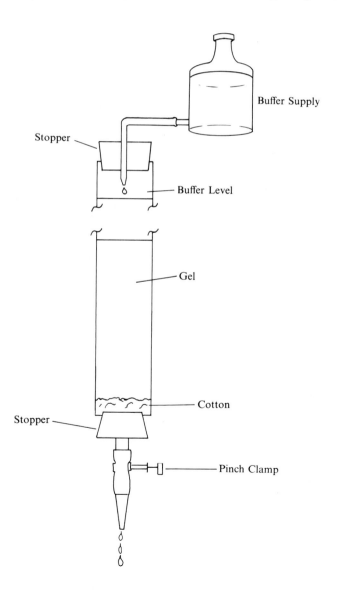

cess, never allow the column to run dry. If more buffer solution is necessary, add it carefully without disturbing the gel. After a bed of desired height has been prepared, continue eluting the column with buffer. Cut a piece of filter paper into a circle with a diameter of 2 to 2.3 cm and put it on top of the Sephadex bed. Be careful not to disturb the gel at the top of the column.

Sample Application

Measure out 4–5 mL of whey for application to the column. If you are using commercial α-lactalbumin, prepare a solution of 3–4 mg of protein in 1–2 mL of Tris buffer. After the eluting solvent has just drained into the top of the G-50 column, add the α-lactalbumin solution dropwise to the top of the gel. Use a pipet and avoid disturbing the top surface of the gel. After the protein solution has just entered the gel, slowly and carefully add 10 mL of buffer to the top of the column without disturbing the gel surface and allow it to penetrate the surface of the column. Then, immediately add buffer to fill the column. Set up an eluting solvent supply as shown in Figure E3.4. This will allow solvent to enter the column at the same rate as solvent elutes from the column.

Column Development, Fraction Collection,
and Analysis of Fractions

The stopper in the top of the column must be airtight. Adjust the screw clamp so that the flow rate of the column is about 5 to 10 mL/hr. Collect 2-mL fractions in test tubes in a fraction collector. Transfer the contents of every other fraction to a quartz cuvette and measure the absorbance at 280 nm. For a blank, use Tris buffer. Record the A_{280} values in your notebook. When you begin to detect protein eluting from the column ($A_{280} > 0$), measure the A_{280} of every fraction. On a single sheet of graph paper, plot A_{280} vs. fraction number. Continue collecting fractions until no more protein is eluted (about 30 fractions). Save all fractions that have A_{280} over 0.2.

When you are finished with the Sephadex column, pour the gel into a container labeled "used Sephadex G-50." The gel may be recycled and used later.

AFFINITY CHROMATOGRAPHY

Obtain a prepacked column and clamp it to a ring stand. If you must prepare your own column of IDA-agarose, use a 1 × 6–8 cm column. Pour in about 2 mL of the IDA-agarose slurry. (Be sure the column outlet is closed.) Allow the gel to settle to a column 1–2 cm high. Protect the surface of the gel by allowing a small circle of filter paper to settle onto the top. Allow most of the solution to pass through the column, close the outlet, and add buffer A to fill the column.

Wash the column (prepacked or handmade) with 10 mL of buffer A. Any flow rate up to 2 mL/min is permissible. Since this eluent will be discarded, it may be collected in a single container. *Do not allow the column to go dry.* To load the column with copper, drain buffer A to a level just slightly above the gel top. Add 0.5 mL of 0.1 M CuSO$_4$ solution to the column, and allow it to enter the gel. Once all the CuSO$_4$ is in the gel, immediately add buffer A, adjust the flow rate to 2 mL/min, and continue to wash the column until all excess Cu(II) has been eluted (about 25 mL). The resin will change from white to a pale blue color.

Attach the column outlet to a UV flow cell, fraction collector, and recorder, if available. Alternatively, collect fractions by hand. Allow buffer A to drain just to the top of the gel, and add 0.5 mL of whey directly to the top of the gel. Begin to collect 1-mL fractions in individual test tubes. Let the whey enter the gel, and add 1.0 mL of buffer A to rinse the inside of the column. Allow this to enter the gel; then fill the column with buffer A, and continue to collect 1-mL fractions. Begin to take A_{280} measurements of each fraction by transferring the contents of each fraction into a quartz cuvette. Buffer A should be used to zero the spectrophotometer. Begin the preparation of a graph of A_{280} (y axis) vs. fraction number (x axis). Continue to wash the column with buffer A until the A_{280} is approximately 0 or until the recorder has returned to the baseline. This will require 20–25 mL of buffer A. To elute α-lactalbumin, pass buffer B through the column. Continue to collect 1-mL fractions, measure the A_{280} of each, and add to the graph. This time use buffer B to zero the spectrophotometer. Elute with a total of 10 mL of buffer B. The gel may be reused after removing Cu(II) with 0.2 M EDTA.

C. Electrophoresis of α-Lactalbumin

PREPARATION OF THE SLAB POLYACRYLAMIDE GEL

 Caution

Acrylamide in the unpolymerized form is a skin irritant and a potential neurotoxin. Wear gloves and a mask while weighing the dry powder. Do not breathe the dust. Prepare all acrylamide solutions in the hood. Do not mouth pipet any solutions used for gel formation or staining.

Do not touch the electrophoresis chamber or wires while the electrophoretic operation is in progress. Voltages may be as high as 300 V and shocks may be fatal.

Wash the glass plates thoroughly with detergent and warm water. Rinse well with distilled water and then 95% ethanol. Wipe dry with tissue. Prepare gels 0.7–1.0 mm thick by following the directions provided with the electrophoresis apparatus (see Chapter 4). Prepare the following working solutions. The letters refer to the stock solutions listed under part A above:

1. Separation gel

   5.0 mL of solution B

   10.0 mL of solution C

   5.0 mL of H_2O

II. Stacking gel

   1.0 mL of solution D

   2.0 mL of solution E

   1.0 mL of solution F

   4.0 mL of solution H

Prepare the separation gel by mixing 20.0 mL of the separation gel, solution I above, and 20.0 mL of ammonium persulfate solution (G). Mix well and immediately add between the glass plates with a Pasteur pipet or a syringe. Fill the space between the plates to a level about 15 mm from the top of the plates. Overlay the gel with a thin layer of water. Allow about 30 minutes for polymerization to occur. Remove H_2O from the top of the gel with a syringe or by pouring.

Form the stacking gel by pouring solution II between the plates on top of the separation gel. Fill the space to the top of the plates. Immediately insert the sample comb. Polymerize for 30 minutes under a fluorescent light. Carefully remove the comb and rinse the sample wells with distilled water from a syringe.

Apply each sample to a well made by the comb. Before application, each sample is mixed with an equal volume of sucrose solution (H) and 0.1 mL of the bromphenol tracking dye solution (I). A typical sample size is approximately 50 μL, containing about 20–100 μg of protein.

Run the slab gel in a vertical dimension. The current should be set at 25–30 mA. Allow the electrophoresis to proceed until the tracking dye is at the bottom of the gel. *Turn the power off* and remove the glass plate from the buffer chambers. Carefully separate the glass plates with a thin spatula and transfer the gel in one piece to a tray containing staining solution.

For Coomassie dye staining, allow the slab to soak for about 30 minutes. Remove the dye solution by decanting and cover the gel with destaining solution. Replace the destaining solution with fresh solution about every 30 minutes until background color is removed. The destaining procedure may require several days. Proteins will show up as dark blue bands on a nearly colorless background.

For silver staining, prepare the gel by soaking in prefixing solution 1 for 30 minutes, then in prefixing solution 2 for 30 minutes. The gel is then fixed with 10% glutaraldehyde for 30 minutes to retard protein movement or diffusion. Rinse the gel with distilled water, making several changes during a 2-hour interval. Soak in dithiothreitol solution for 30 minutes. Decant the solution and soak in 0.1% silver nitrate for 30 minutes. Rinse once with distilled water and twice with formaldehyde developing solution. Allow the gel to soak in the developing solution until the protein bands are readily visible. Stop the staining process with 5.0 mL of 2.3 M citric acid. Finally, rinse the stained gel several times with water. Proteins appear as gray-black bands on light gray background.

Measure the total length of the slab gel and the distance migrated by each protein and the tracking dye from the top. In your notebook, draw a picture of the gel showing protein bands.

D. Bradford Protein Assay

Set up 12 test tubes (10 × 100 mm, colorimetric tubes) and add water and proteins according to the top three rows of Table E3.2. Tube 1 is used as a blank and tubes 2 through 6 are for construction of a standard calibration curve. Tubes 7 to 10 are duplicates of two different concentrations of the isolated α-lactalbumin solution, and tubes 11 and 12 are two concentrations of whey. Water is added to give a final volume of 1.0 mL in each tube. Add 5.0 mL of dilute Bradford dye reagent to each tube and mix well by gentle inversion. After a period of at least 5 minutes, read A_{595} for each tube, using tube 1 as a blank. The tubes should be read within an hour after adding the dye. If absorbance readings are too low (< 0.05) or too high (> 1.0), repeat the assay with more or less protein fraction.

E. Ultraviolet Spectrum of α-Lactalbumin

Turn on the recording spectrophotometer and the UV lamp. Allow the instrument to warm up for about 15 minutes. Set the wavelength to 260 nm. During the warm-up period, prepare samples by transferring 1 or 3 mL of Tris buffer into a quartz cuvette and an equal volume of purified

Table E3.2
Procedure for the Bradford Protein Assay[1]

Reagents	Tube Number											
	1	2	3	4	5	6	7	8	9	10	11	12
Water	1.0	0.9	0.8	0.6	0.4	0.2	0.7	0.7	0.4	0.4	0.95	0.90
Standard												
gamma globulin	—	0.1	0.2	0.4	0.6	0.8	—	—	—	—	—	—
α-Lactalbumin	—	—	—	—	—	—	0.3	0.3	0.6	0.6	—	—
Whey	—	—	—	—	—	—	—	—	—	—	0.05	0.10
Dye solution	5.0 →											

Mix well, allow to sit at least 5 min, and read A_{595}

[1] Units are mL.

α-lactalbumin solution into a matched cell. Place the cuvette with buffer only in the sample beam and adjust the absorbance to zero. Remove the blank cuvette and replace with the cuvette containing α-lactalbumin. Read and record the absorbance at 260 nm. If the A_{260} is greater than 1.0, dilute the sample with a known volume of Tris buffer and repeat the reading.

Using the above procedure, read and record the absorbances at 280 and 290 nm.

Obtain the continuous spectrum of the purified α-lactalbumin sample from 240 to 340 nm. Do this by placing a cuvette containing Tris buffer in the reference beam and α-lactalbumin in the sample beam. Record the UV spectrum. Dilute the solution with a known amount of buffer if the recorder runs off scale. Record the spectrum with standard α-lactalbumin.

IV. Analysis of Results

A. and B. Isolation of α-Lactalbumin

Describe the appearance of the whey before chromatography. Study your experimentally derived plot of A_{280} vs. Sephadex or affinity column fraction number. How many major protein peaks are present? Which peak has the greatest amount of protein? Do you believe that each peak contains a single type of protein, or does it contain a mixture? If each peak represents a mixture of proteins, what one thing does each protein have in common?

The presence of α-lactalbumin in a fraction cannot be confirmed until further characterization. However, if you know that α-lactalbumin is among the smallest proteins in milk (molecular weight is 15,000), can you determine which protein peak contains α-lactalbumin?

C. Electrophoresis

Study the destained gel or diagram and estimate the number of protein components in each sample. Calculate the **relative mobility** of each protein band using Equation E3.1.

▶ $$\text{Relative mobility} = \frac{\text{distance moved by protein}}{\text{distance moved by dye}} \qquad \text{(Equation E3.1)}$$

Prepare a table of relative mobilities of all bands in the gel. Compare the profile for isolated α-lactalbumin to that of standard α-lactalbumin. Is α-lactalbumin the predominant protein in your preparation from milk?

In whey? Try to estimate the percentage of the isolated sample that is α-lactalbumin. Assume that all proteins on the gel stain to the same extent with the dye, even though this is probably not true. Approximately what percent of toal whey proteins is α-lactalbumin?

D. The Bradford Assay

Prepare a standard calibration curve using the data obtained for standard gamma globulin (Table E3.2, tubes 2 to 6). Plot the absorbance on the y axis and mg of standard protein per assay on the x axis. Use the standard curve to calculate the protein concentration in the isolated α-lactalbumin fractions (in units of mg/mL.) Compare your graph with Figure 2.14.

E. Ultraviolet Spectrum of α-Lactalbumin

CALCULATION OF PROTEIN CONCENTRATION

Calculate the ratio A_{280}/A_{260} and estimate the concentration of α-lactalbumin.

▶   Protein concentration (mg/mL) $= 1.55A_{280} - 0.76A_{260}$ (Equation E3.2)

Compare these results with the results from the Bradford assay. Explain any differences.

CALCULATION OF ABSORPTION COEFFICIENT

Use the Beer-Lambert relationship (Equation 3.3) to calculate the absorption coefficient in terms of $E^{1\%}$ at 280 nm.

▶ $A = Elc$ (Equation E3.3)

where

A = absorbance at 280 nm
E = absorption coefficient, $E^{1\%}$
l = path length (usually 1 cm)
c = concentration of protein, g/100 mL

The literature value for $E_{280}^{1\%}$ of purified bovine α-lactalbumin is 20.1.
Calculate the ratio A_{280}/A_{290}, and compare to the standard value of 1.30 for purified α-lactalbumin.

ULTRAVIOLET SPECTRUM

Compare the spectrum of isolated α-lactalbumin with that of standard α-lactalbumin. Describe and explain any differences.

V. Questions and Problems

▶ 1. What other purification techniques could be applied to the purification of α-lactalbumin? Study the following references for methods that have been applied to the isolation and purification of α-lactalbumin: W. G. Gordon and W. F. Semmett, *J. Amer. Chem. Soc.* **75,** 328 (1953). W. G. Gordon and J. Ziegler, *Biochem. Prep.* **4,** 16 (1955). U. Brodbeck, W. L. Denton, N. Tanahashi, and K. E. Ebner, *J. Biol. Chem.* **242,** 1391 (1967). F. J. Castellino and R. L. Hill, *J. Biol. Chem.* **245,** 417 (1970). L. Lindahl and H. Vogel, *Anal. Biochem.* **140,** 394 (1984).

▶ 2. What methods of analysis could you perform on the purified α-lactalbumin in order to determine the extent of purity?

▶ 3. What assumptions must be made about the relative mobility of bromphenol blue when used as a tracking dye?

▶ 4. What would be the effect of each of the following changes on the relative mobility of α-lactalbumin in disc gel electrophoresis?

(a) Lower the pH of all buffers.

(b) Increase the ionic strength of the buffers.

(c) Change the temperature of the electrophoretic operation.

(d) Increase the concentration of α-lactalbumin.

(e) Increase the concentration of bis-acrylamide in the gels.

(f) Decrease the electrophoretic running time.

▶ 5. Describe the chemistry involved in the Bradford assay.

▶ 6. Describe the procedure for obtaining a "difference spectrum" of isolated α-lactalbumin vs. standard α-lactalbumin.

▶ 7. Why do most protein solutions absorb light of 280 nm wavelength?

▶ 8. Comment on the validity of the following statement: "Ideally, the best Bradford calibration curve for this experiment would use purified α-lactalbumin rather than gamma globulin."

▶ 9. Absorbance measurements of column fractions at 280 nm were used in this experiment to detect the presence of protein material. Do you think this method of analysis could lead to a quantitative determination of protein concentration? What other biomolecules might interfere with this measurement?

VI. References

J. Bell and E. Bell, *Proteins and Enzymes* (1988), Prentice-Hall (Englewood Cliffs, NJ). An excellent text for protein study.

R. Boyer, *J. Chem. Educ.* **68,** 430–432 (1991). "Purification of Milk Whey α-Lactalbumin by Immobilized Metal-Ion Affinity Chromatography."

K. Ebner, *Acct. Chem. Res.* **3,** 41–47 (1970). Biological role of α-lactalbumin.

F. Franks, Editor, *Characterization of Proteins* (1988), Humana Press (Clifton, NJ). Analysis of proteins.

L. Hall and P. Campbell, in *Essays in Biochemistry,* Vol. 22, R. Marshall and K. Tipton, Editors (1988), Academic Press (London), pp. 1–26. A review on α-lactalbumin.

E. Harris and S. Angal, Editors, *Protein Purification Methods: A Practical Approach,* Vol. 1 (1989), IRL Press (Oxford). Excellent for theory and applications.

E. Harris and S. Angal, Editors, *Protein Purification Applications: A Practical Approach,* Vol. 2 (1990), IRL Press (Oxford). An excellent book for undergraduates.

J. Janson and L. Ryden, Editors, *Protein Purification: Principles, High Resolution Methods and Applications.* (1989), VCH (New York). An advanced book.

C. Mathews and K. van Holde, *Biochemistry* (1990), Benjamin/Cummings (Redwood City, CA), pp. 156–161. Isolation and purification of proteins.

J. H. Morrissey, *Anal. Biochem.* **117,** 307 (1981). "Silver Stain for Proteins in Polyacrylamide Gels."

J. Porath, J. Carlsson, I. Olsson, and G. Belfrage, *Nature* **258,** 598–599 (1975). "Metal Chelate Affinity Chromatography, a New Approach to Protein Fractionation."

T. Reid and A. Stancavage, *Biochromatography* **4,** 262–265 (1989). Purification of α-lactalbumin by metal-ion affinity chromatography.

J. Robyt and B. White, *Biochemical Techniques: Theory and Practice* (1987), Brooks/Cole (Monterey, CA). An excellent review of techniques for protein isolation and study.

R. Scopes, *Protein Purification: Principles and Practice,* 2nd ed. (1987), Springer-Verlag (New York). Excellent reference for undergraduates.

G. Sofer and V. Britton, *Biotechniques* Nov./Dec. 1983, pp. 198–203. "Designing an Optimal Chromatographic Purification Scheme for Proteins."

L. Stryer, *Biochemistry,* 3rd ed. (1988), W. H. Freeman (New York), pp. 43–67. Isolation, purification, and characterization of proteins.

D. Voet and J. Voet, *Biochemistry* (1990) John Wiley & Sons (New York), pp. 75–122. Study of proteins.

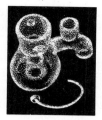

Structural Analysis of a Dipeptide or Tripeptide

■ Recommended Reading
Chapter 3, Sections A, B, G; Experiments 2 and 3.

■ Synopsis
The primary structure of a protein is defined by the sequence of amino acids. In this experiment the procedures that are in common use to determine protein primary structure are applied to an unknown dipeptide or tripeptide. Amino acid composition of the peptide will be determined by acid hydrolysis followed by paper chromatography. The identity of the NH_2-terminal amino acid will be achieved by the dansyl or Edman method followed by thin-layer or high-performance liquid chromatography.

I. Introduction and Theory

Structural elucidation of natural macromolecules is an important step in understanding the relationships between the chemical properties of a biomolecule and its biological function. The techniques used in organic structure determination (NMR, IR, UV, and MS) are somewhat useful when applied to biomolecules, but the unique nature of natural molecules also requires the application of specialized chemical techniques. Proteins, polysaccharides, and nucleic acids are polymeric materials, each composed of hundreds or sometimes thousands of monomeric units (amino acids, monosaccharides, and nucleotides, respectively). But there is only a limited number of these units from which the biomolecules are synthesized. For example, only 20 different amino acids are found in proteins but these different amino acids may appear several times in the same protein molecule. Therefore, the structure of a peptide or protein can be recognized only after the amino acid composition *and* sequence have been determined.

A protein sample must be purified before it can be analyzed for amino acid content and sequence. Methods for protein purification were discussed in Experiment 3. After the isolation and purification of a protein, it must be analyzed for purity. Although several techniques are available, the homogeneity of a protein sample is best tested by electrophoresis (Chapter 4) and/or chromatography (Chapter 3). In Experiment 3, α-lactalbumin isolated from milk whey was analyzed by gel electrophoresis. Once purity of the protein sample has been established by electrophoresis or chromatography, the analysis of amino acid composition and sequence can begin.

Amino Acid Composition

The amide bonds in peptides and proteins can be hydrolyzed in strong acid or base. Treatment of a peptide or protein under either of these conditions yields a mixture of the constituent amino acids. Neither acid- nor base-catalyzed hydrolysis of a protein leads to ideal results because both tend to destroy some constituent amino acids. Acid-catalyzed hydrolysis destroys tryptophan, causes some loss of serine and threonine, and converts asparagine and glutamine to aspartic acid and glutamic acid, respectively. Base-catalyzed hydrolysis leads to destruction of serine, threonine, cysteine, and cystine and also results in racemization of the free amino acids. Because acid-catalyzed hydrolysis is less destructive, it is often the method of choice. The hydrolysis procedure consists of dissolving the protein sample in aqueous acid, usually 6 N HCl, and heating the solution in a sealed, evacuated vial at 110°C for 12 to 24 hours. The time interval necessary for hydrolysis depends on the nature of the amino acid residues. Because of steric interactions, hydrolysis at valine and isoleucine is slower than at other amino acid residues.

Now that the free amino acids present in a peptide or protein have been released by hydrolysis, they must be separated and quantified. The most versatile, economical, and convenient techniques for separation are based on chromatographic methods. Ion-exchange chromatography as described in Chapter 3 and Experiment 2 is probably best. Earlier workers relied on paper chromatography and thin-layer chromatography (TLC); however, more sensitive techniques are now available. Automated ion-exchange chromatography (amino acid analyzers), gas chromatography, and high-performance liquid chromatography (HPLC) are now powerful tools for the qualitative and quantitative analysis of amino acids and derivatives. Because there is still a demand for rapid, qualitative, routine analysis of amino acids, thin-layer and paper chromatographic methods are still being developed and improved. Amino acids and derivatives may be analyzed directly by TLC without further derivatization as required for GC. Several support materials are available, but most analyses are carried out on silica gel or cellulose. The free amino acids can be detected on the developed chromatographic plates by reaction with ninhydrin. A pink-purple color is obtained for all amino acids except proline. A yellow color develops with proline. Since many derivatives of amino acids, including phenylthiohydantoin and dansyl derivatives, contain aromatic moieties, they can be detected on TLC plates by fluorescence under an ultraviolet lamp.

Gas chromatography is used to analyze volatile derivatives of amino acids. Phenylthiohydantoins (products of Edman degradation) may be analyzed directly by GC but are better resolved if converted to their trimethylsilyl derivatives with N,O-bis(trimethylsilyl) acetamide. Free amino acids are generally converted to their N-trifluoroacetyl-n-butyl esters or trimethylsilyl derivatives before GC analysis. For best results, all gas chromatography of amino acid derivatives should be done with a glass column and injection port, as contact with metals causes extensive decomposition of the derivatives.

High-performance liquid chromatographic techniques have been applied with success to the analysis of phenylthiohydantoin, 2,4-dinitrophenyl, and dansyl amino acid derivatives.

If only very small samples of amino acids are available for analysis, fluorescence is used for detection. One of the most sensitive methods of microanalysis is based on the reaction of amino acids with o-phthalaldehyde and β-mercaptoethanol (Equation E4.1). The isoindole derivative is fluorescent and can be measured with amounts as small as 10^{-12} mole.

▶   (Equation E4.1)

o-Phthalalaldehyde   Amino acid β-Mercaptoethanol

Isoindole derivative
of amino acid

End Group Analysis

Sequential analysis of amino acids in purified peptides and proteins is best initiated by analysis of the terminal amino acids. A peptide has one amino acid with a free α-amino group (NH_2-terminus end) and one amino acid with a free α-carboxyl group (COOH-terminus end). Many chemical methods have been developed to selectively tag and identify these terminal amino acids.

NH_2-terminal amino acid analysis is achieved by the use of (1) 2,4-dinitrofluorobenzene (Sanger reagent), (2) 1-dimethylaminonaphthalene-5-sulfonyl chloride (dansyl chloride), or (3) phenylisothiocyanate (Edman reagent). Figure E4.1 shows the structures of these reagents. Although the chemistry is different for each of these reagents, the same general concept is used. These compounds react with the NH_2-terminal amino acid of a peptide or protein to produce covalent derivatives that are stable to acid-catalyzed hydrolysis. The process is illustrated in Figure E4.2. The mixture of modified amino acid and free amino acids is separated, and the NH_2-terminal amino acid is identified by thin-layer, paper, gas, or high-performance liquid chromatography. The first NH_2-terminal method to be widely used was based on the Sanger reagent, which produces yellow-colored 2,4-dinitrophenyl (DNP) derivatives of NH_2-terminal amino acids. The Sanger method has several disadvantages, including poor yield of the DNP-peptide derivative, low sensitivity of analysis, and instability of some DNP–amino acids during acid hydrolysis. The dansyl

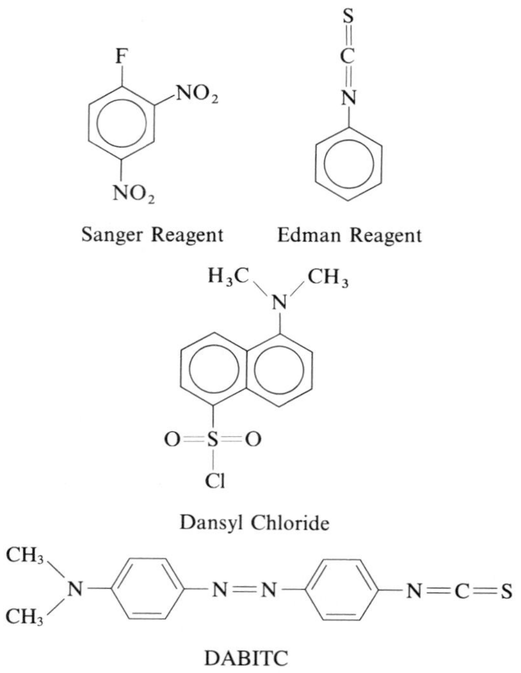

Figure E4.1
Reagents useful in NH_2-terminal analysis.

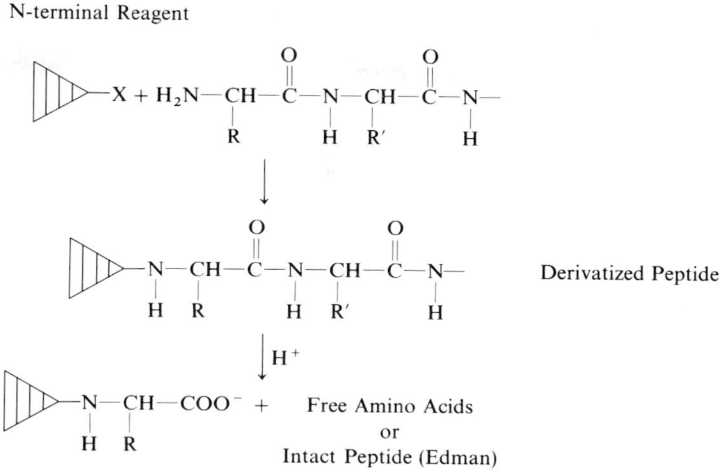

N-terminal Reagent

Figure E4.2
Analysis of NH$_2$-terminal amino acids.

chloride method has largely replaced the Sanger method because very sensitive fluorescence techniques may be used for detection and analysis of the dansyl amino acid derivatives and the derivatives are more stable during acid hydrolysis.

Even more versatile than the dansyl method is the Edman method (Figure E4.3) (Konigsberg, 1972). The NH$_2$-terminal amino acid is removed as its phenylthiohydantoin (PTH) derivative under anhydrous acid conditions, while all other amide bonds in the peptide remain intact. The derivatized amino acid is then extracted from the reaction mixture and identified by paper, thin-layer, gas, or high-performance liquid chromatography. The intact peptide (minus the original NH$_2$-terminal amino acid) may be isolated and recycled by reaction with phenylisothiocyanate. Since this method is nondestructive to the remaining peptide (aqueous acid hydrolysis is not required) and results in good yield, it can be used for stepwise sequential analysis of peptides. The method has now been automated.

Under all NH$_2$-terminal labeling conditions, internal amino acid residues in a peptide may be modified, but in ways different from the NH$_2$-terminal amino acid. For example, if the NH$_2$-terminal amino acid is alanine, only the free α-amino group is reactive to the selective reagent. However, if lysine, serine, or other amino acids with nucleophilic side chains are present at the NH$_2$-terminus, both nucleophilic groups (α-amino and ϵ-amino for lysine; α-amino and β-hydroxyl for serine) have the potential for reaction with the labeling reagent. If lysine or serine residues are located internally, only the side chain nucleophilic group (ϵ-amino for lysine, β-hydroxyl for serine) has the potential for reaction. Chromatographic techniques may be used to separate and identify each of these three types of amino acid modification (α-amino, side chain, and both).

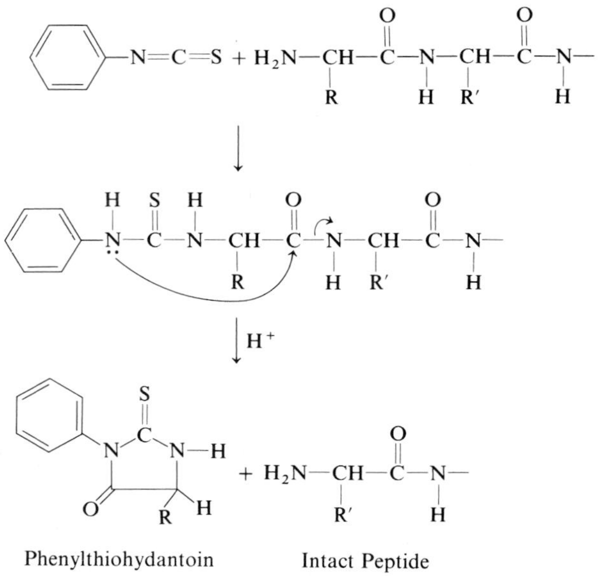

Figure E4.3
The Edman method of NH$_2$-terminal amino acid analysis.

Phenylthiohydantoin   Intact Peptide

The development of new chemical reagents and instrumentation has now made it possible to achieve end group analysis and sequencing on extremely small samples of protein (microsequencing). By using a chromophoric derivative of phenylisothiocyanate, 4-N,N-dimethylaminoazobenzene-4'-isothiocyanate (DABITC; see Figure E4.1), and analyzing the derivatives by HPLC, it is possible to sequence peptides and proteins at the nanomolar level.

The COOH-terminal amino acid of a peptide or protein may be analyzed by either chemical or enzymatic methods. The chemical methods are similar to the procedures for NH$_2$-terminal analysis. COOH-terminal amino acids are identified by hydrazinolysis or are reduced to amino alcohols by lithium borohydride. The modified amino acids are released by acid hydrolysis and identified by chromatography. Both of these chemical methods are difficult, and clear-cut results are not readily obtained. The method of choice is peptide hydrolysis catalyzed by carboxypeptidases A and B (Ambler, 1972). These two enzymes catalyze the hydrolysis of amide bonds at the COOH-terminal end of a peptide (Equation E4.2), since carboxypeptidase action requires the presence of a free α-carboxyl group in the substrate.

Carboxypeptidase A catalyzes the hydrolysis of carboxyl-terminal acidic or neutral amino acids; however, the rate of hydrolysis depends on the structure of the side chain R'. Amino acids with nonpolar aryl or alkyl side chains are cleaved more rapidly. Carboxypeptidase B is specific for the hydrolysis of basic COOH-terminal amino acids (lysine and arginine). Neither peptidase functions if proline occupies the COOH-terminal position or is the next to last amino acid.

(Equation E4.2)

$$-\overset{\overset{\displaystyle O}{\|}}{\underset{\underset{\displaystyle H}{|}}{C}}-N-\underset{\underset{\displaystyle R}{|}}{CH}-\overset{\overset{\displaystyle O}{\|}}{C}-N-\underset{\underset{\displaystyle R'}{|}}{CH}-COO^- + H_2O \xrightarrow{\text{carboxypeptidase}}$$

$$-\overset{\overset{\displaystyle O}{\|}}{\underset{\underset{\displaystyle H}{|}}{C}}-N-\underset{\underset{\displaystyle R}{|}}{CH}-COO^- + H_2N-\underset{\underset{\displaystyle R'}{|}}{CH}-COO^-$$

It should be recognized that the release of the COOH-terminal amino acid by carboxypeptidase A or B uncovers yet another COOH-terminal amino acid that may be cleaved by the enzyme. Cleavage of the second amino acid at a rate greater than the first amino acid may lead to ambiguity in COOH-terminal analysis, since both free amino acids may be present in approximately equal concentrations in the reaction mixture. To avoid this difficulty, it is often possible to incubate the protein with a carboxypeptidase, remove aliquots from the reaction mixture at various times, analyze the free amino acids, and graph the release of individual amino acids as a function of time.

Overview of the Experiment

In this experiment, the sequence of amino acids in a dipeptide or tripeptide will be determined by using some of the techniques just described. The amino acid composition of the unknown peptide will be found by acid-catalyzed hydrolysis followed by chromatographic analysis of the amino acids. Sequence analysis will then be carried out by identification of the NH$_2$ and/or COOH terminus. See Figure E4.4 for a summary of

Figure E4.4
Flowchart for the analysis of a dipeptide.

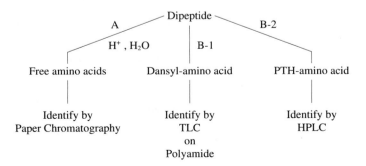

the procedure. The general experimental procedure will be outlined, but the specific analyses you will perform will depend on the unknown peptide. The following schedule is appropriate for this experiment.

DIPEPTIDE

Method I: Total Hydrolysis and Dansyl Analysis by TLC

Period 1: (a) Begin acid hydrolysis of peptide. (b) Complete dansyl chloride reaction and begin acid hydrolysis of dansyl peptide. (c) Prepare paper chromatogram by applying standard amino acids.

Period 2: (a) Work up the dansyl hydrolysate and spot on the TLC plate along with standard dansyl amino acids. Develop the plate in solvents. (b) Work up the peptide acid hydrolysate, spot on the paper chromatogram, and develop.

Method II: Total Hydrolysis and Edman Analysis by HPLC

Period 1: (a) Same as 1 (a and c) above. (b) Begin Edman reaction and proceed through toluene evaporation.

Period 2: (a) Complete Edman reaction and prepare for HPLC analysis. (b) Work up the peptide acid hydrolysate. Spot on the paper chromatogram and develop.

TRIPEPTIDE

A combination of methods can be used for this analysis, but the following is recommended.

Periods 1 and 2: Same as method I or II above.

Period 3: Carboxypeptidase reaction and paper chromatographic analysis.

Many variables must be considered as you plan the execution of this experiment. For example, there is no single paper or thin-layer chromatographic method that will successfully separate and identify every combination of two or three amino acids or derivatives. Successful results in this experiment depend partly on the proper choice of unknown peptides by your instructor. Since each peptide unknown may require a specific set of analyses and reaction conditions, your instructor should give you some hints on how to proceed. If your laboratory is equipped with HPLC instruments, method II is an ideal option.

II. Materials and Supplies

A. *Total Hydrolysis of Unknown Peptide*

Unknown dipeptide or tripeptide, 2 mg
Melting point capillary tube, 1.5 × 100 mm
6 N HCl
Heat lamp

Whatman 3MM paper, 20 × 20 cm

Amino acid standard solutions, 1% in water

Syringe, 50–100 μL

Chromatography chambers (jars or sandwich-type by Kodak)

Chromatography solvent, acetonitrile and 0.1 M ammonium acetate, 60:40, pH 4.0 or 5.0

Oven at 100°C

Ninhydrin spray

Rotary evaporator or tank of N_2 gas

B. NH_2-Terminal Analysis

METHOD 1—DANSYL CHLORIDE

Unknown dipeptide or tripeptide, 1 mg

Dansyl chloride solution, 5 mg/mL in acetone

Conical centrifuge tube, 12 mL

Sodium bicarbonate, 0.2 M

Hydrocarbon foil

Constant-temperature bath at 37°C

Hydrolysis vial ($\frac{1}{2}$ dram with polyvinyl liner, Sargent-Welch)

6 N HCl

Acetone and 6 N HCl (1:1)

Oven at 100°C

Thin-layer plates, polyamide, double-sided, 9 × 9 cm (Schleicher and Schuell)

Chromatography solvents:

  Solvent 1. Formic acid–H_2O, 1.5:100 (v/v)

  Solvent 2. Toluene–acetic acid, 10:1 (v/v)

  Solvent 3. Ethyl acetate–methanol–acetic acid, 2:1:1 (v/v/v)

Standard mixture of dansyl amino acids in acetone, 1 mg/mL each: Phe, Leu, Pro, Gly, Ala, Val.

UV detection lamp

METHOD 2—EDMAN METHOD

Unknown dipeptide or tripeptide, 3 mg

Conical centrifuge tube, 12 mL

Pyridine, 70%

Phenylisothiocyanate

Tank of purified N_2 gas

Constant-temperature water bath at 37°C

Toluene

Phenylthiohydantoin amino acid standards in dichloroethane

UV detection lamp

Anhydrous trifluoroacetic acid

Hydrocarbon foil

Ethyl acetate

For HPLC Analysis:

> HPLC equipment with reverse-phase column, for example, Hi-Pore RP-318 (Bio-Rad), 250 × 4.6 mm
>
> Elution solvents:
>
> > Solvent A. 0.14 M sodium acetate, 0.004 M triethylamine, pH 6.4
> >
> > Solvent B. 60% acetonitrile in H_2O

C. COOH-Terminal Analysis Using Carboxypeptidase A

Unknown peptide, 2 mg

N-Ethylmorpholine acetate buffer, 0.2 M, pH 8.5

Solution of carboxypeptidase A in water, 1 mg/mL (50 EC units per mL)

Constant-temperature bath at 37°C

Acetic acid, 0.1 M

Clinical centrifuge

Heat lamp

Solvent for paper chromatography (see part A)

Ninhydrin spray

Rotary evaporator

III. Experimental Procedure

A. Total Hydrolysis of Unknown Peptide

Place approximately 1 mg of your peptide sample on a small watch glass and dissolve it in 0.05 mL of 6 N HCl delivered with a syringe. With a syringe, transfer the peptide solution into a melting-point capillary, and seal the open end with a Bunsen burner. Put the sealed tube in a small labeled beaker and place in a 110°C oven for 12 to 15 hours, or at room temperature for 1 week. After hydrolysis, open the capillary by scratching one end with a sharp file, and transfer the hydrolysate to a watch glass. Evaporate the liquid by *gentle* heating under a heat lamp (**Hood!**), add 0.1 mL of distilled water, and repeat the evaporation step. Finally, prepare the residue for chromatography by dissolving it in 0.1 mL of distilled water.

For paper chromatography, obtain a 20×20 cm sheet of Whatman 3MM. Wear disposable plastic gloves when you handle the paper in order to avoid transfer of amino acids from your fingers to the paper sheet. Prepare the chromatogram by applying spots of the unknown hydrolysis mixture and amino acid standards along one edge of the paper. The amino acid standards can be applied during period 1 and the hydrolysis mixture at the beginning of period 2. The spot should be no closer than 2 cm from the bottom edge or side edges. You may wish to mark the paper *lightly* 2 cm from one edge with a pencil. Prepare a "key" in your notebook to keep track of the identity of each spot (Figure E4.5). Using small capillary tubes drawn to a fine point or microcapillary pipets, make two separate applications of your unknown solution, one containing approximately 5 μL and the other containing two or three times as much sample. It is best to apply the larger amount in small increments. Make one application and allow it to dry before more is applied to the spot. Apply the amino acid standards at selected spots along the 2-cm line; 5 to 10 μL of each standard should be sufficient. Applications of amino acids should be kept as small as possible (0.2–0.3 cm) so that the developed spots do not overlap. To develop the paper chromatogram, staple the sheet into a cylinder and place it in the chromatography jar containing a 1-cm layer of acetonitrile and 0.1 M ammonium acetate (60:40, pH 4.0 or 5.0) (Heimer, 1972). When preparing the cylinder, handle the paper only with gloves and do not overlap the edges while stapling. Solvent development time is 40 to 60 minutes.

When the developing solvent reaches a level approximately 1 cm from the top of the chromatogram, remove the paper from the jar and allow it to dry in a hood. Remove the staples and spray the paper lightly with ninhydrin solution. Allow it to dry in a hood for 10 minutes. The color may be developed by placing the chromatogram in an oven (110°C for 10 min) or at room temperature for 1 to 2 hours. All the spots should be circled with a pencil, as they will fade with time. Describe in your notebook the color of each spot.

Figure E4.5
Preparation of a thin-layer or paper chromatogram.

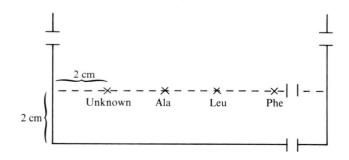

B. NH₂-Terminal Analysis

METHOD 1—DANSYL CHLORIDE

⚠ **Caution**

> Dansyl chloride is a corrosive organic acid halide, which should be used only in a hood. Wear disposable gloves while using the reagent. If it is spilled on the skin, sprinkle with sodium bicarbonate and wash with copious amounts of warm soapy water. Rinse well. For floor or bench spills, cover with sodium bicarbonate and transfer mixture to a beaker of water. Pour the aqueous solution down the drain with excess water.
>
> **Do not pipet solutions of dansyl chloride by mouth!**

Weigh about 1 mg of the unknown peptide and transfer into a conical centrifuge tube. Dissolve the peptide in 0.5 mL of 0.2 M sodium bicarbonate. Add 0.2 mL of the solution of dansyl chloride in acetone and mix well. Cover with hydrocarbon foil and incubate for 1 hour at 37°C or 2 hours at room temperature. After completion of the reaction, evaporate the solution to dryness under vacuum (rotary evaporator) or by a gentle stream of N₂ gas, with the reaction tube held in a beaker of warm water (Figure E4.6). Dissolve the residue in 0.5 mL acetone–6 N HCl solution and transfer it into the hydrolysis vial. Evaporate the acetone from the vial with a gentle stream of nitrogen gas. Seal the hydrolysis vial and heat in a 110°C oven for 10 to 16 hours. Remove the hydrolysate from the vial and evaporate it on a watch glass under a heat lamp. Dissolve the residue in a minimal amount of 50% aqueous pyridine (about 10 μL). Using a microsyringe, spot 1–5 μL of the solution about 1 cm from the corner of a polyamide thin-layer sheet (see Figure E4.7). The diameter of the spot should not exceed 3–4 mm. Dry the spot with warm air. On the reverse side of the plate load 1 μL of the standard dansyl–amino acid mixture. This should be spotted at the same position (1 cm from corner) as the unknown. Develop the plate in solvent 1, 1.5% formic acid, in the direction shown in Figure E4.7 until the solvent reaches 1 cm from the top. This may take up to 15 minutes. Dry the solvent from both sides of the plate in warm air. View the plate under UV light. **Be sure to wear safety goggles!** A blue fluorescent streak with faint green fluorescent spots may be seen. Develop the plate in solvent 2, toluene–acetic acid, at right angles to solvent 1 (see Figure E4.7). The fluorescent blue streak will be along the bottom of the plate. Remove the plate from the solvent, allow to air-dry, and view under UV light. There is now some separation of the blue and green spots (see Figure E4.7). If the unknown cannot be identified, run the plate in solvent 3, ethyl acetate–methanol–acetic acid, in the same direction as solvent 2.

Figure E4.6
Evaporation of solvent by a stream of N₂ gas.

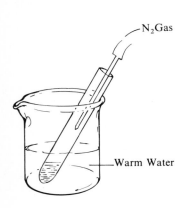

N₂ Gas

Warm Water

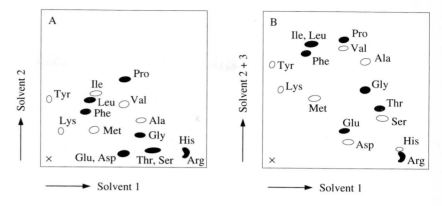

Figure E4.7
Thin-layer chromatography of dansyl-amino acids on a polyamide plate.   (A) After solvents 1 (1.5% formic acid) and 2 (toluene–acetic acid).   (B) After solvents 1, 2, and 3 (ethyl acetate–methanol–acetic acid). DNS-OH = dansic acid; DNS-NH₂ = dansyl amide. See text for further details.

METHOD 2—EDMAN METHOD

  Caution

Phenylisothiocyanate is a lacrimator and should be handled only with gloves in a hood. If it is spilled on the skin, wash immediately with strong soap solution and rinse well. Small spills on the lab bench or floor should be absorbed on paper towels or vermiculite. Sweep the contaminated solid onto paper and allow the phenylisothiocyanate to evaporate in a hood. Check with your instructor for disposal of the paper. For large spills, absorb or mix with vermiculite, sodium bicarbonate, or sand. Package this in a paper carton and have your instructor dispose of it.

   Trifluoroacetic acid is a corrosive liquid. Handle it only with disposable gloves in a hood. Wash skin spills immediately with soap and water and rinse well. Cover floor or bench spills with excess sodium bicarbonate and vermiculite. Mix well and transfer to a beaker of water. After reaction, pour solution down the drain with excess water.

   Pyridine is a toxic, flammable liquid (flash point is 20°C) with a very irritating smell. Use only in a hood with no flames. Gloves should be worn while handling.

   Toluene is a flammable liquid (flash point is 4°C). Use only in a hood with no flames.

   Do not pipet these four reagents—pyridine, trifluoroacetic acid, toluene, or phenylisothiocyanate—by mouth!

To a conical centrifuge tube, add about 2 mg of the unknown peptide and dissolve in 1.0 mL of 70% aqueous pyridine (**Hood**!). Add 0.1 mL of phenylisothiocyanate and mix well. Flush the tube with purified N_2 gas to exclude oxygen, and stopper it with a cork or cover with Parafilm. If O_2 is present, oxygen atoms will replace sulfur in the thiocarbonyl group of the phenylthiocarbamyl peptide. Incubate the reaction mixture in a constant-temperature bath (37°C) for 2 hours. Evaporate the solvent using a rotary evaporator. An alternative method of solvent evaporation is to place the test tube in a beaker of warm water and flush the inside of the tube with N_2 gas as in Figure E4.6. Wash the residue remaining in the tube with two 1-mL portions of toluene to remove unreacted phenylisothiocyanate. Remove residual toluene on the rotary evaporator or in a warm-water bath under a stream of N_2.

Cyclization of the phenylthiocarbamyl peptide and cleavage of the peptide bond are brought about by reaction with anhydrous trifluoroacetic acid. Add 0.5 mL of trifluoroacetic acid to the unknown peptide residue, cover with hydrocarbon foil, and heat for 10 minutes at 60°C or allow the reaction to proceed for 24 hours at room temperature. Remove the excess trifluoroacetic acid by placing the tube in a beaker of warm water with evaporation under a stream of N_2 in a hood. Extract the phenylthiohydantoin amino acid derivative from the residue with ethyl acetate (3 × 1 mL). Combine the ethyl acetate extracts in a small beaker and concentrate to 0.5 mL over a steam bath in a hood.

Have your instructor assist you in the use of the HPLC equipment. The instrument should be equipped with a reverse-phase column and a UV detection system set at 254 nm. Obtain the retention time for each of the standard phenylthiohydantoin–amino acid samples provided and then inject the unknown. For elution, a gradient of 5–40% solvent B (60% acetonitrile) in solvent A (0.14 M sodium acetate, 0.004 M triethylamine, pH 6.4) is recommended. Flow rate should be about 1.5 mL/min and temperature around 50°C. These conditions may vary depending on the HPLC instrument and column used.

C. COOH-Terminal Analysis Using Carboxypeptidase A

Weigh about 2 mg of the unknown peptide into a test tube and dissolve in 1.0 mL of 0.2 N N-ethylmorpholine acetate buffer, pH 8.5. Add 0.1 mL (5 EC units) of the carboxypeptidase solution and mix well, but do not shake. Incubate at 37°C and withdraw small aliquots (0.1 mL) at timed intervals (0, 5, 10, 20, 30, and 40 minutes). Immediately after withdrawal, transfer each aliquot to a test tube containing 0.5 mL of 0.1 M acetic acid. This stops the enzyme-catalyzed hydrolysis and causes precipitation of the enzyme. Remove the precipitate from each acidified aliquot by centrifugation at 1000 rpm for 10 min. Evaporate 0.2 mL of each supernatant on separate watch glasses. Dissolve the residue in 0.1 mL of water and chromatograph against amino acid standards on Whatman 3MM paper as in part A. Since a somewhat "quantitative" analysis will be made of the COOH-terminal amino acid, be sure the same amount of solution is applied to each chromatogram. For detection, spray with ninhydrin solution.

IV. Analysis of Results

A. *Total Hydrolysis of Unknown Peptide*

Prepare a table of R_f values of the standard amino acids. You should also describe the color of the original ninhydrin spot. Since the colors vary slightly with the amino acids, this aids in the identification of an unknown amino acid. Calculate the R_f values for the constituent amino acids of the unknown peptide. From these data you should be able to identify the amino acids present in your unknown.

B. *NH₂-Terminal Analysis*

METHOD 1—DANSYL CHLORIDE

Draw a picture in your notebook of the polyamide thin-layer plate exposed under UV light after each of the three solvent developments. These pictures should look similar to Figure E4.7. Three fluorescent areas should be evident after solvent 2; however, better separation is achieved by solvent 3. A blue fluorescent area at the bottom of the plate is dansic acid (DNS-OH), which is a hydrolysis product of dansyl chloride. A blue-green fluorescent spot about one-third to one-half up the plate is dansyl amide (DNS-NH₂), which is produced by reaction of dansyl chloride with ammonia. A third spot, which usually fluoresces green, is the dansyl derivative of the NH₂-terminus amino acid. Note the positions of the standard dansyl amino acids and compare with the unknown. What is the identity of the NH₂-terminal amino acid? Are any other fluorescent spots evident on the plate? Using polarity or nonpolarity, try to explain the position of each molecule on the thin-layer plate.

METHOD 2—EDMAN METHOD

From the HPLC data, identify the NH₂-terminal amino acid. Is the amino acid one of those identified in part A?

C. *COOH-Terminal Analysis Using Carboxypeptidase A*

Study the chromatograms of the released COOH-terminal amino acids and prepare a graph of intensity of each purple ninhydrin spot vs. time. For early reaction times you will probably see only one amino acid spot, but later aliquots may yield two or even three amino acid spots. If this is the case, make a separate line on the graph for each appearing amino acid. Estimate the intensity of each amino acid spot using such terms as light, medium, and dark or use a scale of 1 to 10 on your graph. Using data from all three parts of the experiment, draw the structure of your unknown tripeptide.

V. Questions and Problems

▶ 1. Compare the advantages and disadvantages of the Edman and dansyl chloride methods of NH_2-terminal analysis.

▶ 2. Why must the layer of solvent in a chromatography jar be below the origin line containing the applied samples on the chromatogram?

▶ 3. Why must the Edman reaction be carried out in a basic solvent such as aqueous pyridine?

▶ 4. An unknown tripeptide was analyzed according to the procedures described in this experiment and the following results were obtained.

Part A: Total hydrolysis and paper chromatography—Phe, Leu, Ala

Part B: Dansyl chloride and thin-layer chromatography—dansyl alanine

Part C: Carboxypeptidase A—relative order of release: Phe > Leu ≈ Ala

What is the sequence of the tripeptide?

▶ 5. Could you determine the amino acid sequence of a dipeptide using two steps, (1) total acid hydrolysis and (2) COOH-terminal analysis with carboxypeptidase?

▶ 6. In the TLC analysis of dansyl-amino acids, why does dansic acid not move during development by solvents 2 and 3?

▶ 7. Using structures, write out the reaction scheme for the Edman-type degradation of your unknown dipeptide using 4-N,N-dimethylamino-azobenzene-4′-isothiocyanate (DABITC).

VI. References

G. Allen, *Laboratory Technique in Biochemistry and Molecular Biology,* 2nd ed. (1989), Elsevier (Amsterdam). Determination of protein primary structure.

R. Ambler, in *Methods in Enzymology,* Vol. XXVB, C. H. W. Hirs and S. N. Timasheff, Editors (1972), Academic Press (New York), pp. 436–445. "Carboxypeptidase."

J. Bell and E. Bell, *Proteins and Enzymes* (1988), Prentice-Hall (Englewood Cliffs, NJ). An excellent text on protein structural analysis.

S. Black and M. Coon, *Anal. Biochem.* **121,** 281–285 (1982). "HPLC of Phenylthiohydantoin Amino Acids."

J. Chang, R. Knecht, and D. Braun, *Biochem. J.* **199,** 547–555 (1981). Amino acid analysis at the picomole level.

Y. Cheng and N. Dovichi, *Science* **242,** 562–564 (1988). "Subattomole Amino Acid Analysis by Capillary Zone Electrophoresis and Laser-Induced Fluorescence."

W. Hancock, Editor, *CRC Handbook of HPLC for the Separation of Amino Acids, Peptides and Proteins* (1984), CRC Press (Boca Raton, FL). Contains many references and methods for analysis.

E. Heimer, *J. Chem. Educ.* **49,** 547 (1972). "Paper Chromatography of Amino Acids."

W. Konigsberg, in *Methods in Enzymology,* Vol. XXVB, C. H. W. Hirs and S. N. Timasheff, Editors (1972), Academic Press (New York), pp. 326–332. "Edman Degradation."

F. Lottspeich and A. Henschen, in *HPLC in Biochemistry,* A. Henschen, K. Hupe, F. Lottspeich, and W. Voelter, Editors (1985), VCH (Weinheim), pp. 139–216. "Amino Acids, Peptides and Proteins."

C. Mathews and K. van Holde, *Biochemistry* (1990), Benjamin/Cummings (Redwood City, CA), pp. 133–170. An introduction to protein sequence analysis.

J. Rawn, *Biochemistry,* 2nd ed. (1989), Neil Patterson Publishers (Burlington, NC), pp. 51–73. Determining amino acid composition and sequence of proteins.

J. Robyt and B. White, *Biochemical Techniques: Theory and Practice* (1987), Brooks/Cole (Belmont, CA), pp. 340–352. Techniques for determination of protein primary structure.

R. Scopes, *Protein Purification: Principles and Practice,* 2nd ed. (1987), Springer-Verlag (New York). Methods for protein purification and analysis.

R. Somack, *Anal. Biochem.* **104,** 464–468 (1980). "Complete Phenylthiohydantoin Amino Acid Analysis by High-Performance Liquid Chromatography on ULTRASPHERE-Octadecyltrimethyloxysilane."

L. Stryer, *Biochemistry,* 3rd ed. (1988), W. H. Freeman (New York), pp. 15–42. A discussion of protein sequence analysis.

D. Voet and J. Voet, *Biochemistry* (1990), John Wiley & Sons (New York), pp. 109–122. Sequence analysis of proteins.

J. Walker, Editor, *Methods in Molecular Biology—Proteins* (1984), Humana Press (Clifton NJ), Vol. I. Detailed experimental techniques for protein analysis.

EXPERIMENT 5

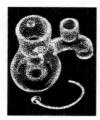

Protein Ligand Interactions: Binding of Coomassie Blue Dye to Ovalbumin

■ Recommended Reading

Chapter 2, Section D; Chapter 5, Sections A and B; Experiment 3.

■ Synopsis

The Bradford protein assay, as used in Experiment 3, is based on the absorbance change that occurs upon binding of Coomassie Blue dye to proteins. This protein-ligand interaction will serve as a model for the many biochemical processes that are dependent on reversible binding of a small molecule to a macromolecule. In this experiment, the dynamics of dye binding to the protein ovalbumin will be investigated in order to gain understanding of the physical and chemical nature of ligand-macromolecule interactions.

I. Introduction and Theory

Most of the processes that occur in an organism are the result of interactions between molecules. The action of hormones is familiar to all. A hormone response is the consequence of a weak, but specific, interaction between the hormone molecule and a receptor protein in the membrane of the target cell. Before a metabolic reaction can occur, a small substrate molecule must physically interact in a certain well-defined manner with a macromolecular catalyst, an enzyme. The biochemical action of a drug also depends on molecular interactions. The drug is first distributed throughout the body via the blood stream. Drugs in the blood stream are often bound to plasma proteins, which act as carriers. When the drug

Table E5.1
Biological Examples of Ligand-Macromolecule Interactions.

Type of Interaction	Biochemical Significance
Lipids-proteins	Cell membranes
Substrate-enzyme	Metabolic reactions
Hormone-receptor protein	Regulation
Inhibitor-enzyme	Metabolic regulation
Ligand-carrier protein	Membrane transport
Antigen-antibody	Immune response
Coenzyme-enzyme	Metabolic reactions

molecules are transported to their site of action, a second molecular interaction is likely to occur. Many drugs elicit their effects by interfering with biochemical processes. This may take the form of enzyme inhibition, where the drug molecule binds to a specific enzyme and prohibits its catalytic action. Table E5.1 lists several other molecular interactions that lead to some dynamic biochemical action.

All of these molecular interactions have at least two common characteristics. (1) The forces that are the basis of these interactions are weak and noncovalent. We usually define the interactions in terms of hydrogen bonding, hydrophobic stabilization, van der Waals forces, and electrostatic interactions. (2) The binding between molecules is close, selective, and specific. Imagine that the interaction brings together two molecular surfaces. Where the surfaces are in contact, the forces must be complementary. If on one surface there is a nonpolar molecular group (phenyl ring, hydrophobic alkyl chain, etc.), the adjacent region on the other surface must also be hydrophobic and nonpolar. If a positive charge exists on one surface, there may be a neutralizing negative charge on the other surface. Simply stated, the two molecules must be compatible, in a chemical sense.

Quantitative Treatment of Binding

A thorough understanding of the biochemical significance of ligand binding to macromolecules comes only from a quantitative analysis of the strength of binding. (In biochemistry, a small molecule that binds to a macromolecule is called a **ligand**.) Binding affinity between two molecules is often expressed as an equilibrium constant, the **formation constant, K_f,** which is derived from the law of mass action. Consider the specific interaction between a small molecule, L (for ligand), and a macromolecule, M (Equation E5.1). These two species combine to form a complex, LM.

$$\blacktriangleright \quad L + M \; \rightleftharpoons \; LM \qquad\qquad \text{(Equation E5.1)}$$

K_f, the formation constant for the complex, is defined by Equation E5.2.

▶ $$K_f = \frac{[LM]}{[L][M]}$$ (Equation E5.2)

Do not confuse K_f with K_d, the dissociation constant. The relationship between K_f and K_d is defined in Equation E5.3

▶ $$K_d = \frac{[L][M]}{[LM]} = \frac{1}{K_f}$$ (Equation E5.3)

The larger the value of K_f, the greater the strength of binding between L and M. (Large K_f implies a high concentration of LM relative to L and M.) Return to Equation E5.2 and note that in order to determine K_f, a method must be developed to measure equilibrium concentrations of L, M, and the complex LM. In a later section, we will describe experimental techniques that are applied to these measurements of binding constants, but first we must reorganize Equation E5.2 into a form that contains more readily measurable terms. We will begin with the assumption that the macromolecule, M, has several binding sites for L and that these sites do not interact with each other. That is, K_f is identical for all binding sites. The following definitions are necessary for the reorganization of Equation E5.2

[L] = equilibrium concentration of free or unbound ligand

[M] = equilibrium concentration of macromolecule with no bound L, or the concentration of unoccupied binding sites

[LM] = equilibrium concentration of ligand-macromolecule complex, or the concentration of occupied sites

$[M]_0$ = total or initial concentration of macromolecule, or total concentration of available binding sites

$[L]_0$ = total concentration of bound and unbound ligand; or initial concentration of ligand

v = fraction of available sites on M that are occupied, or the fraction of M that has L in binding site:

▶ $$v = \frac{[LM]}{[M]_0}$$ (Equation E5.4)

The term v is particularly significant because it can be considered a ratio of the number of occupied sites to the total number of potential binding sites on M. It can be measured experimentally, but first it must be redefined in the following manner: Since $[M]_0 = [LM] + [M]$, then

$$v = \frac{[LM]}{[LM] + [M]}$$

From Equation E5.2, $[LM] = K_f[L][M]$. Therefore,

$$v = \frac{K_f[L][M]}{K_f[L][M] + [M]}$$

Simplifying,

▶ $$v = \frac{K_f[L]}{K_f[L] + 1}$$ (Equation E5.5)

You should recognize the similarity of Equation E5.5 to the Michaelis-Menten equation for enzyme catalysis. A graph of v vs. [L] yields a hyperbolic curve (see Figure E5.1) that approaches a limiting value or saturation level. At this point, all binding sites on M are occupied. Because of the difficulty of measuring the exact point of saturation, this nonlinear curve is seldom used to determine K_f. Linear plots are more desirable, so Equation E5.5 is converted to an equation for a straight line. The equation will now be put into a more general form to account for any number of potential binding sites on M. The symbol $\overline{v}$ will be used to represent

Figure E5.1
A plot illustrating saturation of binding sites with ligand.

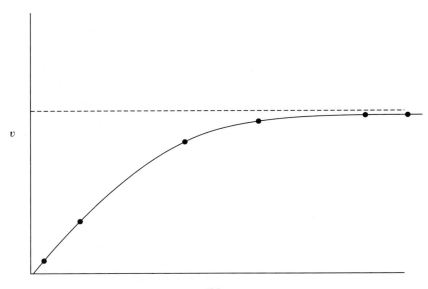

the average number of occupied sites per M, and n will represent the number of potential binding sites per M molecule. Assuming that all the binding sites on M are equivalent, Equation E5.5 becomes

▶ $$\overline{v} = \frac{nK_f[L]}{K_f[L] + 1}$$ (Equation E5.6)

Equation E5.6 contains terms such as [L] that are difficult to determine, so it must be converted to a form amenable to experimental measurements. This will be completed in the next section.

Experimental Detection of Binding and Measurements of K_f and n

A variety of techniques has been developed for measuring the dynamics of ligand-macromolecule interactions. The method most applicable to investigating reversible binding is equilibrium dialysis, a technique that requires specialized equipment, sometimes cumbersome procedures, and the use of radiolabeled ligands (Bell and Bell, 1988; van Holde, 1985). Since this method is not practical for a teaching laboratory, other procedures have been developed.

One of the simplest and most convenient methods for detecting ligand binding is the differential method, which detects and quantifies some measurable change that accompanies the binding of L and M. Most often, this is a change in spectral absorption or fluorescence. Fluorescence measurements are more sensitive, with detection at the level of 10^{-6} to $10^{-9}M$. Fluorescence measurements require, however, a specialized spectrophotometer called a fluorimeter. Because this instrument is not readily available in many laboratories and the fluorescence technique for measuring ligand binding is rather difficult, absorbance measurement using a more common UV-VIS spectrometer is often the method of choice. A spectral method (absorbance or fluorescence) can be used only if ligand binding results in a spectral change for the ligand or macromolecule.

When Coomassie Blue dye binds to proteins, the dye undergoes a significant spectral change in the visible region (Figure E5.2). The spectrum of free dye displays a minimum in the range 575–625 nm. When dye and protein are mixed, a new absorption band appears at a λ_{max} of 596 nm. The increase in absorption at 596 nm is directly related to the concentration of protein, which is the basis of the Bradford protein assay (Chapter 2, Experiment 3). The new absorption band is due to the presence of dye-protein complex. Note that the new band is rather broad and that a range of wavelengths may be used for measurement. Because the maximum wavelength change occurs in the region of 596 nm, that wavelength is chosen for this experiment.

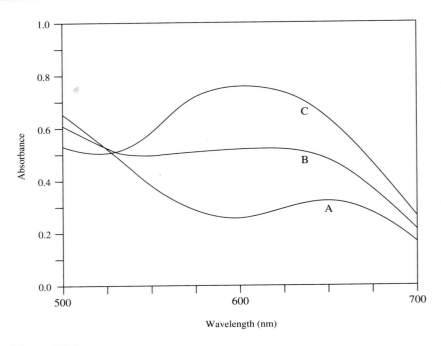

Figure E5.2
The spectral changes occurring during the binding of Coomassie Blue dye to ovalbumin. Curve A, free dye in solution; curve B, dye plus 0.5 mg of ovalbumin; curve C, dye plus 1.0 mg of ovalbumin. The dye concentration is identical in all three curves.

Equation E5.6 for evaluating protein-ligand interactions is easily adaptable to the differential method, but it must be converted into a linear form. Absorption data can be expressed in terms of [bound]/[free] ligands or [occupied]/[unoccupied] sites on the macromolecule. The data analysis described here assumes that L and LM, but not M, contribute to the measured absorbance. The measured absorbance change is proportional to the ratio of bound L to unbound L.

The absorbance observed from a mixture of L and M, A_{obs}, is

$$A_{obs} = A_L + A_{LM}$$

where

A_L = absorbance of free L

A_{LM} = absorbance of bound L

Absorbance data may be used directly to calculate the fraction of ligand bound:

▶   Fraction of ligand bound $= f = \dfrac{A_{obs} - A_L}{A_{max} - A_L}$ (Equation E5.7)

where

A_{max} = total absorbance of LM when all the L molecules in solution
are bound to M. This number is experimentally determined in
part B.

The term $A_{max} - A_L$ is proportional to the total number of binding sites,
and $A_{obs} - A_L$ represents the average number of occupied sites on M
caused by a certain concentration of L.

For graphical analysis of the experimental data, $\overline{v}$ for Equation E5.6
can be defined in terms that can be measured experimentally:

$$\blacktriangleright \quad \overline{v} = f\frac{[L]_0}{[M]_0} \qquad \text{(Equation E5.8)}$$

If Equation E5.8 is combined with Equation E5.6, an equation for a
straight line can be derived in terms that are readily measured.

$$f\frac{[L]_0}{[M]_0} = \frac{nK_f[L]}{K_f[L] + 1}$$

Take the reciprocal of each side and solve for $[M]_0/f$:

$$\frac{[M]_0}{f[L]_0} = \frac{K_f[L] + 1}{nK_f[L]}$$

$$\frac{[M]_0}{f} = \frac{[L]_0}{n} + \frac{[L]_0}{nK_f[L]}$$

The term $[L]/[L]_0 = 1 - f$; therefore:

$$\frac{[M]_0}{f} = \frac{[L]_0}{n} + \frac{1}{nK_f(1 - f)}$$

Rearrange to the form for a straight line, $y = mx + b$:

$$\blacktriangleright \quad \frac{[M]_0}{f} = \frac{1}{nK_f(1 - f)} + \frac{[L]_0}{n} \qquad \text{(Equation E5.9)}$$

This is a form of the Scatchard equation. The values n and K_f can be calculated from a plot of $1/(1 - f)$ vs. $[M]_0/f$. The slope of the line is represented by $1/nK_f$, whereas the y intercept is $[L]_0/n$. If the data points
fall on a straight line as in Figure E5.3; then all the dye binding sites on
the protein M are identical and independent.

The derivation of Equation E5.9 assumes that K_f is identical for all
binding sites; that is, the binding of one molecule of L does not influence
the binding of other L molecules to binding sites on M. However, it is

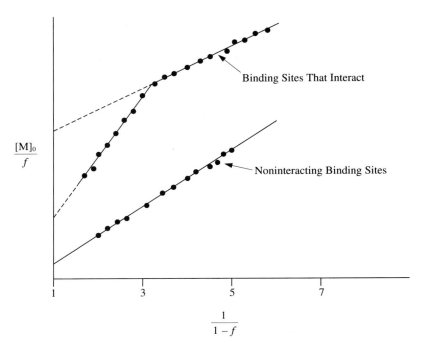

Figure E5.3
Scatchard plot for typical binding data. See text for details.

common for ligand-macromolecule interactions to display such influences. The binding of one L molecule to M may encourage or inhibit the binding of a second L molecule to M. For example, the binding of dioxygen to one of the four subunits of hemoglobin increases the affinity of the other subunits for dioxygen. There is said to be **cooperativity** of sequential binding. If the sites do show cooperative binding, the plot becomes nonlinear as shown by the curved line in Figure. E5.3 This line is resolved into two lines, indicating that two types of binding sites are present on M. The n and K_f for each type of binding may be estimated by resolving the smooth curve into two straight lines. K_f and n may be calculated from the intercepts and slopes. An alternative way to derive the Scatchard equation has been published (Sohl and Splittgerber, 1991).

Overview of the Experiment

In this experiment, you will evaluate the binding of Brilliant Blue dye to ovalbumin. Increasing amounts of protein will be added to a fixed concentration of dye. Absorbance measurements will be made on the solutions and used to prepare a saturation curve and a Scatchard plot.

The time requirements for this experiment are:

Part A: Concentration of ovalbumin solution—30 minutes.

Part B: Binding of dye to protein, the saturation curve—1 hour.

Part C: Binding of dye to protein, the Scatchard plot—2 hours.

Although it is not essential for students to complete part A (concentration of ovalbumin solution), it affords good experience for individual students or it may be used as a class demonstration.

II. Materials and Supplies

Ovalbumin solutions in water, two solutions: I = 5.0 mg/mL; II = 20 mg/mL

Coomassie Brilliant Blue G-250 dye, stock solution, $9.0 \times 10^{-4}\ M$ in ethanol, 85% phosphoric acid, H_2O. Prepare by adding 920 mg of dye in 100 mL of 95% ethanol and 200 mL of 85% phosphoric acid. Dilute to 1200 mL with distilled water

Dye solvent, 100 mL ethanol, 200 mL 85% phosphoric acid, and 900 mL water

Automatic pipets and tips, 20 μL, 200 μL, 1000 μL

3-mL glass or plastic cuvettes

Matched quartz cuvettes, 1 or 3 mL

UV-VIS spectrophotometer

13 × 100 mm test tubes

III. Experimental Procedure

 Caution

> Disposable gloves and protective clothing should be used during this experiment. The Coomassie Brilliant Blue dye solution is very acidic and may cause acid burns if splashed on hands or clothing. In addition, the dye will stain skin and clothing.

A. Concentration of Ovalbumin Solution (Optional)

The exact concentration of the ovalbumin solution may be determined by measuring the A_{280} of the solution. Turn on the spectrometer and UV lamp. Set the wavelength to 280 nm. Allow the instrument to warm up for at least 15 minutes. Pipet 1.0 or 3.0 mL of distilled water into a quartz cuvette. Place the cuvette in the sample position of the spectrometer and adjust the absorbance to 0.0. Carefully pipet 1.0 or 3.0 mL of ovalbumin solution I into the matched cuvette. Read the A_{280}. If the absorbance reading is above 1, dilute the solution with water and again read A_{280}. Calculate the protein concentration as described in Analysis of Results, part A.

B. Binding of Dye to Ovalbumin—The Saturation Curve

Turn on the spectrometer and set the wavelength to 596 nm. Use Table E5.2 to prepare the various solutions for absorbance measurements. All reagents should be at or near room temperature. Prepare each solution in a test tube numbered to correspond to Table E5.2. The dye solution is prepared by mixing 3.0 mL of the dye stock solution with 27.0 mL of the dye solvent. Mix well! Accurately pipet 2.0 mL of the diluted dye solution into each of nine test tubes. Using an automatic pipetting system, transfer the appropriate amount of ovalbumin solution II (20 mg/mL) into each tube. For example, deliver 0.025 mL into tube #2 and 0.05 mL into tube #3. Adjust the total volume in each tube to 3.0 mL by adding the appropriate amount of distilled water. Mix each tube well by holding a small square of hydrocarbon foil over the tube and inverting it several times. *Do not shake as this will cause foaming and may denature the protein!* After about 15 minutes of reaction time, transfer the contents of each tube into a 3-mL cuvette and read the A_{596} to the nearest 0.001 absorbance unit. Use a cuvette of distilled water as a blank. The color of the reaction mixture will remain stable for up to 1 hour.

C. Binding of Dye to Ovalbumin—The Scatchard Plot

This assay is set up in a similar fashion to part B except that the 5 mg/mL ovalbumin solution is used. Prepare the dye solution by mixing 6.0 mL of dye stock solution with 54 mL of dye solvent. Mix well! Accurately pipet 2.0 mL of the diluted dye solution into each of 20 test tubes. Using an automatic pipetting system, carefully transfer variable amounts (0 to 1.0 mL) of a 5 mg/mL solution of ovalbumin. Suggested amounts are 0.0, 0.01, 0.02, 0.03, 0.04, 0.05, 0.06, 0.07, 0.08, 0.09, 0.10, 0.20,

Table E5.2
Preparation of Cuvettes for Part B[1]

Reagent	Tube Number								
	1	2	3	4	5	6	7	8	9
Dye solution	2.0	2.0	2.0	2.0	2.0	2.0	2.0	2.0	2.0
Protein solution (20 mg/mL)	0.0	0.025	0.05	0.10	0.20	0.40	0.60	0.80	1.0
H_2O	1.0	0.975	0.95	0.90	0.80	0.60	0.40	0.20	0.0

[1]Units are milliliters

0.25, 0.30, 0.40, 0.50, 0.60, 0.70, 0.80, and 1.0 mL. Use the most accurate pipet available for each transfer. Adjust the total volume in each tube to 3.0 mL by adding the appropriate amount of distilled water. Mix each tube well by holding a piece of hydrocarbon foil and inverting several times. *Do not shake!* After a 15-minute reaction period, read the A_{596} for each tube. Use a cuvette of water as blank.

IV. Analysis of Results

A. Concentration of Ovalbumin Solution

The molar absorption coefficient, ϵ, for ovalbumin at 280 nm is $2.85 \times 10^4\ M^{-1}cm^{-1}$. Use Beer's Law, $A = \epsilon cl$, to calculate the concentration, c, of the solution (see Chapter 2, Section D). If dilution of the protein was necessary, be sure you account for this in your calculation. Report the protein concentration in M units.

B. Binding of Dye to Ovalbumin—The Saturation Curve

Construct the curve by plotting the A_{596} of each tube vs. the mg of protein present in that tube. Calculate the total amount of ovalbumin in each tube based on the original solution concentration (20 mg/mL). For example, tube 2, which was prepared with 0.025 mL of the stock protein solution, contains 0.5 mg of ovalbumin.

What is the shape of the curve? Explain what is happening on a molecular level. Determine the maximum absorbance (A_{max}). This is observed when enough protein has been added to bind all dye molecules and is indicated on the curve by a leveling of the absorbance readings.

C. Binding of Dye to Ovalbumin—The Scatchard Plot

The curve in part B does not allow a quantitative description of the binding process. To obtain one, it is necessary to plot the data in straight-line form using Equations E5.6, E5.7, and E5.9.

Prepare a table with the following headings and calculate the numerical values for each reaction mixture.

Tube#	A_{obs}	$[M]_0$	f	$\dfrac{1}{1 - f}$	$[M]_0/f$

where $[M]_0$ is the total ovalbumin concentration in molar units. Tube 2 has 0.01 mL of a 5 mg/mL protein solution diluted to a final volume of 3.0 mL. Therefore the concentration is

$$\frac{0.05\ mg}{3\ mL} = \frac{X\ mg}{1000\ mL}$$

$$X = 0.0167 \text{ g}/1000 \text{ mL}$$

$$\frac{0.0167 \text{ g/liter}}{45,000 \text{ g/mole}} = 3.71 \times 10^{-7} M$$

▶ $f = \dfrac{A_{obs} - A_L}{A_{max} - A_L}$ (Equation E5.7)

A_{obs} and A_L are defined in the Introduction and A_{max} was determined experimentally in part B.

Now plot $[M]_0/f$ versus $1/1 - f$ as in Figure E5.3. Calculate $[L]_0$ and use Figure E5.3 to determine the value(s) for n and K_f. Curvature in the graph implies cooperative effects in the binding of dye to ovalbumin. If this is the case, evaluate the constants as shown for the dashed line in Figure E5.3. Explain on a molecular level the shape of the binding curve.

V. Questions and Problems

▶ 1. A drug, X, was studied for its affinity for serum albumin. When X was bound to albumin, an increase in absorbance was noted. The $\bar{v}$ values were determined from these absorbance measurements. Use the data below to determine K_f and n for the interaction between X and albumin. Prepare two types of graphs and compare the results. In one graph plot $\bar{v}$ vs. $[X]$, and in the second plot $\bar{v}/[X]$ vs. $\bar{v}$. Which is the better method? Why?

[X]	$\bar{v}$
0.36	0.43
0.60	0.68
1.2	1.08
2.4	1.63
3.6	1.83
4.8	1.95
6.0	1.98

▶ 2. You are attempting to develop a colorimetric probe for use in binding studies. List several requirements that the probe must meet in order to be effective. For example, it must bind at specific locations of the macromolecule. Can you think of other requirements?

▶ 3. A research project you are working on involves the study of sugar binding to human albumin. The sugars to be tested are not fluorescent, and you do not wish to use a secondary probe such as a dye. Human albumin has only one tryptophan residue, and you know that this amino acid is fluorescent. You find that the tryptophan fluorescence spectrum of human albumin undergoes changes when various sugars are added. Can you explain the results of this experiment and discuss the significance of the finding?

▶ 4. Equation E5.5 in this experiment can be used to determine K_f values, but hyperbolic plots are obtained. Convert Equation E5.5 into an equation that will yield a linear plot without going through all the changes necessary for the Scatchard equation.

VI. References

J. Bell and E. Bell, *Proteins and Enzymes* (1988), Prentice-Hall (Englewood Cliffs, NJ) pp. 379–380. Methods for measurement of ligand binding.

M. Bradford, *Anal. Biochem.* **72**, 248–254 (1976). "A Rapid and Sensitive Method for the Quantitation of Microgram Quantities of Protein Utilizing the Principle of Protein-Dye Binding."

F. Dalhquist, in *Methods in Enzymology,* Vol. 48F, C. Hirs and S. Timasheff, Editors (1978), Academic Press (New York), pp. 270–299. "The Meaning of Scatchard and Hill Plots."

D. Freifelder, *Physical Biochemistry,* 2nd ed. (1982), W. H. Freeman (San Francisco), pp. 654–684. Introduction to ligand binding.

G. Scatchard, *Ann. N.Y. Acad. Sci.* **51**, 660–672 (1949). "The Attraction of Proteins for Small Molecules and Ions."

J. Sohl and A. Splittgerber, *J. Chem. Educ.* **68**, 262–264 (1991). "The Binding of Coomassie Brilliant Blue to Bovine Serum Albumin—A Physical Biochemistry Experiment."

L. Stryer, *Biochemistry,* 3rd ed. (1988), W. H. Freeman (New York), pp. 154–155, 239–240. Introduction to ligand binding.

K. van Holde, *Physical Biochemistry,* 2nd ed. (1985), Prentice-Hall (Englewood Cliffs, NJ), pp. 51–93. Measurement of ligand binding by equilibrium dialysis.

D. Voet and J. Voet, *Biochemistry* (1990), John Wiley & Sons (New York), pp. 1153–1154. Ligand binding.

EXPERIMENT **6**

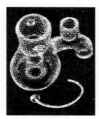

Kinetic Analysis of an Enzyme, Tyrosinase

■ Recommended Reading
Chapter 5, Sections A, B, C.

■ Synopsis
Tyrosinase, a copper-containing oxidoreductase, catalyzes the orthohydroxylation of monophenols and the aerobic oxidation of catechols. The enzyme activity can be assayed by monitoring the oxidation of 3,4-dihydroxyphenylalanine (dopa) to the red-colored dopachrome. The kinetic parameters K_M and V_{max} will be evaluated using Lineweaver-Burk or direct linear plots. Inhibition of tyrosinase will also be studied. Finally, two stereoisomers, L-dopa and D-dopa, will be tested and compared as substrates.

I. Introduction and Theory

Enzymes are biological catalysts. Without their presence in a cell, most biochemical reactions would not proceed at the required rate. The physicochemical and biological properties of enzymes have been investigated since the early 1800s. The unrelenting interest in enzymes is due to several factors—their dynamic and essential role in the cell, their extraordinary catalytic power, and their selectivity. Two of these dynamic characteristics will be evaluated in this experiment, namely a kinetic description of enzyme activity and molecular selectivity.

Enzyme-catalyzed reactions proceed through an ES complex as shown in Equation 6.1. The individual rate constants, k_n, are placed with each arrow.

▶ $$E + S \underset{k_2}{\overset{k_1}{\rightleftharpoons}} ES \underset{k_4}{\overset{k_3}{\rightleftharpoons}} E + P$$

(Equation E6.1)

299

E represents the enzyme, S the substrate or reactant, and P the product. For a specific enzyme, only one or a few different substrate molecules can bind in the proper manner and produce a functional ES complex. The substrate must have a size, shape, and polarity compatible with the active site of the enzyme. Some enzymes catalyze the transformation of many different molecules as long as there is a common type of chemical linkage in the substrate. Others have absolute specificity and can form reactive ES complexes with only one molecular structure. In fact, some enzymes are able to differentiate between D and L isomers of substrates.

Kinetic Properties of Enzymes

The initial reaction velocity, v_0, of an enzyme-catalyzed reaction varies with the substrate concentration, [S], as shown in Figure E6.1. The Michaelis-Menten equation has been derived to account for the kinetic properties of enzymes. (Consult a biochemistry textbook for a derivation of this equation and for a discussion of the conditions under which the equation is valid.) The common form of the equation is

Figure E6.1
Michaelis-Menten plot for an enzyme-catalyzed reaction.

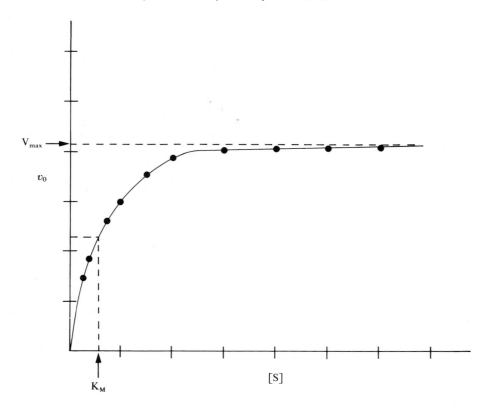

$$\blacktriangleright \quad v_0 = \frac{V_{max}[S]}{K_M + [S]} \qquad \text{(Equation E6.2)}$$

where

v_0 = initial reaction velocity

V_{max} = maximal reaction velocity; attained when all enzyme active sites are filled with substrate molecules

$[S]$ = substrate concentration

K_M = Michaelis constant = $\dfrac{k_2 + k_3}{k_1}$

The important kinetic constants, V_{max} and K_M, can be graphically determined as shown in Figure E6.1. Equation E6.2 and Figure E6.1 have all of the disadvantages of nonlinear kinetic analysis. V_{max} can be estimated only because of the asymptotic nature of the line. The value of K_M, the substrate concentration that results in a reaction velocity of $V_{max}/2$, depends on V_{max}, so both are in error. By taking the reciprocal of both sides of the Michaelis-Menten equation, however, it is converted into the Lineweaver-Burk relationship (Equation E6.3).

$$\blacktriangleright \quad \frac{1}{v_0} = \frac{K_M}{V_{max}} \cdot \frac{1}{[S]} + \frac{1}{V_{max}} \qquad \text{(Equation E6.3)}$$

This equation, which is in the form $y = mx + b$, gives a straight line when $1/v_0$ is plotted against $1/[S]$ (Figure E6.2). The intercept on the $1/v_0$ axis is $1/V_{max}$ and the intercept on the $1/[S]$ axis is $-1/K_M$. A disadvantage of the Lineweaver-Burk plot is that the data points are compressed in the high substrate concentration region. A third type of graphical analysis results from the Eisenthal and Cornish-Bowden modification (Equation E6.4).

$$\blacktriangleright \quad V_{max} = v_0 + \frac{v_0}{[S]} K_M \qquad \text{(Equation E6.4)}$$

This modification is a direct linear graph that results from plotting each pair of experimental values of v_0 and $[S]$. Figure E6.3 is an example of the Eisenthal and Cornish-Bowden analysis. K_M and V_{max} values are read directly from the graph.

Significance of Kinetic Constants

The **Michaelis constant**, K_M, for an enzyme-substrate interaction has two meanings: (1) K_M is the substrate concentration that leads to an initial reaction velocity of $V_{max}/2$ or, in other words, the substrate concentration that results in the filling of one-half of the enzyme active sites, and (2) $K_M = (k_2 + k_3)/k_1$. The second definition of K_M has special

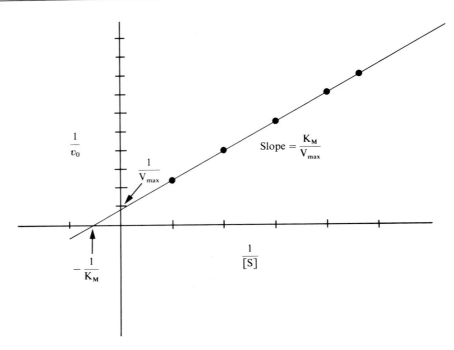

Figure E6.2
Lineweaver-Burk plot for an enzyme-catalyzed reaction.

significance in certain cases. When $k_2 \gg k_3$, then $K_M = k_2/k_1$ so K_M is equivalent to the dissociation constant of the ES complex. When $k_2 \gg k_3$, a large K_M implies weak interaction between E and S, whereas a small K_M indicates strong binding between E and S.

V_{max} is important because it leads to the analysis of another kinetic constant, k_3, **turnover number**. The analysis of k_3 begins with the basic rate law for an enzyme-catalyzed process (Equation E6.5), which is derived from Equation E6.1:

▶ $v_0 = k_3[ES]$ (Equation E6.5)

If all of the enzyme active sites are saturated with substrate, then [ES] in Equation E6.5 is equivalent to $[E_T]$, the total concentration of enzyme, and v_0 becomes V_{max}; hence

▶ $V_{max} = k_3[E_T]$ (Equation E6.6)

or

$$k_3 = \frac{V_{max}}{[E_T]}$$

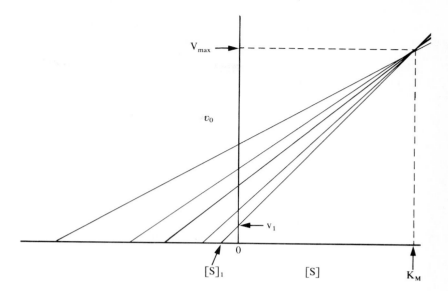

Figure E6.3
Eisenthal and Cornish-Bowden
(direct linear) plot for an
enzyme-catalyzed reaction.

For an enzyme with one active site per molecule, the turnover number, k_3, is the number of substrate molecules transformed to product by one enzyme molecule per unit time, usually in minutes or seconds. The turnover number is a measure of the efficiency of an enzyme.

Inhibition of Enzyme Activity

Nonsubstrate molecules may interact with enzymes, leading to a decrease in enzymatic activity. The study of enzyme inhibition is of interest because it often reveals information about the mechanism of enzyme action. Also, many toxic substances, including drugs, express their action by enzyme inhibition.

The process of reversible inhibition is described by an equilibrium interaction between enzyme and inhibitor. Most inhibition processes can be classified as **competitive** or **noncompetitive,** depending on how the inhibitor impairs enzyme action. A competitive inhibitor is usually similar in structure to the substrate and is capable of reversible binding to the enzyme active site. In contrast to the substrate molecule, the inhibitor molecule cannot undergo chemical transformation to a product; however, it does interfere with substrate binding. A noncompetitive inhibitor does not bind in the active site of an enzyme but binds at some other region of the enzyme molecule. Upon binding of the noncompetitive inhibitor, the enzyme is reversibly converted to a nonfunctional conformational state, and the substrate, which is fully capable of binding to the active site, is not converted to product.

Any complete inhibition study requires the experimental differentiation between competitive and noncompetitive inhibition. The two inhibitory processes are kinetically distinguishable by application of the

Lineweaver-Burk or direct linear plot. For each inhibitor studied, at least two sets of rate experiments are completed. In all sets, the enzyme concentration is identical. In set 1, the substrate concentration is varied and no inhibitor is added. In set 2, the same variable substrate concentrations are used as in set 1, and a constant amount of inhibitor is added to each assay. If additional data are desired, more sets are prepared with variable substrate concentrations as in set 2; but a constant, and different, concentration of inhibitor is present. These data, when plotted on a Lineweaver-Burk graph, lead to three lines as shown in Figure E6.4 and E6.5. The Lineweaver-Burk plot of a competitively inhibited process is shown in Figure E6.4. All three lines intersect at the same point on the $1/v_0$ axis; hence, V_{max} is not altered by the competitive inhibitor. If enough substrate is present, the competitive inhibition can be overcome. The apparent K_M value (measured on the $1/[S]$ axis) changes with each change in inhibitor concentration. Figure E6.4 reveals several kinetic characteristics of the inhibition process: K_M, V_{max}, and K_I. The quantity K_I is the dissociation constant for the EI complex, and its value indicates the strength of binding between enzyme and inhibitor. A large K_I implies relatively weak interaction between E and I.

A Lineweaver-Burk plot of enzyme kinetics in the presence and absence of a noncompetitive inhibitor is shown in Figure E6.5. V_{max} in the presence of a noncompetitive inhibitor is decreased, but K_M is unaffected. The effect of a competitive inhibitor on the direct linear plot is shown in Figure E6.6.

Figure E6.4
Lineweaver-Burk plot for competitive inhibition.

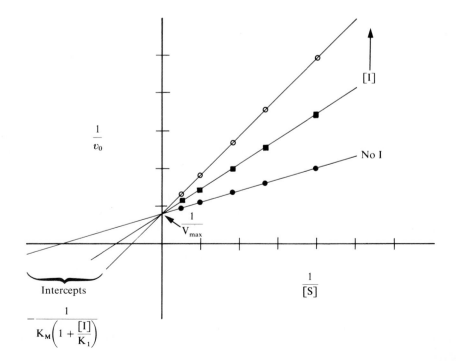

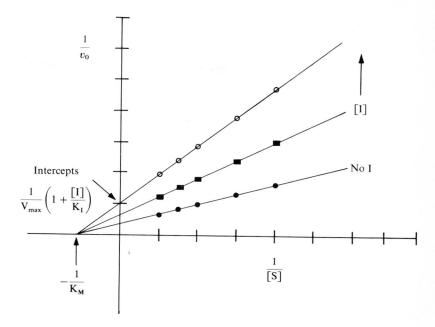

$\dfrac{1}{v_0}$

Intercepts

$$\dfrac{1}{V_{max}}\left(1 + \dfrac{[I]}{K_I}\right)$$

[I]

No I

$-\dfrac{1}{K_M}$

$\dfrac{1}{[S]}$

Figure E6.5
Lineweaver-Burk plot for non-competitive inhibition.

Units of Enzyme Activity

The actual molar concentration of an enzyme in a cell-free extract or purified preparation is seldom known. Only if the enzyme is available in a pure crystalline form, carefully weighed, and dissolved in a solvent can the actual molar concentration be accurately known. It is, however, possible to develop a precise and accurate assay for enzyme activity. Consequently, the amount of a specific enzyme present in solution is most often expressed in units of activity. Three units are in common use, the *international unit* (IU), the *katal,* and *specific activity*. The International Union of Biochemistry Commission on Enzymes has recommended the use of a standard unit, the **international unit,** or just **unit,** of enzyme activity. One IU of enzyme corresponds to the amount that catalyzes the transformation of 1 μmole of substrate to product per minute under specified conditions of pH, temperature, ionic strength, and substrate concentration. If a solution containing enzyme converts 10 μmoles of substrate to product in 5 minutes, the solution contains 2 enzyme units. A new unit of activity, the katal, has been recommended. One **katal** of enzyme activity represents the conversion of 1 mole of substrate to product in 1 second. One international unit is equivalent to 1/60 μkatal or 0.0167 μkatal. One katal, therefore, is equivalent to 6×10^7 international units. It is convenient to represent small amounts of enzyme in millikatals (mkatals), microkatals (μkatals), or nanokatals (nkatals). The enzyme activity in the above example was 2 units which can be converted to katals as follows: Since one katal is 6×10^7 units, 2 units are

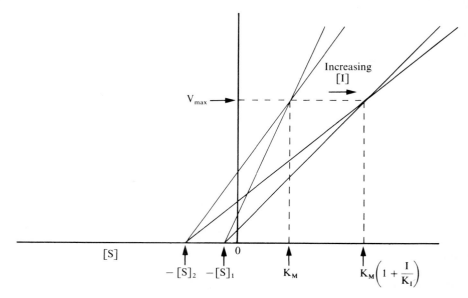

Figure E6.6
Direct linear plot for competitive inhibition.

equivalent to $2/6 \times 10^7$ or 33 nkatals (0.033 μkatals). If the enzyme is an impure preparation in solution, the activity is most often expressed as units/mL or katals/mL.

Another useful quantitative definition of enzyme efficiency is **specific activity**. The specific activity of an enzyme is the number of enzyme units or katals per milligram of protein. This is a measure of the purity of an enzyme. If a solution contains 20 mg of protein that express 2 units of activity (33 nkatals), the specific activity of the enzyme is 2 units/20 mg = 0.1 units/mg or 33 nkatals/20 mg = 1.65 nkatals/mg. As an enzyme is purified, its specific activity increases. That is, during purification, the enzyme concentration increases relative to the total protein concentration until a limit is reached. The maximum specific activity is attained when the enzyme is homogeneous or in a pure form.

The activity of an enzyme depends on several factors, including substrate concentration, cofactor concentration, pH, temperature, and ionic strength. The conditions under which enzyme activity is measured are critical and must be specified when activities are reported.

Design of an Enzyme Assay

Whether an enzyme is obtained commercially or prepared in a multistep procedure, an experimental method must be developed to detect and quantify the specific enzyme activity. During isolation and purification of an enzyme, the assay is necessary to determine the amount and purity of the enzyme and to evaluate its kinetic properties. An assay is also essential for a further study of the mechanism of the catalyzed reaction.

The design of an assay requires certain knowledge of the reaction:

1. The complete stoichiometry.
2. What substances are required (substrate, metal ions, cofactors, etc.) and their kinetic dependence.
3. Effect of pH, temperature, and ionic strength.

In addition to these conditions, a method must be available to identify and monitor any physical, chemical, or biological change that occurs during enzyme-catalyzed conversion of substrate to product. The most straightforward approach is to monitor the change in substrate or product concentration during the reaction. Many techniques, including spectrophotometry, fluorimetry, acid-base titration, and radioactive counting, are used to make these measurements. The choice depends primarily on the structural characteristics of the substrate and/or product and the chemistry that occurs during the reaction.

If an enzyme assay involves continuous monitoring of substrate or product concentration, the assay is said to be **kinetic**. If a single measurement of substrate or product concentration is made after a specified reaction time, a **fixed-time assay** results. The kinetic assay is more desirable because the time course of the reaction is directly observed and any discrepancy from linearity can be immediately detected.

Figure E6.7 displays the kinetic progress curve of a typical enzyme-catalyzed reaction and illustrates the advantage of a kinetic assay. The rate of product formation decreases with time. This may be due to any combination of a decrease in substrate concentration, denaturation of the enzyme, and product inhibition of the reaction. The solid line in Figure E6.7 represents the continuously measured time course of a reaction (kinetic assay). The true rate of the reaction is determined from the slope of the dashed line drawn tangent to the experimental result. From the data given, the rate is 5 μmoles of product formed per minute. Data from a fixed-time assay are also shown on Figure E6.7. If it is assumed that no product is present at the start of the reaction, then only a single measurement after a fixed period is necessary. This is shown by a circle on the experimental rate curve. The measured rate is now 16 μmoles of product formed every 5 minutes or about 3 μmoles/minute, considerably lower than the rate derived from the continuous, kinetic assay. Which rate measurement is correct? Obviously, the kinetic assay gives the true rate because it corrects for the decline in rate with time. The fixed-time assay can be improved by changing the time of the measurement, in this example, to 2 minutes of reaction time, when the experimental rate is still linear. It is possible to obtain true rates with the fixed-time assay, but one must choose the time period very carefully. In the laboratory, this is done by removing aliquots of a reaction mixture at various times and

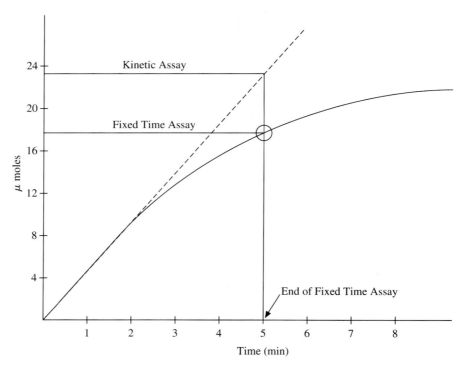

Figure E6.7
Kinetic progress of an enzyme-catalyzed reaction. See text for details.

measuring the concentration of product formed in each aliquot. Figure E6.7 reinforces an assumption used in the derivation of the Michaelis-Menten equation. Only measurements of initial velocities lead to true reaction rates. This avoids the complications of enzyme denaturation, decrease of [S], product inhibition, and reversion of the product to substrate.

Several factors must be considered when the experimental assay conditions are developed. The reaction rate depends on the concentrations of substrate, enzyme, and necessary cofactors. In addition, the reaction rate is under the influence of environmental factors such as pH, temperature, and ionic strength. Enzyme activity increases with increasing temperature until the enzyme becomes denatured. The enzyme activity then decreases until all enzyme molecules are inactivated by denaturation. During kinetic measurement, it is essential that the temperature of all reaction mixtures be maintained constant.

Ionic strength and pH should also be monitored carefully. Although some enzymes show little or no change in activity over a broad pH range, most enzymes display maximum activity in a narrow pH range. Any assay developed to evaluate the kinetic characteristics of an enzyme must be performed in a buffered solution, preferably at the optimal pH. Enzymes vary greatly in their response to inorganic salts. Some enzymes require the presence of anions or cations for activity, whereas others are inhibited. Trial and error using common salts (KCl, NaCl, etc.) is the best approach.

Applications of an Enzyme Assay

The conditions used in an enzyme assay depend on what is to be accomplished by the assay. There are two primary applications of an enzyme assay procedure. First, it may be used to measure the concentration of active enzyme in a preparation. In this circumstance, the measured rate of the enzyme-catalyzed reaction must be proportional to the concentration of enzyme; stated in more kinetic terms, there must be a linear relationship between initial rate and enzyme concentration (the reaction is first order in enzyme concentration). To achieve this, certain conditions must be met: (1) the concentrations of substrates(s), cofactors, and other requirements must be in excess; (2) the reaction mixture must not contain inhibitors of the enzyme; and (3) all environmental factors such as pH, temperature, and ionic strength should be controlled. Under these conditions, a plot of enzyme activity (μmole product formed/minute) vs. enzyme concentration is a straight line and can be used to estimate the concentration of active enzyme in solution.

Second, an enzyme assay may be used to measure the kinetic properties of an enzyme such as K_M, V_{max}, and inhibition characteristics. In this situation, different experimental conditions must be used. If K_M for a substrate is desired, the assay conditions must be such that the measured initial rate is first order in substrate. To determine K_M of a substrate, constant amounts of enzyme are incubated with varying amounts of substrate. A Lineweaver-Burk plot ($1/v_0$ vs. $1/[S]$) or direct linear plot may be used to determine K_M and V_{max}. If a reaction involves two or more substrates, each must be evaluated separately. The concentration of only one substrate is varied while the other is held constant at a saturating level. The same procedure holds for determining the kinetic dependence on cofactors. Substrate(s) and enzyme are held constant and the concentration of the cofactor is varied.

Inhibition kinetics are included in the second category of assay applications. An earlier discussion outlined the kinetic differentiation between competitive and noncompetitive inhibition. The same experimental conditions that pertain to evaluation of K_M and V_{max} hold for K_I estimation. A constant level of inhibitor is added to each assay, but the substrate concentration is varied as for K_M determination. In summary, a study of enzyme kinetics is approached by measuring initial reaction velocities under conditions where only one factor (substrate, enzyme, cofactor) is varied and all others are held constant.

Properties of Tyrosinase

Tyrosinase, also commonly called polyphenol oxidase, has two catalytic activities; *o*-hydroxylation of monophenols and aerobic oxidation of *o*-diphenols (Equations E6.7 and E6.8).

▶   monophenol + O_2 $\longrightarrow$ catechol + H_2O (Equation E6.7)

▶   2 catechol + O_2 $\longrightarrow$ 2 *o*-quinone + 2 H_2O (Equation E6.8)

The official Enzyme Commission classification and number are monophenol, *o*-diphenol: O_2 oxidoreductase, EC 1.14.18.1. Tyrosinases are present in plant and animal cells. The biological function of tyrosinase in plant cells is unknown, but its presence is readily apparent. Upon injury, many plant products such as potatoes, apples, and bananas undergo a browning process. This is due to the formation and further oxidation of natural quinones as shown in Equation E6.8. In mammalian cells, tyrosinase catalyzes two steps in the biosynthesis of melanin pigments from tyrosine. The reaction sequence, shown in Figure E6.8, is localized in melanocytes or pigment cells and produces the melanin substances that impart color to skin, hair, and eyes. Tyrosinase located in skin is activated by the ultraviolet rays of the sun, causing greater melanin production and, ultimately, suntan.

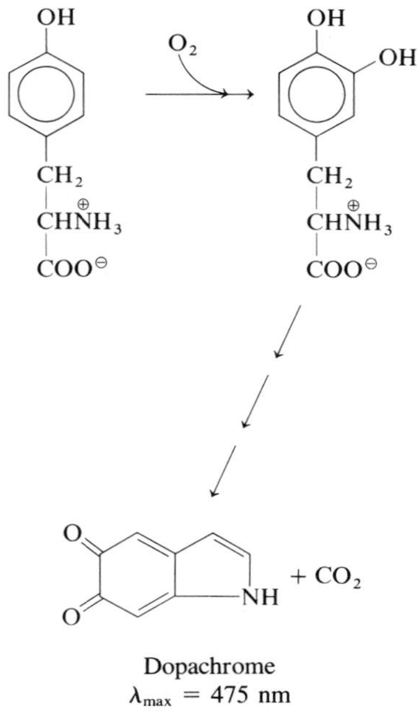

Figure E6.8
The oxidation of tyrosine and dopa as catalyzed by tyrosinase.

Dopachrome
λ_{max} = 475 nm

The many tyrosinases in nature differ in cofactor requirement, metal ion content, substrate specificity, molecular weight, and oligomeric structure. The diverse properties of the tyrosinases will not be discussed; instead, the properties of a single enzyme, mushroom tyrosinase, will be outlined.

Mushroom tyrosinase is tetrameric, with a total molecular weight of 128,000. There are four atoms of Cu^+ associated with the active enzyme. Two types of substrate binding sites exist in the enzyme, one type for the phenolic substrate and one type for the dioxygen molecule. The copper atom is associated with the dioxygen binding site; hence, chemicals that form complexes with copper atoms are potent inhibitors of tyrosinase activity. The inhibitors azide, cyanide, phenylthiourea, and cysteine compete with oxygen binding. Nonphenolic, aromatic compounds compete with the binding and oxidation of phenols and catechols.

Overview of the Experiment

Several kinetic characteristics of mushroom tyrosinase will be examined in this experiment. A spectrophotometric assay of tyrosinase activity will be introduced and applied to the evaluation of substrate specificity, K_M of the natural substrate, 3,4-dihydroxyphenylalanine (L-dopa), and inhibition characteristics.

Many assays for tyrosinase activity have been developed. Procedures in the literature include use of the oxygen electrode, oxidation of tyrosine followed at 280 nm, and oxidation of dopa followed at 475 nm. The most convenient assay involves following the tyrosinase-catalyzed oxidation of dopa by monitoring the initial rate of formation of dopachrome at 475 nm (Figure E6.8).

Two substrates are required in the tyrosinase-catalyzed reaction, phenolic substrate (dopa) and dioxygen. The conditions described in the experiment are such that the reaction mixtures are saturated with dissolved dioxygen. Therefore, when measurements are made for K_M, only the concentration of dopa is limiting, so the rate of the reaction depends on dopa concentration.

The dopachrome assay is extremely flexible, as it can be applied to a variety of studies of tyrosinase. Projects not described here, but of potential interest, are pH optimum studies, inhibition studies, and formation of the apoenzyme by dialysis and reconstitution (Pomerantz, 1966).

Time requirements for this experiment are:

Part A: Tyrosinase concentration—15 minutes.

Part B: Tyrosinase level for kinetic assay—1 hour.

Part C: K_M of dopa—2 hours.

Part D: Inhibition—2 hours.

It is recommended that part A be done as a group project so that time and enzyme are not wasted. If only one 3-hour laboratory period is allotted to this experiment, the following schedule is recommended:

Part A: Instructor or teaching assistant prepares the enzyme solution just before class and A_{280} is reported to students.

Part B: Do only three or four assays with D- or L-dopa.

Part C: Do three or four assays with the same dopa isomer as in part B.

Part D: Students choose one inhibitor and do four assays on the same dopa isomer as in parts B and C.

If 5 or 6 hours are available, students can do the following:

Part A: Measure A_{280} as a group project.

Part B: Do four or five assays with D- or L-dopa.

Part C: Do four assays each with D- and L-dopa.

Part D: Choose one inhibitor and do four assays with L- or D-dopa.

II. Materials and Supplies

Sodium phosphate buffer, 0.1 M, pH 7.0

Mushroom tyrosinase, 100 units/mL

1 unit represents the amount of enzyme that causes an A_{280} change of 0.001 using tyrosine as substrate. Prepare solution in phosphate buffer. A_{280} should be about 0.20. Store in ice bath during laboratory period.

L-Dopa, 2 mg/mL in phosphate buffer

D-Dopa, 2 mg/mL in phosphate buffer

Thiourea, 1×10^{-3} M in phosphate buffer

Benzoic acid, sodium salt, 3×10^{-3} M in phosphate buffer

Cinnamic acid, 1×10^{-3} M in phosphate buffer

Cuvettes, glass and quartz

Stopwatch

Constant-temperature bath at $25-30°C$

UV-VIS spectrophotometer with recorder and constant-temperature circulating bath or colorimeter

III. Experimental Procedure

A. *Measurement of Tyrosinase Concentration*

In order to determine the specific activity of the enzyme, the exact concentration of the enzyme must be known. The concentration of the solution of tyrosinase may be determined as a class project by the following procedure. Turn on the spectrophotometer and the UV lamp. Adjust the wavelength to 280 nm. Allow the instrument and lamp to warm up for 15 to 20 minutes. Transfer 1.0 or 3.0 mL of the phosphate buffer to a 1- or 3-mL quartz cuvette. Place it in the sample position of the spectrophotometer and adjust the balance to zero absorbance. Discard the buffer, and clean and dry the cuvette. Transfer 1.0 or 3.0 mL of the tyrosinase solution into the quartz cuvette. Place in the sample position and record the absorbance at 280 nm. Calculate the tyrosinase concentration as described in the Analysis of Results section.

B. Determination of the Tyrosinase Level for Kinetic Assays

All solutions listed under Materials and Supplies, except tyrosinase, should be stored in a constant-temperature bath. The tyrosinase solution should be stored in an ice bath. If there is no constant-temperature circulating bath on the spectrophotometer, maintain all solutions except tyrosinase at room temperature. Otherwise, set both the constant-temperature bath and the circulating bath at the same temperature. The recommended temperature range is 25 to 30°C. Turn on the spectrophotometer and recorder for warm-up. Adjust the wavelength to 475 nm.

Before kinetic constants can be evaluated, it is critical to find the correct concentration of enzyme to use for the assays. If too little enzyme is used, the overall absorbance change for a reaction time period will be so small that it is difficult to detect differences due to substrate concentration changes or inhibitor action. On the other hand, too much enzyme will allow the reaction to proceed too rapidly, and the leveling off of the time course curve as shown in Figure E6.7 will occur very early because of the rapid disappearance of substrate. A rate that is intermediate between these two extremes is best. For the dopachrome assay, it is desirable to use the level of tyrosinase that gives a linear absorbance change at 475 nm for 3 minutes.

The general procedure for the assay is as follows. Pipet phosphate buffer and L-dopa into a glass cuvette according to the directions in Table E6.1. Mix by inversion (cover the cuvette with hydrocarbon foil), and place the cuvette in the spectrometer sample position. If a recorder is available, adjust it to a full-scale absorbance of 1 and set the pen to 0 or the 0.10 absorbance line with the recorder zero adjust dial. Set the recorder speed to 1 or 2 cm/min and turn on the recorder. Observe whether there is a "blank" rate (no enzyme). Obtain a recorder trace for 3 minutes. Now, pipet the appropriate amount of enzyme into the cuvette. Mix by inversion of the cuvette after covering it with hydrocarbon foil (**Do not shake!**), and immediately place in the spectrophotometer. Turn on the recorder pen and observe the recorder trace for 3 minutes. Be sure you mark, on the recorder paper, the point of tyrosinase addition. With a straight-edge, draw a line tangent to the recorder trace as

Table E6.1
General Procedure for the Tyrosinase Assay[1]

| Reagent | Assay | | | | |
	1	2	3	4	5
Phosphate buffer	1.45	1.40	1.30	1.20	1.10
L-Dopa	1.5	1.5	1.5	1.5	1.5
Tyrosinase	0.005	0.10	0.20	0.30	0.40

[1] Units are milliliters.

shown in Figure E6.7. Is the time course linear for the full 3 minutes? Determine the longest period of linearity and calculate the average $\Delta A/\text{min}$ over this time period. If you observed a "blank" rate with no enzyme present, calculate its $\Delta A/\text{min}$ and subtract it from the enzyme-catalyzed reaction rate.

If no recorder is available, you must take absorbance readings at 475 nm every 30 seconds. Proceed as follows. To the cuvette, add buffer and L-dopa according to Table E6.1. Mix well and place in the photometer. Immediately adjust the absorbance to 0. Begin timing with a stopwatch and record in your notebook an absorbance reading every 30 seconds. Continue for 3 minutes. Now, add the appropriate level of tyrosinase, mix **(Do not shake),** immediately place the cuvette in the photometer, and set the absorbance to 0. Start the stopwatch and record an absorbance reading every 30 seconds for 3 minutes. On graph paper, prepare a plot of A_{475} vs. time. Calculate $\Delta A/\text{min}$ over a period during which the assay is linear. On the same graph paper, plot A_{475} vs. time for the "blank." If there is a significant rate (greater than 5% of the enzyme rate), subtract it from the enzyme-catalyzed rate before preparing the graph.

Repeat the above procedures (with or without recorder) with a different concentration of tyrosinase. If the corrected rate ($\Delta A/\text{min}$ for enzyme minus $\Delta A/\text{min}$ for blank) was less than 0.03, double the amount of enzyme added. If the corrected rate was greater than 0.3/min, reduce the enzyme by one-half. Determine the rate for a total of five different enzyme levels. The enzyme levels chosen should cover a $\Delta A/\text{min}$ range from about 0.025/min to 0.25/min. The tyrosine levels listed in Table E6.1 are only recommended; you may have to try higher or lower amounts of tyrosinase.

C. K_M of L-Dopa and D-Dopa

Now that the appropriate enzyme level has been determined, the kinetic constants may be evaluated. The K_M for L-dopa can be obtained by setting up the same assay as in part B, except that the factor to vary will be the concentration of L-dopa. The concentration of L-dopa in part B was sufficient to saturate all the tyrosinase active sites, so the rate depended only on the enzyme concentration. In part C, L-dopa levels will be varied over a range that is nonsaturating.

Choose the level of enzyme that will give a convenient rate, at saturating levels of L-dopa, of 0.10 to 0.15 $\Delta A/\text{min}$. Prepare a table, for five assays, similar to Table E6.1. The total volume for each assay must be constant at 3.0 mL. Since the enzyme added to each assay remains constant, only the volume of buffer and L-dopa is varied. The recommended levels of L-dopa are 0.10, 0.20, 0.40, 0.80, 1.0 and 1.5 mL. The procedure to follow is identical to part B. The enzyme should be added last, after a "blank" rate is determined for each assay. If the recommended levels of L-dopa do not result in a saturating rate, increase the amount of L-dopa. From the slope of each recorder trace or graph of absorbance vs. time, calculate the $\Delta A/\text{min}$ for each level of L-dopa. In order to determine the K_M of D-dopa, repeat the above procedure using D-dopa rather than L-dopa. Calculate the $\Delta A/\text{min}$ for each concentration level of D-dopa.

D. Inhibition of Tyrosinase Activity

Whether an inhibitor acts in a competitive or noncompetitive manner is deduced from a Lineweaver-Burk or direct linear plot using varying concentrations of inhibitor and substrate. In separate assays, two substances will be added to the dopa-tyrosinase reaction mixture, and the effect on enzyme activity will be quantified. The structures of the potential inhibitors, cinnamic acid, sodium benzoate, and thiourea, are shown in Figure E6.9. The inhibition assays must be done immediately following the K_M studies. To measure inhibition, reaction rates both with and without inhibitor must be used and the tyrosinase activity must not be significantly different. If it is necessary to do the inhibition studies later, the K_M assay for L-dopa must be repeated with freshly prepared tyrosinase solution.

To set up the inhibition assay, prepare a table similar to Table E6.1. Inhibitor should appear in the list of reagents before tyrosinase. Choose a level of tyrosinase that results in a relatively high reaction rate ($\Delta A/\text{min}$ of approximately 0.2). Vary the amount of L-dopa as in part C. A constant amount of inhibitor (cinnamic acid, benzoic acid, or thiourea)

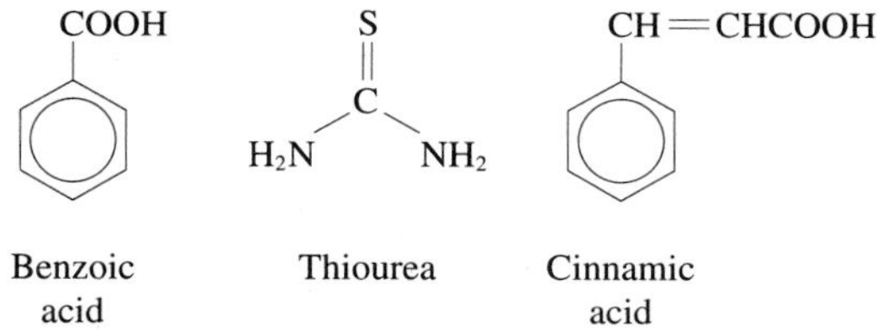

Figure E6.9
Three inhibitors of tyrosinase.

Benzoic acid

Thiourea

Cinnamic acid

should be added to each cuvette. You will have to determine this level of inhibitor by trial and error. The desired inhibition rate with saturating substrate is 10 to 20% of the uninhibited rate. Add all reagents except tyrosinase, mix well, and determine the blank rate, if any. Add tyrosinase, mix, and immediately record A_{475} for 3 minutes. From recorder traces or graphs of A_{475} vs. time, calculate $\Delta A/\text{min}$ for each assay.

IV. Analysis of Results

A. Measurement of Tyrosinase Concentration

If the tyrosinase preparation is homogeneous, the concentration of enzyme may be determined by absorbance measurement at 280 nm. The absorption coefficient, $E_{280}^{1\%}$, for tyrosinase is 24.9. That is, the absorbance of a 1% (w/v) solution of pure tyrosinase in a 1-cm cell at 280 nm is 24.9. Beer's law states that

$$A = Elc$$

where

   A = absorbance
   E = absorption coefficient
   c = concentration of solution
   l = cell path length (usually 1 cm)

For a 1% solution of tyrosinase, A is equal to E if l is 1 cm,

$$A = E_{280}^{1\%}$$

Therefore, the concentration of tyrosinase in solution is calculated from a simple ratio,

$$\frac{c_1}{A_1} = \frac{c_2}{A_2}$$

where

   c_1 = concentration of a standard solution of tyrosinase = 1%
   A_1 = absorption of the standard solution of tyrosinase at 280 nm = 24.9
   c_2 = concentration of unknown solution of tyrosinase in % (w/v)
   A_2 = absorption of unknown solution of tyrosinase at 280 nm

Convert your concentration (now in w/v%) to mg of tyrosinase/mL.

B. Determination of the Tyrosinase Level for Kinetic Assays

Prepare a table of rate (ΔA/min) vs. enzyme concentration (mg per assay). ΔA/min units are converted to a more useful form, amount of product formed/time, usually μmole/minute. The conversion is straightforward when Beer's law is used in the form of Equation E6.9.

$$\blacktriangleright c = \frac{A}{El} \qquad \text{(Equation E6.9)}$$

where A refers to the absorbance change that occurs per minute, l is the light path length through the cuvette (1 cm), E is the molar absorption coefficient for the product dopachrome (3600 M^{-1} cm^{-1}), and c is the concentration of product. According to this relationship, the absorbance change during 1 minute for production of 1 molar dopachrome is 3600:

$$1 \text{ molar} = \frac{A}{3600 \ M^{-1} \ \text{cm}^{-1} \times 1 \ \text{cm}}$$

$$A = 3600$$

To convert the raw data to moles of product formed per minute per liter, each ΔA/min is divided by 3600. For example, a ΔA/min of 0.1 is converted to μmoles of product formed (Δc) in the following way:

$$\Delta c = \frac{\Delta A/\text{min}}{El} = \frac{0.1/\text{min}}{3600 \ M^{-1} \ \text{cm}^{-1} \ \text{cm}} = 2.78 \times 10^{-5} \ \frac{M}{\text{min}}$$

Now convert to moles/min:

$$2.78 \times 10^{-5} \ \frac{\text{moles}}{\text{liter}-\text{min}} \times 0.003 \text{ liters} = 8.33 \times 10^{-8} \ \frac{\text{moles}}{\text{min}}$$

Now convert to μmoles/min:

$$\Delta c = 8.33 \times 10^{-2} \ \frac{\mu\text{moles}}{\text{min}}$$

Add another column to your table, label it μmoles product formed per minute, and calculate the appropriate rate for each enzyme concentration. Prepare a graph of rate (μmole/min on the ordinate, y axis) vs. enzyme concentration in each assay (mg, on the abscissa, x axis). Connect as many of the points as possible with a straight line passing through the origin. If most of the points are on this line, the assay and the standard curve can be used to quantify an unknown level of tyrosinase. The standard curve also provides the experimenter with a choice of enzyme levels to use for further kinetic studies.

C. K_M of L-Dopa and D-Dopa

The treatment of results will be described for L-dopa. The procedure for D-dopa is identical. Prepare a table of L-dopa concentration per assay (mmolar) vs. ΔA/min. Convert all ΔA/min units to μmoles/min as described in part B. Prepare a Michaelis-Menten curve (μmoles/min vs. [S]) as in Figure E6.1 and a Lineweaver-Burk plot ($1/\mu$mole/min vs. $1/$[S]) as in Figure E6.2. Alternatively, you may wish to use the direct linear plot. Estimate K_M and V_{max} from each graph. The intercept on the rate axis of the Lineweaver-Burk plot is equal to $1/V_{max}$. For example, if the line intersects the axis at 0.02, then $V_{max} = 1/0.02$ or 50 μmoles of product formed per minute. The line intersects the $1/$[S] axis at a point equivalent to $-1/K_M$. If the intersection point on the $1/$[S] axis is -0.67, then $K_M = -1/-0.067 = 15$ μmolar. Repeat this procedure for the data obtained for D-dopa. Compare the K_M and V_{max} values and explain any differences.

Some enzymes are able to differentiate between D and L isomers of substrates. Dopa, with a chiral center, exists in enantiomeric pairs, D and L. The naturally occurring dopa molecule has the L configuration, so it might be expected that the enzyme-catalyzed oxidation of L-dopa is more facile than that of D-dopa. The relative reactivity can be determined in a quantitative way by comparing K_M or V_{max} values. Does tyrosinase exhibit stereoselectivity?

D. Inhibition of Tyrosinase Activity

A Lineweaver-Burk plot is prepared for each chemical inhibitor tested. Each plot of $1/v_0$ vs. $1/$[S] consists of two lines, one representing the absence of inhibitor, the other representing a specific concentration of inhibitor.

Convert all ΔA/min data into μmole/min and calculate the reciprocal of each data point. Plot against the reciprocal of the corresponding substrate concentration ($1/$mmolar). One line on each graph is obtained from the set of data acquired from the L-dopa K_M experiment in part C. Draw the two separate lines with a straight-edge, making sure the appropriate data points are used. Compare the two graphs with Figures E6.4 and E6.5 to determine the type of inhibition. Explain the inhibitory effect of the substance.

V. Questions and Problems

▶ 1. One of the many straight-line modifications of the Michaelis-Menten equation is the Eadie-Hofstee equation:

$$v_0 = -K_M \frac{v_0}{[S]} + V_{max}$$

Beginning with the Lineweaver-Burk equation, derive the Eadie-Hofstee equation. Explain how it may be used to plot a straight line from experimental rate data. How are V_{max} and K_M calculated?

▶ 2. The table below gives initial rates of an enzyme-catalyzed reaction along with the corresponding substrate concentration. Use any graphical method to determine K_M and V_{max}.

v_0 (μmole/min)	[S] (mole/liter)
130	65×10^{-5}
116	23×10^{-5}
87	7.9×10^{-5}
63	3.9×10^{-5}
30	1.3×10^{-5}
10	0.37×10^{-5}

▶ 3. The enzyme concentration used in each assay (1000 mL) in Problem 2 is 5×10^{-7} mole/liter. Calculate k_3, the turnover number, in units of sec^{-1}.

4. Using the data in Problems 2 and 3, calculate the specific activity of the enzyme in units/mg and katal/mg. Assume the enzyme has a molecular weight of 55,000 and the reaction mixture for each assay is contained in a total volume of 1.00 mL.

▶ 5. Each of the compounds listed below is known to inhibit tyrosinase activity.

sodium azide, NaN$_3$	sodium cyanide, NaCN
L-phenylalanine	8-hydroxyquinoline
tryptophan	diethyldithiocarbamate
cysteine	thiourea
4-chlororesorcinol	phenylacetate

(a) Study the structure of each and try to predict whether the substance is a competitive or noncompetitive inhibitor of tyrosinase activity as measured by the dopachrome assay. Assume that the inhibition data were evaluated using a Lineweaver-Burk plot of $1/v_0$ vs. $1/[dopa]$.

(b) An alternative method for measuring tyrosinase activity is the use of an oxygen electrode to measure the rate of dioxygen utilization during phenol oxidation (see Equations E6.7 and E6.8). Predict whether each substance is a competitive or noncompetitive inhibitor of tyrosinase as measured by an oxygen electrode. Assume that a Lineweaver-Burk plot of $1/v_0$ vs. $1/[O_2]$ was used to evaluate the rate data.

6. Compound X was tested as an inhibitor of the enzyme in Problem 2. Use the rate data in Problem 2 and the following inhibition data to evaluate compound X. Is it a competitive or noncompetitive inhibitor? Calculate K_I for the inhibitor.

$[X] = 3.7 \times 10^{-4} M$ v_0 (μmole/min)	$[S]$ (mole/liter)	$[X] = 1.58 \times 10^{-3} M$ v_0 (μmole/min)
125	6.5×10^{-4}	110
102	2.3×10^{-4}	80
70	7.9×10^{-5}	40
45	3.9×10^{-5}	20
20	1.3×10^{-5}	—
5	0.37×10^{-5}	—

VI. References

J. Bell and E. Bell, *Proteins and Enzymes* (1988), Prentice-Hall (Englewood Cliffs, NJ), pp. 314–331. A review of enzyme kinetics.

A. Cornish-Bowden, *Fundamentals of Enzyme Kinetics* (1979), Butterworths (London). An entire book devoted to enzyme kinetics.

H. Duckworth and J. Coleman, *J. Biol. Chem.* **245**, 1613–1625 (1970). "Physico-chemical and Kinetic Properties of Mushroom Tyrosinase."

V. Hearing and M. Jiminez, *Int. J. Biochem.* **19**, 1141–1147 (1987). "Mammalian Tyrosinase."

K. Lerch, *Life Chem. Rep.* **5**, 221–234 (1987). "Molecular and Active Site Structure of Tyrosinase."

C. Mathews and K. van Holde, *Biochemistry* (1990), Benjamin/Cummings (Redwood City, CA), pp. 339–380. An introduction to enzyme kinetics.

S. Pomerantz, *J. Biol. Chem.* **241**, 161–168 (1966). "The Tyrosine Hydroxylase Activity of Mammalian Tyrosinase."

J. Rawn, *Biochemistry,* 2nd ed. (1989), Neil Patterson Publishers (Burlington NC), pp. 149–193. A complete chapter on enzyme kinetics.

D. Robb, in *Copper Proteins and Copper Enzymes*, R. Lontie, Editor (1984), CRC Press (Boca Raton, FL), pp. 207–241. "Tyrosinase."

L. Stryer, *Biochemistry*, 3rd ed. (1988), W. H. Freeman (New York), pp. 177–200. Introduction to enzymes and kinetic analysis.

D. Voet and J. Voet, *Biochemistry* (1990), John Wiley & Sons (New York), pp. 316–328. An introduction to enzymes.

W. Wood, J. Wilson, R. Benbow, and L. Hood, *Biochemistry, A Problems Approach,* 2nd ed. (1981), Benjamin/Cummings (Menlo Park, CA), pp. 144–172. Enzyme kinetics with an introduction to the Eisenthal and Cornish-Bowden direct linear plot.

C. Worthington, Editor, *Worthington Enzyme Manual* (1988), Worthington Biochemical Corporation (Freehold, NJ), pp. 288–291. "Tyrosinase."

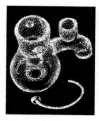

Purification and Characterization of a Lipid Extracted from Nutmeg

■ **Recommended Reading**

Chapter 3, Sections A, B, D.

■ **Synopsis**

Lipids, a group of chemical substances found in both plant and animal cells, are among the most ubiquitous of biomolecules. A lipid extract of ground nutmeg will be purified by adsorption chromatography on a silica gel column. Analysis of the lipid extract by thin-layer chromatography will provide the classification of the unknown lipid. The molecular structures of the major components present in the extract will be elucidated in Experiment 8.

I. Introduction and Theory

Lipids are low-molecular-weight biomolecules that can be extracted from plant and animal tissues by organic solvents. The major classes of compounds represented in the lipid group are triacylglycerols, glycerophosphatides, sphingolipids, glycolipids, and sterols (Figure E7.1). Lipids are found in most cells and tissue, but rarely are they present in a "free" or "uncombined" state. They are usually associated with proteins or carbohydrates, which complicates their extraction and structural identification. Lipids not only have a wide variety of molecular structures but also are involved in a whole array of biological functions, including membrane structural integrity, vitamin activity, cell recognition, and cellular energy.

In this experiment we will approach the world of lipids with an introduction to analytical procedures currently applied to lipid extraction, purification, and characterization.

Lipid Extraction

The extraction procedure used to isolate lipids from biological tissue depends on the class of lipid desired and the nature of the biological source (animal tissue, plant leaf, plant seed, cell membranes, etc.). Because lipids are generally less polar than other cell constituents, they may be selectively extracted with the use of organic solvents. Early studies of lipids used ether, acetone, hexane, and other organic solvents for extraction; however, these solvents extract only lipids bound in a nonpolar or

Figure E7.1
Structures of the major lipid classes.

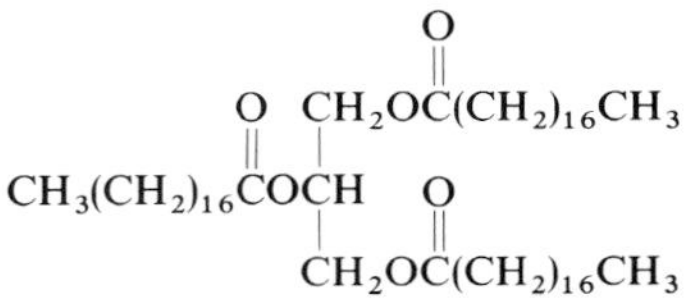

A Triacylglycerol
(tristearin)

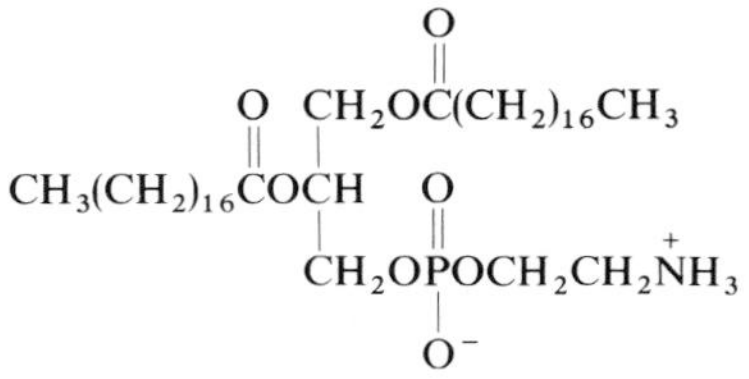

A Glycerophosphatide
(phosphatidyl ethanolamine)

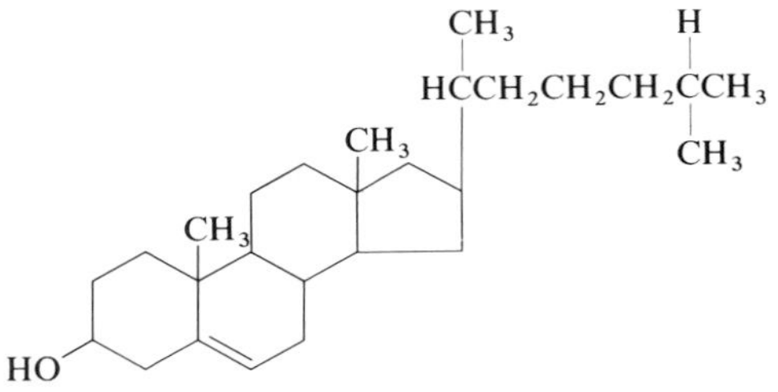

A Steroid
(cholesterol)

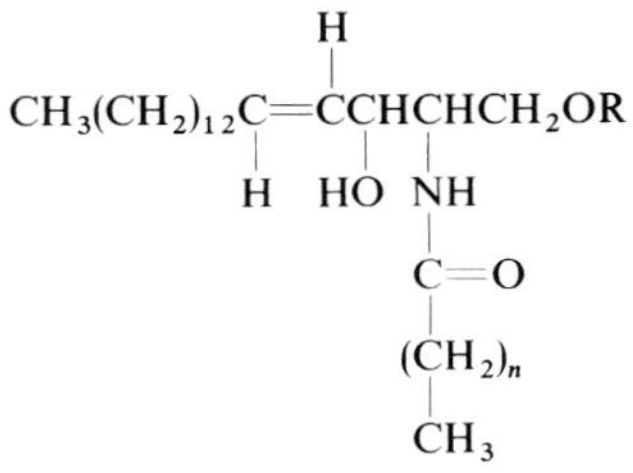

A Sphingolipid or Glycolipid

R = H, Ceramide
R = Phosphocholine, Sphingomyelin
R = galactose, Cerebroside

hydrophobic manner. In the 1950s, Folch's group reported the use of chloroform and methanol (2:1) in lipid extractions (Folch, 1957). This became the solvent system of choice for total lipid extraction because it was especially useful in dissociating lipid-protein complexes of plasma membranes. The only apparent disadvantage of this solvent system was its tendency to dissolve some nonlipids, including proteins. Increasing awareness of safety in the laboratory (both methanol and chloroform are toxic) led to a search for other solvent systems. A more desirable lipid solvent is hexane and isopropanol (3:2) (Radin, 1981). In addition to lower toxicity, it has the advantage over chloroform-methanol that it dissolves very little nonlipid material. Whatever solvent system is used, chloroform-methanol or hexane-isopropanol, a total lipid extraction process yields a complex mixture of organic compounds in which all components have one common characteristic, a relatively nonpolar structure.

Solvent extraction of lipids from biological tissues is best carried out in a homogenizer as described in Experiment 3. More drastic extraction conditions, such as ultrasonic vibration or grinding with sand, may be necessary if the biomaterial contains excessive amounts of connective tissue. After the solvent extraction, several "work-up" steps should be completed before isolation of the crude lipid residue. If chloroform-methanol has been used, it is essential to wash the extracted lipid solution with an aqueous salt solution (0.5 to 1% KCl, NaCl, or $CaCl_2$) to remove water and nonlipid materials. It is generally not necessary to wash hexane-isopropanol extractions, but if desired they may be extracted with 7 to 10% aqueous sodium sulfate. This treatment removes any nonlipid materials that may be present and also reduces the volume of the extract because it removes water-soluble isopropanol without removing desired lipid material. Reducing the volume of the extraction solvent is particularly beneficial if further purification of the lipids by chromatography is under consideration.

The lipid extract, now washed to remove nonlipid materials, must be dried with anhydrous sodium sulfate or magnesium sulfate to remove the last traces of water. Evaporation of the water-free solvent extract is accomplished by rotary evaporation under vacuum. The crude, dry lipid residue may be susceptible to chemical modification by air oxidation, so it should be stored in the cold ($-20°C$) and dark, under an atmosphere of N_2 or argon gas. Lipids containing polyunsaturated fatty acids are especially sensitive to air oxidation, so protective measures should be taken if these substances are of interest. For a more detailed discussion of lipid extraction and analysis, consult the list of references at the end of the experiment.

Lipid Purification

Separation of crude lipid extracts into individual lipid classes is difficult and time-consuming. In some cases a crude separation of lipids can be attained by selective solvent extraction. The more nonpolar neutral lipids (triacylglycerols and steroids) may be efficiently extracted from biological material with chloroform, ether, or acetone. The more polar lipids (phospholipids and glycolipids) may be selectively isolated by initial extraction of the tissue with cold acetone to remove neutral nonpolar lipids, followed by extraction of more polar lipids with hexane-isopropanol. For more extensive purification of lipids, the researcher must turn to chromatography. Chromatographic methods can both resolve a complex lipid mixture into the various classes of lipids and separate the individual components of a single class of lipids. Silica gel chromatography (column and thin-layer) and HPLC are the most frequently used methods for the resolution of lipid extracts. For large-scale separation of lipid classes, column chromatography using various adsorbents (silica gel, Florisil, ion-exchange celluloses, and lipophilic Sephadexes) is most effective.

Adsorption chromatography (Chapter 3) on columns of silica gel is the best preparative method for the separation and purification of neutral lipids (Sweeley, 1969). Silica gel (sometimes called silicic acid) is chemically represented as $SiO_2 \cdot xH_2O$ and is a relatively polar adsorbent. Neutral or nonpolar lipids are not readily adsorbed by the support and can be separated from the more polar lipids, which are preferentially adsorbed. In practice, a glass column is packed with a slurry of silica gel in a hydrocarbon solvent (petroleum ether, pentane, or hexane). The lipid fraction to be purified is dissolved in the same hydrocarbon solvent and applied to the top of the column. To elute the various lipid classes, solvents of increasing polarity are allowed to percolate through the silica gel column and are collected at the bottom of the column. Table E7.1 outlines the order of elution of lipids from silica gel columns. The polarity of the eluting solvent is increased by adding a gradually increasing proportion of ethyl ether. For example, 1% ether in petroleum ether (fraction I) elutes only the least polar lipids—paraffins and fatty acid esters; 4% ethyl ether in petroleum ether has sufficient polarity to elute triacylglycerols, and 25 to 75% ethyl ether in petroleum ether elutes diacylglycerols and monoacylglycerols. For elution of phospholipids and glycolipids, solvent systems of increasing polarity such as chloroform-methanol must be used; however, silica gel columns are most effective for resolution of the neutral lipids.

Once several silica gel column fractions have been collected, the eluted fractions must be analyzed for the presence of lipids. Thin-layer chromatography on silica gel plates will be introduced in this experiment for the detection and identification of neutral lipids.

As discussed in Chapter 3, thin-layer chromatography utilizes a thin coating of silica gel on a glass or plastic plate to separate components of a mixture. The mixture of chemical substances to be analyzed is applied near one edge of the plate, and development of the plate with a solvent system allows the applied samples to move a distance that is related to the polarity of the sample. A solvent system consisting of hexane, diethyl ether, and acetic acid will separate triacylglycerols, cholesterol esters, fatty acid esters, and fatty acids (see Figure E7.2). More polar lipids,

Table E 7.1
Order of Elution of Lipids from Silica Gel Columns

Fraction Number	Solvent	Lipid Component
Neutral		
I	1% Ethyl ether/petroleum ether	Paraffins
II	1% Ethyl ether/petroleum ether	Squalene, waxes, fatty acid esters
III	1% Ethyl ether/petroleum ether	Cholesterol esters
IV	4% Ethyl ether/petroleum ether	Triacylglycerols, fatty acids
V	8% Ethyl ether/petroleum ether	Steroids (cholesterol)
VI	25% Ethyl ether/petroleum ether	Diacylglycerols
VII	100% Ethyl ether	Monoacylglycerols
Polar		
VIII	5% Methanol/chloroform	Ceramides, phosphatidic acid, cardiolipins
IX	20% Methanol/chloroform	Phosphatidylethanolamine, phosphatidylserine phosphatidylinositol
X	30% Methanol/chloroform	Phosphatidylcholine
XI	50% Methanol/chloroform	Sphingomyelin
XII	70% Methanol/chloroform	Lysophosphatidylcholine

such as glycerophosphatides, sphingolipids, and glycolipids, remain at the origin. Pure standard samples of lipids are evaluated on the same TLC plate in order to help identify unknowns. The lipids on the solvent-developed plate are not colored but will react with iodine vapor to produce dark red-brown spots on a yellow background.

Overview of the Experiment

In this experiment you will isolate an unknown lipid mixture from its natural source by solvent extraction. The source of the unknown lipid is *Myristica fragrans* seed (nutmeg). This particular seed is unusual in that it contains an appreciable quantity of a single type of lipid. In fact, a single compound makes up 70 to 80% of the total lipid extract. The lipid is of sufficient homogeneity to allow its isolation as a solid with a relatively sharp melting point. The crude isolated material will be purified by silica gel column chromatography or recrystallization and classified by thin-layer chromatography with standard lipids. For a flowchart of the procedure, see Figure E7.3. Complete identification of the molecular structure of the lipid will be the objective of Experiment 8. Approximate time requirements for each experimental section are:

    A. Isolation of the unknown lipid—1 hour

    B. Purification of the unknown lipid
        1. Recrystallization—$\frac{1}{2}$ hour
        2. Silica gel column—1 hour

    C. Characterization by thin-layer chromatography—2 hours

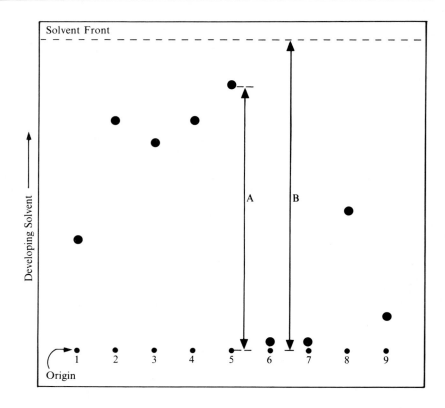

Figure E7.2
Typical thin-layer chromatogram of lipids. The solvent system, hexane–diethyl ether–acetic acid (90:10:1). (1) cholesterol, (2) fatty acid, (3) triacylglycerol, (4) fatty acid methyl ester, (5) cholesterol ester, (6) glycerophosphatide, (7) sphingolipid, (8) 1,3- or 1,2-diacylglycerol, (9) 1- or 2-monocylglycerol. The R_f for (5), cholesterol ester, is calculated as follows:

$$R_f = \frac{\text{distance of migration of cholesterol ester}}{\text{distance of migration of solvent}} = \frac{A}{B} = \frac{72 \text{ mm}}{85 \text{ mm}} = 0.85$$

II. Materials and Supplies

Nutmeg, ground or whole

Hexane-isopropanol, 3:2

Acetone

Diethyl ether

Hexane

Silica gel, 100–200 mesh

Flow Chart of Procedure

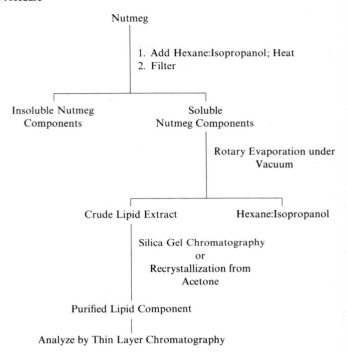

Figure E7.3
Flowchart for isolation of lipid from nutmeg.

Glass column (1 × 25 cm)

Glass wool or cotton

Small funnel and qualitative grade filter paper

Silica gel thin-layer plates (20 × 20 cm and 5 × 10 cm)

Chromatography solvent system, hexane–diethyl ether–acetic acid (80:20:1) in a chromatography jar

Lipid standards (2% solutions in chloroform): a triacylglycerol (triolein), cholesterol ester (cholesterol linoleate), fatty acid (palmitoleic, oleic, etc.), fatty acid methyl ester (linolenic acid, methyl ester), a glycerophosphatide (phosphatidylcholine, phosphatidylethanolamine, etc.), a diacylglycerol (diolein), and a monoacylglycerol (monoolein).

Microcapillaries or tapered capillaries

Iodine chamber. Place a few crystals of iodine in an empty chromatography jar and cover.

Rotary evaporator

III. Experimental Procedure

A. *Isolation of the Unknown Lipid*

⚠ Caution

Acetone, ether, isopropanol, and hexane cause eye and skin irritation. Vapor or mist of these solvents is irritating to the eyes, mucous membranes, and upper respiratory tract. Prolonged exposure should be avoided. Harmful if swallowed.

The solvents, also, are volatile and extremely flammable. Always measure quantities in a hood. No lighted burners or flames should be in the vicinity of your work. In case of fire, use a CO_2 extinguisher.

First aid: eyes—flush with copious amounts of water for at least 15 minutes; skin—wash with soap and copious amounts of water; inhalation—remove to fresh air.

Dispose in the organic waste container.

Place 5 g of finely crushed or ground nutmeg into a 125-mL Erlenmeyer flask. Add 50 mL of hexane-isopropanol (3:2) and warm on a steam bath or hot plate in a hood for 15 minutes. Mix well while heating. Filter the extraction mixture rapidly through fluted qualitative grade filter paper and pour an additional 20 mL of warm hexane-isopropanol through the solid residue in the filter. Remove the solvent from the extract under vacuum on a rotary evaporator to obtain a yellow oil or off-white solid. Alternatively, the solvent may be removed on a steam bath (**Hood**!) while flushing the inside of the flask with N_2 gas. Weigh the crude product and retain a 1- to 2-mg portion for chromatographic analysis (part C).

B. *Purification of the Unknown Lipid*

Partial and rapid purification of the lipid is possible by recrystallization from acetone. If more time is available, and a purer sample of the lipid is desired, a silica gel column should be run.

RECRYSTALLIZATION

To recrystallize the unknown, dissolve the crude residue in a minimum amount (about 20 mL) of warm acetone on a steam bath (**Hood**!). After slow cooling of the acetone solution to room temperature and then in ice, filter by suction, and wash the solid product with ice-cold acetone. A typical yield is about 0.5 g of purified lipid. Save the purified lipid for Experiment 8.

SILICA GEL CHROMATOGRAPHY

To prepare the silica gel column, clamp the glass column to a ring stand. Prepare a slurry of 6 g of silica gel in 20 mL of hexane. Put a small piece of glass wool or cotton into the bottom of the column and pour about 5 mL of hexane into the column. Be sure the stopcock is closed! With a long glass rod, tap the glass wool or cotton to drive out air bubbles and push it tightly into the column. Place a small funnel on top of the column and begin to pour the well-stirred slurry of silica gel into the column. As the silica gel begins to pack into a column, open the stopcock to allow release of solvent. Continue to add the slurry of silica gel until a tightly packed column of 10 to 12 cm is attained. *Do not allow the column to run dry during packing.* Allow 1 to 2 cm of solvent to remain above the level of the silica gel.

Prepare the lipid for chromatography by dissolving 50 to 75 mg of crude lipid in a minimum of hexane (5 to 10 mL). Open the stopcock and allow excess solvent to drain from the column until the level of solvent just reaches the top of the silica gel column. Very carefully add the solution of crude lipid to the top of the column. This should be done by using a Pasteur pipet and allowing drops of the solution to run down the inside of the glass column to the top of the silica gel bed. It is important not to disturb the top of the silica gel column. After addition of the lipid solution, begin to collect a fraction from the bottom of the column into a 25-mL Erlenmeyer flask. Set the flow rate to about 2 drops per second. When the level of solution in the column reaches the top of the silica gel, turn off the stopcock and carefully rinse the inside of the glass column with 1 mL of hexane. Open the stopcock until the solvent just enters the column. This solvent washes the lipid material that may have adhered to the inside glass walls. Now fill the column with hexane and continue to collect the eluting solvent at 2 drops per second. Collect a total of 20 mL of solvent in the first fraction. During this collection, continue to add solvent to the top of the column. Now change the eluting solvent to 90% hexane and 10% diethyl ether. Pour this into the column and collect four 20 mL fractions. Be sure to number the collection flasks. Add solvent to the column as necessary. Finally, elute the column with 80% hexane and 20% diethyl ether and collect two more 20-mL fractions.

Each column fraction is analyzed for lipid material by spotting on a 5×10 cm silica gel thin-layer plate and exposing it to iodine vapor as follows. Prepare seven tapered capillary tubes and use these to place a spot of each solution on the TLC plate. Put at least 10 capillary applications from a single fraction on a spot. Your final plate should then have seven spots, one for each fraction. Set the plate in an iodine chamber and allow it to remain for about 15 minutes or until some spots are yellow or red-brown. The presence of lipid in a fraction is indicated by the red-brown color. Retain all the column fractions that appear to have lipid.

Each of these fractions will be analyzed in part C by thin-layer chromatography.

C. Partial Characterization of the Unknown Lipid

⚠ **Caution**

Chloroform is harmful if inhaled, swallowed, or absorbed through the skin. Vapor or mist is irritating to the eyes, mucous membranes, and upper respiratory tract. Chloroform is suspected to be a cancer-causing agent.

First aid: eyes—prolonged irrigation with copious amounts of water; skin—wash extensively with soap and water; inhalation—remove to fresh air.

Dispose in the organic waste container.

Analysis of the purity of each silica gel column fraction and classification of the unknown lipid can be accomplished by thin-layer chromatography on silica gel plates. On a single plate will be spotted (1) a solution of the crude lipid extract from part A, (2) aliquots from each lipid fraction of the column (or recrystallized lipid), and (3) solutions of standard lipids (listed in the Materials section). On a 20 × 20 cm silica gel plate there is room for nine different analyses. Prepare a 1% solution of the crude lipid from part A in chloroform (10 mg/1 mL). If recrystallized lipid is to be analyzed rather than column-purified lipid, prepare a 1% solution in chloroform as for the crude lipid. Following the procedure you used in Experiment 2 (amino acid analysis), apply each sample in a row across one side of the silica gel plate, 2 cm from the edge (see Figure E7.2). Use small, tapered capillary tubes or microcapillaries. To apply each solution, dip the capillary into the solution, and gently touch the end of the filled capillary on the TLC plate where you desire the spot. Allow the solution from the capillary to spread onto the TLC plate to form a circle no wider than 0.5 cm diameter. Lift the capillary from the plate and allow the solvent on the plate to dry. Again touch the end of the capillary to the same spot and reapply solution. Repeat this application process about 20 times for each sample. Of course, each sample to be analyzed is placed on a different location along the edge of the TLC plate.

Develop the plate in hexane–diethyl ether–acetic acid (80:20:1) by placing the TLC plate in a chamber containing the solvent system, making sure the edge with the applied samples is down. The solvent level in the chamber must not be above the application spots on the plate. (Why?) Leave the chromatogram in the chromatography jar until the solvent front rises to about 1 cm from the top of the plate (45 to 60 min). Remove the plate and make a small scratch at the solvent level. Allow the chromatogram to dry (**Hood**!) and then place it in an iodine chamber for several minutes. Remove the plate and lightly trace, with a pencil or other sharp object, around each red-brown spot. This should be done promptly, as the colors will fade with time. Calculate the mobility of each standard and unknown lipid relative to the solvent front (R_f):

$$R_f = \frac{\text{distance from the origin migrated by a compound}}{\text{distance from origin migrated by solvent}}$$

IV. Analysis of Results

Calculate the approximate percentage lipid composition of the nutmeg (assume the lipid recovery is quantitative). Prepare a table listing the R_f values obtained for each standard lipid and for the crude and purified unknown. By comparing R_f values, identify the general class of lipid to which your purified extract belongs. How effective was your purification step? If impurities are present in the crude or purified lipid, can you identify them? If you recrystallized the crude lipid, calculate the recovery yield.

$$\text{Recovery yield}(\%) = \frac{\text{g purified lipid}}{\text{g crude lipid}} \times 100$$

V. Questions and Problems

▶ 1. If you wished to determine the complete structure of the isolated lipid, what experimental approach would you follow?

▶ 2. Why is acetic acid added as a minor component (1%) of the chromatography solvent system?

▶ 3. Try to explain the relative order of R_f values you obtained for the standard lipids.

▶ 4. How does iodine react to produce the red-brown spots on a developed chromatogram? Why do some lipids give darker spots than others?

5. Calculate the R_f for standard lipid number 2 (a fatty acid) in Figure E7.2.

6. Why should whole nutmeg be ground before the solvent extraction?

VI. References

W. Christie, *HPLC and Lipids—A Practical Guide* (1987), Pergamon Press (Oxford). Introduction to HPLC and practical applications.

J. Folch, M. Lees, and G. Stanley, *J. Biol. Chem.* **226**, 497–509 (1957). Use of chloroform-methanol as a lipid solvent.

F. Gunstone, J. Harwood, and F. Padley, Editors, *The Lipid Handbook* (1986), Chapman & Hall (London). Reference book presenting the chemistry, biochemistry, and technology of lipids.

M. Gurr and J. Harwood, *Lipid Biochemistry,* 4th ed. (1991), Chapman & Hall (New York). An introduction to lipid structure and function.

A. Kuksis, Editor, *Chromatography of Lipids in Biomedical Research and Clinical Diagnosis* (1987), Elsevier (Amsterdam). A complete book on lipid analysis.

C. Mathews and K. van Holde, *Biochemistry* (1990), Benjamin/ Cummings (Redwood City, CA), pp. 298–307. An introduction to lipid structure.

J. Mead, R. Alfin-Slater, D. Howton, and G. Popják, *Lipids: Chemistry, Biochemistry and Nutrition* (1986), Plenum Publishing (New York). An excellent book covering all aspects of lipids.

G. Nelson, in *Analysis of Lipids and Lipoproteins,* E. Perkins, Editor (1975), The American Oil Chemist's Society (Champaign, IL), pp. 1–19. General extraction methods.

N. Radin, in *Methods in Enzymology,* Vol. 72, J. M. Lowenstein, Editor (1981), Academic Press (New York), pp. 5–7. Extraction of lipids with hexane-isopropanol.

J. Rawn, *Biochemistry,* 2nd ed. (1989), Neil Patterson Publishers (Burlington, NC), pp. 209–218. Introduction to lipids.

L. Stryer, *Biochemistry,* 3rd ed. (1988), W. H. Freeman (New York), pp. 283–312. Introductory material on lipid structure and function.

C. Sweeley, in *Methods in Enzymology,* Vol. XIV, J. M. Lowenstein, Editor (1969), Academic Press (New York), pp. 255–267. Silica gel column chromatography.

M. Tevini and D. Steinmuller, in *HPLC in Biochemistry,* A. Henschen, K. Hupe, F. Lottspeich, and W. Voelter, Editors (1985), VCH (Weinheim), pp. 349–412. "Lipids."

D. Voet and J. Voet, *Biochemistry* (1990), John Wiley & Sons (New York), pp. 271–279. A discussion of lipid structure.

G. Zubay, *Biochemistry,* 2nd ed. (1985), Macmillan (New York), pp. 154–170. Structure and function of lipids.

EXPERIMENT **8**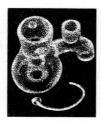

Fatty Acid Content of Triacylglycerols in Natural Oils

■ **Recommended Reading**

Chapter 3, Section C.

■ **Synopsis**

Fatty acids that are used for energy metabolism are stored in triacylglycerols. Three fatty acids are linked through ester bonds to the glycerol skeleton. In this experiment, triacylglycerol samples will be characterized by identifying the fatty acids at carbon 2 and comparing them to the fatty acids at carbons 1 and 3 combined. The methods used include release of the fatty acids by saponification and lipase-catalyzed hydrolysis followed by analysis of the corresponding fatty acid methyl esters with gas chromatography.

I. Introduction and Theory

Triacylglycerols, the neutral, saponifiable lipids found in most organisms, serve as biochemical energy reserves in the cell. Neutral lipids may be isolated from natural sources by extraction with nonpolar solvents (ethyl ether, chloroform, hexane-isopropanol) as in the isolation of the unknown lipid from nutmeg (Experiment 7). Chemically, the triacylglycerols are fatty acid esters of the trihydroxy alcohol glycerol (Figure E8.1). The figure also illustrates the recommended stereospecific numbering (sn) scheme for glycerol derivatives. According to IUPAC (1967), "The carbon atom that appears on top in that Fischer projection that shows a vertical carbon chain with the secondary hydroxyl group to the left is designated as C-1." To differentiate such numbering from conventional numbering conveying no steric information, the prefix "sn" (stereospecific numbering) is used; therefore, the structure in Figure E8.1 is designated 1,2,3-triacyl-*sn*-glycerol.

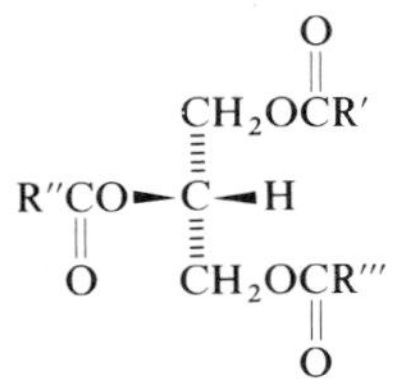

Figure E8.1.
General structure for a triacylglycerol. R represents the alkyl chains of fatty acids.

A variety of saturated and unsaturated fatty acids is present in triacylglycerols. Among the most abundant fatty acids are myristic (14:0), palmitic (16:0), stearic (18:0), and oleic ($18{:}1^{\Delta 9}$). To completely characterize the neutral lipids, one must know the identity of the fatty acids *and* the position that each fatty acid occupies on the glycerol skeleton (carbon number 1, 2, or 3). The biosynthetic distribution of fatty acids in positions 1, 2, or 3 of glycerol was once considered to be random; however, more recent studies indicate some selectivity. For example, the proportion of unsaturated to saturated fatty acids in position 2 is quite often greater than at positions 1 and 3. Analytical techniques are available to determine the identity and position of each fatty acid in a triacylglycerol sample. Since glycerol is a **prochiral** molecule, it should be recognized that the 1 and 3 positions of glycerol are *not* equivalent. Methods have been devised for distinguishing between the fatty acids at positions 1 and 3 (stereospecific analysis of triacylglycerols) but a discussion is beyond the scope of this experiment (Brockerhoff, 1975). Because rather extensive experimentation is required to distinguish between fatty acids at positions 1 and 3, in this experiment you will consider these fatty acids as a group and instead distinguish between the fatty acids at position 2 and those at 1 and 3 combined. This will allow you to compare the amount of unsaturated fatty acids at position 2 relative to 1 and 3 combined.

Characterization of Triacylglycerols

The development of procedures in lipid enzymology, gas chromatography, and thin-layer chromatography now makes it possible to do structural analysis of triacylglycerols. Fatty acid content (identity and percentage composition) of triacylglycerols is most easily determined by complete saponification (NaOH/methanol) followed by esterification of the released fatty acids (Equation E8.1). Gas chromatographic analysis of the fatty acid methyl esters (FAMEs), in conjunction with analysis of the appropriate standard FAMEs, conveniently provides both qualitative and quantitative fatty acid analysis. HPLC methods for the separation of fatty acids and fatty acid methyl esters are now under development, but gas chromatography still provides the best separation and versatility.

(Equation E8.1)

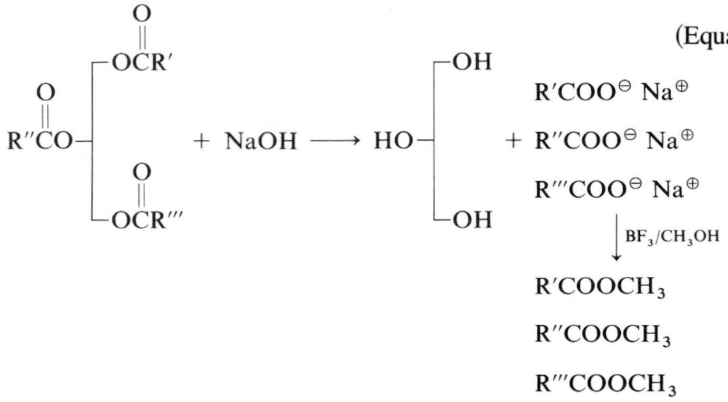

Positional distribution of fatty acids in triacylglycerols is elucidated by exploiting the selectivity of the lipolytic enzymes called lipases. Lipases (e.g., EC 3.1.1.3) are found in animals, microorganisms, and higher plants, and they catalyze ester hydrolysis, with preference for the acyl linkages at positions 1 and 3 of triacylglycerols. The cellular products of lipase-catalyzed hydrolysis of triacylglycerols are 2-monoacylglycerols and the free fatty acids from positions 1 and 3 (Equation E8.2).

▶

(Equation E8.2)

$$
\begin{array}{ccc}
R''CO\begin{bmatrix} -OCR' \\ \\ -OCR''' \end{bmatrix} + 2\,H_2O & \xrightarrow{\text{Lipase}} & R''CO\begin{bmatrix} -OH \\ \\ -OH \end{bmatrix} + \begin{array}{l} R'COOH \\ R'''COOH \end{array}
\end{array}
$$

Lipases act most efficiently in a heterogeneous medium, when the lipid substrate is dispersed as an emulsion in the buffered, aqueous phase containing the enzyme. The free fatty acids retrieved from the 1 and 3 positions of the triacylglycerol are characterized by esterification and gas chromatographic analysis. Information from this analysis leads directly to identification of the fatty acids at the combined positions 1 and 3. The 2-monoacylglycerol can be isolated from the reaction mixture by thin-layer or silica gel column chromatography. Fatty acid composition of the 2-monoacylglycerol is then determined by saponification, esterification, and gas chromatography of the fatty acid methyl esters.

Overview of the Experiment

In this experiment you will characterize an oil or the purified lipid isolated from nutmeg in Experiment 7. You will release the fatty acids by complete saponification and identify and determine the mole percentage composition of each fatty acid by gas chromatography (see parts A and D in Figure E8.2). The positional distribution of the acids is elucidated by lipase action followed by gas chromatography (part B and C in Figure E8.2). Become familiar with Figure E8.2 before you begin the experiment. The time required for each part of the experiment is as follows:

A. Saponification of Triacylglycerols—1 hour

B. Lipase-Catalyzed Hydrolysis of Triacylglycerols—2 hours

C. Isolation and Saponification of 2-Monoacylglycerol—3 hours

D. Gas Chromatography of Fatty Acid Methyl Esters—2 hours

If only one period (3 hours) is available for this experiment, parts A and D may be completed. This will allow students to carry out a percent composition of fatty acids in triacylglycerol samples. If this alternative is used, up to three oil samples may be analyzed.

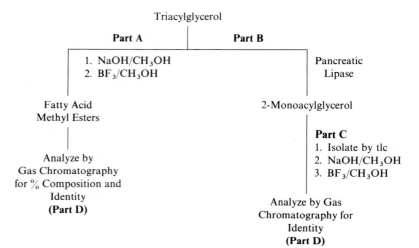

Figure E8.2.
Flowchart for characterization of a triacylglycerol.

II. Materials and Supplies

Several fat and oil samples (lard, butter, vegetable oil, etc.)

Unknown lipid from Experiment 7

Methanolic sodium hydroxide, 0.5 N

BF_3 solution (14% in methanol)

Hexane

Saturated NaCl

Anhydrous $MgSO_4$

Tris buffer, 1 M, pH 8

H_2SO_4, 1M

Aqueous $CaCl_2$, 40%

Porcine pancreatic lipase (Steapsin), 50 triacetin units/mg

Fatty acid methyl ester standards (12:0, 14:0, 16:0, 18:0, 16:1$^{\Delta 9}$, 18:1$^{\Delta 9}$, 18:2$^{\Delta 9, 12}$, 18:3$^{\Delta 9, 12, 15}$, solutions in hexane; 2 mg/mL). Standard samples containing mixtures of FAMES are commercially available. Rapeseed oil is an appropriate standard for this experiment.

Ethyl ether, peroxide free

Methanol containing 1% acetic acid

Silica gel thin-layer plates (2.5 × 10 cm)

Silica gel plates, preparative (20 × 20 cm, 1–2 mm thickness)

Chromatography solvent system, hexane–ethyl ether–acetic acid (80:20:1) in chromatography jar

Centrifuge tubes, 10-mL conical

Iodine chamber. Place a few crystals of iodine in an empty chromatography jar and cover.

Capillary tubes

Constant-temperature water bath at 37°C

Steam bath

Gas chromatograph

Conditions:

gas flow: 50–70 mL/min

$T_{injector}$: 200°C

T_{column}: 185°C

$T_{detector}$: 200°C

Attenuation: 2 or 4

Injection sample size: 5–10 μL

Column: 6 ft, 10% diethylene glycol succinate (DEGS) on Chrom W, 60/80 mesh

⚠ Caution:

All procedures in parts A, B, and C requiring the heating or evaporating of hexane, ethyl ether, and methanol should be carried out in a well-ventilated hood. If hood space is not available, a vent may be constructed according to Figure E8.3. There should be no flames of any kind in the laboratory.

III. Experimental Procedure

A. Saponification of Triacylglycerol for Fatty Acid Composition

Select a triacylglycerol sample for analysis. This may be the unknown lipid that was isolated and purified in Experiment 7 or a fat or oil supplied by your instructor. Weigh 25 to 50 mg of the sample into a small conical test tube. Add 3 mL of 0.5 N methanolic sodium hydroxide. Heat the mixture over a steam bath **(Hood!)** until a homogeneous solution is obtained. Add 5 mL of BF$_3$-methanol to the saponification reaction mixture and boil 2 to 3 minutes. Cool and transfer the solution into a separatory funnel containing 25 mL of hexane and 20 mL of saturated NaCl solution. Shake well, but gently, and allow the layers to separate. Vigorous shaking will give rise to an emulsion. The hexane layer, containing the fatty acid methyl esters, is dried with about 1 g of anhydrous MgSO$_4$ and

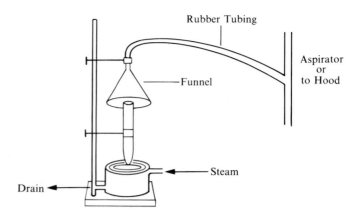

Figure E8.3.
Construction of a vent for removal of noxious vapors from laboratory.

filtered or decanted into a small vial. Concentrate the solution on the steam bath to a volume of about 0.5 mL **(Hood!)**. This solution of fatty acid methyl esters should be capped to prevent air oxidation and stored in a refrigerator until it is to be analyzed by gas chromatography.

B. Lipase-Catalyzed Hydrolysis of Triacylglycerol

In a conical test tube, dissolve or suspend 25 to 50 mg of another sample of the triacylglycerol used in part A in 0.2 mL of hexane. Add the following reagents to the triacylglycerol:

1. 1.5 mL of 1 M Tris buffer, pH 8.0
2. 0.005 mL of 40% aqueous $CaCl_2$
3. 50 mg of pancreatic lipase suspended in 0.3 mL of 1 M Tris buffer, pH 8.0

After a brief (5 second) homogenization with a vortex mixer, or vigorous mixing, place the enzymatic reaction mixture in a constant-temperature bath at 37°C. Shake the mixture every few minutes. The extent of reaction can be monitored by TLC analysis of aliquots removed at various times (1, 30, and 60 minutes). Aliquots of 1 drop are removed and clarified with a few drops of methanol containing 1% acetic acid. The clarified solution is spotted on a small (2.5 × 10 cm) silica gel plate, developed in the hexane–ethyl ether–acetic acid solvent, and placed in the iodine chamber for detection. When the triacylglycerol spot (fastest running) has nearly disappeared (the 60-min aliquot), the maximum yield of 2-monoacylglycerol is obtained. Stop the lipase reaction at this time by adding 1 mL of 1 M H_2SO_4. The spot on the TLC plate near the origin represents the 2-monoacylglycerol, and it should have increased in intensity with time of incubation. The end point of the lipase reaction can also be monitored by noting a turbidity or grainy appearance due to the precipitation of calcium salts of the released free fatty acids.

C. *Isolation and Saponification of 2-Monoacylglycerols*

To isolate the 2-monoacylglycerol from the reaction mixture, extract it three times with 3 mL of ethyl ether. Combine and dry the ether layers with 0.5 g of anhydrous $MgSO_4$ and concentrate on a steam bath (**Hood!**) to about 1 mL. If necessary, the ether extract may be stored until the next lab period. The ether solution is streaked on a preparative silica gel plate (20 × 20 cm) in a 15-cm-long band about 2.5 cm from the bottom edge. Develop the plate in hexane–ethyl ether–acetic acid (80:20:1). The components of the reaction mixture will separate into about four rows of spots. The components may be visualized by placing the plate in an iodine chamber for a few minutes. Mark the monoacylglycerol area (bands closest to the origin; lowest R_f), scrape the silica gel from this area with a razor blade, and suspend in 5 mL of ethyl ether. Mix the suspension for 5 minutes and filter by gravity through fluted paper into a 10-mL test tube. Dry with anhydrous $MgSO_4$ and carefully remove the ether on a steam bath or warm water bath (**Hood!**). Add 1 mL of 0.5 N methanolic NaOH to the residue and heat on the steam bath for 5 minutes or until all the residue is dissolved. Add 2 mL of BF_3-MeOH to the saponification mixture and boil for 2 to 3 minutes (**Hood!**). Cool and transfer the solution into a separatory funnel containing 20 mL of hexane and 15 mL of saturated NaCl solution. Shake gently and allow the layers to separate. The hexane layer, containing the methyl esters of the fatty acids originally at position 2 of the monoacylglycerol, is dried with anhydrous $MgSO_4$ and filtered into a small vial. Concentrate the solution on the steam bath to a volume of about 0.5 mL (**Hood!**). The sample is now ready for gas chromatographic analysis.

D. *Gas Chromatography of Fatty Acid Methyl Esters*

Study the section on gas chromatography in Chapter 3. If you are not familiar with the use of the instrument, have your instructor assist you. Be sure to record all instrumental conditions in your notebook (temperatures of the column, oven, and detector; rate of gas flow (mL/min); attenuation setting; sample size; and recorder chart speed). It is recommended that you begin with the FAME standards, using 5- to 10-μL injection samples with an attenuator setting of 2. Two peaks should appear, one for the hexane solvent and a second representing the FAME. The hexane peak will be off scale; if the FAME peak is also off the recorder paper, reduce the injection sample or increase the attenuation setting to 8 or 16. If the recorder peaks for the FAMEs are too small, decrease attenuation setting (1 or 2). When you are analyzing the standard FAMEs, you may inject a second sample as soon as the FAME from the first sample has been eluted from the column. This occurs when the recorder pen returns to the original baseline.

Use the same instrumental conditions to analyze the samples from parts A and C. The samples will contain a mixture of fatty acid methyl esters, so several recorder peaks will be obtained. Do not inject a second sample until you are sure all the fatty acids have been eluted from the first sample. The sample from part C is sometimes contaminated with an organic binder that is present in the silica gel. Fortunately, this substance has a retention time greater than the C_{18} fatty acid methyl esters. Determine the retention time for each standard FAME and for each FAME in the unknown samples. Retention time is the time interval between injection of sample and maximum response of the recorder.

IV. Analysis of Results

If you completed only parts A and D, it is possible to identify all the fatty acids present and to calculate the composition in weight percentage or mole percentage of each fatty acid present in the triacylglycerol sample. If you completed parts A–D, you are able to calculate the above plus the mole percentage of each fatty acid in the secondary position (carbon 2) and primary positions (carbons 1 and 3).

Prepare a table listing the retention time for each standard FAME. Use this table to identify the fatty acids present in each triacylglycerol you analyzed. Alternatively, plot the log of the retention time against the chain length of each saturated FAME (see Figure E8.4). Unknown saturated fatty acids can be identified from experimental retention times using this plot. Unsaturated fatty acids cannot be identified from the plot of saturated FAMEs. A separate plot of log retention time vs. the chain length must be prepared for each level of saturation (saturated, monounsaturation, diunsaturation, etc.).

The amount of each FAME present is proportional to the area under its peak on the recorder chart. If the peaks are symmetrical, the peak area may be calculated by triangulation:

Peak area = peak height × peak width at half height

or

Peak area = $HW_{1/2}$

Thermal conductivity detectors used in gas chromatographs do not respond equally to all FAMEs. To correct for varying detector sensitivity, peak area for each FAME should be multiplied by the proper **response correction factor (RCF)** (Table E8.1). If extra time is available, you may want to calculate your own response correction factors for the fatty acid methyl esters. The factors are experimentally determined on a gas chromatograph by comparing the area under a GC peak due to a known amount of compound to the area under a GC peak represented by a reference compound.

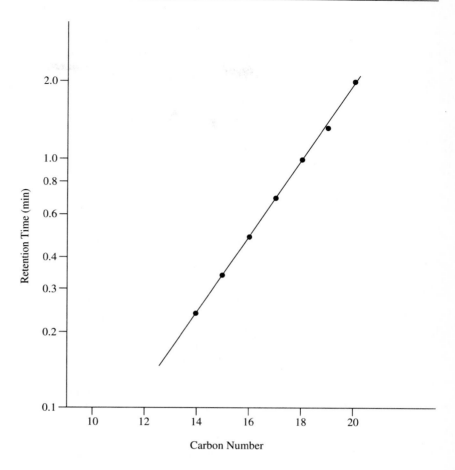

Figure E8.4.
A plot of log retention time vs. carbon number for gas chromatography of a series of saturated fatty acid methyl esters.

$$RCF_1 = \frac{\text{amount of compound 1}}{\text{peak area of 1}} \times \frac{\text{peak area of reference}}{\text{amount of reference}}$$

The response correction factors for compound 1 and other compounds are calculated relative to the response correction factor for the reference, which is assumed to be 1.00.

The mole percentage of each FAME is calculated according to the following sequence of equations.

1. Correction of peak area of $FAME_1$ for varying detector sensitivity.

   Peak area$_1$ $\times$ RCF_1 = corrected peak area$_1$

2. Conversion of corrected peak area (now in weight %) to moles.

   $$\frac{\text{Corrected peak area}_1}{\text{Molecular weight}_1} = \text{mole-corrected peak area}_1$$

3. Conversion of mole-corrected peak area$_1$ to mole % of $FAME_1$.

Table 8.1
Fatty Acid Methyl Ester Response Factors Measured
with a Thermal Conductivity Detector, DEGS Column
at 190°C

FAME	Response Correction Factor (RCF)
Myristic (14:0)	0.908
Palmitic (16:0)	0.954
Stearic (18:0)	1.010
Linoleic (18:2$^{\Delta9,12}$)	1.071
Linolenic (18:3$^{\Delta9,12,15}$)	1.172

$$\blacktriangleright \quad \text{Mole } \%_1 = \frac{\text{mole-corrected peak area}_1}{\text{total of mole-corrected peak area for all FAMEs}} \times 100$$

(Equation E8.3)

The denominator for Equation E8.3 is calculated by performing steps 1 and 2 for all of the FAMEs present in a single triacylglycerol sample. The mole-corrected peak areas are added together to obtain the total of all mole-corrected areas. If we assume that the extent of conversion of free fatty acids to FAMEs is essentially quantitative, or at least equal for all fatty acids in our experiment, the above calculation leads directly to the combined mole percent composition of fatty acids in all positions of the triacylglycerol.

The fatty acid composition (mole %) in positions 1 and 3 is obtained by subtracting the mole % of each fatty acid in position 2 (from analysis of 2-monoacylglycerol) from the mole % of the same fatty acid in the triacylglycerol:

Mole % of FA$_A$ in positions 1 and 3

$$= \frac{3(\text{mole } \% \text{ FA}_A \text{ in TG}) - (\text{mole } \% \text{ FA}_A \text{ in MG})}{2}$$

where

FA$_A$ = a fatty acid present in position 2
TG = triacylglycerol
MG = monoacylglycerol

For the lipid sample that you analyzed, report the fatty acid composition in mole % and the positional distribution of each fatty acid in mole %. Compare your results with those of other students.

V. Questions and Problems

1. Positional distribution analysis of hundreds of naturally occurring tri-acylglycerols has led to the discovery of general trends in fatty acid composition. Position 2 tends to contain unsaturated fatty acids, while positions 1 and 3 contain saturated fatty acids. Did you observe this general trend in your analysis?

▶ 2. Why are FAMEs rather than free fatty acids used for gas chromatographic analysis?

▶ 3. Why was $CaCl_2$ added to the lipase reaction mixture?

4. Your TLC plate of the ether extract of the lipase reaction mixture should have four spots. Identify each spot, and explain the order of spots.

▶ 5. What is the identity of the lipid that was extracted from nutmeg seed (Experiment 7)?

▶ 6. What is the function of BF_3 in the formation of FAMEs?

7. Write out the lipase-catalyzed reaction using your unknown as triacylglycerol substrate.

▶ 8. Using data from your experiment, write the order of gas chromatographic elution (first to last) for each set of fatty acid methyl esters below. Explain your answers.

(a) 14:0, 20:0, 12:0, 16:0, 18:0

(b) $16:1^{\Delta 9}$, 16:0, $16:2^{\Delta 9,12}$

(c) 16:0, 18:0, $16:1^{\Delta 9}$, $18:1^{\Delta 9}$

VI. References

R. Ackman, in *Methods in Enzymology,* Vol, XIV, J. Lowenstein, Editor (1969), Academic Press (New York), pp. 329–381. "Gas-Liquid Chromatography of Fatty Acids and Esters."

H. Brockerhoff, in *Methods in Enzymology,* Vol. XXXVB, J. Lowenstein, Editor (1975), Academic Press (New York), pp. 315–325. "Determination of the Positional Distribution of Fatty Acids in Glycerolipids."

W. Christie, *HPLC and Lipids—A Practical Guide* (1987), Pergamon Press (Oxford). Introduction to HPLC and practical applications.

A. Fallon, R. Booth, and L. Bell, *Applications of HPLC in Biochemistry* (1987), Elsevier (Amsterdam), pp. 193–212. Application of HPLC to lipid analysis.

F. Gunstone, J. Harwood, and F. Padley, Editors, *The Lipid Handbook* (1986), Chapman & Hall (London). Reference book presenting the chemistry, biochemistry, and technology of lipids.

M. Gurr and J. Harwood, *Lipid Biochemistry,* 4th ed. (1991), Chapman & Hall (New York). An introduction to lipid structure and function.

IUPAC—International Union of Biochemistry, *J. Biol. Chem.* **242,** 4845–4849 (1967).

M. Kates, *Techniques of Lipidology* (1986), Elsevier (Amsterdam). Isolation, analysis, and identification of all lipids.

A. Kuksis, Editor, *Chromatography of Lipids in Biomedical Research and Clinical Diagnosis* (1987), Elsevier (Amsterdam). A complete book on lipid analysis.

C. Mathews and K. van Holde, *Biochemistry* (1990), Benjamin-Cummings (Redwood City, CA), pp. 301–302. Analysis of triacylglycerols.

J. Mead, R. Alfin-Slater, D. Howton, and G. Popják, *Lipids: Chemistry, Biochemistry and Nutrition* (1986), Plenum Publishing (New York). An excellent introduction to lipids.

G. Patton, S. Cann, H. Brunengraber, and J. Lowenstein, in *Methods in Enzymology,* Vol. 72, J. Lowenstein, Editor (1981), Academic Press (New York), pp. 8–20. "Separation of Methyl Esters of Fatty Acids by Gas Chromatography on Capillary Columns."

J. Rawn, *Biochemistry,* 2nd ed. (1989), Neil Patterson Publishers (Burlington, NC), pp. 209–218. Lipid structure and function.

L. Stryer, *Biochemistry,* 3rd ed. (1988), W. H. Freeman (New York), pp. 283–312. Introductory material on lipid structure and function.

M. Tevini and D. Steinmüller, in *High Performance Liquid Chromatography in Biochemistry,* A. Henschen, K. P. Hupe, F. Lottspeich, and W. Voelter, Editors (1985), VCH (Weinheim, West Germany), pp. 349–412. HPLC of all lipids.

D. Voet and J. Voet, *Biochemistry* (1990), John Wiley & Sons (New York), pp. 273–274. A discussion of triacylglycerols.

C. Worthington, Editor, *Worthington Enzyme Manual* (1988), Worthington Biochemical Corporation (Freehold, NJ), pp. 212–215. Pancreatic lipase.

G. Zubay, *Biochemistry,* 2nd ed. (1985), Macmillan (New York), pp. 154–160. Triacylglycerol structure and analysis.

EXPERIMENT **9**

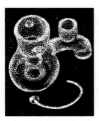

Identification of Carbohydrates by Polarimetry

■ **Recommended Reading**

Chapter 5, Section A.

■ **Synopsis**

Carbohydrates are found in all forms of life. They are unique among biologically significant molecules in that they can exist in many stereochemical forms. The naturally occurring carbohydrates are asymmetric molecules; therefore, they can be detected and identified by measurements of optical rotation. An unknown monosaccharide and disaccharide will be identified by chemical and polarimetric characterization.

I. Introduction and Theory

Carbohydrate Structure and Function

Carbohydrates are naturally occurring compounds that have the basic formula $C_x(H_2O)_x$. We are all familiar with the central role of carbohydrates in energy metabolism, with glucose being the molecule of primary significance. But carbohydrates are also essential components of cell membranes and bacterial cell walls. Here they play a role in structural integrity or a more dynamic role in cellular recognition and communication.

In chemical terms, carbohydrates are polyhydroxy aldehydes or ketones. Two common carbohydrates, glucose and fructose, are shown in Figure E9.1. Glucose is a polyhydroxy aldehyde with six carbons or an aldohexose, whereas fructose is a polyhydroxy ketone or a ketohexose. Glucose and fructose are **monosaccharides**. Larger carbohydrates formed by the combination of two or more monosaccharides are **disac-**

Figure E9.1
Two common carbohydrates,
D-Glucose and D-Fructose.

charides, trisaccharides, and so on; in general, they are called **oligo-saccharides**. Carbohydrates with 10 or more monosaccharides are called **polysaccharides**.

Carbohydrates may contain one or many **asymmetric carbon atoms** or **chiral centers**. Glucose has four chiral centers, and fructose has three. The stereochemistry of glucose, fructose, and other carbohydrates is defined by comparison to one of the simplest carbohydrates, glyceraldehyde. Glyceraldehyde, as shown in Figure E9.2, has one chiral center and thus can exist in two possible arrangements or enantiomers, D-glyceraldehyde and L-glyceraldehyde. Solutions of the isomers rotate the plane of polarized light to the same extent, but in opposite directions. Carbohydrates are classified as D or L depending on how their stereochemical structures compare to D- and L-glyceraldehyde. (In the R,S system, D-glyceraldehyde is R. The stereochemistry of other carbohydrates may also be assigned by the R,S format; however, they are traditionally denoted by D or L.) Since most carbohydrates contain more than one chiral center, convention dictates that the stereochemistry of the asymmetric center most remote (highest in number) from the aldehyde end determines the classification. If the stereochemical configuration around that carbon atom is identical to D-glyceraldehyde, the carbohydrate is said to be in the D family; if the configuration is opposite to D-glyceraldehyde (or identical to L-glyceraldehyde), the carbohydrate is L. Both of the carbohydrates shown in Figure E9.1 are of the D family. Note that the classification of the carbohydrates is not dependent on the direction of the rotation of plane polarized light, but is done strictly on a configurational basis. In fact, D-glucose has a specific rotation, $[\alpha]_D^{20}$, of $+52.2°$, whereas D-fructose has an $[\alpha]_D^{20}$ of $-92.0°$.

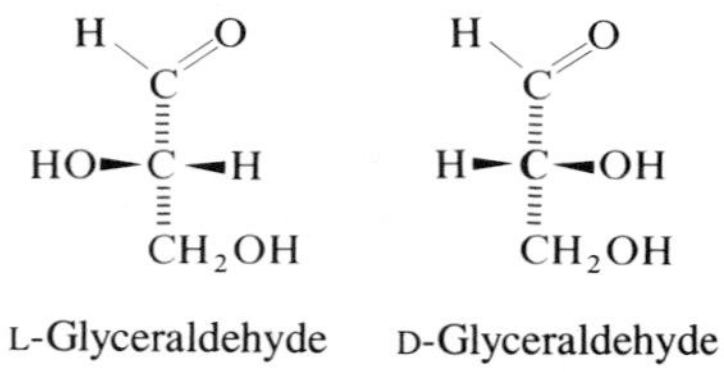

Figure E9.2
The two stereoisomers of glyceraldehyde.

L-Glyceraldehyde   D-Glyceraldehyde

When glucose, fructose, and other carbohydrates are dissolved in water, the open chain form in Figure E9.1 (the Fischer projection structure) is transformed to an aldehyde hydrate, which rearranges to a cyclic structure as shown in Equations E9.1 and E9.2.

(Equation E9.1)

α-ᴅ-Glucose   β-ᴅ-Glucose

(Equation E9.2)

α-ᴅ-Fructose   β-ᴅ-Fructose

Formation of the intramolecular hemiacetals results in two forms of the compound that differ in the stereochemistry at the hemiacetal (C-1 of glucose) or hemiketal (C-2 of fructose) carbon. For glucose, for example, these two forms are called α-ᴅ-glucose and β-ᴅ-glucose.

An aqueous solution of pure β-ᴅ-glucose has an initial $[\alpha]_D^{20}$ of $+19°$. However, if the solution is allowed to sit for several hours, the optical rotation slowly changes, resulting in a final $[\alpha]_D^{20}$ of $+52.2°$. Likewise, a solution of α-ᴅ-glucose has an initial $[\alpha]_D^{20}$ of $+113.4°$, but after several hours the specific rotation is $+52.2°$. In each case, the same equilibrium mixture of α-ᴅ- and β-ᴅ-glucose has been produced. This process of inter-conversion, called mutarotation, has been observed for many carbohydrate solutions when the sugars can easily form a five- or six-membered ring. Fructose, in Equation E9.2, also undergoes mutarotation, resulting in an equilibrium mixture of β-ᴅ-fructose and α-ᴅ-fructose. In this case, a five-membered hemiketal ring is formed. The two predominant forms of the sugars (α and β) are referred to as **anomers**. The mutarotation process is catalyzed by acid or base.

Basic Principles of Polarimetry

Both absorption and fluorescence spectroscopy discussed in Chapter 5, provide information about the molecular structure and conformation of biomolecules. A spectroscopic technique that reveals even more detailed information about molecular structure involves **polarimetry,** the interaction of **plane-polarized light** with biomolecules. Polarimetry is used to measure optical activity or chirality in molecules. A high percentage of biochemically significant molecules display chirality. This section will introduce the principles of polarimetry and discuss its application to the characterization of biomolecules, especially carbohydrates.

Recall, from an earlier discussion, that light is composed of electrical and magnetic vectors. If light is passed through a Polaroid lens or a Nicol prism, only one plane of the electrical vector is transmitted (Figure E9.3). The transmitted light is said to be **plane-polarized**, since it has only a single plane of polarization in the E vector.

Plane-polarized light interacts in a unique and reproducible manner with molecules that are asymmetric. (Asymmetric molecules are those that cannot be superimposed on their mirror images.) Such molecules are said to be optically active because their solutions change the orientation of plane-polarized light. The most often encountered optically active molecules are those with chiral centers, carbon atoms bonded to four different groups. Figure E9.4 shows two chiral molecules, the amino acids D- and L-phenylalanine and the carbohydrates D- and L-glucose. Only one form of each of these occurs in nature (L-phenylalanine, D-glucose); therefore, when they are extracted from biological material they should show optical activity.

When plane-polarized light is transmitted through a solution of a chiral substance, the plane of light exiting the sample is rotated or changed. In polarimetry, the direction and extent of that rotation are measured (see Figure E9.5). The direction ($+$ or $-$) of the rotation depends on the form of the chiral molecule under study. The extent of rotation is expressed in terms of the angle (α) and identified by $[\alpha]_\lambda$ defined by Equation E9.3.

$$\blacktriangleright \quad [\alpha]_\lambda^T = \frac{\alpha_{obs}}{lc} \qquad \text{(Equation E9.3)}$$

where

$[\alpha]_\lambda^T$ = specific rotation of an asymmetric substance at temperature T, with light of wavelength λ

α_{obs} = experimental measurement of rotation in degrees

l = path length of the cell holding the sample, in dm

c = concentration of the optically active sample in g/mL

Figure E9.3
Production of polarized light.

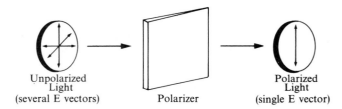

Unpolarized Light (several E vectors)   Polarizer   Polarized Light (single E vector)

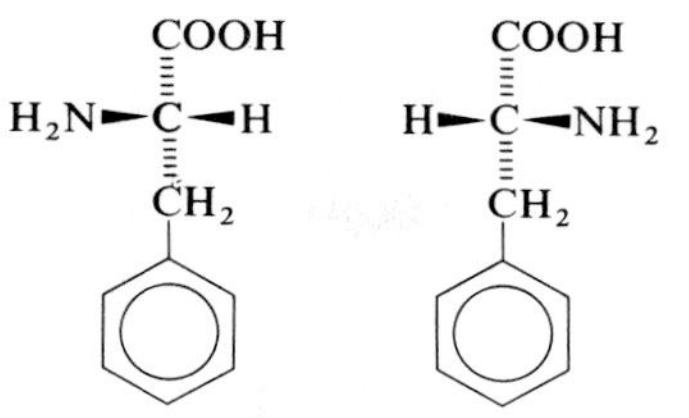

L-Phenylalanine D-Phenylalanine

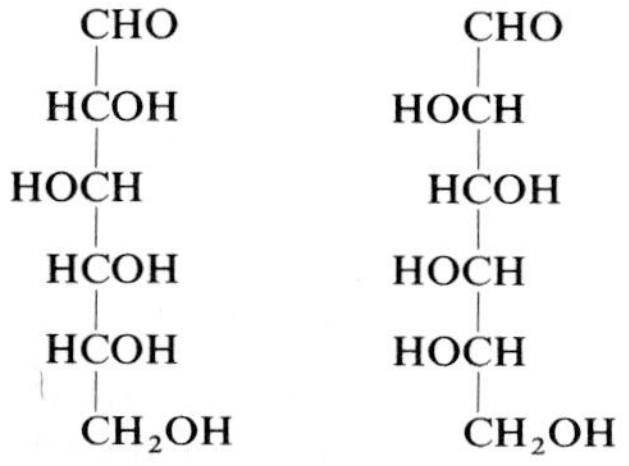

D-Glucose L-Glucose

Figure E9.4
Optical isomers of phenylalanine and glucose.

The magnitude and sign of α_{obs} depend on the nature and concentration of the asymmetric substance, the temperature of the solution, the path length of the plane-polarized light through the sample, and the wavelength of the plane-polarized light. Most polarimetric measurements use the D line from a sodium lamp; hence, the **specific rotation** becomes $[\alpha]_D^T$. Some instruments allow α_{obs} measurements at several different wavelengths. More expensive instruments plot optical rotation as a function of wavelength, which provides extensive information about molecular structure. (This technique, called optical rotatory dispersion, ORD, will not be discussed here.)

The size and sign of the specific rotation can be used to identify unknown asymmetric molecules, because each compound has a characteristic $[\alpha]_D^T$. This is a physical property of that substance, similar to boiling point, refractive index, and density.

Instrumentation for Measurement of $[\alpha]_D$

The essential components of a polarimeter are shown in Figure E9.6. Light from a sodium lamp is transmitted through a Polaroid lens or a Nicol prism. The plane-polarized light is directed through a glass cell

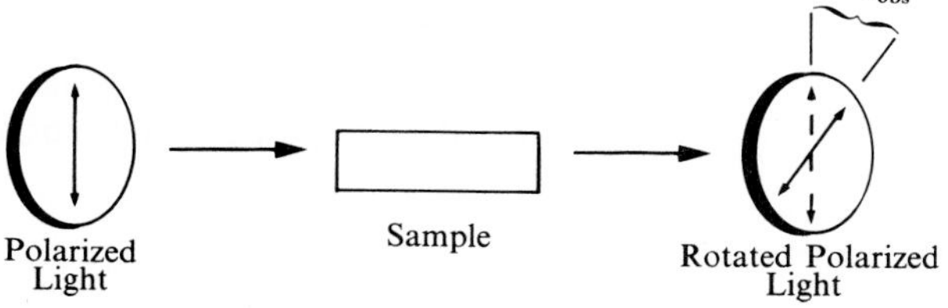

Figure E9.5
Diagram showing the basic principles of polarimetry.

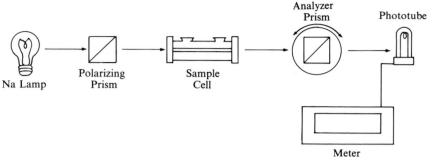

Figure E9.6
A diagram of a typical polarimeter.

containing a solution of the material under study. A second polarizer (called the analyzer) serves to measure the angle of rotation.

In older instruments, the experimenter must adjust the analyzer manually and use the naked eye to detect the angle of rotation. More recent instruments provide for automatic measurement of the angle of rotation with digital readout of the magnitude and sign of α_{obs}.

Applications of Polarimetry

Polarimetry is a valuable tool for identifying and characterizing optically active biomolecules such as amino acids, carbohydrates, nucleic acids, and lipids. Once a molecule has been classified by other techniques (chromatography, titration, chemical reactivity, etc.) into one of these groups, its identity can be confirmed by polarimetric measurements.

The technique is simple, fast, and accurate and requires relatively inexpensive equipment. A standard solution of known concentration of the compound is prepared, and the optical rotation is measured relative to the optical rotation of pure solvent. The observed α is then used to calculate specific rotation, $[\alpha]_D^T$. Standard $[\alpha]_D^T$ values for biomolecules are readily available in common reference books. Since α_{obs} depends on temperature and wavelength, these variables must be known. Most modern polarimeters provide thermostated cells to maintain constant temperature (see Figure E9.7). When specific rotation measurements are reported, they must include the temperature of the solution, the λ of the plane-polarized light, the concentration of the optically active material, and the identity of the solvent.

More detailed information can be obtained from polarimetry experiments conducted under conditions that cause changes in optical rotation measurements (mutarotation). This technique is of particular value in identifying and characterizing carbohydrate structure and is illustrated in this experiment.

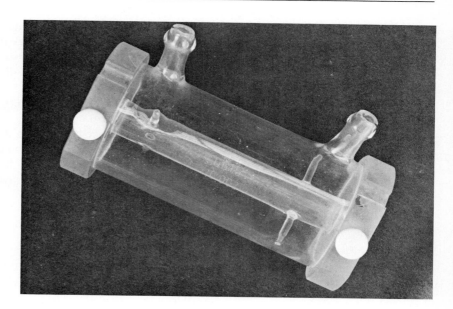

Figure E9.7
A polarimeter cell. Photo courtesy of Mr. Mark Billadeau.

Identification and Characterization of Carbohydrates by Polarimetry

Detection of mutarotation by optical rotation can aid in the identification of carbohydrates. Unknown carbohydrates are often identified by an initial measurement of optical rotation under controlled conditions. Standard values for $[\alpha]_D^{20}$ are readily available for most carbohydrates. Table E9.1 gives several of these values for common monosaccharides. Identification of an unknown carbohydrate by optical rotation measurements is similar to identifying an organic compound by melting point. However, two or more carbohydrates may have similar $[\alpha]_D^{20}$ values, and the absolute identity of the unknown cannot be determined by this measurement alone. Measurement of the specific rotation after mutarotation has occurred to produce an equilibrium mixture will yield a second optical rotation value. The equilibrium mixture attained by mutarotation of a sugar is reproducible; therefore, standard $[\alpha]_D^{20}$ values can be determined for each sugar. The specific optical rotations of several equilibrium mixtures of α and β anomers are given in Table E9.1. Now two optical rotation values can be experimentally measured and used for the identification of unknown sugars.

The method for measuring optical rotations of pure sugars is relatively straightforward. A standard solution of the sugar is prepared and an initial optical rotation measurement is rapidly made. If this measurement is made within a few minutes, the result will be the observed optical rotation for the pure anomer. A drop of dilute acid or base is then added to the carbohydrate sample and the optical rotation is monitored until a constant value is obtained. The final, constant reading corresponds to the equilibrium mixture of α and β anomers of the unknown carbohydrate.

Characterization of Glycosides by Polarimetry

Disaccharides and polysaccharides are formed when two or more monosaccharides are chemically combined. The bond linking the monosaccharides is formed between the hemiacetal or hemiketal functional group of one carbohydrate and a hydroxyl functional group of a second carbohydrate (Equation E9.4).

▶

(Equation E9.4)

The new linkage is called a **glycosidic bond** and the product has the general name **glycoside**. In the example, two glucose units are joined by an $\alpha\,(1 \rightarrow 4)$ glycosidic linkage to produce a disaccharide, maltose. Several variations of glycosidic bonds are observed in nature, and they are shown in Figure E9.8. Milk sugar, lactose, is a disaccharide consisting of galactose and glucose linked by a $\beta\,(1 \rightarrow 4)$ glycosidic bond. Common table sugar, sucrose, consists of glucose and fructose linked by an $\alpha\,(1 \rightarrow 2)$ glycosidic bond. Many identical monosaccharide molecules may combine by glycosidic bonds to form polysaccharides. Starch consists of glucose units linked by $\alpha\,(1 \rightarrow 4)$ glycosidic bonds, whereas cellulose is made up of glucose units linked by $\beta\,(1 \rightarrow 4)$ glycosidic bonds.

Glycosidic bonds are subject to hydrolysis under certain conditions. They hydrolyze readily in the presence of dilute aqueous acid or enzymes called **glycosidases**. Using enzymes for the specific cleavage of these bonds is a valuable technique for structure determination of di- and polysaccharides.

An interesting enzyme that catalyzes the hydrolysis of specific glycosidic bonds is invertase. The overall reaction catalyzed by invertase is shown in Equation E9.5.

▶   Sucrose $\rightleftharpoons$ glucose + fructose   (Equation E9.5)

Table E9.1
Specific Optical Rotation, $[\alpha]_D^{20}$, in Degrees, for Several Carbohydrates in Water

Carbohydrate	α	β	Equilibrium Mixture After Mutarotation
Monosaccharides			
L-Arabinose	+55.4	+190.6	+104.5
D-Fructose	—	−133.5	−92.0
D-Galactose	+150.7	+52.8	+80.2
D-Glucose	+112	+18.7	+52.7
L-Lyxose	+5.8	—	+13.5
D-Mannose	+29.3	−16.3	+14.5
D-Ribose	−23.7	—	−23.7, −21.5
L-Sorbose	—	—	−43.1
D-Xylose	+93.6	—	+18.8
Disaccharides			
Cellobiose	—	+14.2	+34.6
Lactose	+90.0	+34.2	+53.6
Maltose	+173.0	+112.0	+130.4
Sucrose	—	—	+66.5
α, α-Trehalose	—	—	+199, +178
Trisaccharides			
Cellotriose	—	—	+22, +25
Maltotriose	—	—	+160
Raffinose	—	—	+105.2, +123
Polysaccharides			
Glycogen	—	—	+198
Starch	—	—	+220

The final outcome is an equimolar mixture of glucose and fructose. The name of the enzyme is derived from the observation that sucrose is dextrorotatory (plane-polarized light is rotated to the right) and an equimolar mixture of glucose and fructose is slightly levorotatory (plane-polarized light is rotated to the left). Therefore, the enzyme-catalyzed reaction results in an inversion of the sign of the optical rotation. Invertase is specific in its action, preferring sucrose and raffinose as substrates.

In this experiment, one sample of an oligosaccharide will be treated with acid and a second sample with invertase. After a brief incubation period under controlled conditions, an optical rotation measurement will be made on each reaction mixture. The experimental data are then compared to the optical rotation of an aqueous solution of the oligosaccharide. By using proper data analysis, it can be determined whether the oligosaccharide is susceptible to acid- and/or invertase-catalyzed hydrolysis. In addition, if the oligosaccharide is cleaved, the percentage of substrate remaining in the sample can be calculated. Hence, polarimetry can be used to monitor the rate of glycoside bond hydrolysis.

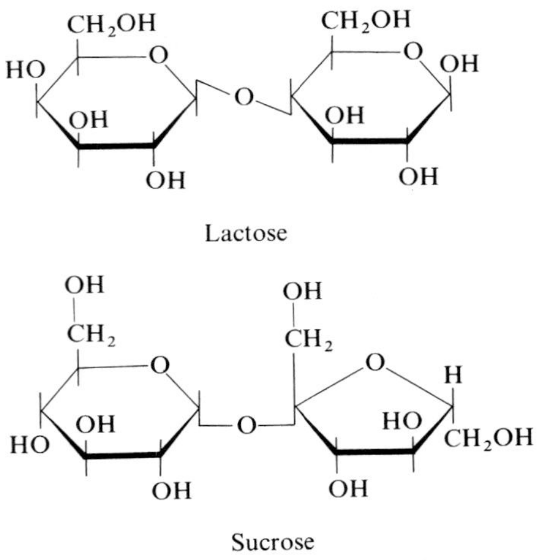

Lactose

Sucrose

Figure E9.8
Two common disaccharides, lactose and sucrose.

There is one difficulty in measuring invertase-catalyzed sucrose inversion by the polarimetric method. The rate of the enzyme-catalyzed reaction is rapid, but the mutarotation of the monosaccharide products is slower. Therefore, the optical rotation of the reaction mixture will slowly change due to mutarotation even after the enzyme is inactivated. The undesirable effect of slow mutarotation is avoided by adding base to the reaction mixture after a suitable reaction period. The added base inactivates the enzyme and catalyzes the mutarotation. If the inversion of sucrose catalyzed by acid is studied, there is no difficulty due to mutarotation because acid acts as a catalyst for both the cleavage of the glycoside and the mutarotation of the products.

Overview of the Experiment

The two major exercises in this experiment are outlined in Figure E.9.9. Part A involves the identification of an unknown carbohydrate by polarimetry measurements before and after mutarotation, and part B consists of the application of polarimetry to characterizing glycosidic bonds and carbohydrate structure. Each part will require approximately $1\frac{1}{2}$ hours. Since most biochemistry laboratories have only one polarimeter available, a schedule of users must be prepared.

II. Materials and Supplies

A. Identification of an Unknown Carbohydrate

Unknown carbohydrates, 10-g samples
10% HCl
Polarimeter and 1-dm cells
Volumetric flasks, 50 mL

B. Characterization of Glycosidic Bonds

Several oligosaccharides: sucrose, maltose, lactose, and raffinose
Concentrated HCl
Invertase solution in acetate buffer, pH 5.0, 200 units/mL
Constant-temperature water bath at 37°C
Polarimeter and 1-dm cells
Volumetric flasks, 50 and 100 mL
Saturated Na_2CO_3 solution

Figure E9.9
Flowchart for analysis of an unknown carbohydrate.

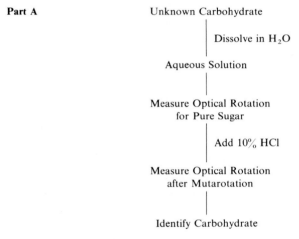

Part A

Unknown Carbohydrate
↓ Dissolve in H_2O
Aqueous Solution
↓
Measure Optical Rotation for Pure Sugar
↓ Add 10% HCl
Measure Optical Rotation after Mutarotation
↓
Identify Carbohydrate

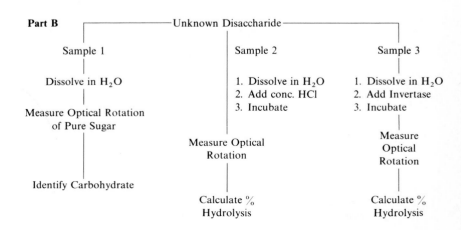

Part B — Unknown Disaccharide —

Sample 1
↓
Dissolve in H_2O
↓
Measure Optical Rotation of Pure Sugar
↓
Identify Carbohydrate

Sample 2
1. Dissolve in H_2O
2. Add conc. HCl
3. Incubate
↓
Measure Optical Rotation
↓
Calculate % Hydrolysis

Sample 3
1. Dissolve in H_2O
2. Add Invertase
3. Incubate
↓
Measure Optical Rotation
↓
Calculate % Hydrolysis

III. Experimental Procedure

A. *Identification of an Unknown Carbohydrate*

Before you begin the experimental procedure, become familiar with the use of the polarimeter. Read the instructions or have your instructor assist you. You may be fortunate enough to use an automatic polarimeter with meter or digital readout of degrees. If not, you will have to operate the polarimeter manually.

Obtain a polarimeter cell and rinse it several times with distilled water. Be careful while handling the cell. Do not scratch or chip the glass lenses at the ends of the cell. Fill the cell to overflowing with distilled water and slide the glass lens over the end. There should be no air bubbles trapped under the lens in the cell. If air bubbles are present, remove the lens and add water to the cell with a disposable pipet. Close the cell and wipe the glass end pieces with lens paper. Place the filled cell into the polarimeter and zero the instrument in the following way. An automatic polarimeter has a "zero adjust" button. Push this button and 00.00 degrees should appear on the readout. A manual polarimeter is adjusted by rotating the analyzer. Look through the eyepiece and note the two halves of the circular field. Rotate the analyzer until the semicircles of the field are equal in illumination, as shown in Figure E9.10. Read the degree setting on the analyzer (sign and number) and record it in your notebook as the zero point. Turn the analyzer a few degrees off the zero point and again adjust to the zero point. Record the second reading. Now, turn the analyzer a few degrees in the opposite direction as before and again adjust to the zero point. Record the third reading. Average the three zero point readings and use this as your zero adjust setting. Pour the water out of the cell and allow it to air dry.

The next step should be completed rapidly to avoid mutarotation of the sugar. Weigh 5.0 ± 0.01 g of your unknown carbohydrate and transfer it to a 50-mL volumetric flask. Add 40 mL of distilled water to dissolve the sugar. Fill to the mark with distilled water. Mix well and fill the polarimeter cell with the sugar solution. Check for air bubbles in the cell and remove them as before. Place the filled cell in the polarimeter and record three optical rotation readings as described above, readjusting the analyzer each time. Average the readings.

Transfer the contents of the polarimeter cell to the original volumetric flask and then add 1 drop of 10% HCl to the flask. Mix well and again fill the polarimeter cell. Record the optical rotation every 10 minutes for 30 minutes or until a constant reading is obtained. The final constant

Figure E9.10
Adjustment of the analyzer for polarimetry measurements.

Unbalanced ⇌ Balanced Zero Point ⇌ Unbalanced

reading is that of the equilibrium mixture after mutarotation. Refer to the Analysis of Results section for instructions to calculate specific rotation, $[\alpha]_D^{20}$.

B. Characterization of Glycosidic Bonds

Obtain three 50-mL volumetric flasks. Label them with numbers 1 to 3. Into each, weigh 2.5 ± 0.01 g of the unknown oligosaccharide. Each sample of unknown will be treated separately. The procedure is completed more rapidly if the samples are started at different times. Begin the preparation of sample 2 immediately after beginning the incubation of sample 1, and so forth.

Sample 1 Dissolve the sugar in about 40 mL of distilled water and then add water to the mark. Mix well, and cool to 20°C in a water bath. Transfer to a polarimeter cell and obtain three optical rotation measurements as described in part A. Use water as a blank for zero point measurement.

Sample 2 Dissolve the sugar in about 40 mL of distilled water and then add 5.0 mL of concentrated HCl. Add water to the mark, mix well, transfer to a 125-mL Erlenmeyer flask, and incubate the mixture at 37°C for 15 minutes. Cool to 20°C, transfer to a polarimeter cell, and obtain three optical rotation measurements. Use water as a blank.

Sample 3 Dissolve the sugar in about 40 mL of distilled water and add 5.0 mL of invertase solution. Add water to the mark, mix well, transfer to a 125-mL Erlenmeyer flask, and incubate the mixture at 37°C for 15 minutes. Add 2.0 mL of saturated Na_2CO_3 to inactivate the invertase and accelerate mutarotation. If a precipitate forms, centrifuge for 20 minutes at 3000 rpm. Incubate an additional 15 minutes and obtain three optical rotation measurements. Use water as a blank.

IV. Analysis of Results

A. Identification of an Unknown Carbohydrate

Calculate average optical rotation measurements for the water blank and the unknown sample. Correct for the water blank by subtracting its rotation from that of the carbohydrate sample. Calculate the specific rotation, $[\alpha]_D^{20}$, for the unknown according to Equation E9.3.

▶ $$[\alpha]_D^{20} = \frac{[\alpha]_{obs}}{lc}$$ (Equation E9.3)

Calculate the $[\alpha]_D^{20}$ for the equilibrium mixture after mutarotation in the same fashion. Compute the statistical significance of the data. Compare the two values with those in Table E9.1 and identify the unknown carbohydrate by name and type of anomer.

B. Characterization of Glycosidic Bonds

Calculate the optical rotation of each of the three samples. Remember to correct the reading for sample 3, which was diluted by addition of 2 mL of Na_2CO_3. [Multiply the observed degrees of rotation (after water blank correction) by 52/50.] The result from sample 1 gives the optical rotation of the pure oligosaccharide before inversion and mutarotation. Samples 2 and 3 give the rotation of mixtures of oligosaccharide and monosaccharide(s) caused by hydrolysis of the glycosidic bonds.

Identify the unknown disaccharide from the specific rotation of sample 1. Using the data from samples 1, 2, and 3, calculate the percentage of disaccharide remaining after treatment with acid or enzyme. Use the following relationships. In each case X = the amount of disaccharide hydrolyzed in gram %. The polarimeter cell path length is assumed to be 1 dm.

Sucrose

$$X = \frac{\alpha_{obs, \text{ sample } 1} - \alpha_{obs, \text{ after inversion}}}{7.7} \times 100$$

Maltose

$$X = \frac{\alpha_{obs, \text{ sample } 1} - \alpha_{obs, \text{ after inversion}}}{8.1} \times 100$$

Lactose

$$X = \frac{\alpha_{obs, \text{ sample } 1} - \alpha_{obs, \text{ after inversion}}}{1.5} \times 100$$

Describe the effectiveness of base and invertase in cleaving your unknown disaccharide.

V. Questions and Problems

▶ 1. Illustrate the structure for the aldopentose D-ribose in the straight chain (Fischer projection) and the cyclic form (Haworth structure).

▶ 2. Why are there no optical rotation values listed in Table 9.1 for α and β forms of sucrose?

▶ 3. Predict the action of invertase on raffinose, shown below.

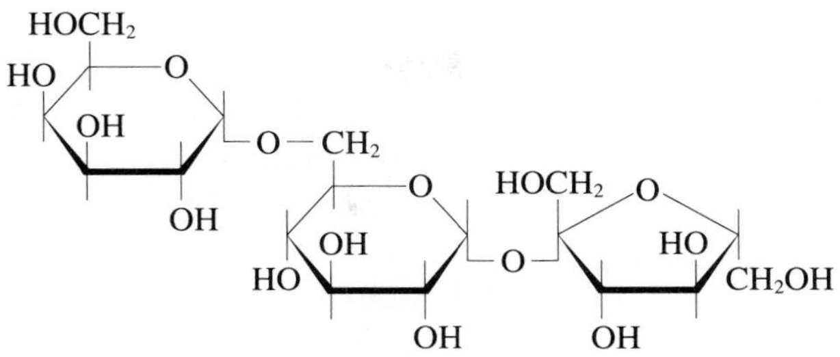

▶ 4. An unknown carbohydrate was dissolved in water (5 g/100 mL) and the optical rotation was immediately measured as −6.75° at 20°C. The carbohydrate solution was allowed to remain in the polarimeter cell for 24 hours, and a second measurement was found to be −4.55° at 20°C. After 24 hours more, the optical rotation was still −4.55°. Study Table E9.1 and identify the unknown. Assume that the "zero point" reading on the polarimeter was 0.01°. The path length of the cell is 1 dm.

▶ 5. The following optical rotation readings were measured for a water blank and an unknown carbohydrate solution.

α_{obs}, Blank	α_{obs}, Carbohydrate
+0.01	+3.24
0.00	+3.15
+0.02	+3.30
−0.01	+3.20
0.00	+3.21
+0.01	+3.19
−0.02	+3.17
0.00	+3.23
+0.01	+3.20
−0.01	+3.25

(a) Calculate the sample mean.

(b) Calculate the standard deviation.

(c) Calculate the 95% confidence levels for the measurements.

VI. References

R. Bentley, *Annu. Rev. Biochem.* **41**, 953–996 (1972). "Configurational and Conformational Aspects of Carbohydrate Biochemistry."

R. Binkley, *Modern Carbohydrate Chemistry* (1988), Marcel Dekker (New York). Excellent introduction for students new to the field.

M. Chaplin and J. Kennedy, Editors, *Carbohydrate Analysis—A Practical Approach* (1986), IRL Press (Oxford). A handbook of laboratory protocols.

C. Mathews and K. van Holde, *Biochemistry* (1990), Benjamin/Cummings (Redwood City, CA), pp. 260–297. Structure and reactions of carbohydrates.

J. Rawn, *Biochemistry,* 2nd ed. (1989), Neil Patterson Publishers (Burlington, NC), pp. 292–295. Introduction to carbohydrates.

J. Robyt and B. White, *Biochemical Techniques: Theory and Practice* (1987), Brooks/Cole (Monterey, CA). Several sections provide methods for carbohydrate analysis.

H. Strobel and W. Heineman, *Chemical Instrumentation: A Systematic Approach,* 3rd ed. (1989), John Wiley & Sons (New York), pp. 664–668. Polarimeters.

L. Stryer, *Biochemistry,* 3rd ed. (1988), W. H. Freeman (New York), pp. 331–348. A brief introduction to carbohydrate structure.

K. van Holde, *Physical Biochemistry,* 2nd ed. (1985), Prentice-Hall (Englewood Cliffs, NJ), pp. 235–252. Theoretical description of optical rotation.

D. Voet and J. Voet, *Biochemistry* (1990), John Wiley & Sons (New York), pp. 245–260. Carbohydrate structure and function.

EXPERIMENT 10

Characterization of Hemoglobin A_{1c} by Affinity Chromatography

■ **Recommended Reading**

Chapter 3, Sections D, E, and H; Chapter 5, Section A.

■ **Synopsis**

Hemoglobin A_{1c}, a glycoprotein, comprises about 5% of the hemoglobin in normal adult red blood cells. In individuals with diabetes mellitus, its concentration may double. In this experiment, HbA_{1c} and other glycohemoproteins will be isolated from human hemolysate by cation exchange chromatography or affinity chromatography. The glycoproteins are characterized by visible spectrophotometry and carbohydrate analysis.

I. Introduction and Theory

Glycoproteins—Structure, Function, and Analysis

Proteins are classified into two broad categories: **simple**, those that consist only of amino acids, and **conjugated**, those containing other moieties, including lipids, metal ions, nucleic acids, coenzymes, and carbohydrates. We are just beginning to realize the universal occurrence and multifaceted biological functions of the conjugated proteins. Perhaps the most diverse group is the **glycoproteins**, those that contain covalently bound carbohydrate groups. A glycoprotein contains one, a few, or several carbohydrate units. The most common sites of carbohydrate attachment are the side chains of asparagine, serine, and threonine, although 5-hydroxyproline and lysine are also important. Figure E10.1 illustrates the manner in which carbohydrates are linked to the common amino acid residues. Figure E10.1A shows a typical N-glycosyl bond between the amide side chain of asparagine and C_1 of *N*-acetylglucosamine. The attachment of a sugar to threonine, shown in Figure E10.1B, consists of an O-glycosyl bond. The most common sugars found in glycoproteins are glucose, mannose, galactose, fucose, xylose, *N*-acetylglucosamine, *N*-acetylgalactosamine, and sialic acid. These sugars are found individually or, more commonly, linked together in oligosaccharide units that are attached to the protein.

The diverse structures of the glycoproteins would be expected to lead to great diversity in biological function. Table E10.1 lists common glycoproteins classified according to function or source. The specific biological functions of many of the glycoproteins are still not completely understood. They are important components of cell membranes, where they coat the extracellular membrane surface. In membranes they may play a role in cellular recognition by immunoglobulins or provide sites for cell-cell interactions. Many plasma proteins, including the blood-group substances, are glycoproteins, but their functions are unclear.

Isolation, separation, and structural analysis of glycoproteins follow the same general procedures common for simple proteins. Ion-exchange chromatography, gel filtration, HPLC, and electrophoresis are routinely used to isolate and purify glycoproteins. One new and very promising technique for the specific isolation of glycoproteins is based on the princi-

Figure E10.1
Linkage of carbohydrates to amino acids in glycoproteins. (A) N-glycosyl bond between asparagine and *N*-acetylglucosamine. (B) O-glycosyl bond between threonine and glucose.

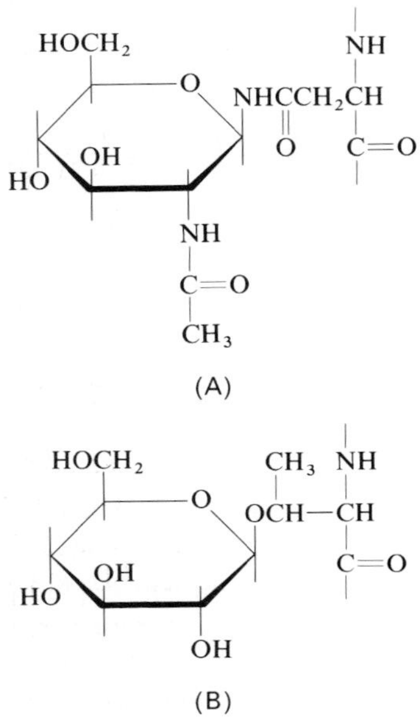

Table E10.1
Representative Examples of Glycoproteins

Blood
Fibrinogen
Immunoglobulins
Blood-group proteins
Glycosylated hemoglobins
"Antifreeze proteins"

Hormones
Chorionic gonadotropin
Follicle-stimulating hormone

Mucus Secretions
Submaxillary mucins
Gastric mucin

Cell Membranes
Glycophorin
Fibronectin

ples of affinity chromatography. A solid support, Glyco-Gel B[1], has been developed that contains covalently linked boronic acid on cross-linked agarose. Any glycoprotein that contains a carbohydrate unit with two neighboring hydroxyl groups is retained by the Glyco-Gel B (see Figure E10.2). A buffer containing sorbitol is used to elute the glycoprotein from the affinity gel.

Identification of the amino acid residues involved in the protein-sugar linkage and characterization of the carbohydrate units require specific removal of the carbohydrate. O-glycosyl linkages between the seryl or threonyl and sugar are susceptible to attack by a mild basic solution containing sodium borohydride. The reducing conditions lead to the production of a sugar alcohol. N-glycosyl linkages between asparagine residues and the carbohydrate unit are stable under these conditions, so the two major types of sugar-protein linkage can be distinguished by this analysis. Carbohydrate units linked to amino acids by N-glycosyl bonds are cleaved by acid-catalyzed hydrolysis. The released carbohydrate unit, which is separated from the protein by ion-exchange chromatography or gel filtration, is analyzed for sugar composition. Sugar analysis is achieved by complete hydrolysis of all internal glycosyl linkages, conversion of the monosaccharides to suitable derivatives, and identification by gas chromatography or high-performance liquid chromatography.

The procedures described above identify the nature of the peptide-sugar linkage and the total sugar composition but do not completely characterize the carbohydrate unit unless the total glycosyl unit on the pro-

[1]Trademark of Pierce Chemical Company.

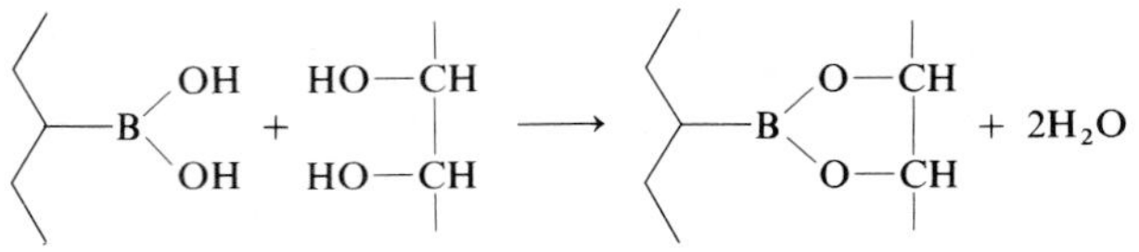

Figure E10.2
Affinity gel showing boronic acid covalently linked to agarose. Attachment of a glycoprotein via two hydroxyl groups is shown on the right side of the equation.

tein is a monosaccharide. Still to be defined are the sequence of sugars in the carbohydrate chain and the type of sugar linkages. The types of sugar linkages must be described in two ways, by determining the anomeric configuration (α or β) and by noting what hydroxyl groups are used in the linkages ($1 \rightarrow 4$, $1 \rightarrow 6$, etc.). These problems are solved by degrading the intact carbohydrate unit with exoglycosidases or endoglycosidases that catalyze the hydrolysis of specific sugar linkages. The released carbohydrates are analyzed by GC or HPLC. The results of such studies are combined, and the original structure is deciphered, using the overlap method in much the same way as for protein sequence determination.

Properties of Hemoglobin A$_{1c}$

The major hemoglobin protein in most adult humans is HbA, which has the subunit structure $\alpha_2\beta_2$. HbA comprises 90 to 95% of adult hemoglobin. Minor components that make up the remaining 5 to 10% are HbA$_2$ (2.5%), HbF (0.5%), HbA$_{1a}$ (0.2%), HbA$_{1b}$ (0.4%), and HbA$_{1c}$ (3–5%). The most abundant minor component, HbA$_{1c}$, has received considerable attention because its concentration is approximately doubled in individuals with diabetes mellitus. Figure E10.3 shows the most likely route of biosynthesis and the partial structure of HbA$_{1c}$. The subunit structure of the hemoglobin is the normal $\alpha_2\beta_2$, but the NH$_2$-terminal amino group (valine) of each of the two β chains is covalently linked to a ketohexose. The synthesis of HbA$_{1c}$ is unusual since it is thought to be nonenzymatic. Under physiological conditions, glucose or glucose-6-phosphate may be the source of the ketohexose. The rate of formation of HbA$_{1c}$ in the red blood cell depends on the glucose concentration; hence, untreated diabetics produce greater amounts of the minor hemoglobin. In the red blood cell, this glucose-modified hemoglobin is able to bind oxygen more strongly than HbA. This is probably due to a decreased reactivity of HbA$_{1c}$ with 2,3,-bisphosphoglycerate. The discovery of glycohemoglobin has stimulated the search for other proteins that are glycosylated in the presence of high glucose concentrations. Several, in addition to glycohemoglobin, have been identified and characterized. These include serum albumin, apoprotein B, which is a component of low-density lipoproteins, erythrocyte membrane proteins, skin collagen, and peripheral nerve proteins. The possible role of these proteins in the development of diabetic complications is being carefully studied.

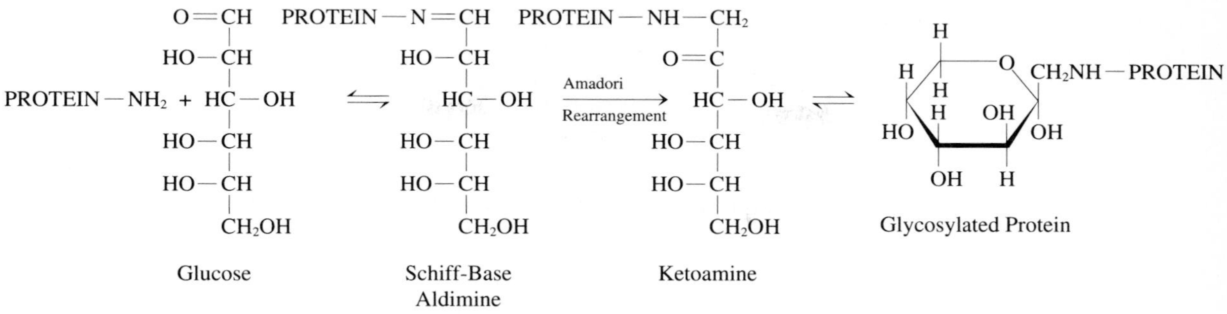

Figure E10.3
Possible route of synthesis of HbA$_{1c}$ and partial structure of the protein. The sugar is covalently attached to the NH$_2$-terminal end of each β chain in hemoglobin.

The link between HbA$_{1c}$ and diabetes mellitus has sparked great interest in the development of methods for separation, characterization and analysis of variant hemoglobins and other proteins. Cation-exchange chromatography on Bio-Rex 70 is one of the most effective methods for separation of the several hemoglobins in human red blood cells. All of the minor hemoglobin components are well resolved and widely separated from normal HbA.

The characteristic absorption spectrum of oxyhemoglobin A, with λ_{max} at 415, 542, 577 nm, which is due to the heme prosthetic group, is well known. What may not be so well understood by students are the spectral changes that occur when hemoglobin is treated with inorganic or organic phosphates. Inorganic phosphate, 2,3-bisphosphoglycerate (2,3-BPG), and inositol hexaphosphate (IHP) bind to hemoglobin in a cleft between the two β chains. When one of these molecules (for example, 2,3-BPG) occupies the small space between the β chains, the hemoglobin undergoes a conformational change. This change, which is primarily quaternary in nature, causes a slight movement of the heme prosthetic groups. This places the heme groups in a different environment and leads to the observed spectral changes of hemoglobin in the presence of phosphates. Recall from your biochemistry lectures that 2,3-BPG and other phosphates also decrease the oxygen affinity of deoxyhemoglobin. When glucose or other small molecules are covalently linked to the NH$_2$-terminal amino acid of the β chains, as in HbA$_{1c}$, the phosphate binding site is blocked. This interferes with the noncovalent binding of 2,3-BPG or IHP. Therefore, smaller spectral changes should occur when phosphate is added to HbA$_{1c}$ than when added to normal HbA.

One of the most effective and convenient methods for detecting and characterizing ligand-hemeprotein interactions is **difference spectroscopy**. This technique was introduced in Chapter 5.

A procedure for partial characterization of glycohemoglobins involves a colorimetric determination of the carbohydrate. When some glycoproteins are treated with oxalic acid at 100°C, the bound carbohydrate is released in the form of 5-hydroxymethyl furfural (5-HMF). 5-HMF produces a colored adduct with thiobarbituric acid (TBA) that has a characteristic absorbance spectrum with $\lambda_{max} = 443$ nm. This test is relatively specific, since 5-HMF is formed only from carbohydrates bound to proteins via ketoamine linkages. If time permits, each of the hemoglobins that you isolate in this experiment will be characterized by the TBA test. A positive test indicates the presence of a ketohexose bound to an amino group as in Figure E10.3.

Glycosylated Hemoglobins and Diabetes Mellitus

Quantitative measurements of glycosylated hemoglobins in hemolysates provide a sensitive and accurate diagnosis in mild cases of diabetes, where glucose tolerance tests are sometimes variable. Clinical tests for glycosylated hemoglobins fall into three categories:

1. Direct analysis of erythrocytes
   (a) Hemolysate is prepared and treated with oxalic acid at 100°C. The 5-HMF released is mixed with thiobarbituric acid and the colored adduct is determined at 443 nm. This method is simple, fast, and reproducible, but it measures the blood content of all glycosylated hemoglobins.
   (b) The differential binding of inositol hexaphosphate by glycosylated hemoglobins compared to normal hemoglobin can be used to measure the concentration of glycohemoglobins. The higher the content of NH_2-terminal glycosylated hemoglobin in a blood sample, the smaller the spectral difference in the visible spectrum. This test has been commercially developed by Abbott Laboratories.
   (c) An antibody against HbA_{1c} has been prepared and used directly on hemolysate; however, the sensitivity is acceptable only at levels of HbA_{1c} of 6% or greater. An advantage is that only HbA_{1c} is detected and measured.

2. Measurements of separated glycohemoglobins
   Most tests based on this method separate the glycosylated hemoglobins as a group, and measurement is of total glycohemoglobins. The glycoproteins are separated by affinity chromatography, ion-exchange chromatography, and HPLC. This measurement is more accurate than those in group 1; however, it measures a mixture of glycohemoglobins rather than HbA_{1c}.

3. Measurement of separated HbA_{1c}
   Cation-exchange chromatography or gel electrofocusing is used to separate HbA_{1c} from the other minor hemoglobins and HbA. The purified HbA_{1c} may be quantified by colorimetry or spectrophotometry.

Overview of the Experiment

In this experiment, animal hemolysate will be subjected to cation-exchange or affinity chromatography. Most of the minor glycohemoglobins can be separated on a Bio-Rex 70 cation-exchange column. Each column fraction containing the separated minor hemoglobins will be characterized by visible spectrophotometry and carbohydrate analysis. This procedure will require two laboratory periods of 3 to 4 hours each.

If time is limited, an alternative procedure may be used in this experiment. Rapid separation of glycoproteins from hemolysate can be accomplished with an affinity column. Nonglycosylated proteins such as HbA pass directly through the column, and glycoproteins are retained by the solid support. The glycoproteins, which are eluted with sorbitol buffer, may be characterized by spectrophotometry and colorimetric analysis of carbohydrate content. The glycoprotein fraction from the affinity column will contain HbA$_{1c}$, but also other glycoproteins that are present in hemolysate. Therefore, your characterization tests will actually be carried out on a mixture of glycoproteins. HbA$_{1c}$ will make up at least 80% of this glycoprotein fraction, so it will have the greatest influence on the results of your analysis. Affinity gel may be purchased or it may be prepared according to Koenst and Edstrom (1985), which requires a period of 1–2 days. The alternative experiment with the affinity column requires about 3 hours, not including the thiobarbituric acid test for carbohydrates.

Figure E10.4 shows the experimental options available.

II. Materials and Supplies

A. *Preparation of Hemolysate*

Fresh whole blood. Because of the possibility of pathogenic contamination, use of human blood is not recommended. Blood from experimental animals (rabbits, dogs, etc.) or a slaughterhouse may be used. Alternatively, blood samples with normal and elevated glycohemoglobin levels are available from Pierce Chemical Co. Blood should be collected in heparinized or EDTA-treated tubes. These reagents prevent coagulation. Blood samples may be stored for up to 1 week at 0–5°C.

NaCl solution, 1% in water

Centrifuge, capable of 12,000 rpm

B. *Bio-Rex 70 Column Chromatography*

Bio-Rex 70 cation-exchange resin, 100–200 mesh, sodium form. Obtained from Bio-Rad Laboratories, Richmond, CA, and prepared as

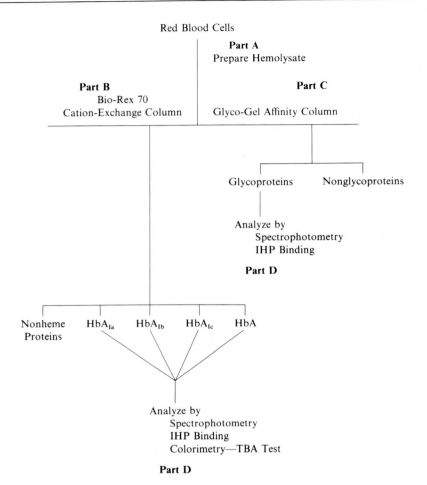

Figure E10.4
Flowchart for characterization
of glycohemoglobins.

described by the supplier. The final buffer for the slurry should be buffer A below.

Glass column, 2.5 × 30 cm

Buffer A, 0.05 *M* potassium phosphate, pH 6.6

Buffer B, 0.05 *M* potassium phosphate, 0.05 *M* NaCl, pH 6.6

Buffer C, 0.05 *M* potassium phosphate, 0.1 *M* NaCl, pH 6.6

Buffer D, 0.05 *M* potassium phosphate, 1.0 *M* NaCl, pH 6.6

Test tubes 50, 10 × 100 mm

Glass cuvettes 2, 3.0 mL

Fraction collector

Spectrophotometer

C. Affinity Chromatography Using Boronic Acid Gel

Glass column, 0.5×5 cm. A disposable pipet will work well.

Glyco-Gel B, obtained from Pierce Chemical Co., Rockford, IL. Prepare as described by the company. Final buffer for slurry should be the wash buffer below. Prepacked columns are also available. Each column can be regenerated twice.

Wash buffer, 0.25 M ammonium acetate, 0.05 M $MgCl_2$, 200 mg/L NaN_3, pH 8.5

Elution buffer, 0.1 M Tris, pH 8.5, containing NaN_3 (200 mg/L) and sorbitol (0.2 M)

Test tubes, ten, 10×100 mm

Hydrocarbon foil

Glass cuvettes 2, 3.0 mL

Spectrophotometer

D. Analysis of Glycoproteins

Double-beam spectrophotometer for difference spectroscopy

Glass cuvettes

0.01 M inositol hexaphosphate (phytic acid)

0.1 M imidazole buffer, pH 6.8, plus 0.2 mM $K_4Fe(CN)_6$

Heating block or constant-temperature bath at 100°C

1.0 M oxalic acid

Trichloroacetic acid, 40% in water

Thiobarbituric acid, 0.025 M in water

Fructose standard solution, 0.001 M in water

Glucose standard solution, 0.001 M in water

Test tubes, ten 10×100 mm

III. Experimental Procedure

A. Preparation of Hemolysate

⚠ **Caution:**

Working with animal blood presents a potential biohazard. Do not pipet any solutions by mouth. Wash hands with hot water and soap before eating, drinking, or smoking. Aseptic techniques must be used.

The red blood cells must be lysed so the hemoglobin can be released and dissolved in solution. Whole blood (2 mL) is centrifuged for 10 minutes at $4000 \times g$ to isolate erythrocytes. Decant and discard the supernatant and wash the erythrocytes by adding 3 mL of 1% NaCl. Centrifuge at $4000 \times g$ for 10 minutes. Discard the supernatant and repeat the wash step two more times. To the isolated and washed red blood cells, add 6 mL of distilled water in order to disrupt the cell membrane by osmotic pressure. Stir and allow the mixture to stand for 20–30 minutes for cell lysis. Centrifuge cell debris for 20 minutes at $30,000 \times g$. The supernatant contains hemoglobin and other red blood cell constituents. The sediment contains fragments of cell membranes and other cell debris.

B. Bio-Rex 70 Column Chromatography

If possible, this column should be run in a cold room or in a refrigerated chamber. Clamp the glass column to a ring stand. Pour a slurry of Bio-Rex 70 into the column and allow the resin to pack to a height of 20 cm. Use the same general procedure that you followed in Experiments 2 and 3. Never allow the column to run dry. When the resin has settled to a final height of 20 cm, cut a small, round piece of filter paper and allow it to settle on top of the column. Place 5 mL of the hemolysate prepared in part A onto the column by adding 1 to 2 mL dropwise with a disposable pipet. Pour the remaining 3 to 4 mL of hemolysate carefully down the inside wall of the column.

Begin to collect 3-mL fractions from the column into test tubes in a fraction collector. The column flow rate should be approximately 2 to 3 mL/min. When all the hemolysate has entered the resin, add dropwise 2 mL of buffer A. Allow the buffer to enter the resin and then elute the column with 200 mL of buffer A, continuing to collect 3-mL fractions. Allow buffer A to enter the column completely and then add buffer B. Elute the column with B, collecting 3-mL fractions, until no more red protein is eluted (approximately 100 mL). Change the eluting buffer to C and collect 3-mL fractions until the column eluent is colorless. Finally, elute the column with buffer D to wash off hemoglobin A. While the column is running, you should be taking absorbance measurements on each fraction. Read A_{415} on each fraction collected with buffers A and B. Read A_{540} for each fraction collected with buffer C and A_{670} for buffer D. The different wavelengths are used because later fractions may be off the absorbance scale at 415 nm. In each case, zero the spectrometer with the proper buffer. Prepare a graph of absorbance (A_{415}, A_{540}, or A_{670}) vs. fraction number. Combine the fractions represented by each peak. Do this by combining three or four of the most intensely colored fractions from each peak. The first major red peak is $HbA_{1a_1} + HbA_{1a_2}$. The second is HbA_{1b} and the third is HbA_{1c}. The largest and last peak is normal hemoglobin A. Label the four fractions I, II, III, and IV. Skip the next section on affinity chromatography and begin part D, Analysis of Glycoproteins.

C. Affinity Chromatography Using Boronic Acid Gel

⚠ **Caution:**

The buffers used in this part of the experiment contain sodium azide, a deadly poison. **Do not pipet any solutions by mouth.** Sodium azide may react with copper or lead plumbing to form explosive azides. For disposal, flush waste buffer down the drain with large volumes of water.

Prepare the small glass column or disposable pipet by placing a very small piece of glass wool or cotton into the constriction. Then plug the constricted end with hydrocarbon foil. Clamp the column to a ring stand and pour 1 mL of the Glyco-Gel B slurry into the column. Allow some buffer to pass out of the column by removing the Parafilm, but the column should not run dry. Close off the column with hydrocarbon foil when most of the buffer is gone. Obtain 0.1 mL of the hemolysate. Allow all the buffer to enter the gel and immediately and carefully add the sample of hemolysate to the top of the column. Allow it to enter the gel and immediately add 0.5 mL of the ammonium acetate buffer to wash the remaining hemoglobin completely into the gel. Begin to collect a fraction from the column and apply 19.5 mL of the wash buffer (ammonium acetate) in a dropwise fashion. Collect total effluent (20.1 mL) in a single test tube. When the wash buffer has just entered the gel, add 5 mL of eluting buffer containing sorbitol and collect all the eluting buffer in another test tube. This last fraction contains glycoproteins. Read the A_{415} of both fractions. Save all fractions for carbohydrate analysis in part D.

D. Analysis of Glycoproteins

Spectrophotometry

Record a visible spectrum (340 to 700 nm) for each pooled column fraction. To do this, dispense 3 mL of a fraction into a glass cuvette and run the spectrum, using as reference the buffer that eluted that particular fraction. For example, for fraction I of the Bio-Rex column, use buffer A in the reference beam. If the spectrometer pen runs off the absorbance scale, dilute the sample with the appropriate buffer and repeat the scan. If you performed the separation using the Bio-Rex column, you should have four fractions (I, II, III, IV) on which to record spectra. For fractions from the affinity column, run spectra on the final fraction (eluted with sorbitol buffer) and on only one of the earlier fractions.

THIOBARBITURIC ACID TEST

Each hemoglobin fraction may be analyzed for carbohydrate content by using the thiobarbituric acid test as described in the introduction. The test is carried out on column fractions and on standard fructose and glucose solutions. The procedure is outlined in Table E10.2.

Tube 1 is a control containing only water and the required chemical reagents. Tubes 2 and 3 contain standard carbohydrates, fructose and glucose. Additional tubes contain column fractions with glyco- or nonglycohemoglobins. After addition of oxalic acid to each tube, mix well and heat at 100°C for 4 hours. Cover each test tube with a marble to avoid evaporation. Add 1.0 mL of 40% trichloroacetic acid to each tube to precipitate the protein. Centrifuge each reaction mixture for 10 minutes at 2000 × g. Transfer 2.0 mL of each supernatant to a separate tube and mix with 2.0 mL of 0.025 M thiobarbituric acid. Incubate the reaction mixtures at 40°C for 1 hour. Allow the tubes to cool to room temperature and record the spectrum of each reaction mixture from 400 to 500 nm.

DIFFERENTIAL BINDING OF INOSITOL HEXAPHOSPHATE

Obtain two matched glass cuvettes for spectral measurements. To check whether the cuvettes are well matched, fill both with water or a buffer and scan from 400 to 600 nm. A straight line should be obtained. Pour out water. Into each cuvette, transfer 0.5 mL of a hemoglobin fraction. This may be one of the two fractions from affinity chromatography or fraction I, II, III, or IV from the Bio-Rex column. Add 2.5 mL of the imidazole buffer containing $K_4Fe(CN)_6$ to each cuvette. To the reference

Table E10.2
Thiobarbituric Acid Test, Part D[1]

Reagent-Procedure	1	2	3	4	5
Distilled water	2.0	1.8	1.8	—	—
Standard fructose	—	0.2	—	—	—
Standard glucose	—	—	0.2	—	—
Hemoglobin column fractions	—	—	—	—	—
I	—	—	—	2.0	—
II, etc.	—	—	—	—	2.0
Oxalic acid, 1.0M	1.0	⟶			
Mix well and heat					
Add TCA, 40%	1.0	⟶			
Centrifuge					
Transfer 2.0 mL of each supernatant to separate tubes					
Add TBA, 0.025 M	2.0	⟶			
Heat and record spectrum					

[1] Units are milliliters.

cuvette, add 0.1 mL of inositol hexaphosphate solution. To the sample cuvette, add 0.1 mL of distilled water. Mix each cuvette by holding a small square of hydrocarbon foil over the opening and inverting several times. Do not shake. Place the cuvettes in the proper compartments of a double-beam spectrometer. With the 0 setting or balance knob, set the recorder pen to midscale (0.5 A) at 400 nm. The recorder pen could move in either direction, so it must have sufficient chart space for movement. Using the above procedure, obtain an IHP difference spectrum for as many of your hemoglobin fractions as time allows.

IV. Analysis of Results

B. and C. Separation of Hemoglobins

If you separated the hemoglobins by ion-exchange chromatography, compare the elution profile with Figure 1 in McDonald et al. (1978). Note, and record in your notebook, any similarities and differences. Identify the hemoglobin fraction represented by each peak in the profile. Can you estimate the quantity of each fraction and calculate the percentage of each type of hemoglobin?

If you used affinity chromatography, define the type of hemoglobin in each of the two fractions. What is the ratio of glycohemoglobins to non-glycohemoglobins? Calculate the percent of glycohemoglobins using Equation E10.1.

$$\blacktriangleright \quad \% \text{Glycohemoglobin} = \frac{5.0 A_{414}(\text{bound Hb}) \times 100}{20.1 A_{414}(\text{bound Hb}) + 5.0 A_{414}(\text{unbound Hb})}$$

A_{414}(bound Hb) = absorbance at 414 nm of the fraction eluted with sorbitol buffer

A_{414}(unbound Hb) = absorbance at 414 nm of the unbound hemoglobins   (Equation E10.1)

D. Analysis of Glycoproteins

SPECTROPHOTOMETRY

Do you note any differences between the absorption spectra of hemoglobin A and the glycohemoglobins? Should there be spectral differences?

THIOBARBITURIC ACID TEST

Explain the results of your TBA test with standard fructose and glucose. Did you obtain a color for each? Which hemoglobin fraction shows the greatest formation of 5-HMF? Are these the expected results?

DIFFERENTIAL BINDING OF IHP

Describe the difference spectrum obtained for each hemoglobin. What does your result imply about the interaction of IHP with hemoglobin? Compare the IHP difference spectrum of hemoglobin A with that of hemoglobin A$_{1c}$ (or the mixture of glycohemoglobins). Explain any differences. Which hemoglobin, A or A$_{1c}$, binds IHP more tightly? Why? Compare your difference spectra with those in Chapter 5, Figure 5.9.

V. Questions and Problems

▶ 1. Describe the principles behind the separation of hemoglobins by Bio-Rex chromatography.

▶ 2. Many of the proteins in the hemolysate are "nonheme" proteins. How would you detect their elution from a Bio-Rex or Glyco-Gel column?

▶ 3. In the early literature describing ion-exchange chromatographic separation of the hemoglobins, glycohemoglobins were called "fast hemoglobins." Why?

4. Did both standard fructose and glucose give an intense color for the TBA test? Explain the results.

▶ 5. Could you substitute Sephadex G-100 for Bio-Rex 70 and expect an effective resolution of the hemoglobins?

▶ 6. Why was it necessary in the difference spectrum experiment to have identical concentrations of protein in both cuvettes?

▶ 7. Why must a "matched pair" of cuvettes be used for difference spectroscopy?

▶ 8. When you monitor the Bio-Rex ion-exchange column at 415 nm, what chromophore are you measuring?

▶ 9. Explain the role of sorbitol in the eluting buffer in part C.

VI. References

E. Abraham, *Glycosylated Hemoglobin: Methods of Analysis and Clinical Applications* (1985), Marcel Dekker (New York). A detailed review.

J. Beeley, in *Laboratory Techniques in Biochemistry and Molecular Biology,* Vol.16, R. Burdon and P. Van Knippenberg, Editors (1985), Elsevier (Amsterdam). "Glycoprotein and Proteoglycan Techniques."

S. Brown and M. Bowes, *Biochem. Educ.* **13**, 2–6 (1985). "Glycosylated Hemoglobins and Their Role in Management of Diabetes Mellitus."

H. Bunn, K. Gabbay, and P. Gallop, *Science* **200**, 21 (1978). "The Glycosylation of Hemoglobin: Relevance to Diabetes Mellitus."

L. Jackman, *J. Chem. Educ.* **58**, 77–78 (1981). "Separation of Glyco-hemoglobins by Ion Exchange Chromatography."

M. Koenst and R. Edstrom, *Biochem. Educ.* **13**, 7–9 (1985). "Determination of Glycosylated Hemoglobin: A Laboratory Experiment Relevant to Diabetes Mellitus."

C. Mathews and K. van Holde, *Biochemistry* (1990), Benjamin/Cummings (Redwood City, CA), pp. 291–297. Structure and function of glycoproteins.

M. McDonald, R. Shapiro, M. Bleichman, J. Solway, and H. Bunn, *J. Biol. Chem.* **253**, 2327 (1978). "Glycosylated Minor Components of Human Adult Hemoglobin."

E. Moore and S. Stroupe, U.S. Patent No. 4,200,435, April 29, 1980. Diagnostics Division of Abbott Laboratories, *Chem. Abstr.* **93**, 110212y (1980). Using differential spectroscopy to measure glycosylated hemoglobin in blood.

J. Rawn, *Biochemistry*, 2nd ed. (1989), Neil Patterson Publishers (Burlington, NC). pp. 228–230, 878–884. Introduction to glycoproteins.

J. Robyt and B. White, *Biochemical Techniques—Theory and Practice* (1987), Brooks/Cole Publishing (Monterey, CA), pp. 50–51. "Difference Spectroscopy."

L. Stryer, *Biochemistry*, 3rd ed. (1988), W. H. Freeman (New York), pp. 298–299, 343–346. Structure and function of glycoproteins.

A. Tarentino, R. Trimble, and T. Plummer, *Methods Cell Biol.* **32**, 111–139 (1989). "Enzymic Approaches for Studying the Structure, Synthesis and Processing of Glycoproteins."

B. Teupe, *Quantitative Determination of Glycated Hemoglobin Using Affinity Chromatography* (1987), Isolab Inc. (Akron, OH).

D. Voet and J. Voet, *Biochemistry* (1990), John Wiley & Sons (New York), pp. 260–268. Structure and function of glycoproteins.

O. Wieland, *Mol. Cell. Endocrinol.* **29**, 125–131 (1983). "Protein Modification by Nonenzymatic Glucosylation: Possible Role in the Development of Diabetic Complications."

K. Winterhalter, in *Methods in Enzymology*, E. Antonini, L. Rossi-Bernardi, and E. Chiancone, Editors Vol. 76 (1981), Academic Press (New York), p. 732. "Determination of Glycosylated Hemoglobin."

EXPERIMENT 11

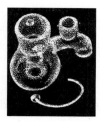

Isolation, Separation, and Spectrophotometric Characterization of Photosynthetic Pigments

■ Recommended Reading
Chapter 3, Section B; Chapter 5, Section A; Experiment 12.

■ Synopsis
Many of the colors associated with higher plants (green leaves in the spring and summer, yellow or red leaves in the fall, the orange color of carrots, some colors in flower petals) are due to the presence of pigment molecules such as chlorophylls and carotenoids. In this experiment a mixture of these pigments will be isolated by solvent extraction of plant tissue, separated by chromatography, and the components identified by visible spectrophotometry.

I. Introduction and Theory

Chlorophylls and Carotenoids

The major photosynthetic pigments of higher plants can be divided into two groups, the *chlorophylls* and the *carotenoids*. Both types of pigments are present in the subcellular organelles called chloroplasts, where they are bound to proteins in the thylakoids, the photochemically active photosynthetic biomembranes (see Experiment 12). Intact pigment-protein complexes, which are held together by weak, noncovalent bonds, can be isolated from chloroplasts and characterized by polyacrylamide gel electrophoresis and isoelectric focusing. The pigments are released in a protein-free form by grinding plant tissue in a solvent such as acetone, methanol, or hexane. Since chlorophyll and the carotenoids are readily soluble in organic solvents, they are biochemically classified as lipids.

The most abundant plant pigments are **chlorophyll a** and **chlorophyll b**, which occur in a ratio (*a:b*) of approximately 3:1. As shown in Figure E11.1, the chlorophylls have a porphyrin ring with a coordinated magnesium atom at its center, a fused five-membered ring, and a C_{20} phytyl side chain. This nonpolar hydrocarbon side chain enhances the solubility of chlorophyll in nonpolar solvents. The only difference between chlorophylls *a* and *b* is the substituent on position 3 (ring 3); chlorophyll *a* has a methyl group, whereas *b* has an aldehyde functional group. The chlorophylls are biosynthesized by condensation of δ-aminolevulinic acid in a process similar to the synthesis of heme in mammals. In the living plant cell, the chlorophylls are the primary photosynthetic pigments. They absorb light in the blue region (450 nm) and the red region (650–700 nm). Absorption of light, which excites an electron in the chlorophyll molecule, provides energy for initiation of the photosynthetic process producing NADPH and ATP to be used for carbohydrate biosynthesis by the plant (see Experiment 12).

The second group of plant pigments, the **carotenoids,** can be divided into two different types: (1) the **carotenes,** which contain only carbon and hydrogen, and (2) the **xanthophylls,** which contain carbon, hydrogen, and oxygen atoms in the form of hydroxyl or epoxide functional groups. The structures of several major carotenoids are shown in Figure E11.2. Note that they all contain 40 carbon atoms and, because of the extensive conjugation, are highly colored. Most carotenoids are yellow, red, or orange; however, some are green, pink, and even black. Many of the bright colors of flower petals are due to the presence of carotenoids. The yellow colors of fall result from the preferential destruction of the green chlorophylls, revealing the carotenoid color (Goodwin, 1958).

The percentage composition of carotenoids in plants depends on growing conditions. The average weight percent ranges are β-carotene, 25–40%; lutein, 40–60%; violaxanthin, 10–20%; and neoxanthin,

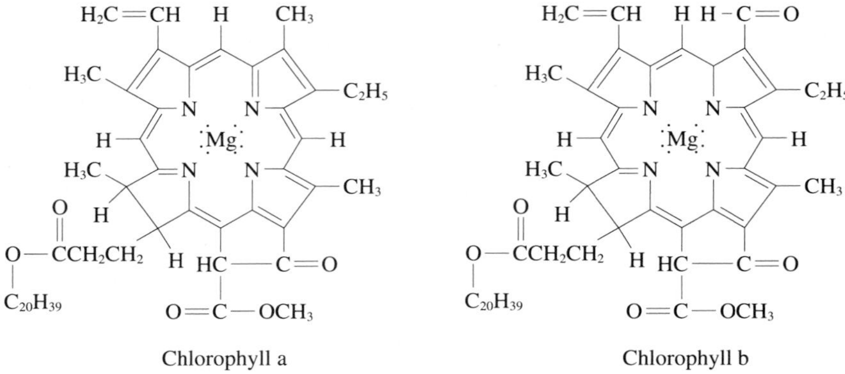

Figure E11.1
Structures of chlorophyll *a* and chlorophyll *b*.

Chlorophyll a

Chlorophyll b

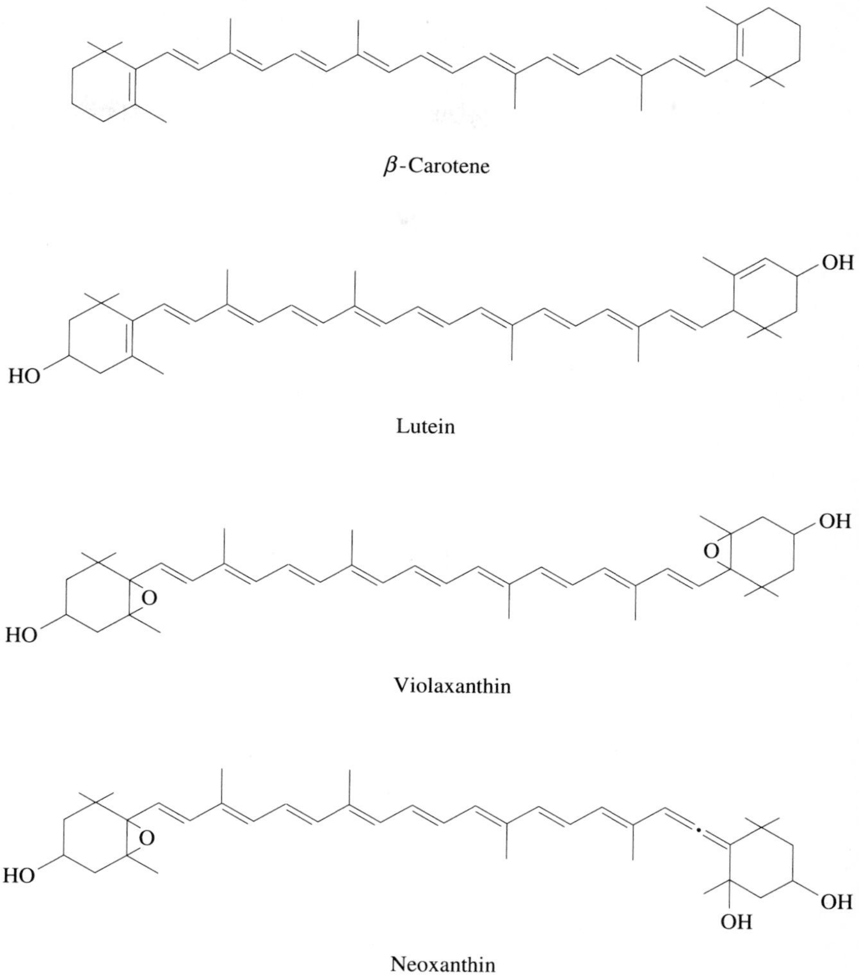

β-Carotene

Lutein

Violaxanthin

Figure E11.2
Structures of major carotenoids:
β-carotene, lutein, violaxanthin,
and neoxanthin.

Neoxanthin

5–13%. Carotenoids are biosynthesized by condensation of acetyl CoA through the mevalonic acid pathway and the Porter-Lincoln pathway.

The physiological function of the carotenoids is not completely understood; however, two probable functions have been described. First, the carotenoids, particularly the xanthophylls, may participate in light absorption for photosynthesis. It has been shown that illuminated xanthophylls can transfer excitation energy directly to chlorophyll *a*. Second, the carotenes may inhibit the photo-oxidative destruction of chlorophyll, a process that consumes oxygen.

General information on photosynthetic pigments can be obtained from standard biochemistry texts and other reference books listed at the end of the experiment.

Spectral Characteristics of Plant Pigments

The intense colors of the chlorophylls and carotenoids make them ideal candidates for absorption spectroscopy studies (Tan and Soderstrom, 1989). In fact, each plant pigment studied in this experiment has a unique visible spectrum that can provide a positive identification. As shown in Figure E11.3, chlorophylls *a* and *b* have absorption maxima in the 600–675 and 400–475 nm ranges. The absorption maximum of each peak depends on solvent polarity. For example, the two most intense ab-

Figure E11.3
Visible absorption spectra of chlorophyll *a* (A) and chlorophyll *b* (B).

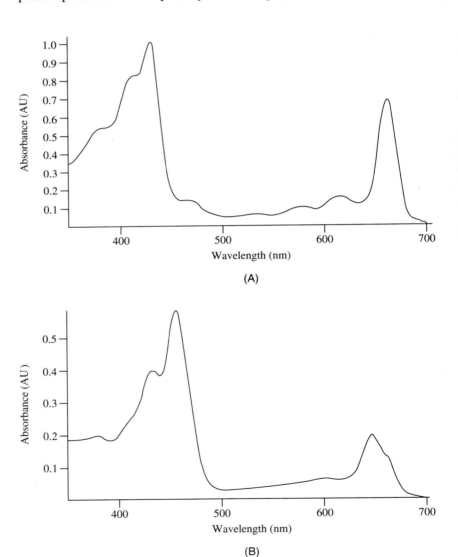

sorption peaks for chlorophyll *a* in diethyl ether are 660 and 428 nm, whereas in more polar methanol the peaks are shifted to 665 and 432 nm.

In contrast to the chlorophylls, which absorb light in two regions of the visible spectrum, the carotenoids exhibit intense absorption in just one, 350–500 nm. Figure E11.4 compares the absorption spectra of four common carotenoids. As with the chlorophylls, the absorption maxima of the carotenoids vary with polarity of the solvent. β-Carotene in diethyl ether has a λ_{max} of 449.8 nm, but in the more polar acetone, the λ_{max} is 454 nm.

In addition to their use in pigment identification, spectrophotometric data can be used to determine the concentration of pigments in a plant extract. Data from spectral analysis were used to calculate absorption coefficients for each pigment in order to derive the following equations (Lichtenthaler, 1987):

Figure E11.4
Visible absorption spectra for four carotenoids: β-carotene (—), lutein (- -), violaxanthin (.._.), and neoxanthin (. . .).

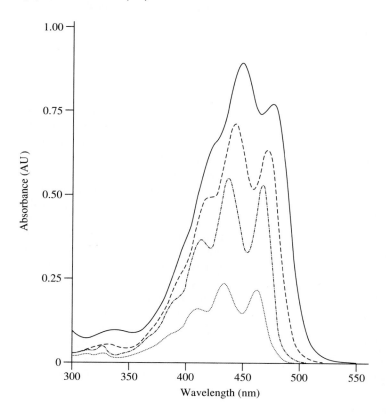

$$C_a = 11.24A_{661.6} - 2.04A_{644.8}$$

$$C_b = 20.13A_{644.8} - 4.19A_{661.6}$$

$$C_{a+b} = 7.05A_{661.6} + 18.09A_{644.8}$$

$$C_{x+c} = \frac{1000A_{470} - 1.90C_a - 63.14C_b}{214}$$

where

C_a = concentration of chlorophyll *a* in units of micrograms per milliliter of plant extract solution (μg/mL)

C_b = concentration of chlorophyll *b* in μg/mL

C_{a+b} = concentration of total chlorophyll in μg/mL

C_{x+c} = concentration of total carotenoids (xanthophylls and carotenes) in μg/mL

A_z = absorbance measurement at wavelength z

These equations are designed for the use of 100% acetone as solvent.

Practical Aspects of the Experiment

Plant pigments will be isolated by careful extraction of plant tissue (green or dried leaves) with 100% acetone. The extraction can be carried out with either a mortar and pestle or a glass homogenizer. Addition of an abrasive agent such as quartz sand enhances the extraction process. Both types of plant pigments, chlorophylls and carotenoids, are extremely light sensitive and are readily destroyed by photobleaching; therefore, the extraction process should be performed rapidly and in subdued lighting. If quantitative determinations are to be made for each pigment, the absorption data must be obtained immediately after extraction.

A wide range of chromatographic techniques, including column, paper, thin-layer, and high-performance liquid chromatography, have been applied to the separation of photosynthetic pigments. Theoretical and practical information on these techniques may be found in Chapter 3. There is great historical interest here, because Tswett, in the first demonstration of chromatography in 1906, separated a plant pigment extract on $CaCO_3$ and MgO columns. In this experiment, you will use either paper or thin-layer chromatography to separate the pigments. Either technique provides excellent resolution of the two chlorophylls and the major carotenoids. It is instructive for students to use both techniques. After solvent elution of the pigments from the paper or silica gel, the visible absorption spectrum of each is determined.

Overview of the Experiment

This experiment provides students with the opportunity to isolate a biomolecule from its natural source, followed by its purification and identification. In addition, students will follow a procedure that is typical of the general extraction and characterization of lipids. However, unlike most lipids, the plant pigments are highly colored and may be characterized and quantified by visible spectrophotometry. Several types of plant tissue may be used. Some recommendations are fresh leaves (tree, plant, grass, spinach), green algae, or mosses. For variety, students may be asked to bring their own samples for analysis.

The approximate time requirements for the experiment are:

A. Extraction of the pigments—30 minutes

B. Determination of pigment concentration—15 minutes

C. Chromatography and elution of pigments—1 hour

D. Measurement of visible absorption spectrum—1 hour

Since the pigment extract deteriorates rapidly by photo-oxidation, it is critical that students work rapidly, but safely and efficiently. Part B must be completed immediately after the extraction. Parts C and D may be completed at a later time, but, in the meantime, the plant extracts must be refrigerated in a concentrated form and under nitrogen gas.

II. Materials and Supplies

Plant tissues (leaves, moss, algae, etc.)

Acetone, 100%

Mortar and pestle or glass homogenizer

Quartz sand

Small funnel and qualitative grade filter paper

Test tubes, 100 × 12 mm

Microcapillaries or tapered capillaries

Anhydrous sodium sulfate, Na_2SO_4

UV-VIS spectrophotometer

Glass cuvettes, 3 or 1 mL

Materials for paper chromatography:

Whatman 3MM paper, 13.5 × 7 cm

Solvent system, petroleum ether–acetone (9:1 v/v)

Chromatography chamber, 250-mL beaker with watch glass

Stapler

Scissors

Materials for thin-layer chromatography:

Silica gel plates, 10 × 10 cm

Solvent system, petroleum ether–dioxane–2-propanol (70:30:10, v/v/v)

Chromatography chamber

Metal spatula or clean razor blade

III. Experimental Procedure

A. *Extraction of the Pigments*

⚠ **Caution**

Exposure to acetone by inhalation, ingestion, or skin absorption is harmful. Contact with the eyes will cause severe irritation.

First aid: eyes—flush with copious amounts of water for at least 15 minutes; skin—flush with water for a few minutes; inhalation—remove to fresh air.

Acetone is volatile and extremely flammable. No open flames should be in the vicinity of your work. In case of fire, use a CO_2 extinguisher.

Dispose of acetone extract and solutions in the organic waste container.

This experiment should be completed in subdued lighting! Obtain five tree leaves or equivalent amount of grass or other plant tissue, tear into small pieces, and place in a mortar or homogenizer. Measure 15 mL of acetone and pour into the container. Add approximately 0.5 g of pure sand and thoroughly grind the mixture until the solution becomes a deep green. The plant cell walls of cellulose are tough, so you need to be persistent. After extraction is complete, add 1–2 g of anhydrous sodium sulfate to remove water from the extract. (What is the source of the water?) Rapidly filter the extract by gravity through fluted filter paper. Measure the volume of the filtered extract and rapidly proceed to part B.

B. *Determination of Pigment Concentration*

In order to determine the chlorophyll and carotenoid content of your extract, you must measure the absorbance at several wavelengths, 661.6, 644.8, and 470 nm. If you are using a single-beam spectrophotometer, use a cuvette of acetone to zero the instrument at each wavelength. A double-beam instrument should have a cuvette of acetone in the reference beam and the acetone extract solution in the sample beam. If desired, it is instructive to obtain a complete spectrum of the extract in the range 400–700 nm. This can be compared to the spectrum obtained for each of the individual pigments in part D.

C. Chromatography and Elution of Pigments

The general procedures for paper and thin-layer chromatography are similar. Obtain a piece of chromatography paper (13.5 × 7 cm) or a thin-layer plate (10 × 10 cm). With a pencil, draw a very light line 1.5 cm from and parallel to the long edge of the paper or TLC plate. Do not scrape silica off the plate. Using a Pasteur pipet or tapered capillary, evenly apply the acetone extract along the line. Begin on the penciled line approximately 1 cm from the edge and stop approximately 1 cm from the opposite edge. The narrower the line, the better your separation will be. Allow the first coating of extract to dry and repeat this procedure with another application of extract. Repeat a third time if the line is not dark green. Allow the paper to dry completely, and roll it into a circle with the pencil line and extract outside and on the circular edge. Use two or three staples to hold the circle in place. (Do not overlap the edges.) Place the paper (extract line down) into a chamber containing no more than 1 cm of appropriate solvent—petroleum ether–acetone (9:1). A chamber may be prepared by pouring 7 mL of solvent into a 250-mL beaker. Cover with a watch glass. If a thin-layer plate was used, place directly in a chamber containing petroleum ether–dioxane–2-propanol (70:30:10) and cover. When the solvent has risen to within 0.5 cm from the top of the paper or TLC plate, remove from the chamber and mark the position of the solvent front. Allow the chromatogram to dry. Measure the position of each pigment band and the solvent front and draw a full-scale picture of the completed chromatogram in your notebook. Since each chromatogram is complete in a few minutes, it is possible to experiment with the amount of extract applied to the paper or plate. If good separation is not attained and pigment bands are very dark, try a chromatogram with less extract. On the other hand, if the bands are light and difficult to distinguish, prepare and develop a chromatogram with more extract.

Each pigment is eluted from the paper or TLC plate using acetone. With the paper chromatogram, use a scissors to cut each colored band from the paper. Cut each paper strip into smaller pieces and place pieces from each band into separate, labeled test tubes containing 4 mL of acetone. Cover the tubes with a cork and allow to sit for 5–10 minutes with gentle mixing at 1- to 2-minute intervals. Proceed directly to part D. To extract the pigments from the TLC plate, use a metal spatula or razor blade to scrap the silica gel at each colored band into labeled test tubes containing 4 mL of acetone. Cover with a stopper and allow to sit for 5–10 minutes with mixing at intervals. Centrifuge or filter each solution using fluted filter paper. Proceed directly to part D.

D. Measurement of Visible Absorption Spectrum

Using glass cuvettes, measure the absorption spectrum of each fraction from 400 to 700 nm. A cuvette containing acetone should be used as reference. A Pasteur pipet may be used to transfer solutions to and from the cuvette. Be careful not to scratch the inside of the glass cuvette with the pipet. After all spectra have been obtained, dispose of the acetone solutions in the waste organic container in the laboratory.

IV. Analysis of Results

A. *Extraction of the Pigments*

Describe the color of the extract. Explain the action of acetone in the extraction process. What is the purpose of the sand?

B. *Determination of Pigment Concentration*

Using the absorbance data obtained on the crude extraction mixture and the equations provided earlier in the experiment, calculate the concentrations of chlorophyll *a*, chlorophyll *b*, and total carotenoids.

C. *Chromatography and Elution of Pigments*

Calculate the mobility of each pigment relative to the solvent front (R_f):

$$R_f = \frac{\text{distance from origin migrated by a compound}}{\text{distance from origin migrated by solvent}}$$

Explain the relative order of the R_f values. Can you make any general statements about the polarity of each pigment? In your notebook, describe the color of each eluted pigment fraction.

D. *Measurement of Visible Absorption Spectrum*

Prepare a table of spectral data (wavelengths of major peaks and absorbance) for each pigment. Using these data and the standard spectra shown earlier in this experiment, try to assign the identity of each pigment isolated. Now that each compound is identified, is there a correlation between the experimental R_f value and polarity as determined from the structure? Which pigments are more polar—those near the top or those near the bottom of the chromatogram?

V. Questions and Problems

▶ 1. What other solvents in addition to acetone could be used for extracting plant photosynthetic pigments?

▶ 2. Predict the relative R_f values for thin-layer chromatography of a mixture of chlorophylls *a* and *b*, β-carotene, lutein, neoxanthin, and violaxanthin. Assume that the solvent is the same as used in this experiment.

▶ 3. Repeat Problem 2 (above), but substitute paper chromatography with petroleum ether–acetone (9:1) as the developing solvent.

▶ 4. What extraction solvent system would you use if you wanted to isolate the intact protein-pigment complexes? Describe experimental techniques that could be used to determine the molecular weight of each protein-pigment complex.

VI. References

R. Boyer, *Biochem. Educ.* **18,** 203–206 (1990), "Isolation and Spectrophotometric Characterization of Photosynthetic Pigments."

H. Dailey, Editor, *The Biosynthesis of Heme and Chlorophylls* (1990), McGraw-Hill (New York). An excellent reference on the chemistry and biochemistry of chlorophyll.

T. Goodwin, *Biochem. J.* **68,** 503–511 (1958), "Studies in Carotenogenesis 24. The Changes in Carotenoid and Chlorophyll Pigments in the Leaves of Deciduous Trees During Autumn Necrosis."

H. Lichtenthaler, in *Methods in Enzymology,* Vol. 148, L. Packer and R. Douce, Editors (1987), Academic Press (New York), pp. 350–382, "Chlorophylls and Carotenoids: Pigments of Photosynthetic Biomembranes."

C. Mathews and K. van Holde, *Biochemistry* (1990), Benjamin/Cummings (Redwood City, CA), pp. 643–669. An introduction to photosynthesis and the photosynthetic pigments.

T. Moore, *Research Experiences in Plant Physiology—A Laboratory Manual,* 2nd ed. (1981), Springer-Verlag (New York), pp. 35–45, "Thin-Layer Chromatography of Chloroplast Pigments and Determination of Pigment Absorption Spectra."

J. Rawn, *Biochemistry,* 2nd ed. (1989), Neil Patterson Publishers (Burlington, NC), pp. 489–531. Fundamentals of photosynthesis.

L. Stryer, *Biochemistry,* 3rd ed. (1988), W. H. Freeman (New York), pp. 517–533. An introduction to photosynthesis.

B. Tan and D. Soderstrom, *J. Chem. Educ.* **66,** 258–260 (1989), "Qualitative Aspects of UV-VIS Spectrophotometry of β-Carotene and Lycopene."

D. Voet and J. Voet, *Biochemistry* (1990), John Wiley & Sons (New York), pp. 586–594. An introduction to the pigments of photosynthesis.

D. Youvan and B. Marrs, *Sci. Am.* **256** (6), 42–48 (1987). "Molecular Mechanisms of Photosynthesis."

G. Zubay, *Biochemistry,* 2nd ed. (1988), Macmillan (New York), pp. 564–588. An introduction to photosynthesis.

EXPERIMENT 12

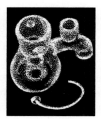

Photoinduced Proton Transport Through Chloroplast Membranes

■ **Recommended Reading**

Chapter 2, Section A; Chapter 5, Section A; Chapter 7, Sections A, B, C; Experiment 11

■ **Synopsis**

Photosynthesis occurs in the plant cell organelle called the chloroplast. During the process of photosynthesis, electrons are transferred from H_2O to $NADP^+$ via an electron carrier system. The energy released by electron transport is converted into the form of a proton gradient and coupled to ADP phosphorylation. In this experiment a method is introduced to demonstrate the formation of the proton gradient across the chloroplast membranes.

I. Introduction and Theory

The Photosynthetic Process

Photosynthesis occurs only in plants, algae, and some bacteria, but all forms of life are dependent on its products. In photosynthesis, electromagnetic energy from the sun is used as the driving force for a thermodynamically unfavorable chemical reaction, the synthesis of carbohydrates from CO_2 and H_2O (Equation E12.1)

▶ $$CO_2 + H_2O \xrightarrow{h\nu} (CH_2O) + O_2 \qquad \text{(Equation E12.1)}$$

Here (CH_2O) represents a carbohydrate molecule. The simplicity of Equation E12.1 is deceiving, because the process of photosynthesis is a complex one; the details are still under investigation. The fundamental photosynthetic event is initiated by the interaction of a photon with a chlorophyll molecule, causing excitation of an electron in chlorophyll. The activated electron is passed through a chain of electron carriers aligned in order of decreasing reduction potential. The electron transport chain is similar to that in mitochondrial respiration. Energy released during electron transport is coupled to the synthesis of ATP from ADP and phosphate, P_i. This process is outlined in Figure E12.1. Two photosystems, I and II, are at work during photosynthesis in plants. Each photosystem contains reaction centers (chlorophylls P680 and P700) that collect the light. The two systems are linked by the electron transport chain, allowing electron flow from H_2O (to produce O_2) ultimately to $NADP^+$. In other words, H_2O is oxidized and the electrons are carried to $NADP^+$,

Figure E12.1
The photosynthetic process, showing coupling of electron transport and ADP phosphorylation. The dashed line shows electron flow in cyclic photophosphorylation. See text for details.

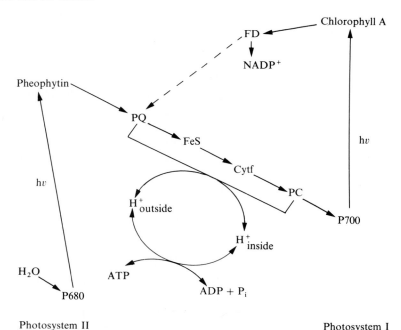

PQ = Plastoquinone
FeS = Iron Sulfur Protein
Cytf = Cytochrome f
PC = Plastocyanin
FD = Ferredoxin

forming the reduced cofactor, NADPH. The overall reaction is shown in Equation E12.2. Note that this is the reverse of the respiratory process occurring in mitochondria, where NADPH is oxidized.

▶ $$H_2O + NADP^+ + n\,ADP + n\,P_i \xrightarrow{h\nu} NADPH + H^+ + n\,ATP + \tfrac{1}{2}O_2$$
(Equation E12.2)

When electrons flow from photosystem I to photosystem II, protons are transported across the chloroplast membranes as indicated in Figure E12.1. This aspect of photosynthesis will be discussed in a later section.

The light reactions illustrated in Figure E12.1 are accompanied by a sequence of "dark reactions" leading to the formation of carbon intermediates and sugars. This sequence of reactions, called the Calvin cycle, incorporates CO_2 into carbohydrate structures.

Figure E12.1 illustrates the photosynthetic process as it occurs in higher plants. This is called **noncyclic photophosphorylation** to distinguish it from **cyclic photophosphorylation** in photosynthetic bacteria. Cyclic photophosphorylation requires only photosystem I and a second series of electron carriers to return electrons to the electron-deficient chlorophyll. The dashed line in Figure E12.1 indicates the flow of electrons in cyclic photophosphorylation. ATP is produced during the cyclic process just as in the noncyclic process, but NADPH is not.

The photosynthetic process in green plants occurs in subcellular organelles called **chloroplasts**. These organelles resemble mitochondria; they have two outer membranes and a folded inner membrane called the **thylakoid**. The apparatus for photosynthesis, including the chlorophyll reaction centers and electron carriers, is in the thylakoid membrane. The chemical reactions of the Calvin cycle take place in the **stroma,** the region around the thylakoid membrane.

Mechanism of Photophosphorylation

One very active area of biochemical investigation is the study of energy transfer in mitochondria and chloroplasts. In both oxidative phosphorylation and photophosphorylation, electron flow through a carrier chain is linked in some way to the generation of ATP. Most research is directed toward determining how the energy provided by electron transport is coupled to the phosphorylation of ADP. In recent years, three theories of energy coupling have received attention: chemical coupling, conformational coupling, and chemiosmotic coupling. The **chemical coupling hypothesis** proposes the formation of high-energy chemical intermediates; the **conformational coupling hypothesis** proposes a high-energy molecular conformation: and **chemiosmotic coupling** involves a charge separation or concentration gradient across a membrane.

The chemiosmotic coupling hypothesis, proposed by P. Mitchell, is the most attractive explanation, and many experimental observations now support this idea. Simply stated, Mitchell's hypothesis suggests that electron transfer is accompanied by transport of protons across the membrane, resulting in a proton concentration gradient. When the proton gradient (a high-energy condition) collapses, the released energy is coupled to the phosphorylation of ADP. In mitochondria, the proton transport is through the inner membrane, from the inside (matrix side) to the outside (cytoplasmic side), whereas in chloroplasts protons are transported across the thylakoid membrane from the outside to the inside.

Experimental results supporting the Mitchell hypothesis were first reported from studies using chloroplasts. Two critical experiments providing unequivocal evidence for linkage between proton transport and ADP phosphorylation were reported. Neumann and Jagendorf (1964) generated a high-energy proton gradient by illuminating chloroplasts in the absence of ADP and phosphate. As one would predict from the Mitchell hypothesis, Jagendorf observed a rise in the pH of the medium indicating proton uptake by the chloroplasts. In later experiments, Jagendorf and Uribe (1966) induced an artificial proton gradient across the thylakoid membrane by incubating chloroplasts under acidic conditions (pH 4) and then rapidly transferring them to pH 8 buffer. When ADP and P_i were added to the chloroplast suspension, the synthesis of ATP was observed, and the previously induced pH gradient disappeared. Similar experiments with mitochondria also confirm the transport of protons in oxidative phosphorylation, but in the opposite direction from photophosphorylation.

Continued study of light-induced proton uptake by chloroplasts has led to the observation that proton uptake is accompanied by Mg^{2+} and K^+ extrusion from the chloroplasts. The specific function of these exchanges, other than ionic balance, is not understood.

Practical Aspects of the Experiment

Chloroplasts will be isolated by careful extraction of spinach leaves, using Tris buffer containing sucrose. The crude extract contains both whole and fragmented chloroplasts, but both contain all the necessary photosynthetic components and are capable of photophosphorylation. The preparation described in this experiment retains almost all of the chlorophyll in the chloroplasts. The total chlorophyll content of the chloroplasts will be determined by extracting the pigment with aqueous acetone and measuring the absorption at $\lambda = 652$ nm. The chlorophyll concentration is calculated according to Equation E12.3 (Arnon, 1949),

$$\blacktriangleright \quad C = \frac{A}{El} = \frac{A_{652}}{34.5 \times l} \qquad \text{(Equation E12.3)}$$

where

C = concentration of chlorophyll in mg/mL

A_{652} = absorption of chlorophyll at 652 nm

E = absorption coefficient for chlorophyll, 34.5 mL/(mg × cm)

l = path length of cuvette

The average range of chlorophyll concentration for the described chloroplast preparation is 0.3 to 1.0 mg/mL. For an alternative method of measuring chlorophyll and other pigment concentrations in plant extracts, see Experiment 11.

The light-induced pH changes are measured by monitoring the pH of a suspension of illuminated chloroplasts. If protons are taken up into the chloroplasts, the pH of the suspension medium should increase. Figure E12.2 shows typical results of such an experiment. The proton transport process is reversed by eliminating the light source (indicated by the dashed line in Figure E12.2). Figure E12.2 shows the result for illuminated chloroplasts alone. It has been found that the presence of a light-sensitive redox dye stimulates the proton transport process. In this experiment phenazine methosulfate (PMS) will be used, although others, including methyl viologen and trimethylhydroquinone, may be used as an electron transport cofactor.

Overview of the Experiment

Chloroplasts will be isolated from spinach leaves and used to explore some intricacies of proton transport and photophosphorylation. The chloroplast preparation should be used as soon as possible (within 2 to 3 hours) since biological activity will decrease with time. The experimental apparatus for measurement of proton uptake should be assembled and all reagents prepared before the chloroplast extract is begun. The time required for the various parts of the experiment is as follows.

A. Preparation of spinach chloroplasts—30 minutes.

B. Determination of chlorophyll content—15 minutes.

C. Measurement of proton uptake—1 to 2 hours, depending on the choice of experiments.

Other optional experiments may be completed if time allows. For example, the effectiveness of various redox dyes may be analyzed. In addition to those listed in the text, FMN, ferricyanide, and dichlorophenolindophenol may be tested (Neumann and Jagendorf, 1964). It has been shown that NH_4Cl and amines stimulate proton uptake (Crofts, 1966). If a potassium ion-specific electrode is available, the light-induced efflux of K^+ from spinach chloroplasts may be studied (Dilley, 1972). Other methods for measuring the rate of ATP formation have been reported (Jagendorf and Uribe, 1966).

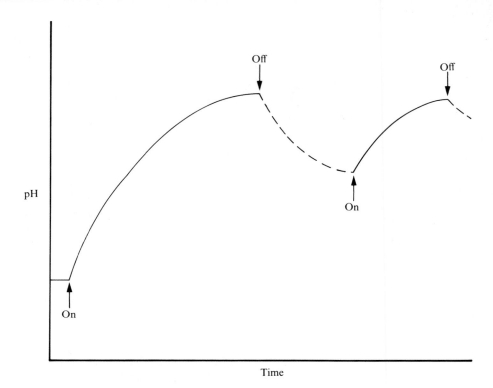

Figure E12.2
Light-induced pH changes in chloroplasts. Light is turned on and off at indicated times.

II. Materials and Supplies

A. Preparation of Spinach Chloroplasts

Fresh spinach leaves, about 50 g. These should be washed well with cold water, placed in plastic bags, and kept cold for several hours before use.

Homogenizing buffer, 0.02 M Tris, pH 8, containing 0.01 M NaCl and 0.4 M sucrose. **Keep cold.**

Sharp knife

Blender

Cheesecloth

Refrigerated centrifuge

Chloroplast suspension solution, 0.4 M sucrose containing 0.01 M NaCl

B. Determination of Chlorophyll Content

Chloroplast preparation from part A

80% acetone solution in water

Conical centrifuge tube

Glass cuvette, one pair

Spectrophotometer for measurement at 652 nm

Clinical or table top centrifuge

C. Measurement of Proton Uptake

pH meter, with recorder if possible. This should have a full-scale pH change of 0.5 pH unit.

500-watt tungsten lamp, with a Corning 1-69 filter or a solution containing a trace of $CuSO_4$ to remove infrared radiation

Magnetic stirrer and small stir bar

Phenazine methosulfate (PMS)

Timer

Large test tube, 2×20 cm

Phosphate solution, 0.01 M K_2HPO_4

Adenosine diphosphate, 0.02 M

III. Experimental Procedure

A. Preparation of Spinach Chloroplasts

Chloroplasts are fragile and unstable; therefore, this part of the experiment should be done as rapidly as possible and in subdued lighting. Maintain all solutions at 4°C or below.

Cut and discard the midribs from 50 g of spinach leaves and tear the leaves into small pieces. Immediately place the leaves into a precooled blender containing 100 mL of ice-cold homogenizing buffer. Grind the leaves at top speed no longer than 5 seconds. Strain the homogenate through eight layers of cheesecloth. Squeeze all the liquid from the cheesecloth. Centrifuge the filtrate in precooled tubes at $1000 \times g$ for 1 minute to remove whole cells. Transfer the supernatant to precooled centrifuge tubes and spin at $6000 \times g$ for about 15 seconds. Decant the supernatant and suspend the chloroplasts (pellet) in 50 mL of Tris homogenizing buffer. Centrifuge the suspension at $6000 \times g$ for about 10 seconds. (Note the color of the supernatant; the deeper the green, the more chlorophyll was lost from the chloroplasts). Pour off and discard the supernatant. Resuspend the isolated chloroplasts by stirring into 50 mL of suspension solution (sucrose containing NaCl). Store the chloroplast preparation in ice until use. It will remain stable for 2 to 4 hours.

B. Determination of Chlorophyll Content

Extract the chlorophyll from the chloroplasts by mixing, in a conical centrifuge tube, 0.05 mL of well-mixed chloroplast suspension with 9.9 mL of 80% acetone in water. Spin in a tabletop centrifuge for 10 minutes. Transfer the supernatant to a glass cuvette and read the absorbance at 652 nm using 80% acetone in water as blank. Calculate the concentration of chlorophyll in the chloroplast suspension using Equation E12.3.

C. Measurement of Proton Uptake (Dilley, 1972)

 Caution:

Phenazine methosulfate is a skin irritant, so it should be handled with care. Avoid contact with skin and do not breathe dust.

The experimental arrangement is shown in Figure E12.3. Maintain the temperature of the water bath at 10°C by adding ice. Add a trace of $CuSO_4$ solution to the water bath to eliminate infrared radiation from the lamp. Obtain a test tube that will hold the pH electrode and 10 to 15 mL of solution. Dilute the chloroplast solution with suspension buffer to a chlorophyll concentration of about 300 μg/mL. Adjust the pH meter with standard pH 7 buffer. The following three experimental conditions are recommended. Each condition should be tested individually and done in duplicate.

1. Chloroplasts with no redox cofactor

    10 mL of chloroplast suspension (about 3 mg chlorophyll)

    1 mL of suspension buffer

2. Chloroplasts with redox cofactor

    10 mL of chloroplast suspension

    5 mg of phenazine methosulfate or other redox dye

    1 mL of suspension buffer

3. Chloroplasts with redox cofactor, ADP, and phosphate

    10 mL of chloroplast suspension

    5 mg of phenazine methosulfate or other redox dye

    0.5 mL of ADP solution

    0.5 mL of phosphate buffer

To complete each experiment, prepare the reaction mixture as described above. Mix well and transfer to the reaction tube. Insert a small

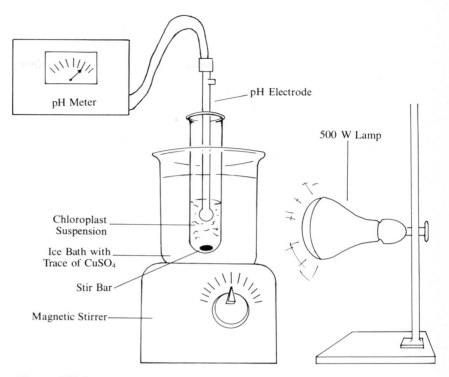

Figure E12.3
Experimental arrangement for measuring proton uptake by illuminated chloroplasts.
See text for details.

magnetic stir bar and place the pH electrode into the reaction mixture.
Turn on the magnetic stirrer to a slow rate and check to see whether the
stirrer affects the pH meter. Adjust the pH of the reaction mixture to a
pH reading in the range of 6 to 6.2 with acid or base. Turn on the pH
meter and recorder, if available, and then turn on the lamp to illuminate
the chloroplasts. Mark the recorder sheet at the time the lamp is turned
on. Allow the recorder to trace the pH reading with time for 60 seconds,
and then turn off the light but continue to record the pH. Turn the lamp
on after a dark interval of 30 seconds. Continue to record the pH for 60
seconds, again turn the lamp off, and continuously record the pH. Repeat the experiment with a fresh portion of chloroplasts.

If no pH recorder is available, pH readings can be taken from the meter or digital readout and a graph of pH vs. time prepared. Record a pH
reading every 10 seconds.

Experimental conditions 2 and 3, with their various additions, are
completed in an identical fashion. If chloroplasts still remain stable and
time permits, other experiments outlined in the overview may be completed.

IV. Analysis of Results

A. Preparation of Spinach Chloroplasts

Write all observations regarding the preparation in your notebook. Record the color of each supernatant and pellet.

B. Determination of Chlorophyll Content

Use Equation E12.3 to calculate the concentration of chlorophyll in the chloroplast suspension. What is the total yield of chlorophyll in mg?

C. Measurement of Proton Uptake

If no recorder tracing is available, prepare a graph of pH vs. time for each experiment. Prepare and complete a table with the following headings:

Addition	Rate of pH Change (sec/0.5 pH)	Relative Rate
None	_____	_____
+ PMS	_____	_____
+ PMS, ADP, P_i	_____	_____

Compare the relative rates of proton uptake for each experimental condition and explain any differences. Did the addition of light-sensitive redox cofactor affect the rate of pH shift? What is the effect of the addition of ADP and phosphate on the rate of pH shift? Explain the effect of alternating light and dark intervals.

V. Questions and Problems

1. Write a flowchart outlining the chloroplast preparation procedure and describe the purpose of each step.

▶ 2. Explain the chlorophyll extraction procedure in chemical terms. Why is acetone effective in the extraction? How could the experiment be modified to increase the yield of chlorophyll extraction?

▶ 3. Describe other methods for measuring the rate of ATP generation during photophosphorylation.

4. Why is the amount of chlorophyll kept constant in all experiments described here?

▶ 5. Describe methods that could be used to measure extrusion of K^+, Mg^{2+}, or Ca^{2+} from chloroplasts.

6. Explain the effect, if any, of the redox cofactor on the rate of proton uptake.

7. Why does the pH of the suspension solution decrease during dark intervals?

VI. References

D. Arnon, *Plant Physiol.* **24,** 1–15 (1949). Determination of chlorophyll content of chloroplasts.

D. Arnon, *Trends Biochem. Sci.* **9,** 258–262 (1984). "The Discovery of Photosynthetic Phosphorylation."

M. Avron, *Annu. Rev. Biochem.* **46,** 143–155 (1977). "Energy Transduction in Chloroplasts."

J. Barber, Editor, *The Light Reactions* (1987), Elsevier (Amsterdam). A detailed look at the chemistry of photosynthesis.

J. Barber and R. Malkin, Editors, *Techniques and New Developments in Photosynthesis Research* (1989), Plenum Publishing (New York). Experimental methods in photosynthesis research.

A. Crofts, *Biochem. Biophys. Res. Commun.* **24,** 127–134 (1966). Uptake of NH_4^+ by illuminated chloroplasts.

R. Dilley, in *Methods in Enzymology,* Vol. 24B, A. San Pietro, Editor (1972), Academic Press (New York), pp. 68–74. Transport of H^+, K^+, and Mg^{2+} in isolated chloroplasts.

Govindjee, Editor, *Molecular Biology of Photosynthesis* (1988), Kluwer Academic Publishers (Dordrecht, The Netherlands). A good mixture of biochemistry and molecular biology.

R. Gregory, *Biochemistry of Photosynthesis,* 3rd ed. (1989), John Wiley & Sons (Chichester). An excellent, up-to-date book on plant biochemistry.

M. Hipkins and N. Baker, Editors, *Photosynthesis—Energy Transduction: A Practical Approach* (1986), IRL Press (Oxford). Experimental techniques for studies in photosynthesis.

A. Jagendorf and E. Uribe, *Proc. Natl. Acad. Sci. USA.* **55,** 170–177 (1966). "ATP Formation Caused by Acid-Base Transition of Spinach Chloroplasts."

J. Jortner and B. Pullman, Editors, *Perspectives in Photosynthesis* (1990), Kluwer Academic Publishers (Dordrecht, The Netherlands). A modern look at photosynthesis.

C. Mathews and K. van Holde, *Biochemistry* (1990), Benjamin/Cummings (Redwood City, CA), pp. 643–669. An introduction to photosynthesis.

P. Mitchell, *Biol. Rev. Camb. Philos. Soc.* **41,** 445–502 (1966). "Chemiosmotic Coupling in Oxidative and Photosynthetic Phosphorylation."

J. Neumann and A. Jagendorf, *Arch. Biochem. Biophys.* **107,** 109–119 (1964). "Light-Induced pH Changes Related to Phosphorylation by Chloroplasts."

W. Nitschke and A. Rutherford, *Trends Biochem. Sci.* **16,** 241–245 (1991). "Photosynthetic Reaction Centres: Variations on a Common Structural Theme."

J. Rawn, *Biochemistry,* 2nd ed. (1989), Neil Patterson Publishers (Burlington, NC), pp. 489–532. An introduction to photosynthesis.

L. Stryer, *Biochemistry,* 3rd ed. (1988), W. H. Freeman (New York), pp. 517–543. An introduction to photosynthesis.

D. Voet and J. Voet, *Biochemistry* (1990), John Wiley & Sons (New York), pp. 586–617. An introduction to photosynthesis.

D. Walker, in *Methods in Enzymology,* Vol. 23A, A. San Pietro, Editor (1971), Academic Press (New York), pp. 211–220. Isolation of chloroplasts and grana using aqueous procedures.

D. Walker, in *Methods in Enzymology,* Vol. 69C, A. San Pietro, Editor (1980), Academic Press (New York), pp. 94–104. "Preparation of Higher Plant Chloroplasts."

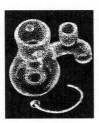

Isolation, Subfractionation, and Enzymatic Analysis of Beef Heart Mitochondria

■ Recommended Reading

Chapter 2, Section D; Chapter 7, Sections A, B, and C; Experiments 3 and 6.

■ Synopsis

Many of the biochemical processes that generate chemical energy for the cell take place in the mitochondria. These organelles contain the biochemical equipment necessary for fatty acid oxidation, di- and tricarboxylic acid oxidation, amino acid oxidation, electron transport, and oxidative phosphorylation. In this experiment, a mitochondrial fraction will be isolated from beef heart muscle. The mitochondria will be analyzed for protein content and fractionated into submitochondrial particles. Each fraction will be analyzed for malate dehydrogenase and monoamine oxidase activities.

I. Introduction and Theory

Mitochondrial Structure and Function

Mitochondria are intracellular centers for aerobic metabolism. They are cell organelles that are identified by well-defined structural and biochemical properties. In morphological terms, mitochondria are relatively large particles that are characterized by the presence of two membranes, a smooth outer membrane that is permeable to most important metabolites and an inner membrane that has unique transport properties. The inner membrane is highly folded, which serves to increase its surface area. Figure E13.1, which shows the structure of a typical mitochondrion, divides the organelle into four major components: inner membrane, outer membrane, intermembrane space, and the matrix. These regions are associated with different and specific biological functions. Mitochondria are characterized by the presence of the proteins and enzymes of respiratory metabolism, many of which are present in the folded inner membrane. Hence, this is the location for the processes of electron transport and oxidative phosphorylation. The enzymes of the tricarboxylic acid cycle, fatty acid oxidation, and amino acid metabolism are present in the matrix region. The outer membrane contains many enzymes, including monoamine oxidase for chemical reactions on neuroactive aromatic amines. The intermembrane space contains enzymes such as adenylate kinase and nucleoside diphosphokinase, which function in nucleotide balance.

Isolation of Intact Mitochondria

Because of the biological significance of these intracellular particles, biochemists have devised several methods for isolating mitochondria. Most isolation methods take advantage of the relatively large size of mitochondria. In animal cells only the nucleus is larger. The standard method for preparing a mitochondrial fraction is differential centrifugation of homogenized tissue (Figure 7.11, Chapter 7.) It should be pointed out that only rarely does one isolate and characterize a preparation of pure mitochondria. Rather, the isolated subcellular fraction is a **mitochondrial fraction**, which indicates that its major components are mitochondria.

Figure E13.1
Structure of the mitochondrion. From *Biology of the Cell,* 2nd ed., p. 135, by Stephen L. Wolfe. ©1981 by Wadsworth, Inc. Reprinted by permission of Wadsworth Publishing Company, Belmont, CA 94002.

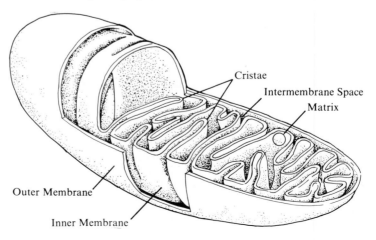

Cristae

Intermembrane Space

Matrix

Outer Membrane

Inner Membrane

Other cellular components that may be present are lysosomes, cell fragments, nuclear fragments, and microbodies (peroxisomes). The purity of the fraction depends on the source of the extract and the method chosen for isolation.

The general procedure for preparation of a mitochondrial fraction consists of the following steps:

1. Select and procure a suitable organ or other biological material. Mitochondrial fractions have been prepared from many animal organs including liver, heart, muscle, and brain, many plant tissues including spinach and avocado, and yeast.

2. The tissue is minced or ground and suspended in sucrose buffer. The choice of buffer conditions is critical in order to avoid loss of protein and other components that may leak from the mitochondria. Cytochrome c is one of the easiest mitochondrial proteins to dislodge, and some is almost always lost. The buffer should be isotonic or hypotonic with a low ionic strength. Sucrose is an ideal buffer component because mitochondria are especially stable and remain intact under these conditions.

3. Homogenize the tissue. This is carried out with a glass-Teflon homogenizer or a blender. This procedure gently breaks open the cells and allows the release of the subcellular organelles into the buffer. The mechanical disruption of cells should be gentle so the fragile mitochondria are not damaged.

4. Centrifuge the homogenate at low speeds. The heavier particles such as whole and fragmented cells, nuclei and nuclear fragments, and membrane fragments are sedimented during this step.

5. Centrifuge the supernatant from step 4 at a higher speed. This step sediments lighter particles including mitochondria. Two distinct layers usually are observed in the pellet: (a) a loosely packed, fluffy upper layer, which consists of damaged mitochondria and is sometimes called **light mitochondria**, and (b) a dark brown layer that consists of **heavy mitochondria**.

6. The heavy mitochondrial fraction is suspended in buffer and homogenized for a brief period in a glass-Teflon homogenizer to complete cell lysis.

7. The mitochondrial fraction is then obtained by centrifugation at relatively high speeds (15,000 *to* 20,000 rpm; 25,000 $\times$ g).

8. The pellet from step 7 now consists primarily of heavy mitochondria suspended in sucrose buffer. Typical yields are approximately 1 mg of protein per gram of starting minced tissue.

Several variations of this procedure may be found in the literature. Often, homogenization is facilitated by adding a proteolytic enzyme to the original suspension in step 2, but the proteolysis must be carefully controlled to avoid degradation of mitochondrial enzymes. The various methods lead to mitochondrial fractions of different quality and biochemical characteristics. The choice of methods depends to a great extent on what is to be done with the mitochondria.

This experiment describes the preparation of a mitochondrial fraction from beef heart muscle. Heart muscle is an excellent choice of tissue because isolated mitochondria are stable and most enzyme activities remain high for a long period of time. Since heart muscle is more fibrous than other tissues, some problems are encountered in homogenizing the tissue. The preparation described here is suitable for the study of characteristic enzymatic activity, electron transport, and oxidative phosphorylation.

Subfractionation of Mitochondria

The discussion to this point has focused on the isolation of intact mitochondria. By various chemical and physical treatments, mitochondria may be separated into their four components. This allows biochemists to study the biological functions of each component. For example, by measuring enzyme activities in each fraction, one can assign the presence of a particular enzyme to a specific region of the mitochondria. Studies of mitochondrial subfractions have resulted in a distribution analysis of enzyme activities in the four locations (Table E13.1). This type of study is often referred to as an **enzyme profile** or **enzyme activity pattern** and the enzyme may be considered a **marker** enzyme. For example, cytochrome oxidase, which is involved in electron transport, is a marker enzyme for the inner membrane.

One of the most instructive fractionation procedures is the preparation of **submitochondrial particles** (SMPs). The particles are produced by sonication (See Experiment 3) and centrifugation. The pellet, which sediments between 12,000 and 100,000 $\times$ g after sonication, defines the submitochondrial particle fraction. Submitochondrial particles are actually chunks of inner membrane that have undergone circularization and

Table E13.1
Average Percent Distribution of Marker Enzyme Activities in
Submitochondrial Fractions

| Fraction | Marker Enzyme[1] | | |
	Cytochrome c Oxidase	Monoamine Oxidase	Malate Dehydrogenase
Outer membrane	10	82	1
Inner membrane	90	11	15
Matrix	0	0	80
Intermembrane space	0	7	4

[1] The numbers represent the % total activity. For example, 82% of the total monoamine oxidase activity of mitochondria is found in the outer membrane.

inversion. In other words, the membrane has been turned "inside-out." Essentially all of the components for electron transport are still present; however, matrix enzymes are largely removed. Some remnants of the outer membrane remain in the isolated fraction.

Characterization of Mitochondria and SMPs

Mitochondrial fractions may be characterized by testing for the presence of known enzyme activities as previously discussed. The relative purity of each fraction can be estimated by measuring the specific activity of marker enzymes. Table E13.1 identifies marker enzymes for the matrix and membranes. Malate dehydrogenase, the tricarboxylic acid cycle enzyme that catalyzes the interconversion of malate and oxaloacetate (Equation E13.1), serves as a marker for the matrix enzymes. Malate dehydrogenase

▶   malate + NAD^+ $\rightleftharpoons$ oxaloacetate + NADH + H^+

(Equation E13.1)

activity would be expected in intact mitochondria, but not in SMPs. The activity of this enzyme in mitochondrial fractions may be estimated by a spectrophotometric assay. Oxaloacetate and NADH are incubated, and the disappearance of NADH is monitored at 340 nm. NAD^+ does not have strong absorption at this wavelength. Note that the reverse reaction is studied because the reaction as shown above is very unfavorable in thermodynamic terms ($\Delta G^{\circ\prime} = +7.1$ kcal/mole).

Monoamine oxidase (MAO) serves as a marker enzyme for outer membrane. There is some MAO activity in the inner membrane and therefore also in SMPs; however, a high level of monoamine oxidase in the SMP preparation indicates a large contamination by outer membrane. Mitochondrial monoamine oxidase is an FAD-dependent enzyme that catalyzes the oxidation of amines to aldehydes (Equation E13.2). A convenient assay for this enzyme uses benzylamine as substrate and monitors the rate of benzaldehyde production at 250 nm.

▶ $RCH_2NH_2 + O_2 + H_2O$ $\rightleftharpoons$ $RCHO + NH_3 + H_2O_2$

(Equation E13.2)

Overview of the Experiment

Students will isolate intact mitochondria from beef heart and fractionate them to prepare submitochondrial particles. Each fraction will be characterized by protein estimation by the biuret method and measurement of malate dehydrogenase and monoamine oxidase activity.

The experiment has the following time requirements:

A. Isolation of the mitochondrial fraction—2 hours. It is recommended that the isolation be done as a class project.

B. Determination of protein—30 minutes.

C. Preparation of SMPs—1–2 hours beginning with intact mitochondria.

D. Measurement of enzyme activities—1 hour.

Completion of all parts will require 4–5 hours. If time is limited (3 hours), it is recommended that students complete parts A and B and then test only for the presence of malate dehydrogenase in the intact mitochondria (part D).

II. Materials and Supplies

A. *Preparation of the Mitochondrial Fraction*

Beef heart, from a local slaughterhouse. It should be obtained 1 to 2 hours after the animal is slaughtered and transported in ice to the laboratory. A relatively fresh organ from a meat market is a suitable source for this experiment.

Sharp knife

Meat grinder, prechilled; plate holes: 4 to 5 mm

Waring blender

Sucrose-Tris homogenizing solution, 0.25 M sucrose and 0.01 M Tris-Cl buffer, pH 7.8. Keep ice-cold.

2 M Tris base solution, unneutralized, pH 10.8

Cheesecloth

Sucrose-Tris isolation solution, 0.25 M sucrose, 0.01 M Tris-Cl buffer, pH 7.8, 0.001 M succinic acid, and 0.2 mM EDTA. Keep ice-cold.

Glass-Teflon homogenizer, manual or motorized

Centrifuge, capable of 1200 $\times$ g and 26,000 $\times$ g

B. *Determination of Protein*

Mitochondrial fraction from part A

Sucrose-Tris isolation solution

10% sodium deoxycholate

Bovine serum albumin solution, 10 mg/mL

Biuret reagent

Spectrophotometer

C. *Preparation of SMPs*

Mitochondrial fraction from part A

Sucrose-Tris isolation solution

Sonicator
Ice-salt bath
Centrifuge (high-speed and ultracentrifuge)

D. *Measurement of Enzyme Activities*

MALATE DEHYDROGENASE

SMP fraction from part C
Mitochondrial fraction from part A
Phosphate buffer, 0.2 M, pH 7.4
Oxaloacetic acid, 0.006 M in phosphate buffer, freshly prepared
NADH, 0.00375 M in phosphate buffer, freshly prepared
Spectrophotometer, suitable for measurements at 340 nm
Cuvettes, 3-mL quartz

MONOAMINE OXIDASE

SMP fraction from part C
Mitochondrial fraction from part A
Phosphate buffer, 0.2 M, pH 7.4
Benzylamine solution, 0.10 M in phosphate buffer
Spectrophotometer, suitable for measurements at 250 nm
Cuvettes, 3-mL quartz

III. Experimental Procedure

A. *Preparation of the Mitochondrial Fraction*

All of the following procedures must be carried out at 2–4°C.

1. With a sharp knife, trim all fat and connective tissue from the ice-cold beef heart.
2. Cut the tissue into cubes 4 to 5 cm wide.
3. Quickly weigh 200 to 300 g of the cubes and pass them through a prechilled meat grinder.
4. Suspend the minced tissue in 400 mL of ice-cold sucrose-Tris homogenizing solution.
5. Adjust the pH to 7.5 ± 0.1 by adding 2 M unneutralized Tris.
6. Pour the neutralized heart mince through two layers of cheesecloth and squeeze out the sucrose solution.
7. Suspend 200 g of the solid mince in 400 mL of ice-cold sucrose-Tris isolation solution.

8. (a) Homogenize 25- to 50-mL portions of the suspension in a glass homogenizing vessel with a motorized pestle. Homogenize each portion for about 10 seconds and then twice more for 5 seconds each. Combine all the homogenate solutions and adjust to pH 7.8 with 2 M Tris base. (b) Alternatively, the minced tissue may be homogenized in a blender. For this procedure, suspend 200 g of mince in 400 mL of sucrose-Tris isolation solution and add 3 mL of 2 M Tris base. Turn on the blender at high speed for 15 seconds. Add 3 mL of 2 M Tris base and blend for another 5 seconds. Adjust the pH of the solution to 7.8 with 2 M Tris base.

9. Centrifuge the homogenate for 20 minutes at 1200 $\times$ g. The pellet consists of unfragmented cells and nuclei.

10. Carefully decant the supernatant without disturbing the loosely packed pellet.

11. Filter the supernatant through two layers of cheesecloth. This process removes lipid granules.

12. Adjust the pH of the filtrate to 7.8 with 2 M Tris base.

13. Centrifuge the pH-adjusted suspension for 15 minutes at 26,000 $\times$ g. At least two distinct layers will be observed in the pellet. The top layer consists of loosely packed light mitochondria (damaged mitochondria). The bottom layer of the pellet is brown and consists of heavy mitochondria. Occasionally a very small black pellet is present beneath the heavy mitochondria.

14. Remove and discard the light mitochondria by pouring off about half of the supernatant and gently shaking the centrifuge tube to release the top layer. Pour off and discard the suspension of light mitochondria.

15. If no tiny black pellet is present, add 10 mL of sucrose-Tris isolation solution and stir the heavy mitochondria with a glass stirring rod. If a black pellet is present, remove the brown heavy mitochondria from each tube with a glass rod and suspend in 10 mL of ice-cold sucrose-Tris isolation solution.

16. Homogenize the mitochondrial suspension in a motorized glass-Teflon homogenizer. Make two passes of the rotating pestle through the suspension, 5 seconds each.

17. Adjust the pH of the homogenate to 7.8 with 2 M Tris base and add sucrose-Tris isolation solution to a total volume of 180 mL.

18. Centrifuge the suspension at 26,000 $\times$ g for 15 minutes. The pellet this time consists primarily of a dark brown layer of heavy mitochondria. If an upper layer of light mitochondria is present, remove as in step 14.

19. Suspend the heavy mitochondria in about 60 mL of sucrose-Tris isolation solution and adjust the pH to 7.8 with 2 M Tris.

20. Store the mitochondrial fraction on cracked ice and begin part B.

B. Determination of Protein

Before the mitochondrial fraction can be biochemically characterized, the protein content must be measured.

1. Obtain five small test tubes and set up the protein assay according to Table E13.2. Tubes 1 and 2 contain two different concentrations of mitochondrial protein. Tubes 3 and 4 are duplicates of a standard protein, bovine serum albumin (BSA). Tube 5 is used as a blank for the spectrophotometer. The purpose of sodium deoxycholate is to disrupt the mitochondria and release the protein material into solution. For a discussion of the biuret analysis of proteins, see Chapter 2.

2. Add the appropriate amount of each of the first four reagents (sucrose solution, mitochondria, sodium deoxycholate, and BSA) to each of the five test tubes. If the solutions are not clear, add 10% sodium deoxycholate with a graduated pipet. Be sure you note the exact amount of deoxycholate added. Do not add more than a total of 0.4 mL.

3. Add sufficient water to each tube so that the total volume of all liquids is 1.5 mL. Table E13.2 is set up assuming that 0.2 mL of deoxycholate solution is sufficient. If, for example, tube 1 requires a total of 0.4 mL of deoxycholate, then only 0.1 mL of water should be added to the tube.

4. Add 1.5 mL of biuret reagent to each tube and mix well by inverting several times while holding a piece of hydrocarbon foil over the opening.

5. Incubate the tubes for 15 minutes at 37°C.

6. Measure and record the absorbance at 540 nm of tubes 1 through 4, using tube 5 to adjust the A_{540} to 0.0 absorbance.

7. If the absorbance readings of tubes 1 and 2 are greater than twice the A_{540} of tubes 3 or 4, repeat the assay using less mitochondrial fraction in tubes 1 and 2. Be sure the new assays contain a total of 3.0 mL of liquid in each tube.

Table E13.2

Preparation of Tubes for the Biuret Protein Assay on Mitochondrial Fractions[1]

Reagents	Tube 1	2	3	4	5
Sucrose-Tris isolation solution	0.5	0.5	0.5	0.5	0.5
Mitochondrial fraction	0.5	0.25	—	—	—
10% Sodium deoxycholate	0.2	0.2	0.2	0.2	0.2
Bovine serum albumin	—	—	0.1	0.1	—
H$_2$O	0.3	0.55	0.70	0.70	0.80
Biuret reagent	1.5	1.5	1.5	1.5	1.5

[1] Numbers are milliliters

C. Preparation of SMPs

1. Dilute (with isolation solution) the appropriate amount of mitochondrial fraction from part A so that you have a 10-mL suspension at a concentration of 15 mg of protein per mL.

2. Transfer the suspension to a 10-mL stainless steel or glass beaker. Chill in an ice-salt bath. With continued cooling in the salt bath, sonicate the suspension at maximum energy for six 5-second bursts. The suspension will warm during sonication. Therefore, after each 5-second sonication, allow the suspension to cool for 30 seconds. The temperature of the suspension should not be allowed to rise above 10°C during sonication. The mitochondrial suspension will become less opaque during the sonication.

3. Add 10 mL of ice-cold isolation buffer and centrifuge for 10 minutes at $10,000 \times g$ and 5°C.

4. Decant the supernatant into another tube and centrifuge at $100,000 \times g$ for 20 minutes.

5. Repeat step 4 two more times.

6. Resuspend the pellet in 15 mL of isolation buffer.

7. Estimate the protein concentration using the biuret assay. If necessary, dilute to 1 mg protein/mL.

D. Measurement of Enzyme Activities

MALATE DEHYDROGENASE

1. Turn on the spectrophotometer and UV lamp and allow to warm up for 15 minutes.

2. Obtain two quartz cuvettes and add the following reagents to each.

Cuvette 1	Reagent	Cuvette 2
1.3 mL	0.2 M phosphate, pH 7.4	1.3 mL
—	NADH	0.2 mL
0.1 mL	Oxaloacetic acid solution	0.1 mL
1.5 mL	H_2O	1.3 mL

3. Add 0.1 mL (0.1 mg protein) of the diluted mitochondrial fraction (1.0 mg/mL) from part B to cuvette 1. Cover the cuvette with hydrocarbon foil and gently invert two or three times.

4. Insert the cuvette into the sample beam of the spectrophotometer and adjust the absorbance at 340 nm to 0.00.

5. Add 0.1 mL (0.1 mg protein) of the diluted mitochondrial fraction to cuvette 2, mix as before, and immediately place it in the spectrophotometer.

6. If a recorder is available, monitor the change in absorbance at 340 nm for 5 minutes. If no recorder is available, read and record in your

notebook the A_{340} at 30-second intervals for 5 minutes. If the reaction is too slow or fast, repeat assay with more or less enzyme fraction. Final volume of assay must be 3.0 mL.

7. Prepare a plot of A_{340} (y axis) vs. time (x axis).

8. Repeat the above steps with the SMP fraction from part C.

MONOAMINE OXIDASE

1. Adjust the spectrophotometer to 250 nm.

2. Add the following to a 3-mL quartz cuvette:

   2.75 mL 0.2 M phosphate buffer, pH 7.4

   0.1 mL 0.1 M benzylamine

Mix well, place in the spectrometer, and record the baseline at 250 nm. Use a range of 0.5 absorbance units or less.

3. Initiate the enzyme-catalyzed reaction by adding 0.15 mL of the 1.0 mg/mL intact mitochondria or SMP suspension. Mix well and record the ΔA_{250} for 3–5 minutes. If the reaction rate is too slow or fast to measure, repeat using more or less protein. Be sure the total volume of each assay is 3.0 mL.

IV. Analysis of Results

A. Preparation of the Mitochondrial Fraction

Prepare a flowchart of the procedures followed in the isolation of mitochondria. Briefly explain the purpose of each step. Describe the appearance of the mitochondrial suspension.

B. Determination of Protein

Calculate the protein concentration in the mitochondrial fractions using Equation E13.3.

▶
$$\frac{C_{std}}{A_{std}} = \frac{C_{unk}}{A_{unk}}$$
(Equation E13.3)

where

A_{std} = absorbance at 540 nm of the standard bovine albumin (tubes 3 and 4)

A_{unk} = absorbance at 540 nm of the protein in the mitochondrial fractions

C_{std} = concentration of standard bovine serum albumin in mg/mL

C_{unk} = concentration of protein in the mitochondrial fractions in mg/mL

The biuret assay is linear up to protein concentrations of about 2 mg/mL.

C. Preparation of SMPs

Describe the appearance of the SMPs. Use a diagram to show disruption of the mitochondria by sonication. Why are the SMPs washed several times (steps 4 and 5)? What proteins are present in the supernatant?

D. Measurement of Enzyme Activities

MALATE DEHYDROGENASE

Use the plot of A_{340} vs. time to calculate $\Delta A/\min$ over the linear portion of the curve. Convert the rate in absorbance terms to activity units. One enzyme unit is the amount of malate dehydrogenase that catalyzes the reduction of 1 micromole of oxaloacetate to L-malate in 1 minute under the described assay conditions. The reduction of 1 micromole of oxaloacetate leads to the oxidation of 1 micromole of NADH; therefore, Equation E13.4 may be used to calculate the specific activity of malate dehydrogenase.

$$\blacktriangleright \quad \text{Specific activity} = \frac{\Delta A_{340}/\min}{6.2 \times \text{mg protein/mL reaction mixture}}$$

(Equation E13.4)

The millimolar absorption coefficient of NADH is 6.2 mM^{-1} cm^{-1}.

MONOAMINE OXIDASE

Convert the rate in ΔA_{250} units to activity units by using Equation E13.5.

$$\blacktriangleright \quad \text{Specific activity} = \frac{\Delta A_{250}/\min}{13 \times \text{mg protein/mL reaction mixture}}$$

(Equation E13.5)

The millimolar absorption coefficient for benzaldehyde is 13 mM^{-1} cm^{-1}.

Now calculate the proportion of total mitochondrial protein (malate dehydrogenase and monoamine oxidase) present in the submitochondrial fraction.

% of total mitochondrial protein in submitochondrial fraction

$$= 100 \times \frac{\text{specific activity of marker enzyme in mitochondria}}{\text{specific activity of marker enzyme in submitochondrial fraction}}$$

This can be calculated separately for malate dehydrogenase and MAO activity. Do your results agree with those in Table E13.1? Explain.

V. Questions and Problems

▶ 1. Why is sucrose used in the isolation buffer?

▶ 2. Describe the function and mode of action of sodium deoxycholate in the protein assay in part B.

3. Explain the derivation of Equation E13.4 for malate dehydrogenase activity calculation.

▶ 4. Briefly explain how the mitochondrial fraction prepared here could be used to study electron transport and oxidative phosphorylation. Could the SMP fraction also be used for these studies?

▶ 5. Using structures, write the reaction for the monoamine oxidase–catalyzed oxidation of benzylamine.

▶ 6. How would you prepare more highly purified mitochondria than described in this experiment?

7. Using a flowchart format, show how you would purify monoamine oxidase from beef heart mitochondria.

▶ 8. What subfraction of mitochondria would be the best source of the pyruvate dehydrogenase complex?

VI. References

M. Ator and P. Ortiz de Montellano, in *The Enzymes,* 3rd ed., Vol. XIX, D. Sigman and P. Boyer, Editors (1990), Academic Press (San Diego), pp. 234–240. Mechanistic discussion of monoamine oxidase.

V. Darley-Usmar, D. Rickwood, and M. Wilson, Editors, *Mitochondria: A Practical Approach* (1987), IRL Press (Oxford). An excellent source for laboratory techniques.

S. Fleischer and L. Packer, Editors, *Methods in Enzymology,* Vol. 55 (1979), Academic Press (New York). Complete volume on bioenergetics and oxidative phosphorylation.

Y. Hatefi, *Annu. Rev. Biochem.* **54,** 1015–1069 (1985). "The Mitochondrial Electron Transport and Oxidative Phosphorylation System."

C. Heisler, *Biochem. Educ.* **19,** 35–38 (1991). "Mitochondria from Rat Liver: Method for Rapid Preparation and Study."

T. King, H. Mason, and M. Morrison, Editors, *Oxidases and Related Redox Systems* (1988), Alan R. Liss (New York). Proceedings of the 4th International Symposium on Oxidases.

C. Lee, in *Methods in Enzymology.* Vol. 55, S. Fleischer and L. Packer, Editors (1979), Academic Press (New York), pp. 105–112. "Tightly Coupled Beef Heart Submitochondrial Particles."

C. Mathews and K. van Holde, *Biochemistry* (1990), Benjamin/Cummings (Redwood City, CA), pp. 504–506. An introduction to mitochondrial structure and function.

B. Mondovi, Editor, *Structure and Functions of Amine Oxidases* (1985), CRC Press (Boca Raton, FL). Excellent review of all known amine oxidases.

J. Rawn, *Biochemistry,* 2nd ed. (1989), Neil Patterson Publishers (Burlington, NC), pp. 12–19, 365–366. An introduction to structure and function of mitochondria.

A. Smith, in *Methods in Enzymology,* Vol. X, R. Estabrook and M. Pullman, Editors (1967), Academic Press (New York), pp. 81–86. "Preparation, Properties and Conditions for Assay of Mitochondria: Slaughterhouse Material, Small-Scale."

L. Stryer, *Biochemistry,* 3rd ed. (1988), W. H. Freeman (New York), pp. 397–413. Mitochondrial structure and function.

D. Voet and J. Voet, *Biochemistry* (1990), John Wiley & Sons (New York), pp. 529–532. Structure and function of mitochondria.

C. Worthington, Editor, *Worthington Enzyme Manual* (1988), Worthington Biochemical Corporation (Freehold, NJ), pp. 224–228, 285–287. Malate dehydrogenase and amine oxidases.

EXPERIMENT 14

Cholesterol (Total and HDL) and Uric Acid Content of Blood Serum: Enzymes as Reagents in Clinical Chemistry

■ **Recommended Reading**

Chapter 5, Section A; Experiment 6.

■ **Synopsis**

The specificity and efficiency of enzymes can be used to great advantage in the quantitative analysis of biological fluids for medical diagnosis. Two analyses that are common in the clinical laboratory are the measurements of serum cholesterol and uric acid. Cholesterol is measured by coupling the enzyme-catalyzed oxidation of cholesterol (generation of H_2O_2) to the peroxidase-catalyzed formation of a chromogen. Uricase-catalyzed oxidation of uric acid is the basis of urate determination.

I. Introduction and Theory

One of the most beneficial and interesting applications of biochemical methods is in the diagnosis of disease states. Most procedures in the clinical laboratory are used to measure the concentrations of various constituents in biological fluids and tissues. An abnormally high or low concentration of a biochemical (enzyme, metabolite, etc.) in a patient's blood or urine specimen is often a signal to the patient's physician that a pathologic condition may exist. Biochemical measurements aid clinicians in at least two ways: (1) they assist in the diagnosis (recognition and identification) of a diseased condition, or (2) they may be used for the confirmation of a suspected diseased condition.

Enzymes as Diagnostic Reagents

The availability of purified enzyme preparations and their unique substrate specificity make enzymes particularly attractive as diagnostic reagents. Since most enzymes act on a single type of reactant and the extent of an enzyme-catalyzed reaction depends on substrate concentration, it is possible to estimate the concentration of a single molecular species (the substrate). It is, of course, essential that the desired reaction be carried out under optimal conditions, that is, proper pH, temperature, and ionic strength, and with assured absence of interfering substances. Of equal importance are the concentrations of enzyme, cofactor, substrate to be measured, and other required reagents. **All reagents, except the substance to be measured, must be present in excess, so that the rate and extent of the enzyme-catalyzed reaction depend on only the concentration of the substance to be determined.** The initial velocity of an enzyme-catalyzed reaction is expressed by the Michaelis-Menten equation (Equation E14.1).

▶
$$v_0 = \frac{V_{max}[S]}{[S] + K_M} \qquad \text{(Equation E14.1)}$$

The maximum rate of a reaction (V_{max}) is attained when all the enzyme active sites are saturated with substrate molecules. For the rate to approach V_{max}, the substrate concentration must be high; in fact, it must be much greater than K_M. When $[S] \gg K_M$, the Michaelis-Menten equation becomes

▶
$$[S] \gg K_M; \qquad v_0 = \frac{V_{max}[S]}{[S]} = V_{max} \qquad \text{(Equation E14.2)}$$

In words, under substrate-saturating conditions, the initial rate of the enzyme-catalyzed reaction (v_0) is independent of substrate concentration, [S] (Equation E14.2). However, when the substrate is present in much less than saturating amounts ($[S] \ll K_M$), the Michaelis-Menten equation becomes

▶
$$[S] \ll K_M; \qquad v_0 = \frac{V_{max}[S]}{K_M} \qquad \text{(Equation E14.3)}$$

The initial rate of the enzyme-catalyzed reaction is directly proportional to [S] (Equation E14.3). Most clinical assays using enzymes are performed under the conditions of Equation E14.3. From further study of this equation, you will note that v_0 also depends on enzyme concentration, since there is an enzyme concentration term hidden in V_{max}. (If you have forgotten this, review the derivation of the Michaelis-Menten equation in your biochemistry textbook.) This can be used to advantage, because if a reaction used for a clinical analysis is very slow (it probably will be, since [S] is low), extra enzyme can be used so that the reaction will proceed to completion in a reasonable period of time.

In general, the principle behind the clinical measurement of cholesterol and uric acid is the following. A biological fluid (serum, urine, etc.) containing the substance to be measured is incubated with an enzyme system that will interact only with that substance. The extent of the enzyme-catalyzed reaction is determined by monitoring some change associated with conversion of substrate to product. Ideally, there is a change in UV or VIS absorbance that can be measured at a convenient wavelength with a spectrophotometer. The spectrophotometric method might involve measurement of a product that appears in solution or the disappearance of substrate. The substance to be determined is incubated for a time interval sufficient to transform it completely into product or to attain equilibrium between substrate and product. Here you should recognize the importance of the enzyme concentration, a point raised earlier. Sufficient enzyme must be present so that an end point (complete conversion of substrate to product or attainment of equilibrium) is reached in the desired amount of time. Once the extent of reaction has been quantitatively measured, knowledge of the stoichiometry of the conversion of substrate to product allows one to calculate directly the initial concentration of the reactant.

Measurement of Cholesterol

The measurement of serum cholesterol is one of the most common tests performed in the clinical laboratory. Hypercholesterolemia (high blood cholesterol levels) can be the result of a variety of medical conditions. Among the conditions implicated are diabetes mellitus, atherosclerosis, and diseases of the endocrine system, liver, or kidney. High blood cholesterol levels do not point to a specific disease; determination of cholesterol is used in conjunction with other clinical measurements mainly for confirmation of a particular diseased condition, rather than for diagnosis of a specific ailment.

Of current interest is the positive correlation between cholesterol levels and heart disease. Cholesterol in the blood is found associated with various lipoproteins. There are two major types of cholesterol-carrying lipoproteins, high-density lipoprotein (HDL) and low-density lipoprotein (LDL). High serum levels of cholesterol-bearing LDLs are positively correlated with the development of atherosclerosis. In contrast, high levels of HDL cholesterol are inversely related to a predisposition to coronary artery disease. In general, the higher the ratio of HDL-cholesterol to LDL-cholesterol, the lower the incidence of heart disease. In order to characterize the LDLs and HDLs and determine the amount of cholesterol associated with each, it is essential to separate them physically. They differ in density and size and may be separated by ultracentrifugation, but they can also be separated by electrophoresis and selective precipitation of LDLs with divalent cations and polyanions.

Clinical measurements of **total cholesterol** in serum or plasma detect cholesterol esters in addition to cholesterol. Between 60 and 70% of the cholesterol transported in blood is in an esterified form, where the β-3-OH group on the steroid skeleton is covalently linked to a naturally occurring fatty acid, shown as an R group in Figure E14.1. A sensitive and reproducible analysis of cholesterol and cholesterol esters is based on the three reactions shown below.

▶ Cholesterol esters + H_2O $\xrightarrow[\substack{\text{or} \\ \text{cholesterol esterase} \\ \text{(EC 3.1.1.13)}}]{H^+ \text{ or } OH^-}$ cholesterol + fatty acids

(Equation E14.4)

▶ Cholesterol + O_2 $\xrightarrow[\substack{\text{cholesterol} \\ \text{oxidase}}]{}$ cholest-4-ene-3-one + H_2O_2

(Equation E14.5)

▶ H_2O_2 + phenol + 4-aminoantipyrine $\xrightarrow[\text{peroxidase}]{}$ quinoneimine chromogen

($\lambda_{max} = 510$ nm)

(Equation E14.6)

Equation E14.4, catalyzed by cholesterol esterase, shows the hydrolysis of cholesterol esters in the sample. In Equation E14.5, cholesterol is then oxidized to cholest-4-ene-3-one. Unfortunately, this reaction does not lead to a major absorbance change at an accessible wavelength, so the rate of the reaction cannot be directly measured. When it is not convenient or possible to monitor directly the progress of a reaction, it may be possible to "couple" it to another reaction that offers a measurable absorbance change. The oxidation of cholesterol in Equation E14.5 is accompanied by production of hydrogen peroxide. In a coupled reaction, H_2O_2 rapidly reacts with phenol and 4-aminoantipyrine in the presence of horseradish peroxidase to produce a quinoneimine chromogen, which has a maximum absorbance at 510 nm (Equation E14.6, Figure E14.2). For Equation E14.5 and E14.6 to be a properly coupled system, the coupled reaction must be fast enough to decompose H_2O_2 as rapidly as it is

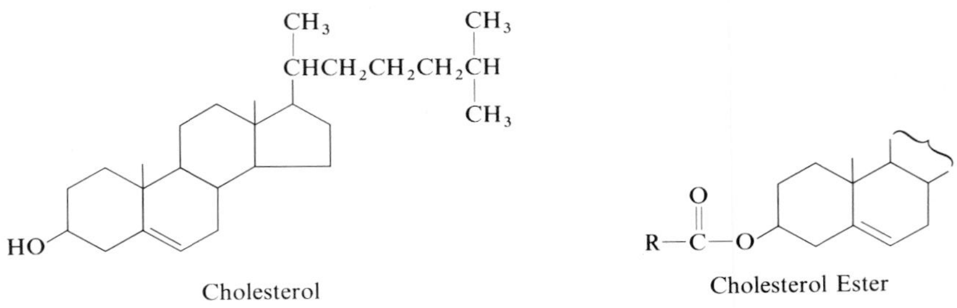

Figure E14.1
Structures of cholesterol and a cholesterol ester.

Cholesterol

Cholesterol Ester

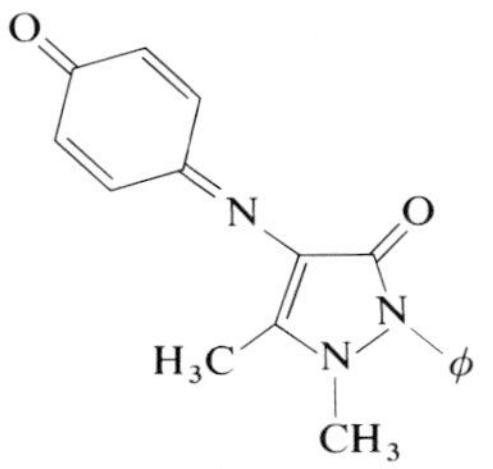

Figure E14.2
Structure of quinoneimine chromogen formed in the assay of cholesterol.

Quinoneimine
Chromogen

$\lambda_{\max} = 510$ nm

produced in Equation E14.5. This condition exists for the coupled reactions, so the absorbance change due to the production of quinoneimine is directly proportional to the concentration of total cholesterol in the sample analyzed. A more detailed kinetic analysis of coupled assays is beyond the scope of this chapter. Interested students should refer to specialized books on enzyme kinetics.

In this experiment, two cholesterol measurements will be made: (1) total serum cholesterol and (2) HDL serum cholesterol, the amount of cholesterol associated with the HDL fraction. The following relationship leads to an estimate of LDL (Equation E14.7).

▶ $\quad C_{\text{LDL}} = C_{\text{serum}} - C_{\text{HDL}}$ $\qquad$ (Equation E14.7)

where

$\quad C_{\text{LDL}}$ = concentration of cholesterol in the low-density lipoproteins and very low-density lipoproteins

$\quad C_{\text{serum}}$ = total concentration of serum cholesterol

$\quad C_{\text{HDL}}$ = concentration of cholesterol in the HDL fraction

C_{serum} will be determined by performing the described cholesterol assay directly on serum (Sigma Chemical Co., 1985). C_{HDL} will be determined on a separated, soluble HDL fraction of serum. Very low-density lipoproteins and low-density lipoproteins are selectively removed from serum by precipitation with magnesium-phosphotungstate reagent.

The raw data collected in the experiment are in the form of absorbance measurements at 510 nm. These numbers are then converted to serum cholesterol concentration, which is reported in mg/100 mL serum.

▶ $\quad$ Cholesterol concentration (mg/100 mL) $= \dfrac{A_{510(x)}}{A_{510(s)}} \times C_{\text{s}}$

(Equation E14.8)

where

$\quad A_{510(x)}$ = absorbance at 510 nm obtained with unknown serum sample

$\quad A_{510(s)}$ = absorbance at 510 nm obtained with standard cholesterol solution

$\quad C_{\text{s}}$ = concentration of cholesterol in standard (mg/100 mL)

Normal levels of serum cholesterol vary widely. For males, total cholesterol is in the range 130 to 320 mg/100 mL and HDL cholesterol ranges from 30 to 70 mg/100 mL. For females, the corresponding ranges are 130–295 and 35–80 mg/100 mL, respectively. Cholesterol levels depend on such factors as sex, age, diet, and emotional stress. Typical ranges at various age intervals are shown in Table E14.1.

Table E14.1
Average Total Human Serum Cholesterol Levels as a
Function of Age

Age (years)	Total Cholesterol (mg/100 mL serum)
0–19	120–230
20–29	120–240
30–39	140–270
40–49	150–310
50–59	160–330

Measurement of Uric Acid

Normal levels of human serum uric acid are 3.5 to 7.2 mg/100 mL for men and 2.6 to 6.0 mg/100 mL for women. However, these levels are influenced by diet, age, and disease states. Uric acid is derived from the metabolism of purines and is normally eliminated by urinary excretion. A diet of purine-rich foods (beef liver, kidney, anchovies) increases serum uric acid levels. Increased levels of serum uric acid are associated with many disease states, including acute and chronic nephritis (renal failure), urinary obstruction, uncompensated hypertension, gout, leukemia, malignant tumors with extensive necrosis, acute infections, and Lesch-Nyhan syndrome. Although the classic medical condition associated with elevated serum levels of uric acid is gout, patients with renal failure show much higher levels. Serum uric acid levels in gouty patients are usually between 7 and 10 mg/100 mL, whereas in chronic renal failure levels may reach 25 to 30 mg/100 mL. The biochemical characteristic of gout is elevated levels of urate in serum, which results in precipitation of sodium urate crystals in joints, causing them to become inflamed.

Uric acid is the end product of purine metabolism in humans, some apes, and the dalmatian. Most other mammals have the enzyme uricase (urate: O_2 oxidoreductase, EC 1.7.3.3), which catalyzes the oxidation of urate to allantoin (Equation E14.9).

Uric acid   Allantoin   (Equation E14.9)

This reaction forms the basis of a quantitative determination of uric acid in biological fluids. Uric acid absorbs strongly in the ultraviolet region ($\lambda_{max} = 292$ nm, $\epsilon = 12,300$), whereas the product allantoin has minimal absorption at this wavelength. The conversion of uric acid to allantoin is therefore associated with a significant change in absorbance, so no

Here

coupling reaction is necessary for assay of uric acid. Uricase and the sample containing uric acid are incubated in constant pH buffer, and the absorbance change at 292 nm is monitored. The overall decrease in absorbance ($A_{initial} - A_{final}$) is directly proportional to the original concentration of uric acid in the specimen. This assay is particularly advantageous because uricase has high specificity for uric acid; however, the enzyme also catalyzes the oxidation of 6-thiouric acid, a metabolic product of an anticancer drug, 6-mercaptopurine.

The data collected in the analysis of serum uric acid will consist of absorbance changes at 292 nm or $\Delta A_{292}(A_{initial} - A_{final})$. In order to relate absorbance changes to concentration of uric acid, Beer's law is used:

$$A = \epsilon l c$$

where

A = absorbance

ϵ = molar absorption coefficient of uric acid, 12,300 M^{-1} cm^{-1}

l = cell path length, usually 1 cm

c = concentration of uric acid in mmole/mL

Rearranging, we have the relationship

$$c = \frac{A}{l\epsilon}$$

This leads to the concentration of uric acid in mmole/mL. This must be converted to mg/100 mL serum (Sigma Chemical Co., 1976).

$$\blacktriangleright \quad c' = \frac{\Delta A_{292} \times 168.1 \times 100 \times 3.05}{12,300 \times 0.0833} \qquad \text{(Equation E14.10)}$$

where

c' = concentration of uric acid in mg/100 mL

ΔA_{292} = absorbance change for reaction, $A_{initial} - A_{final}$

168.1 = molecular weight of uric acid

100 = factor to convert mg/mL of uric acid to mg/100 mL serum

3.05 = correction for reaction volume

0.0833 = volume of serum used in analysis

12,300 = molar absorption coefficient of uric acid

Equation E14.10 illustrates the relationship between the decrease in absorbance at 292 nm associated with Equation E14.9 and the concentration of uric acid in mg/100 mL.

Overview of the Experiment

Clinical methods based on enzyme reactions are now used routinely for the determination of glucose, cholesterol, urea, uric acid, and many other metabolites in blood, urine, and other biological fluids as well as tissue specimens. Diagnostic kits containing all the required reaction components are commercially available at reasonable cost. These kits make it possible for measurements to be made in a simple, accurate, rapid, and reproducible manner. In modern hospital laboratories, much of this testing is completely automated, which greatly improves the speed and reproducibility, but not necessarily the accuracy.

In this experiment, two measurements that are common in the clinical laboratory will be performed. Both procedures rely on spectrophotometric techniques, but they differ in several regards. The measurement of cholesterol introduces coupled enzyme assays. Uric acid determination illustrates the direct measurement of a molecular species and use of Beer's law for the calculation of concentration. Study Figures E14.3, E14.4, and E14.5 for a summary of the procedures. All parts of this experiment (total cholesterol, HDL-cholesterol, and uric acid) can be completed in 3 hours.

Because of the possible transmission of infectious diseases during the use of human blood samples, it is recommended that students use bovine, porcine, or other animal serum samples. Sera from a wide variety of animals, including humans, are available in a lyophilized form from Sigma Chemical Co. When reconstituted with water, these provide convenient unknowns.

Figure E14.3
The flowchart for measurement of total serum cholesterol.

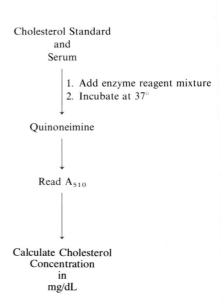

Cholesterol Standard
and
Serum

1. Add enzyme reagent mixture
2. Incubate at 37°

Quinoneimine

Read A_{510}

Calculate Cholesterol
Concentration
in
mg/dL

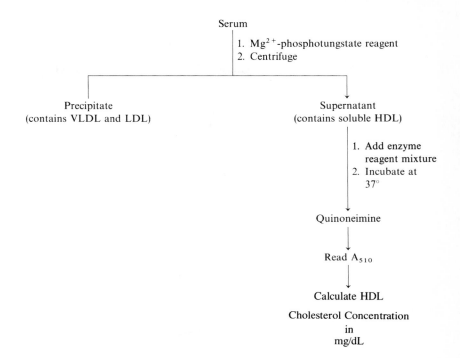

Figure E14.4
The flowchart for measurement of total serum HDL-cholesterol.

The flowchart shows:

Serum
1. Mg^{2+}-phosphotungstate reagent
2. Centrifuge

Precipitate
(contains VLDL and LDL)

Supernatant
(contains soluble HDL)
1. Add enzyme reagent mixture
2. Incubate at $37°$

Quinoneimine

Read A_{510}

Calculate HDL
Cholesterol Concentration in mg/dL

II. Materials and Supplies

A. *Measurement of Cholesterol*

Serum samples. Bovine or porcine blood samples can be obtained from a local slaughterhouse or serum samples may be purchased.

Cholesterol aqueous standard I. This contains cholesterol (200 mg/100 mL) in water containing stabilizers and sodium azide as preservative.

Cholesterol aqueous standard II. This contains cholesterol (50 mg/100 mL) in water solution containing stabilizers and sodium azide as preservative.

LDL precipitating reagent. This solution contains phosphotungstate, magnesium ions, and sodium azide.

Cholesterol assay solution. This may be obtained from Calbiochem-Behring or Sigma Chemical Co. The stock reagent contains pancreatic cholesterol esterase, microbial cholesterol oxidase, horseradish peroxidase, 4-aminoantipyrine, and phenol. Your instructor will reconstitute the stock reagent by addition of water.

Spectrophotometer and cuvettes. Any spectrometer that measures in the range 510 ± 10 nm is suitable.

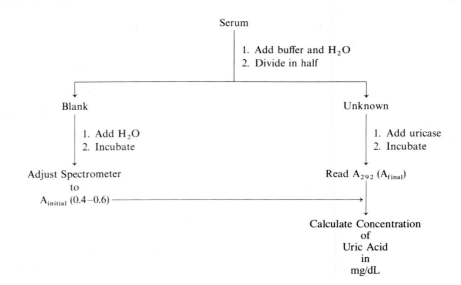

Figure E14.5
The flowchart for measurement of serum uric acid.

Constant-temperature bath at 37°C

Centrifuge, capable of speeds up to 2000 rpm

Centrifuge tubes, 10 mL, conical

Saline solution, 0.15 M NaCl in water

Hydrocarbon foil

B. Measurement of Uric Acid

Blood serum. Use human sera in a lyophilized form available from Sigma Chemical Co. Reconstitute with water.

Glycine buffer, 0.7 M, pH 9.4

Uricase solution. Activity should be between 0.2 and 0.4 units/mL. A uric acid diagnostic kit is available from Sigma Chemical Co.

Spectrophotometer for measurements in the ultraviolet

Quartz cuvettes, matched set

III. Experimental Procedure

 Caution

Several reagents used in this experiment contain sodium azide, a deadly poison. **Never pipet any solutions by mouth.**

Sodium azide may react with lead and copper plumbing to form highly explosive metal azides. For disposal, flush with a large volume of water to prevent azide accumulation.

A. *Measurement of Cholesterol*

MEASUREMENT OF TOTAL CHOLESTEROL

The procedure will be described for the analysis of two different serum samples. Turn on the spectrometer and allow warm-up for 15 to 20 minutes. Set the wavelength to 510 nm. Obtain two 3-mL cuvettes or Bausch and Lomb colorimeter tubes for the blank and standard. In addition, one cuvette will be needed for each serum sample to be tested. Label the cuvettes Blank, Standard, $Serum_1$, and $Serum_2$. Make the following additions:

Blank: 0.02 mL of water

Standard: 0.02 mL of cholesterol standard I

$Serum_1$: 0.02 mL of $Serum_1$

$Serum_2$: 0.02 mL of $Serum_2$

Pipet 1.0 mL of cholesterol enzyme reagent into each tube. Mix well, but do not shake. Shaking will cause foaming and protein denaturation. Mixing is best done by tightly covering the cuvette with hydrocarbon foil and gently inverting 3 to 5 times. Incubate the cuvettes at 37°C in a constant-temperature water bath for 10–15 minutes. Remove all the cuvettes from the bath and wipe dry with a tissue. Add 2.0 mL of saline water to each cuvette and mix well. Place the blank in the spectrometer and adjust A_{510} to 0.0. Read and record in your notebook A_{510} for the Standard, $Serum_1$, and $Serum_2$. Take the absorbance readings within 30 minutes after removal from the bath to avoid color fading.

This procedure does not take into account any absorbance due to the serum. If the serum is turbid, a correction should be made by measuring the absorbance at 510 nm of a 0.02-mL sample of blood serum in 3.0 mL of saline water. Read the A_{510} of this solution using saline water as blank. Record this reading in your notebook as A_c for correction. The calculation for cholesterol concentration will be described in the Analysis of Results.

MEASUREMENT OF HDL-CHOLESTEROL

Obtain a conical centrifuge tube for each serum sample you wish to analyze. To each tube add 0.4 mL of serum and 0.05 mL of phosphotungstate precipitating reagent. Mix well. Centrifuge the tubes for 10 minutes at 2000 rpm. Obtain four cuvettes, one each for Blank, Standard, $Serum_1$, and $Serum_2$. Label the cuvettes and add the following reagents:

Blank: 0.05 mL of water

Standard: 0.05 mL of cholesterol aqueous standard II (50 mg/100 mL)

$Serum_1$: 0.05 mL of $supernatant_1$ from centrifugation

$Serum_2$: 0.05 mL of $supernatant_2$ from centrifugation

Add 1.0 mL of cholesterol enzymatic reagent to each cuvette, cover with hydrocarbon foil, and mix well. Incubate for 10–15 minutes at 37°. Add 2.0 mL of saline water to each cuvette. Adjust spectrometer to zero with Blank and read A_{510} for each sample. Record in your notebook for further calculations.

B. Measurement of Uric Acid

Turn on the spectrophotometer and UV lamp. Pipet into a clean test tube 0.2 mL of serum, 1.0 mL of glycine buffer, pH 9.4, and 6.0 mL of water. Mix well. Obtain two test tubes: label one tube Blank and the other Unknown. Add reagents to the two labeled tubes as follows:

Blank: 3.0 mL of serum in glycine buffer prepared above, plus 0.05 mL of water

Unknown: 3.0 mL of serum in glycine buffer prepared above, plus 0.05 mL of uricase solution

Gently mix each tube well and let stand at room temperature for 15 minutes. Transfer the two solutions to two separate matched quartz cuvettes. Set the spectrophotometer to 292 nm and, with the Blank cuvette in the light path, set the meter to an initial absorbance setting of about 0.40 by adjusting slit width or energy level. The actual $A_{initial}$ is not critical; it may be 0.30, 0.40, or 0.50. However, it should be high enough so the A_{final} is not less than 0. Be sure to record the $A_{initial}$ in your notebook. Remove the Blank cuvette from the light path and read A_{292} for the Unknown. After 5 more minutes of incubation, adjust the spectrophotometer with Blank to $A_{initial}$ as before and read the absorbance of the Unknown cuvette. If this absorbance has decreased, wait another 5 minutes and then read A_{292} of the unknown. Be sure to adjust to $A_{initial}$ with Blank. Continue taking A_{292} readings for the unknown until the uricase reaction is complete (until A_{292} readings are constant). Read and record the final absorbance of the Unknown cuvette.

The uricase solution may contribute some absorbance to the reaction mixture that is not associated with the water blank. To correct for this, pipet 0.05 mL of uricase solution into 3.0 mL of glycine buffer, pH 9.4, in a quartz cuvette. Mix well and read the A_{292} for this solution, using glycine buffer in a quartz cuvette as reference. The A_{292} ($A_{uricase}$) will be approximately 0.005.

In summary, the above procedure should result in three absorbance readings,

1. $A_{initial}$

2. A_{final} depends on uric acid concentration

3. $A_{uricase}$ = approximately 0.005

The calculations for uric acid determination are described in the Analysis of Results.

IV. Analysis of Results

A. *Measurement of Cholesterol*

CALCULATION OF TOTAL SERUM CHOLESTEROL

The cholesterol assay as used in this experiment is linear up to 500 mg/100 mL serum. A calibration curve is not essential and a single standard can be used. Use Equation E14.8 to calculate the cholesterol concentration. If the serum sample was turbid, you should have determined A_{510} of serum in saline water, A_c. If so, subtract this from $A_{510(x)}$.

$$\text{Total cholesterol concentration (mg/100 mL)} = \frac{A_{510(x)} - A_c}{A_{510(s)}} \times C_s$$

where A_c is the absorbance of diluted serum at 510 nm. The concentration of standard cholesterol (C_s) is 200 mg/100 mL. Calculate total cholesterol in the serum samples you tested. Compare your results with normal values of serum cholesterol.

CALCULATION OF HDL-CHOLESTEROL CONCENTRATION

The calculation is identical to that of total serum cholesterol except that the standard cholesterol solution has a concentration of 50 mg/mL. Calculate the concentration of serum HDL-cholesterol in the samples you tested. Also, calculate the concentration of cholesterol associated with LDL for each serum sample.

B. *Measurement of Uric Acid*

The relationship derived from Beer's law was described in the Introduction. This may be simplified to

$$c' = \Delta A_{292} \times 50.04$$

where

c' = concentration of uric acid in mg/100 mL serum

$\Delta A_{292} = A_{\text{initial}} - A_{\text{final}}$

If you corrected for uricase absorption, then this term becomes $(A_{\text{initial}} - A_{\text{final}}) + A_{\text{uricase}}$. Calculate the concentration of uric acid in each serum sample. Compare these values to normal values.

V. Questions and Problems

▶ 1. What is the structure of uric acid under physiological conditions?

2. Derive the equation used for the calculation of uric acid concentration. Hint: Begin with the Beer's law relationship.

▶ 3. Assume you are to determine the blood serum level of cholesterol by the single standard method. The absorbance data obtained are: $A_{standard}$ = 0.350 (250 mg/100 mL); A_{sample} = 0.390; and $A_{correction}$ = 0.02. Calculate the concentration of cholesterol in mg/100 mL. What would be the % error if $A_{correction}$ were not used?

▶ 4. What is the primary assumption made in the use of a single standard as in Problem 3?

▶ 5. The cholesterol assay is linear to a level of 500 mg/100 mL. What would you do if a determination for cholesterol showed the level to be about 650 mg/100 mL?

▶ 6. Study Equation E14.9 (uricase-catalyzed reaction) and suggest an alternative method for the determination of uric acid.

References

A. Cornish-Bowden, *Fundamentals of Enzyme Kinetics* (1979), Butterworths (London), pp. 42–45. An excellent text for advanced reading in enzyme kinetics.

N. den Boer, C. van der Heiden, B. Leijnse, and J. H. M. Souverijm, Editors, *Clinical Chemistry—An Overview* (1989), Plenum Publishing (New York). An excellent textbook on all aspects of modern clinical chemistry.

R. Ellefson and W. Caraway, in *Fundamentals of Clinical Chemistry*, N. Tietz, Editor, 2nd ed. (1976), W. B. Saunders (Philadelphia), pp. 506–512. "Cholesterol."

W. Faulkner and J. King, in *Fundamentals of Clinical Chemistry*, N. Tietz, Editor, 2nd ed. (1976), W. B. Saunders (Philadelphia), pp. 999–1002. "Uric Acid."

G. Feuer and F. de la Iglesia, *Molecular Biochemistry of Human Disease* Vol I (1985), Vol. II (1986), Vol. III (1990), CRC Press (Boca Raton, FL). Three volumes written for students in biochemistry, medicine, and pharmacy.

L. Kaplan and A. Pesce, *Clinical Chemistry—Theory, Analysis and Correlation* (1989), C. V. Mosby (St. Louis). A teaching text for medical technologists and laboratory technicians.

J. Luzio and R. Thompson, *Molecular Medical Biochemistry* (1990), Cambridge University Press (Cambridge). Clinical biochemistry for medical students.

R. Montgomery, T. Conway, and A. Spector, *Biochemistry—A Case-Oriented Approach*, 5th ed. (1990), C. V. Mosby (St. Louis), pp. 465–496, 575–583. A good mix of biochemistry and medicine.

Sigma Chemical Co., Cholesterol, Total and HDL, Technical Bulletin, Procedure No. 351 (1985), Sigma Diagnostics (P.O. Box 14508, St. Louis, MO 63178).

Sigma Chemical Co., HDL Cholesterol, Technical Bulletin, Procedure No. 352-3 (1987), Sigma Diagnostics (P.O. Box 14508, St. Louis, MO 63178).

Sigma Chemical Co., The Ultraviolet Determination of Uric Acid, Technical Bulletin, Procedure No. 292-UV (1976), Sigma Diagnostics (P.O. Box 14508, St. Louis, MO 63178).

E. Taylor, Editor, *Clinical Chemistry* (1989), Wiley-Interscience (New York). An introduction to clinical chemistry written for chemists.

N. Tietz, Editor, *Fundamentals of Clinical Chemistry*, 2nd ed. (1976), W. B. Saunders (Philadelphia), pp. 506–514; 999–1002. A classic reference in the area of clinical chemistry.

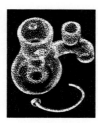

Vitamin C Content of Fruits, Vegetables, and Other Foods

■ Recommended Reading

Chapter 1, Section F.

■ Synopsis

Vitamin C is a nutritional factor that must be present in our diet. Its presence in dietary materials can be detected and quantified by titration with 2,6-dichlorophenolindophenol. In this experiment several fruits, vegetables, and other foods are analyzed for vitamin C content.

I. Introduction and Theory

Complete lack of vitamin C (ascorbic acid) in the diets of humans and other primates leads to a classic disease, scurvy. This nutritional disease, which was probably the first to be recognized, was widespread in Europe during the fifteenth and sixteenth centuries, but it is rare today.

Ascorbic acid is widely distributed in nature, but it occurs in especially high concentration in citrus fruits and green plants such as green peppers and spinach. Ascorbic acid can be synthesized by all plants and animals with the exception of humans, other primates, and guinea pigs. Therefore, vitamin C must be present in our dietary substances.

The fundamental role of ascorbic acid in metabolic processes is not well understood. There is some evidence that it may be involved in metabolic hydroxylation reactions of tyrosine, proline, and some steroid hormones, and in the cleavage-oxidation of homogentisic acid. Its function in these metabolic processes appears to be related to the ability of vitamin C to act as a reducing agent.

Although there is much controversy about the exact requirement, the adult Recommended Daily Allowance of vitamin C is 70 mg per day. Some scientists and physicians have suggested doses up to 1 to 3 grams per day in order to help resist the common cold. Deficiency of vitamin C results in swollen joints, abnormal development and maintenance of tissue structures, and eventually scurvy.

Chemically, ascorbic acid is a water-soluble, slightly acidic carbohydrate that exists in an oxidized or reduced form (Equation E15.1).

(Equation E15.1)

Ascorbic acid
(reduced form)

Dehydroascorbic acid
(oxidized form)

Both forms are biologically active. Ascorbic acid (the reduced form) is relatively stable to heat; however, dehydroascorbic acid (the oxidized form) is unstable. The lactone ring is easily hydrolyzed to diketogulonic acid, which has no antiscurvy activity. When fruits, vegetables, and other foods are heated, there is some loss of active vitamin C by conversion to diketogulonic acid.

Because of the clinical significance of vitamin C, it is essential to be able to detect and quantify its presence in various biological materials. Analytical methods have been developed to determine the amount of ascorbic acid in foods and in biological fluids such as blood and urine. Ascorbic acid may be assayed by titration with iodine, reaction with 2,4-dinitrophenylhydrazine, or titration with a redox indicator, 2,6-dichlorophenolindophenol (DCIP) in acid solution. The latter method will be used in this experiment because it is reasonably accurate, rapid, and convenient and can be applied to many different types of samples.

The reaction of DCIP with ascorbic acid is shown in Figure E15.1. Ascorbic acid reduces the indicator dye from an oxidized form (red in acid) to a reduced form (colorless in acid). The procedure is simple, beginning with dissolution of the sample to be tested in metaphosphoric acid. An aliquot of the sample is then titrated directly with a solution of DCIP. Although the original DCIP solution is blue, it becomes light red in the acid solution. Upon reaction with ascorbic acid in the sample, the dye becomes colorless. Titration is continued until there is a very slight excess of dye added (faint pink color remains in the acid solution).

Samples for analysis often contain traces of other compounds, in addition to ascorbic acid, that reduce DCIP. One way to minimize the interference of other substances is to analyze two identical aliquots of the sample. One aliquot is titrated directly and the total content of all reducing substances present is determined. The second aliquot is treated with

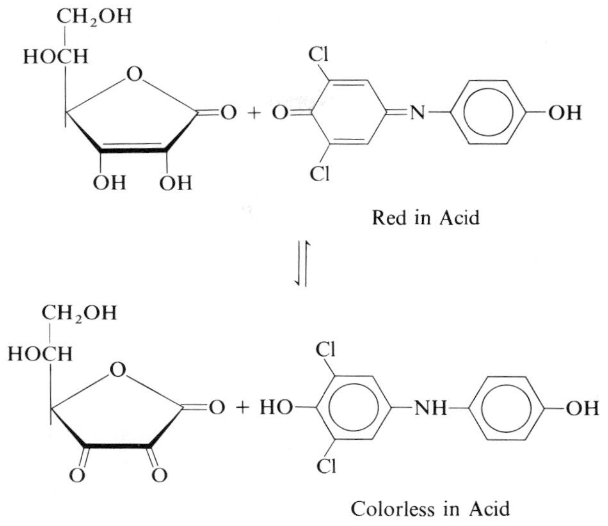

Figure E15.1
Reaction of the redox dye
DCIP with ascorbic acid.

ascorbic acid oxidase to destroy ascorbic acid and then titrated with DCIP. The second titration allows determination of reducing substances other than ascorbic acid. A second way to reduce the effect of other reducing substances is to perform the titration in the pH range 1 to 3. The interfering agents react very slowly with DCIP under these conditions.

Determination of vitamin C in biological fluids such as blood and urine is more difficult because only small amounts of the vitamin are present and many interfering reducing agents are present. Substances containing sulfhydryl groups, sulfite, and thiosulfate are common in biological fluids and react with DCIP, but much more slowly than ascorbic acid. The interference by sulfhydryl is often minimized by the addition of *p*-chloromercuribenzoic acid.

Overview of the Experiment

In this experiment several samples will be prepared, and the amount of vitamin C in each will be determined. Each sample will be titrated with standardized redox indicator, DCIP. The vitamin C content will be calculated in terms of milligrams per milliliter (mg/mL) of sample.

This experiment is flexible in that students may choose the samples they wish to analyze. It is suggested that several fruits and vegetables as well as an unknown ascorbic acid sample be available for analysis. The vitamin C content of raw and cooked vegetables can also be measured and compared. Commercially available vitamin C tablets or multivitamin pills provide interesting samples for analysis.

The time required for this experiment can be adjusted by controlling the number and type of samples.

II. Materials and Supplies

Fruit juices—orange, grapefruit, lemon, or lime.

Whole fruits and vegetables—oranges, grapefruit, lemons, limes, green peppers, tomatoes, potatoes, spinach, lettuce, cabbage, and others

Vitamin C tablets or multivitamin pills

Metaphosphoric acid/acetic acid solution, 4%

Unknown ascorbic acid in metaphosphoric acid/acetic acid solution, 0.5 to 3.0 mg/mL

Standard ascorbic acid in metaphosphoric acid/acetic acid solution, 0.50 mg/mL

2,6-Dichlorophenolindophenol solution in H_2O, 25 mg/100 mL
Ascorbic acid oxidase, lyophilized powder
Buret; use a 10-mL microburet
Mortar and pestle
Knife
Filter paper, fast flow
Glass funnel

III. Experimental Procedure

A. Standard Ascorbic Acid Solution

Fill a microburet with DCIP solution. The top of the solution should be at or slightly below the zero mark of the buret, and the glass tip must be full of solution. Using a pipet, transfer 1.0 mL of the ascorbic acid standard solution to a 50-mL Erlenmeyer flask containing 5 mL of 4% metaphosphoric acid/acetic acid solution. Read and record the initial reading on the buret. Titrate by rapid, dropwise addition of DCIP from the buret while mixing the contents of the flask. Add DCIP solution until a distinct rose-pink color persists for 15 to 20 seconds. Record the final reading on the buret. Repeat this procedure twice more, each time with a fresh 1.0-mL sample of ascorbic acid standard. In a similar fashion, titrate three blanks, each containing 5.0 mL of 4% metaphosphoric acid/acetic acid solution and 1.0 mL of water. Average the results for each series of measurements.

B. Unknown Ascorbic Acid Solution

Obtain a sample containing an unknown amount of ascorbic acid from your instructor. Place 1.0 mL of the unknown in a 50-mL Erlenmeyer flask. Add 5.0 mL of metaphosphoric acid solution to the flask and titrate as before with DCIP. Repeat with two more samples of the unknown.

C. Fruit Juice

Mix the stock solution of juice well and pour about 15 mL into a small beaker. Dilute 10.0 mL of this juice to 50 mL with the metaphosphoric acid solution. Filter this solution through a rapid flow, fluted filter paper. Pipet 10.0 mL of the filtrate into a 50-mL Erlenmeyer and titrate rapidly to a persistent rose-pink color with the standard DCIP solution as previously described. Repeat the titration with two 10.0-mL samples of the diluted juice. For a blank, add 10.0 mL of water to 40 mL of metaphosphoric acid solution and titrate three 10.0-mL samples with DCIP. Record the volume of DCIP necessary to titrate each juice sample and each blank.

To test for the presence of interfering substances in the juice, pipet a 10.0-mL sample of the fresh, undiluted juice into a 50-mL Erlenmeyer. Add a few crystals of ascorbic acid oxidase to destroy the ascorbic acid. Let stand, after gentle mixing, for 10 minutes. Add 40 mL of metaphosphoric acid solution. Titrate three 10.0-mL portions of this diluted juice with DCIP.

D. Whole Fruit

Peel the fruit, weigh to the nearest 0.1 g, and cut a sample of 25 to 50 g from the edible portion. Weigh the sample to the nearest 0.1 g. Slice the sample into many small pieces, transfer to a mortar, and grind with 25 mL of the metaphosphoric acid solution. Pour the liquid from the extract into a 100-mL volumetric flask. Grind the solid residue in the mortar twice using 25 mL of the metaphosphoric acid solution. Add the liquid extract each time to the 100-mL volumetric flask. Add metaphosphoric acid solution to the volumetric flask to a total volume of 100 mL. Filter the solution through a rapid flow, fluted filter paper. Transfer 10.0 mL of the filtrate into a 50-mL Erlenmeyer flask. Titrate rapidly with the DCIP solution. Repeat the titration on two more 10.0-mL samples of the filtrate. Titrate 10.0 mL of metaphosphoric acid solution as a blank.

E. Vitamin C Content in Raw and Boiled Vegetable

Cut two samples of a vegetable (green pepper, lettuce, cabbage, etc.) so that each weighs exactly the same amount (10–15g). Cut one sample into smaller pieces and grind in a mortar with 25 mL of the metaphosphoric acid solution. Pour the liquid from the extract into a 100-mL volumetric flask. Grind the solid residue in the mortar two more times using 25 mL of the metaphosphoric acid solution. Each time, combine all the extracts in the 100-mL flask. Add metaphosphoric acid solution to the volumetric flask to a total volume of 100 mL. Filter the solution through a rapid flow, fluted filter paper. Transfer 10.0 mL of the filtrate into a 50-mL Erlenmeyer flask and titrate, as before, with the DCIP solution. Repeat the titration on two more 10.0-mL samples of the filtrate. Titrate 10.0 mL of metaphosphoric acid solution as a blank.

Place the other vegetable sample in a beaker containing 25 mL of water. Heat the water to boiling and continue to boil for 10 minutes. Pour off and save the water, cut vegetable into small pieces, and transfer to a mortar. Grind with three 25-mL portions of metaphosphoric acid solution as before. Combine all the metaphosphoric acid extracts in a 100-mL volumetric flask. Filter the extract and titrate with DCIP as before. Mix 5.0 mL of the cooking water with 5.0 mL of metaphosphoric acid solution. Titrate with DCIP.

F. Vitamin C Tablets or Multivitamin Pills

Design your own procedure for analysis.

IV. Analysis of Results

A. Standard Ascorbic Acid Solution

Subtract the volume of titrant required for the blank from the titrant required for the sample. The difference represents the volume of DCIP that is equivalent to 0.50 mg of vitamin C. Calculate the standard deviation of your answer. What are the limits for 95% confidence? Review Chapter 1, section F for statistical analysis.

B. Unknown Ascorbic Acid Solution

Calculate the concentration of ascorbic acid in the unknown sample in units of mg/mL. Express your answer with confidence limits at the 95% confidence level.

C. Fruit Juice

Calculate the amount of ascorbic acid in the juices and report in terms of mg/100 mL of pure juice. Again, calculate confidence limits at the 95% confidence level. Remember that you started with 10 mL of pure juice, diluted it with 40 mL of metaphosphoric acid solution, and titrated 10 mL of the diluted juice. Are there reducing substances in addition to ascorbic acid in the juice?

D. Whole Fruit

Calculate the total amount of ascorbic acid in the whole fruit. Use proper statistical analysis.

E. Vitamin C Content in Raw and Boiled Vegetable

Calculate the amount of ascorbic acid in each sample. Was any lost during the boiling process? Explain. Could vitamin C be detected in the boiled water? Typical values for vitamin C content are shown below.

Orange juice	20–80 mg/100 mL
	Average = 45 mg/100 mL
Grapefruit juice	35–65 mg/100 mL
	Average = 40 mg/100 mL
Lemon juice	30–70 mg/100 mL
	Average = 45 mg/100 mL
Lime juice	5–40 mg/100 mL
	Average = 15 mg/100 mL
Plasma	0.2–2 mg/100 mL

F. Vitamin C Tablets or Multivitamin Pills

Calculate the amount of ascorbic acid in each sample. Does it agree with the label on the bottle?

V. Questions and Problems

1. Write the reaction for the oxidation of ascorbic acid catalyzed by ascorbic acid oxidase.

▶ 2. A 10-mL sample of pure orange juice was diluted with 40 mL of metaphosphoric acid solution. Then 10 mL of the diluted juice was titrated to an end point with 9.27 mL of DCIP. A blank required 0.25 mL of DCIP. A 1-mg sample of pure ascorbic acid required 6.52 mL (after blank correction) of DCIP. What is the concentration of ascorbic acid in the orange juice in mg/100 mL?

▶ 3. How does p-chloromercuribenzoic acid function to remove sulfhydryl group interference during the DCIP titration?

▶ 4. A whole, peeled orange weighed 100 g. Three sections (30 g) were extracted with metaphosphoric acid and the total extract filtered and diluted to 100 mL. A sample of 10 mL of the filtered extract required 4.10 mL of DCIP after blank correction. Also, 1 mg of standard ascorbic acid required 7.2 mL of DCIP (blank corrected). How many milligrams of ascorbic acid are present in the whole, peeled orange?

▶ 5. To what class of biomolecules does vitamin C belong?

VI. References

S. Ashoor, W. Monte, and J. Welty, *J. Assoc. Offic. Anal. Chem.* **67,** 78–80 (1984). "Liquid Chromatographic Determination of Ascorbic Acid in Foods."

J. T. Baker Chemical Co., *Product Information Bulletin,* Commodity No. H114, (1971), Phillipsburg, N.J. 08865. Analytical uses of 2,6-DCIP.

M. Bui-Nguyen, in *Modern Chromatographic Analysis of the Vitamins,* A. De Leenheer, W. Lambert, and M. De Ruyter, Editors (1985), Marcel Dekker (New York). pp. 267–301 "Ascorbic Acid and Related Compounds."

A. Floridi, R. Coli, A. Fidanza, C. Bourgeois, and R. Wiggins, *Int. J. Vitam. Nutr. Res.* **52,** 193–196 (1982). "HPLC Determination of Ascorbic Acid in Food: Comparison with Other Methods."

W. Friedrich, *Vitamins* (1988), Walter de Gruyter (Berlin), pp. 929–1001, "Vitamin C."

L. Machlin, *Handbook of Vitamins,* 2nd ed. (1991), Marcel Dekker (New York). A complete guide to the vitamins.

J. Orten and O. Neuhaus, *Human Biochemistry,* 9th ed. (1975), C. V. Mosby (St. Louis), pp. 588–593. Biochemical properties of ascorbic acid.

J. Rawn, *Biochemistry,* 2nd ed. (1989), p. 87–88. Neil Patterson Publishers (Burlington, NC). Biological function of ascorbic acid.

N. Tietz, Editor, *Fundamentals of Clinical Chemistry,* 2nd ed. (1976), W. B. Saunders (Philadelphia), pp. 547–551. A detailed discussion of chemical, biological, and clinical aspects of ascorbic acid.

EXPERIMENT 16

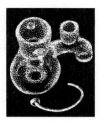

Activity and Thermal Stability of Acrylamide Gel–Immobilized Peroxidase: An Experiment in Biotechnology

■ **Recommended Reading**

Chapter 4, Section B; Experiment 6.

■ **Synopsis**

Immobilized enzymes are becoming increasingly important in commercial processes. In this experiment, students will trap molecules of the enzyme horseradish peroxidase within a polyacrylamide gel matrix. The reaction kinetics and thermal stability of the immobilized enzyme will be measured. This experiment introduces students to the use of enzymes in biotechnology.

I. Introduction and Theory

The principles of biotechnology, the application of biological cells or cell components to technically useful operations, are seldom introduced to students in today's biology, chemistry, and biochemistry classes. With the increasing impact of modern biology on all areas of science and industry, it is imperative that students have some understanding of biotechnology operations.

The use of immobilized enzymes in biotechnology is becoming widespread and encompasses several areas, including the chemical industry, pharmaceuticals, food production, medicine, environmental control, clinical chemistry, and agriculture. Specific applications include winemaking, cheesemaking, production of alcohol for fuel, treatment of wastewater, metal mining, measurement of glucose and other constituents in body fluids, and enzyme replacement in individuals with genetic disorders.

Immobilized Enzymes

Enzymes are valuable and versatile commercial reagents for the following reasons: (1) they have high catalytic activities, (2) they catalyze a great variety of reactions, (3) they provide the potential for stereospecific reactions, (4) they function under generally mild reaction conditions, and (5) side reactions and secondary products are rare. However, along with these advantages as catalysts, enzymes also have some disadvantages: (1) they are generally available only in small quantities, (2) they are fragile, unstable molecules, and (3) they are expensive from a commercial point of view. Therefore, their commercial use is practical only if methods can be found to increase their stability and if enzymes can be recovered and reused after the desired process is complete. In other words, enzymes used in commercial processes must be recycled. One of the best methods for ensuring recovery of active enzyme is immobilization of the catalyst molecules. Immobilization may be defined as any process that limits the movement or free diffusion of the enzyme molecule and usually involves attachment of the enzyme to an inert, water-insoluble support or entrapment within a water-insoluble matrix or microsphere.

In addition to recovery and reuse of immobilized enzymes, there are several other advantages to their use: (1) since they are in an insoluble form, they can be separated from the reaction mixture and do not contaminate the product; (2) an immobilized enzyme is often more stable to heat, pH change, organic solvents, and other adverse reaction conditions than a free enzyme; (3) they can be used in a continuous manner, for example, in a column flow reactor; and (4) the end point of the reaction can be controlled simply by physically separating the enzyme from the solution. Immobilized enzymes are also becoming important in basic biochemical research. Many enzymes in their natural environment in organisms are not free but are immobilized in cells, tissues, and membranes. Even cytoplasmic enzymes are held somewhat rigidly in a gelatinous matrix or cytoskeleton. Therefore, studying the characteristics of immobilized enzymes will provide additional insight into how enzymes function in the living organism.

Methods of Immobilization

Four methods have been developed for enzyme immobilization: (1) physical adsorption onto an inert, insoluble, solid support such as a polymer; (2) chemical covalent attachment to an insoluble polymeric support; (3) encapsulation within a membranous microsphere such as a liposome; and

(4) entrapment within a gel matrix. The choice of immobilization method is dependent on several factors, including the enzyme used, the process to be carried out, and the reaction conditions. Trial and error are often necessary to find the most effective method. In this experiment, an enzyme, horseradish peroxidase (donor: H_2O_2 oxidoreductase; EC 1.11.1.7), will be imprisoned within a polyacrylamide gel matrix. This method of entrapment has been chosen because it is rapid, inexpensive, and allows kinetic characterization of the immobilized enzyme. Immobilized peroxidase catalyzes a reaction that has commercial potential and interest, the reductive cleavage of hydrogen peroxide, H_2O_2, by an electron donor, AH_2:

▶ $$H_2O_2 + AH_2 \xrightleftharpoons{\text{peroxidase}} 2H_2O + A \qquad \text{(Equation E16.1)}$$

The most commonly used matrix for entrapment of enzymes is crosslinked polyacrylamide. When acrylamide and methylene bisacrylamide (a cross-linking agent) are allowed to polymerize, a cross-linked polymer is produced (Figure E16.1). If the proper ratio of acrylamide and methylene bisacrylamide is used, the extent of cross-linking is such that relatively large enzyme molecules can be trapped within the polymer cage, while still allowing relatively small substrate and product molecules to diffuse readily in and out of the matrix. Acrylamide gel has another advantage in that it is nonionic; therefore, charged reactant and product molecules are not retained inside the gel. To produce an entrapped enzyme, a solution of the desired enzyme is mixed with the monomer acrylamide, cross-linking agent methylene bisacrylamide, and a catalyst added for polymer initiation. Within a few minutes, a solid gel mass is produced, which is washed to remove free enzyme that has not been trapped. The gel containing the trapped enzyme is now ready for further investigation and analysis.

Figure E16.1
Entrapment of an enzyme molecule (◯) by cross-linked polyacrylamide gel prepared by mixing enzyme with acrylamide, methylene bisacrylamide, and catalyst.

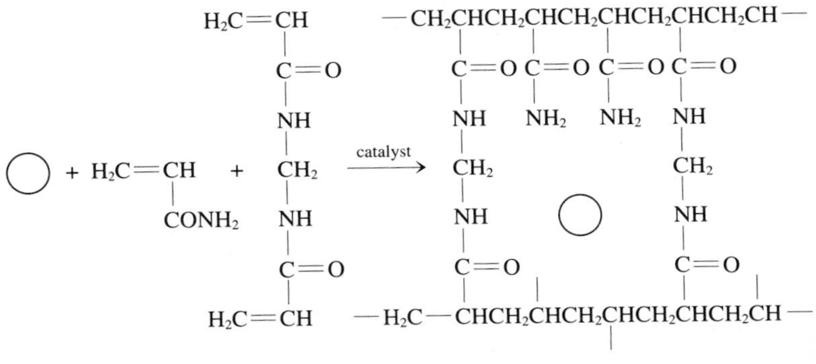

Characterization of Immobilized Enzymes

Before an immobilized enzyme can be used for an industrial process, it is essential to characterize it in terms of its catalytic and kinetic properties. A quantitative assay must be developed to measure the activity, kinetic parameters, and stability of the enzyme. In a coupling reaction, H_2O_2 rapidly reacts with phenol and 4-aminoantipyrine (electron donor) in the presence of peroxidase to produce a quinoneimine chromogen (Equation E16.2, Figure E14.2), which is intensely colored with a maximum absorbance at 510 nm. (This is the same as the product formed in the analysis of cholesterol in Experiment 14.)

$$\blacktriangleright \quad H_2O_2 + \text{phenol} + \text{4-aminoantipyrine} \; \underset{\longleftarrow}{\overset{\text{peroxidase}}{\rightleftharpoons}} \; \text{quinoneimine} + 2H_2O$$
$$\text{(Equation E16.2)}$$

The amount of peroxidase present will influence the amount of quinoneimine product formed. In fact, there is a direct and linear relationship between the quantity of peroxidase present in the gel or solution and the intensity of the color produced. The intensity of color or absorbance can be measured in a spectrophotometer. Therefore, this assay can be used to measure the rate of the enzyme-catalyzed reaction, which allows calculation of important kinetic constants including the number of enzyme activity units (U), the Michaelis constant (K_M), and the maximum velocity (V_{max}). An enzyme supported within a matrix may not behave kinetically in the same way as one free in solution. There are several reasons for this: (1) immobilization may force the enzyme molecule to take on a different conformation, (2) the chemical microenvironment around the immobilized enzyme may be different, and (3) the rate of the reaction may be more greatly influenced by the diffusion of substrate molecules into the gel. The stability of an enzyme is also often changed upon immobilization. This characteristic may be increased in some instances due to stabilizing effects of the surrounding matrix environment, or it may be decreased because of a denaturing microenvironment inside the gel. It is difficult to predict whether an enzyme will be stabilized upon immobilization, so trial and error are still necessary in order to find optimum conditions. In this experiment, you will measure and compare the thermal stability of free and immobilized peroxidase.

Overview of the Experiment

Several reaction characteristics of gel-immobilized peroxidase, including activity and thermal stability, will be examined in this experiment. A spectrophotometric assay for peroxidase is introduced. The literature reports many assays for measurement of peroxidase activity. The aminoantipyrine-phenol assay is selected because it is rapid, convenient, and accurate and requires less toxic reagents than other assays.

Time requirements for this experiment are:

Part A: Preparation of immobilized peroxidase—45 minutes.

Part B: Assay of immobilized peroxidase—1 hour.

Part C: Thermal stability of immobilized peroxidase—1 hour.

II. Materials and Supplies

⚠ Caution

Phenol is a chronic poison. It may be fatal if inhaled, swallowed, or absorbed through skin.

First Aid: In case of contact, immediately flush eyes or skin with copious amounts of water for at least 15 minutes. If inhaled, remove to fresh air.

Acrylamide in the unpolymerized form is a skin irritant and a potential neurotoxin. Wear gloves and a mask while weighing the dry powder. Do not breathe the dust. Prepare all acrylamide solutions in the hood. Do not mouth pipet any solutions used for gel formation.

First Aid: In case of contact, immediately flush with copious amounts of water for at least 15 minutes. If inhaled, remove to fresh air.

Disposal of Gels: Allow to dry in air and throw in a special container provided by the instructor.

Potassium phosphate buffer, 0.2 M, pH 7.0

Screw cap scintillation or similar type of vials, 20 mL

Stock solution for gel: 100 mg of electrophoresis grade acrylamide and 1.1 g of methylene bisacrylamide in 100 mL of 0.2 M, pH 7.0, phosphate buffer

Riboflavin solution, 10 mg/100 mL in glass-distilled water

Fluorescent lamp or sunlight

Ammonium persulfate solution, 0.1 g/10 mL of phosphate buffer (prepare fresh just before use)

Tank of purified nitrogen

Horseradish peroxidase, 0.1 mg/mL in glass-distilled H_2O (150–200 units/mg)

Spectrophotometer and cuvettes

Clinical centrifuge

4-Aminoantipyrine–phenol solution. Prepare by dissolving 810 mg phenol (**Caution**) in 40 mL of glass-distilled water and adding 25 mg of 4-aminoantipyrine. Dilute to a final volume of 50 mL with glass-distilled water.

Hydrogen peroxide, 0.0017 M in water. Prepare by mixing 1 mL of 30% H_2O_2 with 99 mL of glass-distilled water. Further dilute 1 mL of this solution to 50 mL with 0.2 M phosphate buffer, pH 7.0 (prepare fresh just before use).

Syringe (5 mL) with 0.8 μ filter system

Constant-temperature water bath, 60°C

III. Experimental Procedure

A. *Preparation of Immobilized Peroxidase*

Mix the following reagents in a 20-mL vial: 10 mL acrylamide–methylene bisacrylamide stock solution, 1.0 mL of 0.1 mg/mL peroxidase solution, 0.5 mL of 10 mg/mL ammonium persulfate solution, and 1.0 mL of 0.1 mg/mL riboflavin solution. Mix well using a vortex mixer, and gently bubble N_2 gas through the mixture for approximately 2 minutes. This procedure removes O_2, a polymerization inhibitor, from the reaction mixture. Blow N_2 gas over the surface of the reaction mixture, seal tightly, and place under a fluorescent lamp or in sunlight. The solution should become opaque within a few minutes and completely polymerize within 20–30 minutes. With a spatula, transfer the gel to a vacuum filtration system to remove most of the solution. To wash the gel, transfer it to a test tube containing 5 mL of water. Break up the gel by aspirating with a Pasteur pipet. Centrifuge the gel mixture for 5 minutes at 1000–1200 rpm. Decant supernatant and add 10 mL of water to the gel. Again, break up the gel by sucking in and out of a Pasteur pipet and centrifuge as before. Repeat this washing process two more times. The final centrifugation should be done for 5 minutes at 1200–1500 rpm. Dry the gel by vacuum filtration for a few minutes and weigh on a balance. Approximately 2–3 g of semiwet gel will be obtained. Proceed directly to part B.

B. *Assay of Immobilized Peroxidase*

A 3-minute fixed-time assay will be used to measure peroxidase activity. Two tubes will be set up for each assay, one tube for the zero point and the other tube for the 3-minute assay. The amount of gel-immobilized peroxidase will be different for each assay. Obtain six test tubes for reaction vessels and label them 1–6. Weigh two samples of 0.05 g of immobilized peroxidase and transfer one sample to tube 1 and the other sample to tube 2. To tubes 1 and 2, add 2.5 mL of the aminoantipyrine-phenol solution. Mix well. To tube 1 (zero point), add 2.5 mL of the H_2O_2 solution, rapidly mix well, and immediately quench the reaction by pouring into the barrel of a syringe with filter system. Use the plunger to force the solution through the filter into a glass cuvette. This whole procedure from mixing to transfer should take no more than 10 seconds. Read and record in your notebook the absorbance of the reaction mixture at 510 nm.

To tube 2 (3-minute point), add 2.5 mL of the H_2O_2 solution, immediately mix, and note the time. Gently and continuously mix the reaction mixture. At the end of exactly 3 minutes, transfer the reaction mixture to the barrel of a syringe with filter system. Force the solution through the filter into a glass cuvette. Read the absorbance of the reaction mixture at 510 nm and record in your notebook.

In tubes 3 and 4, 0.1 g of immobilized peroxidase will be used. Tube 3 will represent the zero point and tube 4 the 3-minute point. Weigh two samples of 0.1 g of immobilized peroxidase and transfer to tubes 3 and 4. Add 2.5 mL of the aminoantipyrine-phenol solution to each tube and mix well. Measure the peroxidase activity in each tube by adding 2.5 mL of the H_2O_2 solution as before; tube 3 will be treated like tube 1 and tube 4 like tube 2. Read and record the A_{510} for tubes 3 and 4. Repeat the procedure with tubes 5 (zero point) and 6 (3-minute point), each containing 0.2 g of immobilized peroxidase.

C. Thermal Stability of Immobilized Peroxidase

In this section, the thermal stability of acrylamide gel–immobilized peroxidase will be compared to that of the free enzyme. The free enzyme is assayed in the following manner: Dilute 1 mL of the stock horseradish peroxidase (0.1 mg/mL, 15 units/mL) with 299 mL of glass-distilled water. Add 1.0 mL of this diluted enzyme to each of two test tubes. Place one of the tubes in a 60°C water bath for exactly 4 minutes. Allow the other tube to sit at room temperature for the same time interval. Cool the higher-temperature tube to room temperature by placing in a water bath. To each tube add 2.0 mL of the aminoantipyrine-phenol stock solution and 2.0 mL of the H_2O_2 solution and mix well. Allow the tubes to sit at room temperature for exactly 3 minutes; then immediately read the A_{510} for each. Record the results in your notebook.

The immobilized enzyme is assayed in the following manner: Weigh out two samples of 0.1 g of the gel and transfer to two separate test tubes each containing 0.5 mL of phosphate buffer. Place one of the tubes in the 60°C water bath and allow the other to remain at room temperature. At the end of 4 minutes, remove the tube from the 60°C bath, cool to room temperature in a water bath, and add 2.25 mL of aminoantipyrine-phenol solution and 2.25 mL of H_2O_2 solution. Gently mix the reaction mixture for exactly 3 minutes and remove the gel by passing the mixture through the syringe-filter system. Read the A_{510}. To the room-temperature tube, add 2.25 mL of aminoantipyrine-phenol solution and 2.25 mL of H_2O_2. Gently mix for exactly 3 minutes and separate the gel with a syringe-filter system. Read the A_{510} of the reaction mixture.

IV. Analysis of Results

A. Preparation of Immobilized Peroxidase

Describe your observations of the polymerization process. How many grams of acrylamide gel–immobilized peroxidase did you obtain?

B. Assay of Immobilized Peroxidase

The activity of the immobilized peroxidase can be calculated from the absorbance change for each reaction mixture, The absorbance change is calculated as follows:

$$\Delta A = A_{3\ min} - A_{0\ min}$$

where

ΔA = overall absorbance change

$A_{3\ min}$ = absorbance at 510 nm of tubes 2, 4, and 6

$A_{0\ min}$ = absorbance at 510 nm of tubes 1, 3, and 5

Calculate the ΔA/min for each of the three sets of conditions (0.05 g, 0.10 g, and 0.2 g of gel). Prepare a plot of ΔA/min (y axis) vs. mg of gel (x axis). Describe and explain the shape of the graph.

Use the equation below to calculate the units of activity per mg of gel. The number 6.58 in the denominator is the absorption coefficient for the quinoneimine chromogen assay product.

$$\text{Units/mg} = \frac{\Delta A/\text{min}}{6.58 \times \text{mg of gel}}$$

Repeat this calculation for all three amounts of gel (0.05, 0.10, 0.20 g). Compare the three results.

C. Thermal Stability of Immobilized Peroxidase

Determine the reaction rate (ΔA/min) for each of the four reaction conditions. Assume that $A_{0\ min} = 0$ for each assay.

ΔA_1 = absorbance change for free peroxidase at room temperature

ΔA_2 = absorbance change for free peroxidase at 60°C

ΔA_3 = absorbance change for immobilized peroxidase at room temperature

ΔA_4 = absorbance change for immobilized peroxidase at 60°C

Calculate the % activity remaining after heating the free and immobilized enzyme:

$$\% \text{ Activity remaining}_{\text{free}} = \frac{\Delta A_2}{\Delta A_1} \times 100$$

$$\% \text{ Activity remaining}_{\text{immobilized}} = \frac{\Delta A_4}{\Delta A_3} \times 100$$

Compare the two final results. Which is more stable to heat, free or immobilized enzyme?

Calculate the specific activity of the free enzyme in units/mg using the following equation:

$$\text{Units/mg} = \frac{\Delta A_{510}/\text{min}}{6.58 \times \text{mg enzyme/mL reaction mixture}}$$

Compare this number with known specific activity of the enzyme listed on the reagent bottle.

V. Questions and Problems

▶ 1. What is the purpose of washing the acrylamide gel several times with water as described in the experimental section, part A? How could you experimentally determine when washing is complete?

▶ 2. Design an experiment using the immobilized peroxidase to calculate the Michaelis constant, K_M, and the maximum velocity, V_{max}.

▶ 3. How could you determine if the gel is "leaking" peroxidase? What experimental modifications could be made to prevent this?

▶ 4. What should be the shape of a plot of $\Delta A/\text{min}$ (immobilized enzyme activity) vs. milligrams of gel? Explain.

▶ 5. What is the purpose of riboflavin, ammonium persulfate, and sunlight in this experiment?

VI. References

R. Boyer and J. McArthur, *Biotech. Educ.* **2,** 17–20 (1991). "Activity and Thermal Stability of an Immobilized Enzyme."

J. BúLock and B. Kristiansen, Editors, *Basic Biotechnology* (1987), Academic Press (London). A complete text with theory and applications.

L. Josefsson, *Biochem. Educ.* **15,** 141-143 (1987). "Biochemistry in the Service of Man."

J. Kennedy and C. White, in *Handbook of Enzyme Biotechnology*, A. Wiseman, Editor (1985), Ellis Horwood (Chichester), pp. 147–207. "Principles of Immobilization of Enzymes."

M. Kierstan and M. Coughlan, in *Immobilised Cells and Enzymes—A Practical Approach,* J. Woodward, Editor (1985), IRL Press (Oxford), pp. 39–48. "Immobilisation of Cells and Enzymes by Gel Entrapment."

K. Laidler and P. Bunting, in *Methods in Enzymology,* Vol. 64, D. Purich, Editor (1980), Academic Press (New York), pp. 227-248. "The Kinetics of Immobilized Enzyme Systems."

K. Mosbach, Editor, *Methods in Enzymology,* Vol. 137 (1988), Academic Press (San Diego). An entire volume on immobilized enzymes and cells.

M. Roig, F. Estevez, F. Velasco, N. Ghais, and J. Silverio, *Biochem. Educ.* **14,** 180–184 (1986). "Methods for Immobilizing Enzymes."

C. Worthington, Editor, Worthington Biochemical Corporation, *Worthington Enzyme Manual* (1988), Freehold, NJ 07728, pp. 254-260. A book that provides data on many enzymes, including peroxidase.

EXPERIMENT 17

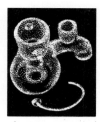

Extraction of Chromosomal DNA from Bacterial Cells

■ **Recommended Reading**

Chapter 2, Sections D, E; Chapter 5, Section A.

■ **Synopsis**

DNA, the genetic substance in biological cells, is a large polymer of de-oxyribonucleotide monomers. This experiment introduces the student to a general method for isolation and partial purification of nucleic acids from microorganisms. The procedure consists of disrupting the cell wall or membrane, dissociating bound proteins, and separating the DNA from other soluble compounds. The isolated DNA is characterized and quantified by ultraviolet spectroscopy under native and denaturing conditions.

I. Introduction and Theory

DNA Structure and Function

DNA in all forms of life is a polymer made up of nucleotides containing four major types of heterocyclic nitrogen bases, adenine, thymine, guanine, and cytosine. The nucleotides are held together by 3′, 5′- phosphodiester bonds (Figure E17.1). The quantitative ratio and sequence of bases vary with the source of the DNA. Native DNA exists as two complementary strands held together by hydrogen bonds and arranged in a double helix.

DNA was first isolated from biological material in 1869, but its participation in the transfer of genetic information was not recognized until the

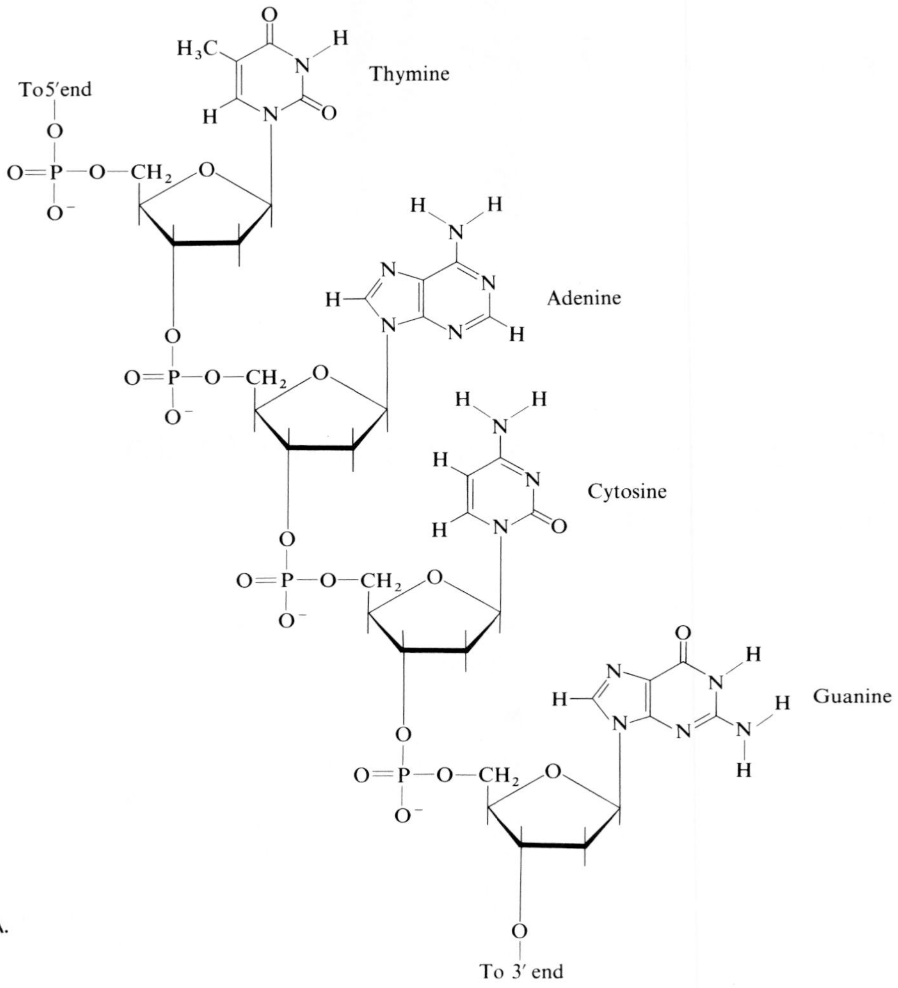

Figure E17.1
The covalent structure of DNA.

mid-1940s. Since that time, DNA has been the subject of thousands of physical, chemical, and biological investigations. A landmark discovery was the elucidation of the three-dimensional structure of DNA by Watson and Crick in 1953. The **double helix** as envisioned by Watson and Crick is now recognized as a significant structural form of native DNA (Figure E17.2). Other discoveries of importance include methods for sequential analysis of the nucleotide bases, regulation of gene expression, cleavage of DNA by specific restriction endonucleases, and recombinant DNA.

DNA in prokaryotic cells (simple cells with no major organelles and a single chromosome) exists as a single molecule in a circular, double helix form with a molecular weight of at least 2×10^9. Eukaryotic cells (cells with major organelles) contain several chromosomes and, thus, several very large DNA molecules.

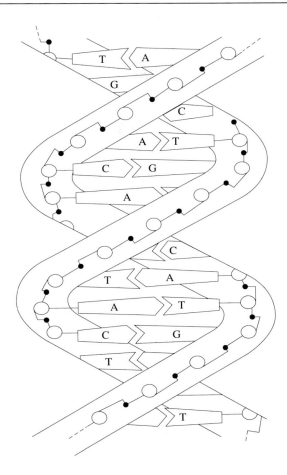

Figure E17.2
The helix conformation of DNA showing base pairs. Reprinted with the permission of Macmillan Publishing Company from *Biochemistry* by Zubay. Copyright © 1988 by Macmillan Publishing Company.

Isolation of DNA

Because of the large size and the fragile nature of chromosomal DNA, it is very difficult to isolate in an intact, undamaged form. Several isolation procedures have been developed that provide DNA in a biologically active form, but this does not mean it is completely undamaged. These DNA preparations are stable, of high molecular weight, and relatively free of RNA and protein. Here, a general method will be described for the isolation of DNA in a stable, biologically active form from microorganisms. The procedure outlined is applicable to many microorganisms and can be modified as necessary.

Designing an isolation procedure for DNA requires extensive knowledge of the chemical stability of DNA as well as its condition in the cellular environment. Figures E17.1 and E17.2 illustrate several chemical bonds that may be susceptible to cleavage during the extraction process. The experimental factors that must be considered and their effects on various structural aspects of intact DNA are outlined below.

1. pH
   (a) Hydrogen bonding between the complementary strands is stable between pH 4 and 10.
   (b) The phosphodiester linkages in the DNA backbone are stable between pH 3 and 12.
   (c) N-glycosyl bonds to purine bases (adenine and guanine) are hydrolyzed at pH values of 3 and less.

2. Temperature
   (a) There is considerable variation in the temperature stability of the hydrogen bonds in the double helix, but most DNA begins to unwind in the range 80–90°C.
   (b) Phosphodiester linkages and N-glycosyl bonds are stable up to 100°C.

3. Ionic Strength
   (a) DNA is most stable and soluble in salt solutions. Salt concentrations of less than 0.1 M weaken the hydrogen bonding between complementary strands.

4. Cellular Conditions
   (a) Before the DNA can be released, the bacterial cell wall must be lysed. The ease with which the cell wall is disrupted varies from organism to organism. In some cases (yeast), extensive grinding or sonic treatment is required, whereas in others (*B. subtilis*), enzymatic hydrolysis of the cell wall is possible.
   (b) Several enzymes are present in the cell that may act to degrade DNA, but the most serious damage is caused by the deoxyribonucleases. These enzymes catalyze the hydrolysis of phosphodiester linkages.
   (c) Native DNA is present in the cell as DNA-protein complexes. The proteins (basic proteins called **histones**) must be dissociated during the extraction process.

5. Mechanical Stress on the DNA
   (a) Gentle manipulations may not always be possible during the isolation process. Grinding, shaking, stirring, and other disruptive procedures may cause cleavage (shearing or scission) of the DNA chains. This usually does not cause damage to the secondary structure of the DNA, but it does reduce the length of the molecules.

Now that these factors are understood, a general procedure of DNA extraction will be outlined.

Step 1. Disruption of the cell membrane and release of the DNA into a medium in which it is soluble and protected from degradation The isolation procedure described here calls for the use of an enzyme, lysozyme, to disrupt the cell membrane. Lysozyme catalyzes the hydrolysis of glycosidic bonds in cell wall peptidoglycans, thus causing destruction of the cell wall and release of DNA and other cellular components. The medium for solution of DNA is a buffered saline solution containing EDTA. DNA, because it is ionic, is more soluble and stable in salt solution than in distilled water. The EDTA serves at least two purposes. First, it binds divalent metal ions (Cd^{2+}, Mg^{2+}, Mn^{2+}) that could form salts with the anionic phosphate groups of the DNA. Second, it inhibits deoxyribonucleases that have a requirement for Mg^{2+} or Mn^{2+}. Citrate has occasionally been used as a chelating agent for DNA extraction; however, it is not an effective agent for binding Mn^{2+}. The mildly alkaline medium (pH 8) acts to reduce electrostatic interaction between DNA and the basic histones and the polycationic amines, spermine and spermidine (see Experiment 18). The relatively high pH also tends to diminish nuclease activity and denature other proteins.

Step 2. Dissociation of the protein-DNA complexes Detergents are used at this stage to disrupt the ionic interactions between positively charged histones and the negatively charged backbone of DNA. Sodium dodecyl sulfate (SDS), an anionic detergent, binds to proteins and gives them extensive anionic character. A secondary action of SDS is to act as a denaturant of deoxyribonucleases and other proteins. Also favoring dissociation of protein-DNA complexes is the alkaline pH, which reduces the positive character of the histones. To ensure complete dissociation of the DNA-protein complex and to remove bound cationic amines, a high concentration of a salt (NaCl or sodium perchlorate) is added. The salt acts by diminishing the ionic interactions between DNA and cations.

Step 3. Separation of the DNA from other soluble cellular components Before DNA is precipitated, the solution must be deproteinized. This is brought about by treatment with chloroform–isoamyl alcohol and followed by centrifugation. Upon centrifugation, three layers are produced: an upper aqueous phase, a lower organic layer, and a compact band of denatured protein at the interface between the aqueous and organic phases. Chloroform causes surface denaturation of proteins. Isoamyl alcohol reduces foaming and stabilizes the interface between the aqueous phase and the organic phases where the protein collects.

The upper aqueous phase containing nucleic acids is then separated and the DNA precipitated by addition of ethanol. Because of the ionic nature of DNA, it becomes insoluble if the aqueous medium is made less polar by addition of an organic solvent. The DNA forms a threadlike precipitate that can be collected by "spooling" onto a glass rod. The isolated DNA may still be contaminated with protein and RNA. Protein can be removed by dissolving the spooled DNA in saline medium and repeating the chloroform–isoamyl alcohol treatment until no more denatured protein collects at the interface.

RNA does not normally precipitate like DNA, but it could still be a minor contaminant. RNA may be degraded during the procedure by treatment with ribonuclease after the first or second deproteinization steps. Alternatively, DNA may be precipitated with isopropanol, which leaves RNA in solution. Removal of RNA sometimes makes it possible to denature more protein using chloroform–isoamyl alcohol. If DNA in a highly purified state is required, several deproteinization and alcohol precipitation steps may be carried out. It is estimated that up to 50% of the cellular DNA is isolated by this procedure. The average yield is 1 to 2 mg per gram of wet packed cells. The isolated DNA has a molecular weight on the order of 10×10^6.

Characterization of DNA

The DNA isolated in this experiment is of sufficient purity for characterization studies. DNA has significant absorption in the UV range because of the presence of the aromatic bases, adenine, guanine, cytosine, and thymine. This provides a useful probe into DNA structure because structural changes such as helix unwinding affect the extent of absorption. In addition, absorption measurements are used as an indication of DNA purity. The major absorption band for purified DNA peaks at about 260 nm. Protein material, the primary contaminant in DNA, has a peak absorption at 280 nm. The ratio A_{260}/A_{280} is often used as a relative measure of the nucleic acid/protein content of a DNA sample. The typical A_{260}/A_{280} for isolated DNA is 1.9. A smaller ratio indicates increased contamination by protein. Compare this measurement of DNA purity with the spectrophotometric method for protein concentration in Chapter 2.

If DNA solutions are treated with denaturing agents (heat, alkali, organic solvents) their ultraviolet absorbing properties are strikingly changed. Figure E17.3 shows the effect of temperature on the UV absorption of DNA. The curve is obtained by plotting $A_{260(T)}/A_{260(25°)}$ vs. temperature (T). Heating through the temperature range of 25°C to about 80°C results only in minor increases in absorption. However, as the temperature is increased, there is a sudden increase in UV absorption followed by a constant A_{260}. The total increase in absorption is usually on the order of 40% and occurs over a small temperature range. Figure E17.3 is called a **thermal denaturation curve,** a **temperature profile,** or a **melting curve.** The temperature corresponding to the midpoint of each absorption increase is defined as T_m, the **transition temperature** or **melting temperature.** (This should not be confused with melting point, transformation of a substance from solid to liquid, as was routinely studied in organic laboratory.) Each species of DNA has a characteristic T_m value that can be used for identification and characterization purposes.

The origin of the absorption increase, called a **hyperchromic effect,** is well understood. The absorption changes are those that result from the transition of an ordered double-helix DNA structure to a denatured state or random, unpaired DNA strands. Native DNA in solution exists in the

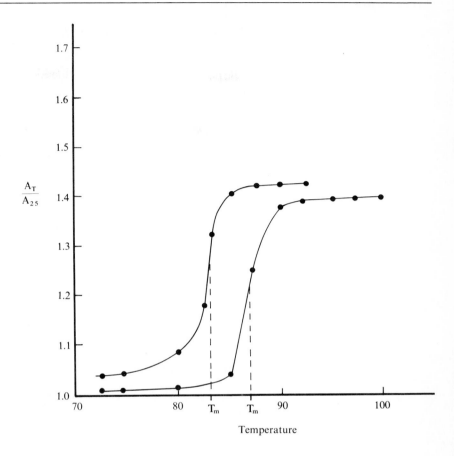

Figure E17.3
A typical thermal denaturation curve for DNA. T_m is measured as the temperature at the midpoint of the absorbance increase.

double helix held together primarily by hydrogen bonding between complementary base pairs on each strand (see Figure E17.2). Hydrophobic and $\pi-\pi$ interactions between stacked base pairs also strengthen the double helix. Agents that disrupt these forces (hydrogen bonding, hydrophobic and $\pi-\pi$ interactions) cause dissociation or unwinding of the double helix. In a random coil arrangement, base-base interactions are at a minimum; this alters the resonance behavior of the aromatic rings, causing an increase in absorption. The process of DNA dissociation can, therefore, be characterized by monitoring the UV-absorbing properties of DNA under various conditions.

Many agents have been evaluated as potential denaturants of DNA. Heat has received the most attention, but chemical agents such as organic solvents, urea, salts, and alkali are known to transform the double helix of DNA to a random coil. In addition, the action of deoxyribonuclease, an enzyme that catalyzes the hydrolysis of phosphodiester bonds, can be assayed by monitoring the change in A_{260}.

Using the Hyperchromic Effect to Characterize DNA

The hyperchromic effect is a relatively simple process to measure, but it can be very useful in characterizing DNA. Preparation of a temperature profile is one of the most convenient methods for measuring the increase in UV absorption. The results may be used for several purposes.

DETERMINATION OF T_M

As noted earlier in this experiment, T_m is a valuable physical constant that can be used in the identification and characterization of DNA. It can be measured as the temperature corresponding to the midpoint of the absorption increase. The T_m for most DNA is the range 80 to 90°C.

CALCULATION OF BASE COMPOSITION

The stability of the DNA double helix depends on the base composition of the DNA. The greater the guanine plus cytosine content (% GC base pairs), the more stable the double helix and the higher the T_m. GC pairs are held together by three hydrogen bonds; in contrast, AT pairs have just two hydrogen bonds. A linear relationship has been described that correlates the T_m value for a DNA sample with the % GC content (Equation E17.1).

▶ $$\%GC = 2.44 \ (T_m - 69.3) \qquad \text{(Equation E17.1)}$$

This has been derived by measuring the T_m for numerous DNA samples for which the base composition was known. Equation E17.1 is valid for GC amounts of 30 to 70%.

PURITY OF DNA

The greater the purity of a DNA sample, the greater the extent of hyperchromicity. Highly purified DNA shows a 40% increase in absorbance upon complete thermal denaturation. Although a quantitative estimate of purity is not possible, a hyperchromic effect of significantly less than 40% indicates relatively impure DNA. This could also mean that the DNA is already partially denatured or contaminated with UV-absorbing materials. The shape of the temperature profile can also be used as an indication of purity. Impure or partially denatured DNA often yields a broad temperature profile, with the hyperchromic effect occurring over a wide temperature range.

Overview of the Experiment

This experiment describes an isolation procedure for DNA that can be applied to most microorganisms. A flowchart of the procedure is shown in Figure E17.4. The general procedure was first reported by Marmur (1961, 1963). Recommended organisms are *Bacillus subtilus, Escherichia coli*, and *Clostridium welchii*. Other microorganisms may be

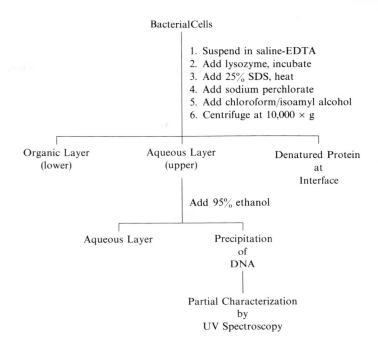

Figure 17.4
Flowchart for isolation of DNA from bacterial cells.

used, but modifications in the procedure may have to be made as described by Marmur. The isolation method used here was first described in the literature over 30 years ago; however, it is still one of the best for extracting larger quantities (mg) of DNA. The isolated DNA is usually recovered in relatively good yield (about 50%) and there is only limited shearing of the product.

After extraction of the DNA, the product will be analyzed by spectral measurement in the ultraviolet range. UV measurements will be made to obtain an estimate of DNA purity and to study thermal denaturation.

The isolation and analysis of DNA may be completed during a 3-hour laboratory period. If more time is available, additional extractions by chloroform–isoamyl alcohol may be completed to remove contaminating protein and achieve a highly purified form of DNA. The DNA product from this experiment may be used for further studies in Experiment 18.

II. Materials and Supplies

Bacterial cells, wet packed, 2 to 3 g. Use *B. subtilus, E. coli,* or *C. welchii.* If lyophilized cells are available, use 0.5 g.

Saline-EDTA, 0.15 *M* NaCl plus 0.1 *M* ethylenediaminetetraacetate, pH 8

Sodium dodecyl sulfate (SDS), 25% in water

Lysozyme solution, 10 mg/mL in water

Sodium perchlorate, 5 M

Chloroform–isoamyl alcohol, 24:1

95% ethanol

Saline-citrate, 0.15 M NaCl plus 0.015 M citrate, pH 7.0

Quartz cuvettes

Centrifuges, one capable of speeds up to 10,000 rpm and a clinical centrifuge

Spectrophotometer for UV measurements

Magnetic stirrer

Water bath at 60°C

Water bath at 90°C

III. Experimental Procedure (Marmur, 1961, 1963)

A. *Isolation of DNA*

 Caution

> Experimental work with bacterial cells presents a potential biohazard. Some strains of the bacteria recommended for use may cause gastroenteritis (abdominal pain and diarrhea). Do not pipet any solutions by mouth. Wash your hands well with hot water and soap before eating, drinking, or smoking.
>
> Ethanol is volatile and flammable. There should be no flames in the laboratory during this experiment.

Suspend 2 to 3 g of wet packed bacterial cells or 0.5 g of lyophilized cells in 25 mL of saline-EDTA solution in a 125-mL Erlenmeyer flask. Add 1.0 mL of the lysozyme solution (10 mg of lysozyme) and incubate the mixture at 37°C for 30 to 45 min. To bring about complete cell lysis, add 2.0 mL of 25% SDS solution and heat the mixture in a water bath (60°C) for 10 minutes. To avoid excessive foaming, stir the mixture very gently. As the nucleic acid is released from the lysed cells, the solution will show an increase in viscosity and a decrease in cloudiness.

After the heating period, allow the mixture to cool to room temperature or cool the flask in cold running water. Add 9.0 mL of sodium perchlorate solution (5 M), which brings the final salt concentration to about 1 M. Mix well, but gently, and then add a volume of chloroform–isoamyl alcohol (24:1) equal to the total volume of the extraction mixture (40 to 45 mL). Shake the solution in a separatory funnel, a tightly stoppered flask, or a screwcap centrifuge tube for 20 to 30 min. Then, centrifuge the emulsion for 5 minutes at 10,000 × g. Three layers should be visible: a bottom organic phase, a middle band of denatured protein at the interface, and an upper aqueous layer containing the nucleic acids. Carefully transfer the aqueous layer to a 125-mL beaker or large test tube. Precipitate the nucleic acids by carefully layering about 2 volumes of 95% ethanol over the aqueous phase. The ethanol should be poured down the side of the beaker or tube using a glass stirring rod. Mix the two layers very gently with a circular motion of the stirring rod and "spool" all the fibers of nucleic acid onto the rod. Press the spooled DNA against the inside of the glass vessel to remove solvent. (If the nucleic acid does not form fibrous threads upon addition of ethanol, centrifuge the mixture for 10 min at 2000 to 3000 rpm in a clinical centrifuge. Transfer the sediment to about 10 mL of saline-EDTA solution and reprecipitate with 95% ethanol.)

If desired, the isolated nucleic acid sample may be treated again with chloroform–isoamyl alcohol for further deproteinization. However, if time does not allow this, dissolve the spooled nucleic acid in 10 to 15 mL of saline-citrate. Save the DNA solution for part B and Experiment 18.

B. Ultraviolet Measurement and Denaturation of Isolated DNA

Dilute a 0.5-mL aliquot of the DNA solution with 4.5 mL of saline-citrate. Transfer some of the solution to a quartz cuvette and determine the A_{260}, using a cuvette containing saline-citrate as reference. If the A_{260} is greater than 1, quantitatively dilute the sample until the absorbance reading is between 0.5 and 1.0. Determine and record the A_{280} on the same DNA sample.

Prepare 10 mL of a DNA solution in standard saline buffer at a DNA concentration of about 20 μg/mL. Measure and record the A_{260}. It should be around 0.4 absorbance units. Transfer 3.0 mL of the DNA solution into each of three test tubes. Place a marble over the top of each tube. Maintain one tube at room temperature and place the other two in a 90°C water bath for 15 minutes. After the incubation period, remove the tubes. Quick-cool one heated tube in an ice bath and allow the other heated tube to cool slowly to room temperature over the period of about 1 hour. Measure and record final A_{260} readings on each of the three tubes.

IV. Analysis of Results

A. Isolation of DNA

Write all observations of your nucleic acid preparation in your notebook. Explain the purpose of each step and the role played by each reagent. Speculate on any damage that may have altered the molecular structure of the DNA.

B. Ultraviolet Measurement and Denaturation of Isolated DNA

Calculate the A_{260}/A_{280} ratio. Extensively purified DNA has a ratio of about 1.9. Comment on the protein content of your isolated nucleic acid. Calculate the DNA concentration (μg/mL) in the saline-citrate solution and determine the total yield of isolated DNA. Assume the $E_{260}^{1\%}$ of native DNA is about 200. Information available in Chapter 2, Section E may also be used to calculate the concentration of DNA.

Calculate the ratio $A_{260(T)}/A_{260(25°)}$ for each of the three tubes. Why are the three ratios not the same? Explain the experimental observation in terms of the molecular structure of DNA. What was the percentage increase in the overall hyperchromic effect? Give a qualitative answer regarding the purity of the DNA sample you used.

V. Questions and Problems

1. Why does the solution become more viscous during lysozyme-SDS treatment?
▶ 2. How could RNA be removed from the nucleic acid preparation?
▶ 3. How does SDS help disrupt the cell membrane?
▶ 4. If extremely high purity DNA is desired, what further steps could be followed in addition to those described here?
▶ 5. A solution of purified DNA gave an A_{260} of 0.55 when measured in a quartz cuvette of 1-cm path length. What is the concentration of DNA in μg/mL?
▶ 6. What techniques could be used to estimate the molecular weight of the isolated DNA?
▶ 7. How could you determine if the DNA were damaged during the isolation procedure?
▶ 8. The concentration of a purified DNA solution is 35 μg/mL. Predict the value of A_{260}.
▶ 9. Predict and explain the action of each of the following conditions on the T_m of native DNA.

   (a) Measure T_m in pH 12 buffer.
   (b) Measure T_m in distilled water.
   (c) Measure T_m in 50% methanol-water.
   (d) Measure T_m in standard saline-citrate solution containing sodium dodecyl sulfate.
▶ 10. A sample of highly purified DNA gave an A_{260} at 25°C of 0.45. The sample was divided into two samples and each was treated as follows:

Sample I	Sample II
1. Heat to 100°C $A_{260(100°C)} = 0.65$	1. Heat to 100°C $A_{260(100°C)} = 0.65$
2. Quick-cool in ice water $A_{260(25°C)} = 0.55$	2. Very slow cool to 25°C $A_{260(25°C)} = 0.46$

Explain why different final A_{260} readings at 25°C are obtained.

▶ 11. A sample of DNA was chemically degraded and found to have an AT content of 60%. Estimate the T_m for the DNA sample.

▶ 12. A sample of purified DNA was incubated with deoxyribonuclease at 37°C. An aliquot was removed from the reaction mixture every minute for 5 minutes and the A_{260} recorded. The following data were obtained.

Time (min)	A_{260}
0	0.60
1	0.64
2	0.67
3	0.70
4	0.72
5	0.73

Describe the action of DNAase on DNA and explain the increase in A_{260}.

VI. References

B. Alberts, D. Bray, J. Lewis, M. Raff, K. Roberts, and J. Watson, *Molecular Biology of the Cell* (1989), Garland (New York), pp. 98–111. Biological approach to the structure and function of DNA.

G. Blackburn and M. Gait, *Nucleic Acids in Chemistry and Biology* (1991), Oxford University Press (New York). An excellent up-to-date book on general aspects of nucleic acids.

M. Mandel and J. Marmur, in *Methods in Enzymology*, Vol. XIIB, L. Grossman and K. Moldave, Editors (1968), Academic Press (New York), pp. 195–206. "Use of Ultraviolet Absorbance-Temperature Profile for Determining the Guanine plus Cytosine Content of DNA."

J. Marmur, in *Methods in Enzymology*, Vol. VI, S. Colowick and N. Kaplan, Editors (1963), Academic Press (New York), pp. 726–738. "A Procedure for the Isolation of Deoxyribonucleic Acid from Microorganisms."

J. Marmur, *J. Mol. Biol.* **3**, 208 (1961). "A Procedure for the Isolation of Deoxyribonucleic Acid from Microorganisms."

C. Mathews and K. van Holde, *Biochemistry* (1990), Benjamin/Cummings (Redwood City, CA), pp. 91–132. An introduction to the structure and function of DNA.

J. Rawn, *Biochemistry* (1989), Neil Patterson Publishers, (Burlington, NC), pp. 665–688; 693–701. The structure and function of DNA.

J. Sambrook, E. Fritsch, and T. Maniatis, *Molecular Cloning: A Laboratory Manual,* 2nd ed. (1989), Cold Spring Harbor Laboratory (Cold Spring Harbor, NY), Vols. I, II, and III. A complete guide to laboratory procedures in molecular biology.

L. Stryer, *Biochemistry,* 3rd ed. (1988), W. H. Freeman (New York), pp. 71–90. An introductory chapter on the structure and function of DNA.

D. Voet and J. Voet, *Biochemistry* (1990), John Wiley & Sons (New York), pp. 786–824. Structure and function of DNA.

W. Wood, J. Wilson, R. Benbow, and L. Hood, *Biochemistry, a Problems Approach,* 2nd ed. (1981), Benjamin/Cummings (Menlo Park, CA), pp. 315–343. A short discussion of DNA structure with excellent problems.

EXPERIMENT 18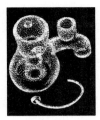

Characterization of Nucleic Acids by Fluorescence of Bound Ethidium Bromide

■ **Recommended Reading**

Chapter 5, Sections A and B; Experiments 17, 19, 20, and 21.

■ **Synopsis**

The central role of the nucleic acids in biochemistry has prompted the development of innumerable techniques for their physical, chemical, and biological characterization. The binding of fluorescent dyes to double-helix DNA forms the basis for many of the analytical methods. In this experiment, ethidium bromide interaction with duplex DNA, which results in enhanced fluorescence of the dye, is applied to (1) measurement of the concentration of DNA in solution and (2) ligand (polyamine) binding to DNA.

I. Introduction and Theory

The nucleic acids are among the most complex molecules that you will encounter in your biochemical studies. When the dynamic role that is played by DNA in the life of a cell is realized, the complexity is understandable. It is difficult to comprehend all the structural characteristics that are inherent in the DNA molecules, but most biochemists are familiar with and accept the double-helix model of Watson and Crick. In fact, the discovery of the double-helical structure of DNA is one of the most significant breakthroughs in our understanding of the chemistry of life. Experiment 17 introduced you to the basic structural characteristics of the DNA molecule and to the forces that help establish the complementary interactions between the two polynucleotide strands. In this experiment, we will be concerned with a more dynamic property of DNA, the binding of natural polyamines.

Circular Supercoiled DNA

Most diagrams of DNA illustrate the linear nature of the double helix and imply that DNA molecules are extended rods with two ends. The chromosomal and plasmid DNA of many microorganisms, however, is a closed circle, a result of covalent joining of the two ends of a linear double helix. Circular DNA has been discovered in Simian virus 40 (SV40), bacteriophages (for example, ϕX174), bacteria (*E. coli,* for example), and several species of animals.

Closed, circular, duplex DNA has a unique structural feature. It is found twisted into a new configuration, **supercoiled DNA**. Other terms used to define this form are **superhelical DNA** and **supertwisted DNA**. Figure E18.1 illustrates both relaxed (A) and supercoiled (B) DNA. Form (B) is not simply a coiling of the completed, circular DNA but is the result of extra twisting in the linear duplex form just before covalent joining of the two strands. The supercoiled form is less stable than the relaxed form; however, there is sufficient evidence to conclude that supercoiled DNA may exist *in vivo*. The most convincing evidence is the isolation and characterization of proteins that catalyze the interconversion of relaxed and supercoiled DNA. These catalytic proteins have received various names in the literature, including DNA relaxing enzyme (Keller, 1975), ω protein (Wang, 1971), and the nicking-closing enzyme (Champoux and Dulbecco, 1972); they are now classified as **topoisomerases** because they catalyze changes in the topology of DNA.

The origin of supercoiling is not completely understood, but it may be caused by the presence of more than the standard number of bases per unit length of a DNA helix. Supercoiling may have biological significance, in that the DNA molecule becomes more compact; hence, is

Figure E18.1.
Structures of closed, circular duplex DNA. (A) Relaxed. (B) Supercoiled.

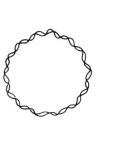

(A)　　　　　　　　(B)

more easily stored in the cell. Supercoiling may also play a role in DNA replication. Supercoiled DNA has less intrinsic viscosity than relaxed or linear DNA; hence, it sediments at a higher rate in the ultracentrifuge and can be readily separated and characterized by this technique.

Methods for the Characterization of DNA

A complete understanding of the biochemical functions of DNA requires a clear picture of its structural and physical characteristics. Nature has built into the genetic molecules many features that make them easy to characterize. (1) The purine-pyrimidine bases absorb light strongly at 260 nm, and they also display weak fluorescence spectra. Minor physical changes in DNA structure modify the extent of absorption and fluorescence. Hence, the bases are natural spectrophotometric probes for the study of DNA structure. (2) The inherent polarization properties of the double helix make optical rotation and circular dichroism measurements convenient. (3) The presence of unique phosphoester linkages in the polynucleotide backbone opens the door for selective chemical cleavage using specific enzymes. (4) Finally, the compact, close-packed nature of DNA leads to molecules of high density, compared to proteins and other biomolecules.

With so many natural characteristics that lend themselves to physical measurement, how does one choose a technique for evaluating the structural properties of DNA? The most important criteria to use in choosing an assay are sensitivity, accuracy, and convenience. Most of the aforementioned methods fit the criteria, except that some require sophisticated techniques, specialized instrumentation, and long periods of time.

ETHIDIUM FLUORESCENCE ASSAYS OF DNA

Fluorescence assays are considered among the most convenient, sensitive, and versatile of all techniques. However, as previously mentioned, the purine-pyrimidine bases yield only weak fluorescence spectra. Le Pecq and Paoletti (1967) showed that the fluorescence of a dye, ethidium bromide, is enhanced about 25-fold when it interacts with DNA. The ethidium bromide, which is a relatively small planar molecule (Figure E18.2), binds to DNA by insertion between stacked base pairs (intercalation). The process of intercalation is especially significant for aromatic dyes, antibiotics, and other drugs. Some dyes, when intercalated into DNA, show an enhanced fluorescence that can be used to detect DNA molecules after gel electrophoresis measurements (see Experiment 21 and Chapter 4) and to characterize the physical structure of DNA. Intercalation of dyes into a supercoiled DNA molecule causes unwinding of the superhelix; if a large excess of dye is used, the DNA begins to twist in the opposite direction (negative to positive coiling) (Bauer and Vinograd, 1968).

Figure E18.2.
Structure of the fluorescent intercalation dye ethidium bromide.

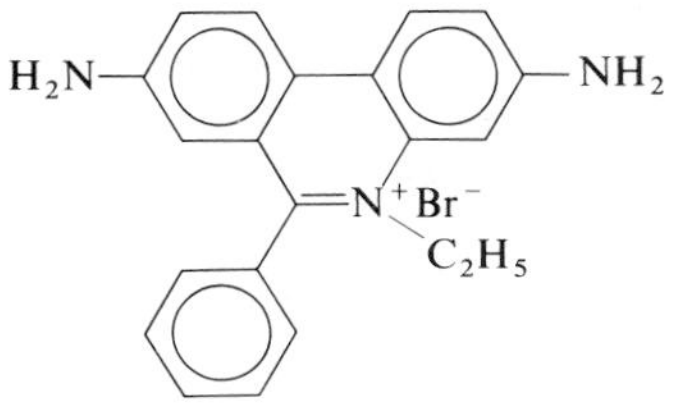

The enhanced fluorescence of ethidium bromide–DNA complexes has been applied to the measurement of structural, physicochemical, and biochemical properties of DNA and RNA (Morgan et al., 1979 a, b). In this experiment two measurements on DNA will be described, (1) determination of the concentration of DNA and RNA in solution and (2) polyamine binding to DNA.

MEASUREMENT OF THE CONCENTRATION OF DNA AND RNA IN SOLUTION

Most assays measure both double-stranded and single-stranded DNA. Ethidium interaction with single-stranded DNA does not lead to increased fluorescence, so duplex DNA can be quantified in the presence of dissociated DNA.

Solutions of purified DNA are commonly contaminated with RNA. Since single-stranded RNA can form hairpin loops with base pairing and duplex formation (as in tRNA), ethidium also binds with enhanced fluorescence to these duplex regions of RNA. Addition of ribonuclease A results in digestion of RNA and loss of fluorescence due to ethidium binding of RNA. The amount of fluorescence lost is proportional to the concentration of RNA. The ethidium fluorescence remaining after ribonuclease treatment is directly proportional to the concentration of duplex DNA (Equation E18.1).

$$F_{\text{total}} = F_{\text{DNA}} + F_{\text{RNA}} \qquad \text{(Equation E18.1)}$$

where

F = fluorescence yield due to each type of nucleic acid

When the F_{RNA} term is reduced to zero, the total fluorescence is a direct measure of DNA concentration. The actual concentration of DNA in solution can be calculated by use of a standard solution of DNA or from a standard curve.

POLYAMINE BINDING TO DNA

The site of action of many antibiotics is the DNA molecule of an invading organism. In fact, the physiological action of many drugs depends on interaction with DNA, which often leads to an intercalation complex. It is possible to demonstrate binding and to estimate the association constants for ligand-DNA complexes using the ethidium assay (Morgan et al., 1979 a, b). A solution of DNA treated with excess ethidium bromide gives a maximum fluorescence yield. If a ligand is added that can compete with ethidium for the intercalation sites on DNA, the ethidium dissociates from the DNA. The fluorescence yield of ethidium decreases at a rate that is directly proportional to the concentration of added ligand. That is, the greater the concentration of bound ligand, the greater the extent of ethidium dissociation and the greater the decrease in ethidium fluorescence. Of course, it must be shown that the ligand does not fluoresce under the conditions of the experiment and that it does not interfere with ethidium fluorescence. An estimate of the binding constant can be obtained from the binding curve of % fluorescence vs. ligand concentration. The ligand concentration that causes a 50% decrease in ethidium concentration is approximately inversely proportional to the binding constant, k_a.

The binding of both synthetic drugs and naturally occurring molecules to DNA is of biochemical significance. There is particularly strong interest in the interactions between polyamines and DNA. Four polyamines, 1,4-diaminobutane (putrescine), 1,5-diaminopentane (cadaverine), spermidine, and spermine (Figure E18.3), are metabolic products in many cells and are found in relatively high concentrations. Putrescine and cadaverine are derived from the decarboxylation of ornithine and lysine, respectively. The other two polyamines, spermidine and spermine, are synthesized in rather complex processes requiring S-adenosylmethionine. The biochemical functions of these unusual compounds are not completely clear. Polyamines are cationic, so they bind strongly to nucleic acids. They interact by forming salt bridges between their positively charged amino groups and negatively charged phosphate anions of the nucleotides. In fact, this interaction may stabilize supercoiled DNA. The polyamines may help control the progression of cells through the cell cycle by regulating the rates of nucleic acid and protein synthesis. The ethidium assay is particularly convenient for demonstrating polyamine binding to DNA because the analysis is rapid; also, since the ligands are not aromatic, they display no measurable interference of ethidium fluorescence. However, the measurements lead only to approximate values for the binding constants. For more precise measurements of binding, Scatchard analysis as described in Experiment 5 must be completed.

Overview of the Experiment

This experiment illustrates two of the many applications of the ethidium fluorescence assay to studies of DNA. The concentrations of DNA solu-

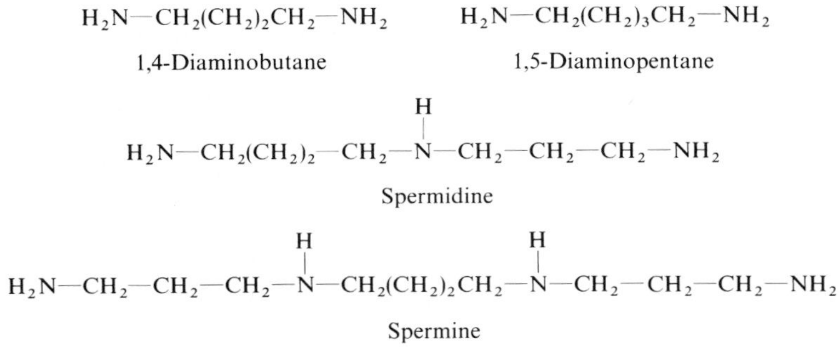

Figure E18.3.
Structure of several significant polyamines.

$$H_2N-CH_2(CH_2)_2CH_2-NH_2 \qquad H_2N-CH_2(CH_2)_3CH_2-NH_2$$

1,4-Diaminobutane   1,5-Diaminopentane

$$H_2N-CH_2(CH_2)_2-CH_2-\overset{\displaystyle H}{\underset{\displaystyle |}{N}}-CH_2-CH_2-CH_2-NH_2$$

Spermidine

$$H_2N-CH_2-CH_2-CH_2-\overset{\displaystyle H}{\underset{\displaystyle |}{N}}-CH_2(CH_2)_2CH_2-\overset{\displaystyle H}{\underset{\displaystyle |}{N}}-CH_2-CH_2-CH_2-NH_2$$

Spermine

tions from Experiment 17 may be determined, or unknown DNA solutions may be prepared. The average time required for part A, measurement of DNA concentration, is 30 to 45 minutes; that for Part B, binding of polyamines, is $1\frac{1}{2}$ to 2 hours.

Several additional experiments can be completed if time allows. Some of the more interesting are (1) kinetics of DNA reassociation (Britten and Kohne, 1968), (2) measurement of the percentage of supercoiled DNA in a sample (Morgan et al., 1979a; Kowalski, 1979), (3) the effect of covalent cross-linking on reassociation kinetics (Morgan et al., 1979a), (4) assay of DNA in whole cells (Morgan et al., 1979b), and (5) various enzyme assays including DNA polymerase, nucleases, and topoisomerases (Morgan et al., 1979b).

II. Materials and Supplies

DNA standard solution, 50 μg/mL in 0.01 M Tris-HCl and 0.05 M NaCl, pH 8.0

Unknown DNA solution, prepared in Tris-HCl buffer

Ethidium bromide solution, pH 8.0, 5 mM Tris-HCl, 0.5 μg/mL-ethidium bromide, and 0.5 mM EDTA

Ribonuclease A, 20 mg/mL in 0.10 M Tris-HCl, pH 8

Constant-temperature bath at 37°C

Spermine solution, $1 \times 10^{-4}M$ in Tris-HCl buffer, pH 8.0

Spectrofluorimeter and fluorescence cuvettes:

Excitation wavelength	525 nm
Emission wavelength	590 nm
Slit width	20 nm
Scale	$\times$ 100

III. Experimental Procedure

A relative fluorescence intensity scale will be defined in part A of the experiment. The cuvette holder should be provided with a water jacket to maintain the temperature within $\pm$ 0.5°. It is recommended that you use the same cuvette for all measurements unless a pair is available that is very well matched. Fluorescence measurements are temperature dependent. It is necessary to maintain all reagents, except the ribonuclease solution, at 37°C, the same as the setting for the cuvette holder in the fluorimeter.

⚠ Caution

Ethidium bromide is a mutagen and a potential carcinogen. Always wear gloves and a face mask while weighing or handling the pure chemical. Gloves should always be worn while using solutions of the dye. Do not allow the solutions of ethidium bromide to come into contact with your skin. **Do not mouth pipet any ethidium bromide solutions**. Save all assay mixtures and excess ethidium bromide solutions after measurements are taken, and pour them into a container marked "Waste Ethidium."

A. Concentration of DNA Solutions

Turn on the fluorimeter lamp and allow it to warm up for 15 minutes. Set the fluorescence scale for the standard DNA in the following manner. Pipet 3 mL of pH 8 ethidium bromide buffer solution into a fluorescence cuvette. With a micropipet, transfer 15 μL of the standard DNA into the cuvette and mix well. Place the cuvette in the spectrofluorimeter and adjust the fluorescence intensity to "100," F_{std}. This is, of course, an arbitrary setting. The actual setting that is used is not especially important, but you must know what the setting is. The number 100 is convenient for comparison purposes and is customary in fluorescence measurements. Pour the contents of the cuvette into the waste ethidium container provided by your instructor. Clean the cuvette carefully with warm water, rinse several times with distilled water, and dry.

Again, transfer 3 mL of the pH 8 ethidium bromide buffer solution into the cuvette. Add 15 μL of the unknown DNA solution. Mix well and place in the spectrofluorimeter. Record the fluorescence intensity, F_{total}. The measured fluorescence is due to both DNA and RNA that may be present in the isolated DNA. The amount of fluorescence due to ethidium-RNA can be eliminated by digesting the RNA with ribonuclease. To the cuvette containing pH 8 ethidium bromide and unknown DNA, add 2 μL of ribonuclease solution. Mix well and incubate the cuvette at 37°C for 20 minutes in a water bath. After incubation, measure the fluorescence intensity, F_{total}. Compare this with the original measurement taken on the unknown DNA (F_{total}). Is RNA present in your DNA sample? Calculate the concentration of DNA in your unknown solution as described in the Analysis of Results. The possibility exists that the solution of ribonuclease may contain a fluorescent impurity. What control experiment should be done to correct this situation?

B. Binding of Spermine to DNA

Standardize the spectrofluorimeter in the following way. Pipet 2 mL of the pH 8 ethidium bromide buffer into a cuvette. Add 1.0 mL of Tris-HCl buffer and 20 μL of standard DNA solution. Mix and place in the fluorimeter. Adjust the fluorescence intensity to 100. Clean the cuvette as described in part A and repeat the assay using various concentrations of spermine. Prepare a table displaying the amount of each component to be added. Four reagents must be in the table: pH 8 ethidium bromide buffer, DNA solution, spermine, and pH 8.0 Tris buffer. Maintain the volume of DNA at 20 μL and ethidium bromide solution at 2.0 mL for all assays. Use 0.05, 0.1, 0.2, 0.4, 0.6, 0.8, and 1.0 mL of spermine in the assays. Remember that the total volume of all constituents in the cuvette must remain constant at 3.02 mL for all the assays. Therefore, the amount of Tris-HCl buffer must change with the amount of spermine added. Prepare each assay separately by adding the proper amount of each component to the cuvette. Mix well and record the fluorescence intensity. Clean the fluorescence cuvette carefully after each assay.

IV. Analysis of Results

A. Concentration of DNA Solutions

The concentration of the unknown DNA solution can be calculated by a simple comparison of the fluorescence intensity obtained from the standard DNA reaction mixture (F_{std}) and the fluorescence of the unknown DNA mixture (Equation E18.2).

$$\blacktriangleright \quad [DNA]_x = \frac{[DNA]_{std} F_{DNA-x}}{F_{DNA-std}} \qquad \text{(Equation E18.2)}$$

where

$[DNA]_x$ = concentration of the unknown solution of DNA in μg/mL

$[DNA]_{std}$ = concentration of the standard solution of DNA in μg/mL

$F_{DNA-std}$ = fluorescence yield of the standard DNA solution

F_{DNA-x} = fluorescence yield of unknown DNA solution after incubation with ribonuclease

To correct for possible fluorescent contamination in the RNAase solution, place 3 mL of pH 8 ethidium bromide buffer in a cuvette. Add 15 μL of Tris-HCl buffer, pH 8, and 2 μL of ribonuclease. Mix well and incubate at 37°C for 20 minutes. Record the fluorescence intensity, if any. If the fluorescence is significant (> 1 intensity unit), subtract from F_{DNA-x} before calculation of unknown DNA. Calculate the concentration of DNA in your sample in units of μg/mL. Was RNA present in your unknown DNA solution?

B. Binding of Spermine to DNA

In Experiment 5, ligand binding to protein was evaluated using Scatchard plots. A reasonably accurate value of the formation constant for the complex, k_a, was determined, but extensive experimentation and data analysis were required. For rapid screening of drug and ligand binding to

DNA, the ethidium assay is especially good. The main disadvantage of this technique is that the value for k_a is just an approximation. However, the convenient and rapid assay will demonstrate the ligands that deserve more detailed binding studies by Scatchard analysis.

Prepare a table of fluorescence reading vs. drug concentration in the cuvette in μM. If the fluorescence of untreated DNA was set to 100 on the fluorimeter, each fluorescence reading taken can be considered a percentage of the maximum possible. Plot the fluorescence reading (y axis) against the spermine concentration (x axis) that caused that particular fluorescence reading. Connect the points with a straight line. The binding constant, k_a, is inversely proportional to the spermine concentration that produces 50% of the fluorescence of untreated DNA.

V. Questions and Problems

1. Outline a procedure for quantitative analysis of DNA in a crude cell extract. Remember that nucleases present in the extract will continue to digest the DNA present and lead to a low result for DNA concentration. What should be done to decrease this error? DNA and RNA both give enhanced fluorescence with ethidium.

▶ 2. Design an ethidium assay for the nicking-closing enzyme, a topoisomerase. (For answer, see Kowalski, 1979.)

▶ 3. Compare the structure of ethidium bromide (Figure E18.2) with those of the polyamines (Figure E18.3). Would you expect the binding of spermine or spermidine to DNA to be competitive with the binding of ethidium? Explain.

▶ 4. Explain how this experiment could be modified to measure the concentration of an RNA solution.

VI. References

U. Bachrach and Y. Heimer, Editors, *The Physiology of Polyamines*, Vols. I and II (1989), CRC Press (Boca Raton, FL). Complete series on the biochemistry of polyamines.

W. Bauer and J. Vinograd, *J. Mol. Biol.* **33**, 141 (1968). "The Interaction of Closed Circular DNA with Intercalative Dyes."

W. Bauer, F. Crick, and J. White, *Sci. Am.* July 1980, pp. 118–133. "Supercoiled DNA."

R. Britten and D. Kohne, *Science* **161**, 529 (1968). "Repeated Sequences in DNA."

J. Champoux and R. Dulbecco, *Proc. Natl. Acad. Sci. USA* **69**, 143 (1972). "An Activity from Mammalian Cells that Untwists Superhelical DNA."

W. Keller, *Proc. Natl. Acad. Sci. USA* **72**, 2550 (1975). "Characterization of Purified DNA-Relaxing Enzyme from Tissue Culture Cells."

D. Kowalski, *Anal. Biochem.* **93**, 346 (1979). "A Procedure for the Quantitation of Relaxed Closed Circular DNA in the Presence of Superhelical DNA: An Improved Fluorometric Assay for Nicking-Closing Enzyme."

J. Le Pecq and C. Paoletti, *J. Mol. Biol.* **27**, 87 (1967). "A Fluorescent Complex Between Ethidium Bromide and Nucleic Acids."

C. Mathews and K. van Holde, *Biochemistry* (1990), Benjamin/Cummings (Redwood City, CA), pp. 91–121, 714–715, 817–859. DNA structure and function and the possible role of polyamines.

A. Morgan, D. Evans, J. Lee, and D. Pulleyblank, *Nucleic Acids Res.* **7**, 571 (1979b). "Review: Ethidium Fluorescence Assay: Part II. Enzymatic Studies and DNA-Protein Interactions."

A. Morgan, J. Lee, D. Pulleyblank, N. Murray, and D. Evans, *Nucleic Acids Res.* **7**, 547 (1979a). "Review: Ethidium Fluorescence Assays. Part 1. Physicochemical Studies."

J. Ramstein and M. Leng, *Biochim. Biophys. Acta* **281**, 18–32 (1972). Interaction between proflavine and DNA.

T. Sakai and S. Cohen, in *Progress in Nucleic Acid Research and Molecular Biology,* Vol. 17, W. Cohn, Editor (1976), Academic Press (New York), pp. 15–41. "Effects of Polyamines on the Structure and Reactivity of t-RNA."

L. Stryer, *Biochemistry,* 3rd ed. (1988), W. H. Freeman (New York), pp. 71–90, 649–685. DNA structure and function.

H. Tabor and C. White Tabor, Editors, *Methods in Enzymology,* Vol. 94 (1983), Academic Press (New York). The entire volume is on polyamines.

D. Voet and J. Voet, *Biochemistry* (1990), John Wiley & Sons (New York), pp. 786–824. Structure and function of DNA.

J. Wang, *J. Mol. Biol.* **55**, 523 (1971). "Interaction Between DNA and an *Escherichia coli* Protein ω."

EXPERIMENT 19

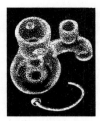

Preparation, Isolation, and Purification of Bacterial Plasmid DNA

■ Recommended Reading

Chapter 7, Sections A, B, and C; Experiments 17, 20, and 21.

■ Synopsis

Extrachromosomal DNA molecules called plasmids are harbored in some strains of *E. coli*. The normal copy number of the plasmids is small, between 20 and 30; however, if these strains of *E. coli* are grown in the presence of chloramphenicol, up to 3000 copies may be replicated per cell. Plasmid DNA has been demonstrated to be a useful vehicle in molecular cloning. This experiment describes a method for the growth of *E. coli* and amplification of the ColE1 plasmids. In addition, the plasmids will be isolated from *E. coli* cells and purified by density gradient ultracentrifugation or ribonuclease treatment.

I. Introduction and Theory

Our knowledge of DNA structure and function has increased at a gradual pace over the past 50 years, but some discoveries have had a special impact on the progress and direction of DNA research. Although DNA was first discovered in cell nuclei in 1869, it was not confirmed as the carrier of genetic information until 1944. This major discovery was closely followed by the announcement of the double-helix model of DNA in 1953. The more recent development of technology that allows manipulation by insertion of "foreign" DNA fragments into the natural, replicating DNA of an organism may well have a greater impact on the direction of DNA research than the previous two discoveries. In the short time since the first construction and replication of plasmid "recombinant DNA," several scientific and medical applications of the new technology have been announced. This new era of "genetic engineering" has captivated the general public and students alike. Some of the predicted achievements in recombinant DNA research have been slow in coming; however, future workers in biochemistry and related fields must be familiar with the principles and techniques of DNA recombination.

Recombinant DNA

DNA recombination or **molecular cloning** consists of the covalent insertion of DNA fragments from one type of cell or organism into the replicating DNA of another type of cell. Many copies of the hybrid DNA may be produced by the progeny of the recipient cells; hence, the DNA molecule is **cloned**. If the inserted fragment is a functional gene carrying the code for a specific protein, many copies of that gene and translated protein will be produced in the host cell. This process has become important for the large-scale production of proteins (insulin, somatostatin, and other hormones) that are of value in medicine and basic science but are difficult and expensive to obtain by other methods.

The basic steps involved in performing a recombinant DNA experiment are listed below and are outlined in Figure E19.1.

1. Select and isolate a natural DNA to serve as the carrier (sometimes called **vehicle** or **vector**) for the foreign DNA.
2. Cleave both strands of the DNA carrier with a restriction endonuclease.
3. Prepare and insert the foreign DNA fragment into the vehicle. This produces a **recombinant** or **hybrid** DNA molecule.
4. Introduce the hybrid DNA into a host organism (usually a bacterial cell), where it can be replicated. This process is called **transformation**.
5. Develop a method for identifying and screening the host cells that have accepted and are replicating the hybrid DNA.

Our current state of knowledge of recombinant DNA is the result of several recent technological advances. The first major breakthrough was the isolation of mutant strains of *E. coli* that are not able to degrade or restrict foreign DNA. These strains are now used as host organisms for the replication of recombinant DNA. The second advance was the development of bacterial extrachromosomal DNA (plasmids) and bacteriophage DNA as cloning vehicles. The final, necessary advance was the development of methods for insertion of the foreign DNA into the natural vehicle. The discovery of **restriction endonucleases,** enzymes that catalyze the hydrolytic cleavage of DNA at selective sites, provided a gentle

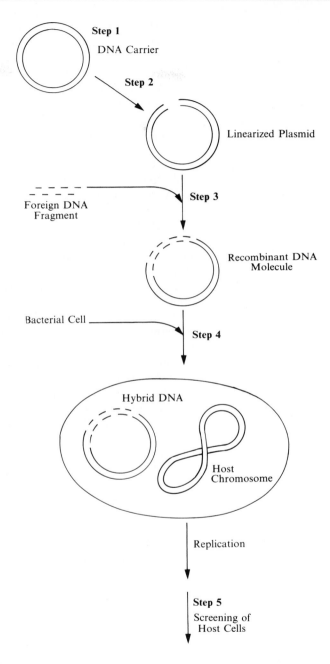

Figure E19.1
Outline of a typical procedure for a DNA recombination experiment.

and specific method for opening (linearizing) circular vehicles. (See Experiment 21.) Methods for covalent joining of the foreign DNA to the vehicle ends and final closure of the circular hybrid plasmid were then developed. The enzyme **DNA ligase,** which can catalyze the ATP-dependent formation of phosphodiester linkages at the insertion sites, is used for final closure. (See Figure E19.2.)

Plasmids and Other Cloning Vehicles

Three types of molecular vehicles are in common use for the preparation and replication of hybrid DNA in *E. coli* cells: plasmids, bacteriophage DNA, and simian virus 40 (SV 40).

PLASMIDS

Many bacterial cells contain self-replicating, extrachromosomal DNA molecules called **plasmids.** This form of DNA is closed circular, double-stranded, and much smaller than chromosomal DNA; its molecular weight ranges from 2×10^6 to 20×10^6, which corresponds to between 3000 and 30,000 base pairs. Bacterial plasmids normally contain genetic information for the translation of proteins that confer a specialized and sometimes protective characteristic (phenotype) on the organism. Examples of these characteristics are enzyme systems necessary for the production of antibiotics, enzymes that degrade antibiotics, and enzymes for the production of toxins. Plasmids are replicated in the cell by one of two possible modes. **Stringent replicated** plasmids are present in only a few copies and **relaxed replicated** plasmids are present in many copies, sometimes up to 200. In addition, some relaxed plasmids continue to be

Figure E19.2
Insertion of a DNA fragment into a linearized plasmid using cohesive ends and DNA ligase.

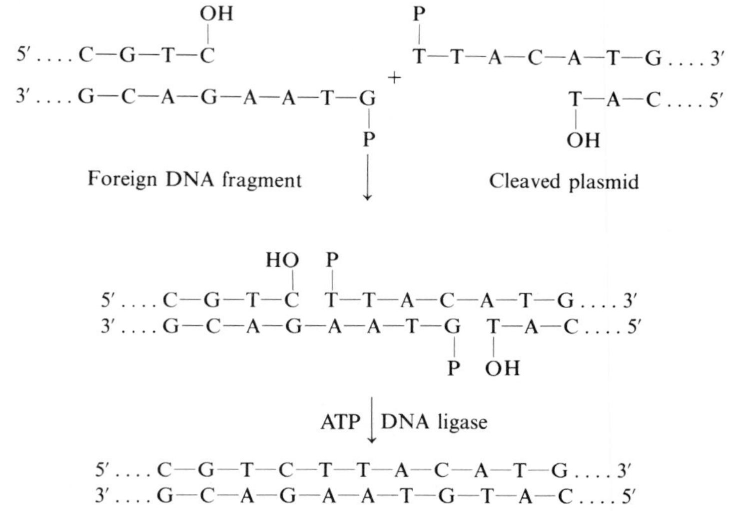

$$
\begin{array}{l}
\overset{\displaystyle \text{OH}}{\mid} \overset{\displaystyle \text{P}}{\mid} \\
5'\ldots C-G-T-C T-T-A-C-A-T-G\ldots 3' \\
 + \\
3'\ldots G-C-A-G-A-A-T-G T-A-C\ldots 5' \\
\underset{\displaystyle \text{P}}{\mid} \underset{\displaystyle \text{OH}}{\mid}
\end{array}
$$

Foreign DNA fragment $\qquad\downarrow\qquad$ Cleaved plasmid

$$
\begin{array}{l}
\overset{\displaystyle \text{HO}}{\mid}\ \overset{\displaystyle \text{P}}{\mid} \\
5'\ldots C-G-T-C\ \ T-T-A-C-A-T-G\ldots 3' \\
3'\ldots G-C-A-G-A-A-T-G\ \ T-A-C\ldots 5' \\
\underset{\displaystyle \text{P}}{\mid}\ \underset{\displaystyle \text{OH}}{\mid}
\end{array}
$$

ATP $\Big\downarrow$ DNA ligase

$$
\begin{array}{l}
5'\ldots C-G-T-C-T-T-A-C-A-T-G\ldots 3' \\
3'\ldots G-C-A-G-A-A-T-G-T-A-C\ldots 5'
\end{array}
$$

produced even after the antibiotic chloramphenicol is used to inhibit chromosomal DNA synthesis in the host cell. Under these conditions, many copies of the plasmid DNA may be produced (up to 2000 or 3000) and may accumulate to 30 to 40% of the total cellular DNA.

The ideal plasmid cloning vehicle has the following properties:

1. The plasmid should replicate in a relaxed fashion so that many copies are produced.

2. The plasmid should be small; then it is easier to separate from the larger chromosomal DNA, easier to handle without physical damage, and probably contains very few sites for attack by restriction nucleases.

3. The plasmid should contain identifiable markers so that it is possible to screen progeny for the presence of the plasmid. At least two selective markers are desirable, a primary one to confirm the presence of the plasmid and a secondary marker to confirm the insertion of foreign DNA. Resistance to antibiotics is a convenient type of marker.

4. The plasmid should have only one cleavage site for a restriction endonuclease. This provides only two "ends" to which the foreign DNA can be attached. Ideally, the single restriction site should be within a gene, so that insertion of the foreign DNA will inactivate the gene (called insertional marker inactivation).

Among the most widely used *E. coli* plasmids are derivatives of the replicon plasmid ColE1. This plasmid carries a resistance gene against the antibiotic colicin E. The plasmid is under relaxed control, and up to 3000 copies may be produced when the proper *E. coli* strain is grown in the presence of chloramphenicol. One especially useful derivative plasmid of ColE1 is pBR322. It has all the properties previously outlined; in addition, its nucleotide sequence of 4363 base pairs is known, and it contains several different restriction endonuclease cleavage sites where foreign DNA can be inserted. For example, pBR322 has a single restriction site for the restriction enzyme *Eco*RI. *In vitro* insertion of a foreign DNA fragment into the *Eco*RI cleavage site and incorporation into a host cell (transformation) lead to immunity of the host cell to colicin E1, but the cell is unable to produce colicin.

BACTERIOPHAGE DNA

Two types of phage DNA have been widely used as cloning vehicles, λ phage and M13 phage DNA. λ DNA is a double-stranded molecule with approximately 50,000 base pairs. It has several advantages as a cloning vehicle:

1. Millions of copies of recombinant phage DNA can be replicated in the host cell.

2. The recombinant phage DNA may be efficiently packaged in the phage particle, which can be used to infect the host bacteria.

3. Recombinant phage DNA is easily screened for identification purposes.

4. λ phage DNA is larger than plasmid DNA; therefore, it is especially useful for insertion of larger fragments of eukaryotic DNA.

M13 phage DNA is single-stranded and contains approximately 6500 nucleotide bases. It is especially useful when DNA sequencing by the Sanger method is to be carried out on the recombinant DNA.

SIMIAN VIRUS 40

The cloning vehicle most often used to introduce recombinant DNA in eukaryotic cells is SV40. Its capacity is rather small; it can accept foreign DNA fragments of only about 4300 base pairs. In contrast, plasmids can be used effectively to carry foreign DNA of up to 10,000 base pairs.

New cloning vehicles are continually being developed and studied. However, at present, plasmids are the most versatile. Methods for the production, isolation, purification, and analysis of plasmid DNA will be described in Experiments 19, 20, and 21.

Growth of Bacteria

Most plasmids currently used as cloning vehicles are of *E. coli* origin. Strains of *E. coli* are relatively easy to grow and maintain. The rate of growth is exponential and can be monitored by measuring the absorbance of a culture sample at 600 nm. One absorbance unit corresponds to a cell density of approximately 8×10^8 cells/mL.

Some *E. coli* strains that harbor the ColE1 plasmids are RR1, HB101, GM48, 294, SK1592, JC411Thy$^-$/ColE1, and CR34/ColE1. The typical procedure for growth and amplification of plasmids is, first, to establish the cells in normal medium for several hours. An aliquot of this culture is then used to inoculate medium containing the appropriate antibiotic. After overnight growth, a new portion of medium containing the antibiotic is inoculated with an aliquot of overnight culture. After the culture has been firmly established, a solution of chloramphenicol is added to inhibit chromosomal DNA synthesis. The ColE1 plasmids continue to replicate. The culture is then incubated for 12 to 18 hours and harvested by centrifugation.

Isolation of Plasmid DNA

The presence of ColE1 and other plasmids in a bacterial cell may be confirmed by genetic screening of antibiotic resistance. However, it is sometimes necessary to isolate plasmid DNA for further characterization and manipulation. Isolation and purification of plasmids are usually carried out for one of the following reasons:

1. Construction of new recombinant plasmids.
2. Analysis of molecular weight by agarose gel electrophoresis.
3. Electrophoretic analyses of restriction enzyme digests (see Experiment 21).
4. Construction of a restriction enzyme map (see Experiment 21).
5. Sequence analysis of nucleotides by the Sanger or Maxam-Gilbert method.
6. Hybridization experiments.

Several methods for isolating plasmid DNA have been developed; some lead to a more highly purified product than others. All isolation methods have the same objective, separation of plasmid DNA from chromosomal DNA. Plasmid DNA has two major structural differences from chromosomal DNA:

1. Plasmid DNA is almost always extracted in a covalently closed circular configuration, whereas isolated chromosomal DNA usually consists of sheared linear fragments.
2. ColE1 plasmids and other potential vehicles are much smaller than chromosomal DNA.

The structural differences cause physicochemical differences that can be exploited to separate the two types of molecules. Methods for isolating plasmid DNA fall into three major categories:

1. Methods that rely on specific interaction between plasmid DNA and a solid support. Examples are adsorption to nitrocellulose microfilters and hydroxyapatite columns.
2. Methods that cause selective precipitation of chromosomal DNA by various agents. These methods exploit the relative resistance of covalently closed circular DNA to extremes of pH, temperature, or other denaturing agents.
3. Methods based on differences in sedimentation behavior between the two types of DNA. This is the approach of choice if highly purified plasmid DNA is required.

An isolation procedure based on a method in category 2 will be described in this experiment. This procedure yields plasmid DNA that is sufficiently pure for digestion by restriction enzymes and analysis by agarose electrophoresis as described in Experiment 21. Two optional methods for further purifying the plasmid are also described: (1) purification on a cesium chloride gradient and (2) removal of contaminating RNA by ribonuclease treatment.

Separation of Plasmid DNA by Boiling (Holmes and Quigley, 1981)

The total cellular DNA must first be released by lysis of the bacterial cells. This is brought about by incubation with the enzyme lysozyme in the presence of reagents that inhibit nucleases (see Experiment 17). Chromosomal DNA is then separated from the plasmid DNA by boiling the lysis mixture for a brief period, followed by centrifugation. In contrast to closed circular plasmid DNA, linear chromosomal DNA becomes irreversibly denatured by heating and forms an insoluble gel, which sediments during centrifugation. Even though plasmid DNA may become partially denatured during boiling, the closed circular double helix reforms on cooling. The boiling serves a second purpose, that of denaturing deoxyribonucleases and other proteins.

Purification of Plasmid DNA by Ultracentrifugation or Ribonuclease Treatment

The plasmids obtained as described above are sufficiently pure for some studies; however, they are contaminated by RNA and a small amount of linear DNA. The RNA may be removed by treatment of the lysate with ribonuclease, but ultracentrifugation is more often the method of choice.

Ethidium bromide and other intercalating dyes bind to double-stranded nucleic acids, causing partial unwinding of the helix and other topologic changes. Closed circular DNA binds less dye than linear DNA, which leads to a difference in the buoyant density of plasmid DNA–dye vs. linear DNA–dye. The density of the DNA–dye complex is related inversely to the amount of bound dye; therefore, the plasmid DNA–dye complex sediments to a region of higher density on a cesium chloride density gradient. Figure E19.3 shows the results of density gradient ultracentrifugation of a cleared lysate. Two red-orange bands are observed; the upper corresponds to linear DNA, the lower to plasmid DNA. Ribonucleic acids sediment as a pellet in the tube, and any contaminating protein floats to the top of the gradient. The two DNA bands can be collected by piercing the tube with a syringe needle and dripping the gradient. The ethidium bromide dye is removed from the plasmid solution before restriction enzyme digestion by an ion-exchange column or extraction with an organic solvent.

If an ultracentrifuge is not available and highly purified plasmid DNA is desired, contaminating RNA may be removed by treatment of the lysate with ribonuclease as described in part D.

Overview of the Experiment

This experiment is divided into three parts: growth of bacteria harboring a plasmid, isolation of plasmid DNA, and purification by density gradient ultracentrifugation or RNAase treatment. Part A of this experiment in-

Figure E19.3
Ultracentrifugation of plasmid DNA on a cesium chloride gradient.

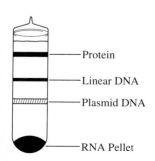

Protein

Linear DNA

Plasmid DNA

RNA Pellet

troduces students to some of the microbiological procedures required in molecular cloning. All media and supplies, except antibiotics, must be autoclaved before use. Sterile transfers are required at all stages. The extensive time requirements and the unusual schedule commitments may make this experiment cumbersome to organize. The total amount of actual working time is small, but several hours of incubation time are required for a suitable culture.

The isolation of plasmid DNA by the boiling method (part B) yields a product of sufficient purity for analysis in Experiment 21. The procedure works well with chloramphenicol-amplified or untreated cell cultures. If cell cultures are available, the isolation method can easily be completed in about 2 hours.

If an ultracentrifuge is available, it is recommended that students further purify the plasmid preparation on a cesium chloride gradient (part C). Only about 1 hour is necessary to prepare the sample for centrifugation; however, 16 to 24 hours of ultracentrifugation are required to separate the plasmid DNA from any remaining chromosomal DNA and other impurities.

Part D, digestion of RNA by ribonuclease, can be done directly on the cleared lysate from part B. Part D can be completed in approximately 1 hour.

The experiment may be carried out in one of the following ways:

1. Part A completed as a class project, preparing enough cell culture for all members to use in parts B, C, and D.
2. Completed in groups of 4 to 6 students. At least one student in each group should be experienced in standard microbiological procedures so that part A can be completed correctly.
3. The instructor or teaching assistant can complete part A and provide cell cultures to students for plasmid isolation.

This experiment illustrates the large-scale isolation and purification of plasmid DNA. If a rapid, microscale method of isolation is desired, see Experiment 20, which describes a method for isolation of plasmid DNA starting with only 5 mL of cell culture. This provides sufficient material for analysis in Experiment 21.

II. Materials and Supplies

A. Growth of Bacteria

Escherichia coli slant, slabs, plate or liquid culture. Recommended strains are RR1, HB101, GM48, 294, SK1591, JC411Thy⁻/ColE1, or CR34/ColE1. The strain must contain a plasmid such as ColE1, pBR322, pUR290, or the pUC series of plasmids.

Luria-Bertani (LB) medium. Each liter contains 10 g Bacto-tryptone, 5 g Bactoyeast extract, and 10 g NaCl. Adjust to pH 7.5 with NaOH.

Ampicillin. A solution of the sodium salt is prepared in water (25 mg/mL) and sterilized by passage through a microfilter (e.g., a 0.22 μ Millipore filter). Store the solution in a freezer.

Chloramphenicol, 25 mg/mL in 100% ethanol

Autoclave

Flask shaker in environmental room at 37°C or a shaker water bath at 37°C

Spectrophotometer and cuvettes for absorbance measurements at 600 nm

Standard microbiological equipment: sterile flasks, pipets, test tubes, etc.

Centrifuge, capable of 4000 $\times$ g

Standard Tris buffer, 0.01 M Tris-HCl, 0.001 M EDTA, and 0.1 M NaCl, pH 7.8. Keep cold.

B. Isolation of Plasmid DNA

E. coli cells, 100 mL of culture grown as described in part A

Standard Tris buffer, 0.01 M Tris-HCl, containing 0.1 M NaCl and 0.1 mM EDTA, 5% Triton X-100, pH 8.0. Keep cold.

Lysozyme solution, 10 mg/mL in 0.01 M Tris-HCl, pH 8.0. Must be freshly prepared.

Bunsen burner

Boiling-water bath in 1-liter beaker

Centrifuge, capable of 12,000 $\times$ g

C. Purification of the Plasmid by Centrifugation on a CsCl Gradient (Optional)

Cleared lysate from part B

Cesium chloride (solid)

Tris-EDTA buffer, 0.01 M Tris-HCl containing 1 mM EDTA, pH 8.0

Ethidium bromide, 10 mg/mL in H_2O

Quick-seal polyallomer ultracentrifuge tubes

Ultracentrifuge and rotor. The recommended rotors are the TLS-55, TLV-100, or the TLA for the Beckman TL-100 instrument, the SW55 or 70Ti for the Beckman L8-70M instrument, or the TV865 for the Sorval instrument.

Hypodermic needles, 18 or 21 gauge

Isoamyl alcohol, saturated with water and solid CsCl

Dialysis tubing, 1 $\times$ 20 cm

D. *Purification of the Plasmid by Ribonuclease (Optional)*

Cleared lysate from part B

Ribonuclease A, 1 mg/mL in 5 mM Tris-HCl, pH 8.0. This must be heated in an 80°C bath for 10 minutes to denature deoxyribonucleases.

Ribonuclease T1, 500 units/mL in 0.05 M Tris-HCl, pH 8.0. This must be heated as described for ribonuclease A.

Sodium dodecyl sulfate (SDS), 10% in water

Acetate-MOPS buffer, 0.1 M Na acetate, 0.05 M MOPS, pH 8.0

Isopropanol

Ethanol

Tris-EDTA, 0.01 M Tris-HCl, 0.01 M EDTA, pH 7.5

III. Experimental Procedure

A. *Growth of Bacteria*

 Caution

Experimentation with bacterial cells presents a potential biohazard. Do not pipet any solutions by mouth. Wash your hands well with hot water and soap before eating, drinking, or smoking.

1. Using sterile techniques, transfer the appropriate bacterial cells from slant to a test tube or flask containing 10 mL of sterilized LB medium and 40 μL of the ampicillin solution. Incubate the culture with vigorous shaking at 37°C for 12 to 15 hours or overnight.

2. Prepare a 125-mL Erlenmeyer flask for transfer by adding 25 mL of sterile LB medium and 0.10 mL of the ampicillin solution. Transfer 0.1 mL of the overnight culture (from step 1) to the flask. Incubate at 37°C with vigorous shaking. At various time periods, remove 1-mL aliquots of the culture, transfer to a cuvette, and determine the cell density by measuring the absorbance at 600 nm.

3. When A_{600} reaches 0.6, transfer the entire culture from step 2 to a flask with 500 mL of sterile LB medium containing 2 mL of the ampicillin solution. Incubate at 37°C for 2.5 hours with vigorous shaking.

4. When A_{600} reaches about 0.4, add 4.0 mL of the chloramphenicol-ethanol solution to the growing culture.

5. Finally, incubate, with shaking, at 37°C for 12 to 16 hours or overnight and proceed to part B.

B. Isolation of Plasmid DNA

The procedure outlined here can be used for bacterial culture ranging from 10 mL to 1 liter. Instructions will be given for 100 mL of cell culture. If only very small cultures (less than 10 mL) are available, follow the procedure in Experiment 20.

1. Obtain 100 mL of cell culture and centrifuge at 4000 × *g* for 10 minutes at 4°C. Discard the supernatant.
2. Wash the cells by adding 100 mL of standard Tris buffer containing NaCl and EDTA. Centrifuge as in step 1 and discard the supernatant.
3. Suspend the pellet in 15 mL of standard Tris buffer and transfer to a 50-mL Erlenmeyer flask.
4. Add 2.0 mL of lysozyme solution.
5. With a clamp, hold the flask over a Bunsen burner flame until the solution begins to boil. Shake constantly.
6. Incubate the flask in a boiling-water bath for 40 seconds. The bath should be set up with a 1-L beaker of water and a hot plate or flame.
7. Cool the flask in ice water.
8. Transfer the viscous lysate to a centrifuge tube and centrifuge at 12,000 × *g* for 20 minutes at 4°C.
9. Remove the cleared lysate (supernatant) and transfer it to a plastic test tube for storage. The present lysate containing plasmids may be analyzed directly without further purification (Experiment 21). Further purification of the lysate is described in part C and part D.

C. Purification of the Plasmid by Centrifugation on a CsCl Gradient (Optional)

 Caution

> Ethidium bromide is a potential carcinogen. Always wear gloves when handling solutions or reaction mixtures containing the substance. Dispose of all solutions containing ethidium bromide in the special container provided in the laboratory.

1. Measure the volume of the cleared lysate from part B.
2. Add 1.0 g of solid CsCl for each milliliter of lysate. If necessary, warm the solution in a 30°C water bath to dissolve CsCl.
3. Add 0.6 mL of ethidium bromide for each 10 mL of CsCl-lysate solution.
4. Mix the contents thoroughly and transfer to a Quick-seal centrifuge tube. Remove bubbles and add standard Tris buffer to fill the neck of the tube. Balance tubes for equal weight. Seal the tubes and load in

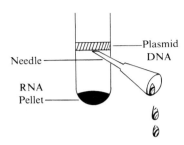

Figure E19.4
Collection of the plasmid fraction after centrifugation. The plastic tube is punctured with a needle and the plasmid collected in a tube.

the rotor. Centrifuge at 40,000 rpm for 16–24 hours. Since several ultracentrifuge models are available, specific instructions will not be outlined here. Have your instructor assist in the operation of the instrument or consult the instrument manual.

5. After the centrifugation, recover the purified plasmid DNA by piercing the side of the tube with an 18- or 21-gauge needle. Insert the needle at a location just below the lower red band and collect the plasmid fraction as shown in Figure E19.4. The upper red band contains chromosomal DNA and nicked circular plasmid DNA. The red pellet consists of ethidium bromide–RNA complexes.

6. Remove the ethidium bromide by extracting with isoamyl alcohol. Mix the plasmid fraction with an equal volume of isoamyl alcohol that has been equilibrated with water and solid cesium chloride. The two phases may be readily separated by centrifugation at $2000 \times g$ for 5 minutes.

7. Transfer the lower aqueous phase containing the plasmid to a new test tube and repeat the extraction process at least three times with fresh isoamyl alcohol. There should no longer be a pink color associated with the aqueous layer containing the plasmid.

8. Finally, transfer the aqueous phase into dialysis tubing and dialyze against standard Tris-HCl–EDTA buffer, pH 8.0. Dialyze at least 12 to 15 hours, with buffer changes every 4 to 5 hours. This step removes CsCl from the plasmid preparation.

9. Transfer the dialyzed extract to a plastic tube and store for Experiment 21.

D. Purification of the Plasmid by Ribonuclease (Optional)

1. Treat the cleared lysate from part B as follows. Add 0.2 mL of ribonuclease A and 0.2 mL of ribonuclease T1.

2. Incubate the reaction mixture in a water bath at 37°C for 15 minutes.

3. Add 0.04 mL of 10% SDS and 2.0 mL of acetate-MOPS buffer, pH 8.0.

4. Precipitate the plasmid DNA by dropwise addition of 4.0 mL of isopropanol.

5. Allow the solution to sit at room temperature for 15 minutes, and then centrifuge at $7500 \times g$ for 15 minutes.

6. Dissolve the pellet (plasmid DNA) in 2.0 mL of acetate-MOPS buffer and precipitate the plasmid DNA with two volumes of ethanol.

7. Centrifuge as in step 5 and dissolve the pellet in 2.0 mL of Tris-HCl–EDTA buffer, pH 7.5.

IV. Analysis of Results

A. Growth of Bacteria

Prepare a flowchart showing each step carried out in the growth of the bacteria. Construct a graph of A_{600} vs. time, and plot on it the data obtained in step 2. Identify any lag, log, and stationary phases of growth. Indicate on the curve where chloramphenicol was added.

B. Isolation of Plasmid DNA

Prepare a flowchart outlining each isolation step. Write a brief explanation of the purpose of each reagent and procedural step.

C. Purification of the Plasmid by Centrifugation on a CsCl Gradient

Draw a picture of the centrifuge tube after centrifugation. Identify the components present in each region of the tube. Compare with Figure E19.3.

D. Purification of the Plasmid by Ribonuclease

Prepare a flowchart for the procedures and outline the purpose of each reagent.

V. Questions and Problems

▶ 1. Explain the action of ampicillin as an inducer of plasmid replication.
2. How does chloramphenicol inhibit protein synthesis?
3. Compare the isolation procedure used in this experiment to that used for chromosomal DNA in Experiment 17.
4. Why are the ribonuclease solutions heated before use in this experiment?
5. Why is a wavelength of 600 nm used to measure growth of bacteria? Could other wavelengths be used?

VI. References

F. Bolivar and K. Backman, in *Methods in Enzymology,* Vol. 68, R. Wu, Editor (1979), Academic Press (New York), pp. 245–267. "Plasmids of *E. coli* as Cloning Vectors."

D. Clewell, *J. Bacteriol.* **110,** 667–676 (1972). "Nature of ColE1 Plasmid Replication in *Escherichia coli* in Presence of Chloramphenicol."

V. Hershfield, H. Boyer, C. Yanofsky, M. Lovett, and D. Helinski, *Proc. Natl. Acad. Sci. USA* **71**, 3455–3459 (1974). "Plasmid ColE1 as a Molecular Vehicle for Cloning and Amplification of DNA."

D. Holmes and M. Quigley, *Anal. Biochem.* **114,** 193–197 (1981). "A Rapid Boiling Method for the Preparation of Bacterial Plasmids."

C. Mathews and K. van Holde, *Biochemistry* (1990), Benjamin/Cummings (Redwood City, CA), pp. 860–884. Plasmids and recombinant DNA.

D. Micklos and G. Freyer, *DNA Science: A First Course in Recombinant DNA Technology* (1990), Cold Spring Harbor Laboratory Press (Cold Spring Harbor, NY). An excellent reference book for beginning students.

H. Miller, in *Methods in Enzymology,* Vol. 152, S. Berger and A. Kimmel, Editors (1987), Academic Press (Orlando, FL), pp. 145–172. "Practical Aspects of Preparing Phage and Plasmid DNA: Growth, Maintenance and Storage of Bacteria and Bacteriophage."

B. Perbal, *A Practical Guide to Molecular Cloning,* 2nd ed. (1988), John Wiley & Sons (New York). Complete laboratory guide to recombinant DNA.

R. Radloff, W. Bauer, and J. Vinograd, *Proc. Natl. Acad. Sci. USA* **57**, 1514–1521 (1967). "A Dye-Buoyant-Density Method for the Detection and Isolation of Closed Circular Duplex DNA."

J. Rawn, *Biochemistry,* 2nd ed. (1989), Neil Patterson Publishers (Burlington, NC), pp. 975–1019. Recombinant DNA.

J. Sambrook, E. Fritsch, and T. Maniatis, *Molecular Cloning, A Laboratory Manual,* 2nd ed. (1989), Cold Spring Harbor Laboratory (Cold Spring Harbor, NY), Vols. I, II and III, pp. 1.21–1.52. Growth of bacteria and amplification of plasmid; extraction and purification of plasmid DNA.

L. Stryer, *Biochemistry,* 3rd ed. (1988), W. H. Freeman (New York), pp. 117–140. Recombinant DNA.

D. Voet and J. Voet, *Biochemistry* (1990), John Wiley & Sons (New York), pp. 837–846. Recombinant DNA.

J. Zyskind and S. Bernstein, *Recombinant DNA Laboratory Manual* (1989), Academic Press (San Diego, CA), pp. 1–10; 31–38. Growth of bacteria and isolation of plasmids.

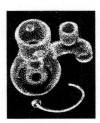

Rapid, Microscale Isolation of Plasmid DNA

■ **Recommended Reading**
Chapter 4; Experiments 17, 19, and 21.

■ **Synopsis**
This experiment describes a simple and rapid method for isolating microgram quantities of plasmid DNA. Boiling a bacterial culture denatures chromosomal DNA, which then precipitates as a gel leaving plasmid DNA in solution. Sufficient quantities of plasmid DNA are isolated for analysis by agarose gel electrophoresis and for restriction endonuclease experiments in Experiment 21.

I. Introduction and Theory

Microscale Isolation of Plasmids

Often it is necessary to detect and analyze plasmid DNA in a large number of small bacterial samples. Rapid screening of single bacterial colonies or small cultures may have several objectives:

1. Detect the presence of plasmids.
2. Estimate the size of plasmids.
3. Analyze plasmids for restriction enzyme cleavage.
4. Perform restriction mapping of plasmids.
5. Analyze nucleotide sequence of plasmids.
6. Determine optimal conditions for the preparation of hybrid DNA.
7. Search for recombinant DNA inserts.

The procedures of plasmid isolation and purification outlined in Experiment 19 are suitable for small and large samples (10-mL to 1-L cultures). The purification methods of density gradient ultracentrifugation and ribonuclease treatment are time-consuming, require costly instrumentation, and are not applicable for surveying many samples. However, they must be completed if a relatively large quantity of highly purified plasmid DNA is required. Typical yields from the large-scale procedures are 2 to 3 mg of plasmid DNA per liter of culture.

Several procedures have been developed for the rapid, small-scale isolation of plasmid DNA. Many of these provide plasmids of sufficient quantity and quality for characterization by agarose gel electrophoresis and restriction enzyme digestion. Besides providing rapid screening of plasmids from several bacterial samples, small-scale procedures require only minimal handling of potentially hazardous plasmids.

One isolation procedure will be outlined in this experiment. The method, which is a micro version of Experiment 19, produces good yields of relatively clean DNA. Bacterial cells containing the desired plasmid grown in small cultures (1 to 5 mL) or in single colonies on agar plates are boiled for a brief period in the presence of reagents that disrupt the cell wall and inhibit nuclease activity. Chromosomal DNA and protein are denatured under these conditions, and they precipitate as a gel that can be centrifuged from the supernatant containing plasmid DNA and bacterial RNA. Because of its appearance, the supernatant is sometimes called a "cleared lysate." The RNA, which may interfere with electrophoretic analysis of the plasmids, is removed by isopropanol precipitation of the plasmid DNA and treatment with ribonuclease (RNAase). The plasmid DNA is then ready for direct analysis by agarose gel electrophoresis or restriction enzyme digestion followed by analysis of the fragments by electrophoresis.

A second, widely used method is the alkaline lysis extraction procedure. For this miniprep, host bacterial cells harboring the plasmid are grown in small culture volumes (1–5 mL) or in single colonies on agar plates. The cells are lysed and their contents denatured by alkaline sodium dodecyl sulfate (SDS). Proteins and high-molecular-weight chromosomal DNA denatured under these conditions precipitate as a gel that can be centrifuged from the supernatant, which contains plasmid DNA and bacterial RNA. Following neutralization and renaturation of the plasmid DNA, the RNA, which may interfere with electrophoretic analysis of the plasmids, is removed by treatment with RNAase. The plasmid DNA is then ready for restriction enzyme digestion followed by analysis and mapping of the fragments by electrophoresis.

Analysis of Plasmid DNA

Several techniques for the characterization of nucleic acids have been introduced in this manual, but the standard method for separation and analysis of plasmid and other smaller DNA molecules is agarose gel electrophoresis. The method has several advantages, including ease of operation, sensitive staining procedures, high resolution, and a wide range of molecular sizes $(0.6-100 \times 10^6)$ that can be analyzed. These characteristics are described in Chapter 4 and in Experiment 21.

One ultimate goal of plasmid isolation and analysis is to construct a **restriction enzyme map** for the plasmid (see Figure E20.1). The map displays the sites of cleavage by restriction endonucleases and the number of fragments obtained after digestion with each enzyme.

A map is constructed by first digesting plasmid DNA with restriction endonucleases that yield only a few fragments. Each digest of DNA obtained with a single nuclease is analyzed by agarose gel electrophoresis,

Figure E20.1
A restriction enzyme map for the plasmid pBR322. From *Molecular Cloning, A Laboratory Manual,* by T. Maniatis, E. Fritsch, and J. Sambrook. Cold Spring Harbor Laboratory, 1982.

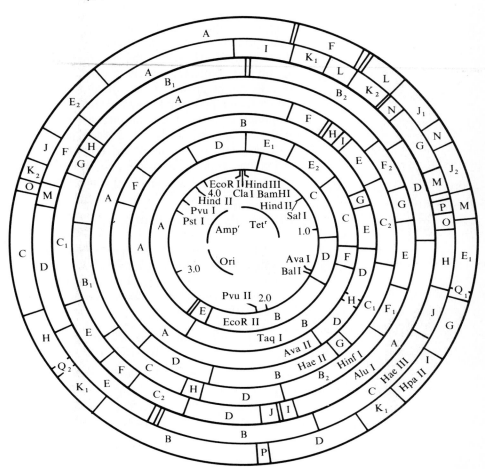

using standards for molecular weight determination. These are referred to as the primary digests. Second, each of the primary digests is treated with a series of additional restriction enzymes, and the digests are analyzed by agarose gel electrophoresis. The restriction map is then constructed by combining the various fragments by trial and error and logic, much as in sequencing a protein by proteolytic digestion by several enzymes and searching for overlap regions in the fragments. In the case of the restriction endonuclease digests, two characteristics of the DNA fragments are known for: the approximate molecular weights (from electrophoresis) and the nature of the fragment ends (from the known selectivity of the individual restriction enzymes). The fragments may also be sequenced by the Sanger or Maxam-Gilbert method. Construction of a restriction map is detailed in Experiment 21.

Overview of the Experiment

This experiment provides an alternative to Experiment 19. Students completing this modification will be exposed to essentially all of the same principles and techniques except ultracentrifugation and density gradient analysis. However, the microscale version is more convenient to set up and complete, especially if facilities are limited. A single culture of cells can be grown for the whole class as described in Experiment 19. The isolation procedure works well with unamplified or chloramphenicol-amplified cultures. Each milliliter of an unamplified, overnight culture will yield approximately 1 to 2 μg of plasmid DNA.

The isolation procedure requires $1\frac{1}{2}$ to 2 hours of student lab time if a cell culture is prepared in advance. The boiling method yields, after centrifugation, a cleared lysate (supernatant) that contains the plasmid DNA.

II. Materials and Supplies

Bacterial culture (5.0 mL) or a colony on an agar plate (4 mm in diameter). The strains of bacteria recommended for this experiment have been listed in Experiment 19. Amplified or unamplified cell cultures are appropriate. Follow directions for bacterial growth in Experiment 19.

STET buffer, containing 8% sucrose, 0.5% Triton X-100, 0.05 *M* EDTA, and 0.05 *M* Tris-HCl, pH 8.0; autoclave and store at 4°C.

Lysozyme solution, 10 mg/mL in 0.01 *M* Tris-HCl, pH 8.0, freshly prepared

Boiling-water bath in a 150- to 200-mL beaker

Microcentrifuge, capable of 12,000 $\times$ *g*, at 4°C

Sodium acetate, 2.5 *M*

Isopropanol

Dry ice–acetone bath, approximately −18°C

Tris-EDTA-RNAase solution, 0.01 M Tris-HCl, 0.001 M EDTA, pH 8.0, containing RNAase (50 μg/mL). Heat this solution in an 80°C bath for 10 minutes to denature deoxyribonuclease.

III. Experimental Procedure

⚠ Caution

Experimentation with bacterial cells presents a potential biohazard. Do not mouth pipet any solutions. Wash hands with hot water and soap before eating, drinking, or smoking.

Ethidium bromide is a mutagen. Always wear gloves when handling solutions or gels. Dispose of solutions as directed by the instructor.

1. Transfer 5 mL of an overnight culture to a small centrifuge tube. Alternatively, a single colony (about 4 mm in diameter) of bacteria can be removed from an agar plate with a flat toothpick and suspended in 5.0 mL of STET buffer in a conical tube. Centrifuge the tube at 4000 × g for 5 minutes.

2. Remove as much of the supernatant as possible with a disposable pipet attached through rubber tubing to an aspirator.

3. Resuspend the cells in 0.35 mL of STET buffer.

4. Add 25 μL of freshly prepared lysozyme solution and mix well.

5. Place the suspension in a boiling-water bath for 40 seconds.

6. Centrifuge immediately at 12,000 × g for 10 minutes at 4°C to sediment gelatinous chromosomal DNA and denatured proteins.

7. Remove the supernatant and transfer to a small conical tube or an Eppendorf tube. Alternatively, the pellet may be removed and discarded with a flat toothpick.

8. To the supernatant, add an equal volume of isopropanol (approximately 0.4 to 0.5 mL) and 40 μL of 2.5 M sodium acetate. Mix well.

9. Cool the mixture in a dry ice–acetone bath at −18°C for 15 minutes.

10. Centrifuge the mixture at 12,000 × g for 5 minutes to sediment the plasmid DNA.

11. Carefully and completely remove and discard the supernatant. Allow the precipitate to air dry for a few minutes.

12. Resuspend the pellet in 0.05 mL of Tris-EDTA-RNAase solution and incubate for 10 minutes at 37°C.

13. The plasmid preparation may be used directly in Experiment 21 or stored in a freezer.

IV. Analysis of Results

Prepare a flowchart showing each step carried out in the growth of the bacteria. Construct a growth curve of A_{600} vs. time, and identify any lag, log, and stationary phases of growth. Identify on the curve where chloramphenicol was added.

Prepare a flowchart outlining the isolation procedure. Explain the function of each reagent and step.

V. Questions and Problems

▶ 1. What is the purpose of the isopropanol precipitation step in the preparation of plasmids?

2. Compare the isolation procedures used in this experiment to that used for chromosomal DNA (Experiment 17). Explain the differences.

3. Compare the boiling method used in this experiment with the alkaline lysis method described in the Introduction.

VI. References

H. Birnboim, in *Methods in Enzymology,* Vol. 100, R. Wu, L. Grossman, and K. Moldave, Editors (1983), Academic Press (New York), pp. 243–255. "A Rapid Alkaline Extraction Method for the Isolation of Plasmid DNA."

H. Birnboim and J. Doly, *Nucleic Acids Res.* **7**, 1513–1523 (1979). "A Rapid Alkaline Extraction Procedure for Screening Recombinant Plasmid DNA."

D. Holmes and M. Quigley, *Anal. Biochem.* **114**, 193–197 (1981). "A Rapid Boiling Method for the Preparation of Bacterial Plasmids."

See also the references in Experiment 19.

EXPERIMENT 21

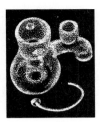

The Action of Restriction Endonucleases on Plasmid or Viral DNA

■ **Recommended Reading**

Chapter 4; Experiments 17, 19, and 20.

■ **Synopsis**

Restriction endonucleases catalyze the hydrolysis of specific phosphodiester bonds in double-stranded DNA. These enzymes are often used to linearize a circular plasmid for hybrid DNA construction. They have also found use in the analysis of DNA and the construction of restriction maps. In this experiment, students will incubate various restriction enzymes with plasmid or viral DNA and analyze the product DNA fragments by agarose gel electrophoresis.

I. Introduction and Theory

Experiments 19 and 20 introduced students to some principles and techniques involved in recombinant DNA research. Specifically, the experiments outlined the replication, isolation, and purification of bacterial plasmid vehicles for molecular cloning experiments. This experiment describes two more tools that are essential in hybrid plasmid construction and analysis—restriction enzyme action and agarose gel electrophoresis. The procedures introduced here have also found widespread use in the analysis and characterization of all DNA molecules.

The Action of Restriction Endonucleases

Bacterial cells produce many enzymes that act to degrade various forms of DNA. Of special interest are the **restriction endonucleases,** enzymes that recognize specific base sequences in double-stranded DNA and catalyze hydrolytic cleavage of the two strands in or near that specific region. The biological function of these enzymes is to degrade or restrict foreign DNA molecules. Host DNA is protected from hydrolysis because some bases near the cleavage sites are methylated. The action of the restriction enzyme *Eco*RI is shown in Equation E21.1.

(Equation E21.1)

$$5'...G \xrightarrow{\downarrow} G-A-T-C-C...3' \quad \xrightarrow[\text{H}_2\text{O}]{\textit{Eco}\text{RI}} \quad 5'...G \qquad\qquad G-A-T-C-C...3'$$
$$3'...C-C-T-A-G \underset{\uparrow}{\,} G...5' \qquad\qquad 3'...C-C-T-A-G \quad + \quad G...5'$$

The site of action of *Eco*RI is a specific hexanucleotide sequence. Two phosphodiester bonds are hydrolyzed (see arrows), resulting in fragmentation of both strands. Note the twofold rotational symmetry feature at the recognition site and the formation of cohesive ends. The weak base pairing between the cohesive ends is not sufficient to hold the two fragments together.

More than 200 restriction enzymes have been isolated and characterized. Nomenclature for the enzymes consists of a three-letter abbreviation representing the source (*Eco* = *E. coli*), a letter representing the strain (R), and a roman numeral designating the order of discovery. *Eco*RI is the first to be isolated from *E. coli* (strain R) and characterized. Table E21.1 lists several other restriction enzymes, their recognition sequence for cleavage, and optimum reaction conditions.

Restriction enzymes are used extensively in nucleic acid chemistry. They may be used to cleave large DNA molecules into smaller fragments that are more amenable to analysis. For example, λ phage DNA, a linear, double-stranded molecule of 48,502 base pairs (molecular weight 31×10^6), is cleaved into six fragments by *Eco*RI or into more than 50 fragments by *Hin*fI (*Haemophilus influenzae,* serotype f). The base sequence recognized by a restriction enzyme is likely to occur only a very few times in any particular DNA molecule; therefore, the smaller the DNA molecule, the fewer the number of specific sites. The λ phage DNA is cleaved into 0 to 50 or more fragments, depending on the restriction enzyme used, whereas larger bacterial or animal DNA will most likely have many recognition sites and be cleaved into hundreds of fragments. Smaller DNA molecules, therefore, have a much greater chance of producing a unique set of fragments with a particular restriction enzyme. It is unlikely that this set of fragments will be the same for any two different DNA molecules, so the fragmentation pattern is unique and can be considered a "fingerprint" of the DNA substrate. The fragments are readily separated and sized by agarose gel electrophoresis as described later in this experiment.

Restriction endonucleases are also valuable tools in the construction of hybrid DNA molecules. Several restriction enzymes act at a single site on

Table E21.1
Specificity and Optimal Conditions for Several Restriction Endonucleases.

Name	Recognition Sequence 5′.......3′	T (°C)	pH	Tris (mM)	NaCl (mM)	MgCl$_2$ (mM)	DTT[2] (mM)
*Aat*II	G—A—C—G—T↓C	37	7.9	20	50	10	10
*Alu*I	A—G↓C—T	37	7.5	10	50	10	10
*Bal*I	T—G—G↓C—C—A	37	7.9	6	—	6	6
*Bam*HI	G↓G—A—T—C—C	37	8.0	20	100	0.7	1
*Bcl*I	T↓G—A—T—C—A	60	7.5	10	50	10	1
*Eco*RI	G↓A—A—T—T—C	37	7.5	10	100	10	1
*Hae*II	Pu[1]—G—C—G—C↓Py[1]	37	7.5	10	50	10	10
*Hind*III	A↓A—G—C—T—T	37–55	7.5	10	60	10	1
*Hpa*I	G—T—T↓A—A—C	37	7.5	10	50	10	1
*Mse*I	T↓T—A—A	37	7.9	10	50	10	1
*Not*I	G—C↓G—G—C—C—G—C	37	7.9	10	150	10	—
*Sal*I	G↓T—C—G—A—C	37	8.0	10	150	10	1
*Sca*I	A—G—T↓A—C—T	37	7.4	10	100	10	1
*Taq*I	T↓C—G—A	65	8.4	10	100	10	10

[1] Pu = a purine base; Py = a pyrimidine base.
[2] DTT = dithiothreitol.

bacterial plasmid vehicles. This linearizes the circular plasmid and allows the insertion of a foreign DNA fragment. For example, the popular plasmid vehicle pBR322 has a single restriction site for *Bam*HI (*Bacillus amyloliquefaciens,* H) that is within the tetracycline resistance gene. The enzyme not only opens the plasmid for insertion of a DNA fragment but also destroys a phenotype; this fact aids in the selection of transformed bacteria.

Restriction enzymes can be used to obtain physical maps of DNA molecules. Important information can be obtained from maps of DNA, whether the molecules are small and contain only a few genes, such as viral or plasmid DNA, or large and complex, such as bacterial or eukaryotic chromosomal DNA molecules. An understanding of the genetics, metabolism, and regulation of an organism requires knowledge of the precise arrangement of its genetic material. Genetic and physical mapping provides this information. Physical mapping of a genome can be conducted at two levels of resolution: (1) initial mapping begins by dividing the overall genome into pieces of manageable size by cleaving with restriction endonucleases and arranging the fragments into a restriction map; (2) subsequent, higher-resolution mapping requires determining the exact nucleotide sequence of each fragment using the DNA sequencing methodology described in Chapter 4. A restriction enzyme map for the plasmid pBR322 is shown in Figure E20.1.

Practical Aspects of Restriction Enzyme Use

Restriction enzymes are heat-labile and expensive biochemical reagents. Their use requires considerable planning and care. Each restriction nuclease has been examined for optimal reaction conditions in regard to specific pH range, buffer composition, and incubation temperature. This information for each enzyme is readily available from the commercial supplier of the enzyme or from the literature. Table E21.1 gives important reaction information for several enzymes. The temperature range and pH optima for most restriction nucleases are similar (37°C, 7.5–8.0); however, optimal buffer composition is variable. Typical buffer components are Tris, NaCl, $MgCl_2$, and a sulfhydryl reagent (β-mercaptoethanol or dithiothreitol). Proper reaction conditions are crucial for optimal reaction rate; but, more important, changing reaction conditions have been shown to alter the specificity of some restriction enzymes.

Although restriction enzymes are very unstable reagents, they can be stored at $-20°C$ in buffer containing 50% glycerol. They are usually prepared in an appropriate buffer and shipped in packages containing dry ice.

Disposable gloves should be worn when you are handling the enzyme container. Remove the enzyme from the freezer just before you need it. Store the enzyme in an ice bucket when it is outside the freezer. The enzyme should never be stored at room temperature. Because of high cost, digestion by restriction enzymes is carried out on a microscale level. A typical reaction mixture will contain about 1 μg or less of DNA and 1 unit of enzyme in the appropriate incubation buffer. One unit is the amount of enzyme that will degrade 1 μg of λ phage DNA in 1 hour at the optimal temperature and pH. The total reaction volume is usually between 20 and 50 μL. Incubation is most often carried out at the recommended temperature for about 1 hour. The reaction is stopped by adding EDTA solution, which complexes divalent metal ions essential for nuclease activity.

Reaction mixtures from restriction enzyme digestion may be analyzed directly by agarose gel electrophoresis. This technique combines high resolving power and sensitive detection to allow the analysis of minute amounts of DNA fragments.

Characterization of DNA by Agarose Gel Electrophoresis

Agarose gels are the standard media for separating and characterizing DNA molecules. As discussed in Chapter 4, the mobility of nucleic acids in agarose gels is influenced by the agarose concentration, the molecular size of the DNA, and the molecular shape of the DNA. In general, the lower the agarose concentration in the gels, the larger the DNA that can be analyzed. A practical lower limit of agarose concentration is reached at 0.3% agarose, below which gels become too fragile for ordinary use. This lower limit of 0.3% agarose in the gel allows analysis of linear double-stranded DNA within the range of 5 and 60 kilobase pairs (up to 150×10^6 in molecular weight). Gels with an agarose concentration of 0.8% can separate DNA in the range of 0.5–10 kilobase pairs and 2% agarose gels are used to separate smaller DNA fragments (0.1–3 kilobase pairs). For most routine analysis of restriction enzyme fragments, agarose concentrations in the range 0.7 to 1% are most appropriate.

Nucleic acids migrate in an agarose medium at a rate that is inversely proportional to their size (kilobase pairs or molecular weights). In fact, a linear relationship exists between mobility and the logarithm of kilobase pairs (or molecular weight) of a DNA fragment. After electrophoresis of a restriction enzyme digest containing DNA fragments of known size, a standard curve may be prepared (see Figure E21.1). The standard reaction mixture shown here is the *Eco*RI digestion of λ phage DNA. The mixture contains six DNA fragments with the following sizes: 21,226, 7421, 5804, 5643, 4878, and 3530 base pairs.

The influence of molecular shape on electrophoretic mobility is not a critical factor to consider in the experiment. The action of restriction en-

Figure E21.1
Standard curve obtained by electrophoresis of λ phage DNA fragments from *Eco*RI cleavage.

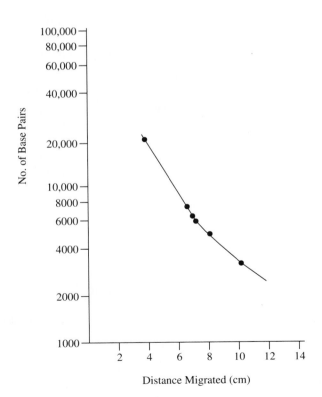

zymes on both covalent closed circular DNA and linear DNA results in the formation of linear fragments.

Agarose gel electrophoresis is an ideal technique for analysis of DNA fragments. In addition to the positive characteristics discussed, the technique is simple, rapid, and relatively inexpensive. Fragments that differ in molecular weight by as little as 1% can be resolved on agarose gels, and as little as 1 ng of DNA can be detected on a gel. Nucleic acids are visualized after electrophoresis by soaking the gel in ethidium bromide or by performing the electrophoresis with ethidium bromide incorporated in the gel and buffer.

Overview of the Experiment

The objective of the experiment is to evaluate the action of restriction enzymes on bacterial plasmids, λ phage DNA, or viral DNA. The DNA will be incubated under the appropriate conditions with selected restriction enzymes. The reaction mixtures will be subjected to agarose gel electrophoresis in order to determine the number and molecular size of the restriction fragments.

The following options are available for this experiment:

1. Students may use isolated and purified plasmids from Experiments 19 and 20. Alternatively, several purified plasmids, including ColE1, pBR322, and pUB110, are available from commercial sources. The plasmids may be cleaved with various restriction enzymes and the products analyzed by agarose gel electrophoresis.

2. Students may characterize λ phage or adenovirus 2 DNA. Both are standard substrates for assays of restriction endonucleases.

It may be more convenient to supply students with DNA rather than have them isolate and purify plasmids as described in Experiments 19 and 20.

Approximately 2 hours are required to prepare and incubate the restriction enzyme reaction mixtures. The agarose slab gel may be prepared during the incubation period. Up to 1 to 2 hours are required to prepare the gel slab. The electrophoresis may be completed as a group project. Most electrophoresis chambers accommodate slab gels with 15 to 20 sample wells. Relatively inexpensive slab gel electrophoresis chambers are available commercially, or they may be constructed from readily available materials.

II. Materials and Supplies

A. Digestion of DNA with Restriction Enzymes

DNA: Plasmid, commercial or from Experiments 19 and 20; λ phage; adenovirus (Ad 2); approximately 0.2–0.4 mg/mL in Tris-HCl, NaCl, EDTA, pH 7.0.

Restriction enzymes, 1 unit/μL. Keep stored in freezer until ready to use. Recommended enzymes are *Bam*HI and *Eco*RI for λ phage DNA, *Eco*RI and *Taq*I for plasmids (pBR322 or ColE1), and *Hpa*I for Ad 2 DNA

Sterilized H_2O

Incubation buffers

*Bam*HI: 0.2 M Tris-HCl, pH 8.0, 0.07 M MgCl$_2$, 1 M NaCl, 0.01 M dithiothreitol

*Eco*RI: 0.1 M Tris-HCl, pH 7.5, 0.1 M MgCl$_2$, 1 M NaCl, 0.01 M dithiothreitol

*Hpa*I: 0.1 M Tris-HCl, pH 7.5, 0.1 M MgCl$_2$, 0.5 M NaCl, 0.01 M dithiothreitol

*Taq*I: 0.1 M Tris-HCl, pH 7.4, 0.1 M MgCl$_2$, 1 M NaCl, and 0.01 M dithiothreitol

Quench buffer, 0.1 M EDTA, pH 7

B. Agarose Gel Electrophoresis

Agarose powder, electrophoresis grade

Agarose slurry buffer, 0.04 M Tris-acetate, 0.002 M EDTA, pH 7.8

Ethidium bromide solution, 10 mg/mL in H_2O

Electrophoresis buffer, Tris-acetate buffer listed above containing ethidium bromide, 0.5 μg/mL

Gel-loading buffer, 0.04 M Tris-acetate containing 50% glycerol and 0.25% bromphenol blue tracking dye, pH 8.0.

Electrophoresis chamber. A setup for horizontal slab gel is recommended.

Power supply

UV lamp

Standard restriction enzyme digest, λ phage DNA digested by *Eco*RI

Constant-temperature water baths

III. Experimental Procedure

A. Digestion of DNA with Restriction Enzymes

The reaction conditions are specific for each restriction enzyme. The commercial supplier of the enzyme will provide an information sheet giving details of the reaction conditions. For a detailed listing of optimal conditions for many restriction enzymes, see Brooks (1987) or Table E21.1.

BamHI

1. Obtain a small microcentrifuge tube for each reaction mixture and add the following components:

    15 μL of sterile H_2O

    2 μL of DNA solution

    2 μL of BamHI incubation buffer; mix well by tapping.

    1 unit of BamHI restriction enzyme. The actual volume depends on the concentration of enzyme. In most cases this will be 1 μL of solution. Mix the tube by gentle tapping.

2. Incubate the reaction mixture at 37°C for 1 hour.

3. Stop the reaction by adding 2 μL of EDTA quench buffer.

4. Store the reaction mixture on ice until agarose gel electrophoresis.

EcoRI

1. Obtain a small test tube for each reaction mixture and add the following components:

    15 μL of sterile H_2O

    2 μL of DNA solution

    2 μL of EcoRI incubation buffer; mix well by tapping.

    Add 1 unit of EcoRI restriction enzyme. The volume depends on the concentration of enzyme. For most enzymes this will be 1 μL of solution. Mix the tube with gentle tapping.

2. Incubate the reaction mixture at 37°C for 1 hour.

3. Add 2 μL of EDTA quench buffer.

4. Store the reaction mixture on ice until agarose gel electrophoresis.

HpaI

1. Obtain a small test tube for each reaction mixture and add the following components:

    15 μL of sterile H_2O

    2 μL of the DNA solution

    2 μL of HpaI incubation buffer; mix well by tapping.

    1 unit of HpaI restriction enzyme. In most cases this will be 1 μL. Mix well by gentle tapping.

2. Incubate the reaction mixture at 37°C for 1 hour.

3. Add 2 μL of EDTA quench buffer.

4. Store the reaction mixture on ice until agarose gel electrophoresis.

TaqI

1. Obtain a small microcentrifuge tube for each reaction mixture and add each of the following components:

15 μL of sterile H_2O

2 μL of DNA solution

2 μL of *Taq*I incubation buffer; mix well by tapping

1 unit of *Taq*I restriction enzyme

2. Incubate the reaction mixture at 37°C for 1 hour.

3. Stop the reaction by adding 2 μL of EDTA quench buffer.

4. Store the reaction mixture on ice until gel electrophoresis.

B. Agarose Gel Electrophoresis

Many varieties of electrophoresis chambers are available. Specific details for the use of the electrophoresis unit should be obtained from the accompanying instructions or from your instructor.

⚠ **Caution**

Do not touch the electrophoresis chamber or wires while the electrophoretic operation is in progress. High voltages are required and shocks may be fatal.

Ethidium bromide is a mutagen. Always wear gloves when handling solutions or agarose gels containing the compound. Dispose of the final electrophoresis buffer containing ethidium bromide according to directions from your instructor. Avoid looking into UV light during the detection of DNA fragments on the agarose gel. Wear goggles!

1. Prepare a 1% (w/v) slurry of agarose in the agarose slurry buffer. The actual amount of solution to use depends on the size of the slab gel to be prepared. Follow the instructions given by your instructor.

2. Heat the slurry in a boiling-water bath until the agarose dissolves.

3. After the agarose solution has cooled to 50°C, add ethidium bromide solution to give a final dye concentration of 0.5 μg/mL.

4. Quickly pour the agarose solution over the glass plate to a depth of 2 to 4 mm. Insert the comb to make the sample wells.

5. Allow the gel to set for approximately 30 minutes and carefully remove the comb.

6. Clamp the agarose slab into the electrophoresis chamber and allow it to set for another 30 minutes.

7. Load one sample into each of the sample wells in the following manner. Mix each 20 μL of reaction mixture from the restriction enzyme with 10 μL of gel-loading buffer and apply one 10 μL sample to each well. Put 10 μL of *Eco*RI standard digest of λ DNA into one sample well.

8. Add electrophoresis buffer to the reservoirs and connect the power supply.

9. Carry out the electrophoresis at approximately 3 V/cm or about 50 to 70 volts total.

10. Turn off the power supply when the tracking dye has migrated to the bottom edge of the gel.

11. Remove the gel from the chamber and glass plate, and examine it under a UV light. DNA fragments will appear as red-orange fluorescent bands.

12. Take a photograph or draw a picture of the slab gel, showing the position of each DNA band.

IV. Analysis of Results

Examine the picture of the slab gel after electrophoresis. Note the number of fragments as represented by red-orange fluorescent bands. (Describe the action of ethidium bromide in formation of the color.) Measure the distance from the origin migrated by each band. Use the data from the *Eco*RI standard digest of λ phage DNA to prepare a standard curve for size determination. Using semilog graph paper, plot the size of each fragment on the log axis vs. the distance migrated by each fragment. The sizes of the fragments are 21,226, 7421, 5804, 5643, 4878, and 3530 base pairs. Use the standard curve to estimate the molecular weight of each restriction enzyme fragment.

V. Questions and Problems

▶ 1. How does the EDTA quench buffer stop the restriction enzyme reaction?

▶ 2. Why are glycerol and bromphenol blue dye added to the gel-loading buffer?

3. Describe the pattern obtained from electrophoresis of a standard *Eco*RI digest of λ DNA on 2% agarose gel.

▶ 4. Why is polyacrylamide electrophoresis not suitable for analysis of restriction enzyme fragments of DNA?

VI. References

J. Brooks, in *Methods in Enzymology*, Vol. 152, S. Berger and A. Kimmel, Editors (1987), Academic Press (Orlando, FL), pp. 113–129. "Properties and Uses of Restriction Endonucleases."

N. Cunningham, J. Tomlinson, and F. Jurnak, in *Practical Biochemistry for Colleges*, E. Wood, Editor (1989), Pergamon Press (Oxford), pp. 149–151. "Restriction Enzyme Mapping of Bacteriophage Lambda DNA."

R. Dryer and G. Lata, *Experimental Biochemistry* (1989), Oxford University Press (Oxford), pp. 241–259. "DNA Restriction Mapping and Sequence Analysis."

R. Fuchs and R. Blakesley, in *Methods in Enzymology*, Vol. 100, R. Wu, L. Grossman, and K. Moldave, Editors (1983), Academic Press (New York), pp. 3–38. "Guide to the Use of Type II Restriction Endonuclease."

B. Lewin, *Genes IV* (1990), Oxford University Press (Oxford), Chapter 6. Various mapping techniques are described.

C. Mathews and K. van Holde, *Biochemistry* (1990), Benjamin/Cummings (Redwood City, CA), pp. 863–867. Introduction to restriction enzymes.

D. Micklos and G. Freyer, *DNA Science: A First Course in Recombinant DNA Technology* (1990), Cold Spring Harbor Laboratory (Cold Spring Harbor, NY). An excellent reference book for beginning students.

R. Ogden and D. Adams, in *Methods in Enzymology*, Vol. 152, S. Berger and A. Kimmel, Editors (1987), Academic Press (Orlando, FL), pp. 61–90. "Electrophoresis in Agarose and Acrylamide Gels."

B. Perbal, *A Practical Guide to Molecular Cloning,* 2nd ed. (1988), John Wiley & Sons (New York). Complete laboratory guide to recombinant DNA.

J. Rawn, *Biochemistry,* 2nd ed. (1989), Neil Patterson Publishers (Burlington, NC), pp. 975–1019. Recombinant DNA.

J. Sambrook, E. Fritsch, and T. Maniatis, *Molecular Cloning, A Laboratory Manual,* 2nd ed. (1989), Cold Spring Harbor Laboratory (Cold Spring Harbor, NY), Vols. I, II, and III. An excellent lab manual for recombinant DNA.

L. Stryer, *Biochemistry,* 3rd ed. (1988), W. H. Freeman (New York), pp. 118–120, 858–859. Introduction to restriction enzymes.

D. Voet and J. Voet, *Biochemistry* (1990), John Wiley & Sons (New York), pp. 825–830. Restriction enzymes.

J. Zyskind and S. Bernstein, *Recombinant DNA Laboratory Manual* (1989), Academic Press (San Diego), pp. 25–30. "Agarose Gel of Chromosomal DNA Restriction Endonuclease Digestions."

EXPERIMENT **22**

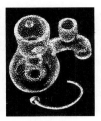

Specific Cleavage of an Oligomeric Ribonucleotide Substrate by a Ribozyme (RNA Enzyme)

■ **Recommended Reading**

Chapters 4, 6; Experiments 6, 17.

■ **Synopsis**

Ribozymes are RNA molecules that catalyze a wide variety of intra- and intermolecular reactions on RNA substrates. They differ in structure from the more well-known biological catalysts, which are protein molecules. In this experiment the action of one type of ribozyme, a sequence-specific endoribonuclease, will be investigated. A radioactively labeled RNA substrate is cleaved by the ribozyme and the products are analyzed by polyacrylamide gel electrophoresis.

I. Introduction and Theory

RNA Enzymes

Prior to 1981, biochemistry students heard in lectures and read in text-books the statement: "All enzymes are proteins, but not all proteins are enzymes." From the early discovery and study of biological catalysts (enzymes) in the mid to late 1800s to the 1980s, hundreds of enzymes were purified and characterized. All were found to be protein molecules. The discovery of RNA molecules that act like enzymes by two independent research groups in 1981 and 1982 began a revolution in biochemistry. The term "ribozyme" was coined for this new form of biological catalyst. When research results on RNA catalysts were first announced, many bio-chemists were skeptical, because for over 75 years it had been thought that all enzymes were protein molecules. More extensive studies of RNA enzymes and the discovery of several different kinds in different organisms finally convinced even the most skeptical biochemist that, indeed, RNA molecules can serve as biological catalysts. The true significance of the discovery of catalytic RNA was finally acknowledged by the awarding of the 1989 Nobel Prize in Chemistry to the two discoverers of these unique catalysts—Thomas Cech (University of Colorado-Boulder) and Sidney Altman (Yale University). Because of its major impact on biochemistry, a brief history of the work that led to the discovery of catalytic RNA is appropriate.

Discovery of Ribozymes

It was discovered in 1977 that eukaryotic genes are discontinuous; that is, coding sequences (exons) are interrupted by stretches of noncoding DNA called intervening sequences (IVSs) or introns. The RNA polymerases acting on the DNA transcribe both the exons and the introns to produce precursors of messenger RNAs that are subsequently spliced to remove the IVSs or introns. The process for nuclear mRNA splicing, which is very complex, is initiated by recognition and cleavage of splice sites (specific nucleotide sequences), followed by formation of a novel lariat RNA with branches and ligation to generate the final mRNA. These reactions are catalyzed by small nuclear ribonucleoproteins. In at least one organism studied, *Tetrahymena thermophila* (a ciliated protozoan), RNA splicing was found by Cech and co-workers to be catalyzed by the RNA intron itself, without the assistance of protein catalysts. The intramolecular excision reaction occurs when isolated pre-rRNA is incubated with guanosine (G-OH) and Mg^{2+}. The splitting process for *Tetrahymena* (A) is outlined and compared to that for nuclear pre-mRNA (B) in Figure E22.1. In chemical terms, splicing process (A) consists of two transesterification steps. After the initial splicing events, the linear excised intron, which consists of 414 ribonucleotide bases, undergoes a series of cyclization and hydrolysis reactions mediated by RNA. The final product is L-19 IVS-RNA, a linear molecule lacking the first 19 nucleotides of the original excised IVS-RNA. Although these splicing processes are mediated by the RNA molecule, the RNA in this series of reactions does not represent a true enzyme because it is not regenerated and does not process several substrate molecules; that is, it does not display multiple turnover like protein enzymes.

The involvement of RNA even as a mediator of splicing processes was, indeed, surprising; however, more revolutionary discoveries were yet uncovered. Cech and his co-workers have since shown that the L-19 IVS-RNA, which can no longer carry out reactions on itself, can catalyze cleavage-ligation reactions, with multiple turnover, on other RNA molecules and therefore can be considered a true enzyme. Several specific enzymatic activities of this shortened form of the *Tetrahymena*

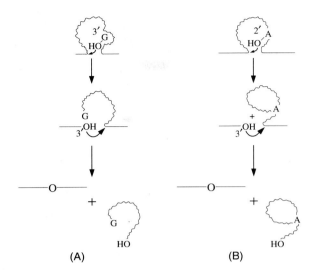

Figure E22.1
Processing of RNA. (A) *Tetrahymena;* (B) nuclear pre-mRNA. G-OH represents guanosine.

self-splicing IVS-RNA have since been discovered and examined in detail: (1) the cleavage and rejoining (ligation) of oligonucleotide substrates, (2) phosphodiesterase activity, (3) acid phosphatase activity, and finally (4) cleavage of large or small RNA molecules at sequence-specific sites (that is, the ribozyme acts like an RNA endonuclease). This last reaction is extremely important because it is the first example of a restriction enzyme that acts on RNA substrates. DNA restriction endonucleases were discovered years earlier, and their use in sequence-specific cleavage of DNA has had a major impact on biochemistry and molecular biology (see Experiment 21). The sequence specificity of ribozymes approaches that of DNA restriction endonucleases. As shown in Figure E22.2, single-stranded RNA substrate molecules are cleaved at a site next to a tetranucleotide recognition sequence (-N-N-N-U$^{\downarrow}$-X). The arrow represents the site of cleavage. Four nucleotides preceding the cleavage site of the substrate form Watson-Crick base pairs with part of the active site (-N'-N'-N'-G-) of the catalytic RNA. The specificity of cleavage can be changed by modifying the sequence of nucleotides at the active site by site-specific mutagenesis. Note that the other substrate, GTP, becomes attached to the newly generated 5' end.

The kinetic parameters and optimal reaction conditions displayed by ribozymes are similar to those of the more common protein catalysts. The temperature and pH optimum for cleavage are 50°C and 7.5–8.0, respectively. The turnover number is about 1.0 min^{-1} and K_M is about 1 μM. There is a requirement for a metal ion; Mg^{2+} is preferred. The de-

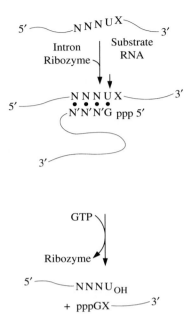

Figure E22.2
The action of a ribozyme on an RNA substrate showing Watson-Crick base pairing and role of GTP. The ribozyme is designed to cleave after the sequence NNNU. N and N′ represent any nucleotide; N is complementary to N′.

naturant urea is sometimes added to ribozyme-cleavage reaction mixtures because it results in greater specificity of cleavage. The action of urea on ribozyme function is not well understood.

Studies of the sequence-specific cleavage of RNA have led to the first commercially available RNA restriction enzyme. This ribozyme, called RNAzyme Tet 1.0 and marketed by United States Biochemical Corporation (USBC), is designed to cleave after the second 3′U of the sequence C-U-C-U. That is, it has at the active site a sequence guide of -G-A-G-G to hold the substrate RNA in place by Watson-Crick base pairing. The enzyme should be of great value as a tool in the analysis of RNA structure and function. For example, it will assist in direct physical mapping, nucleotide sequencing using base-specific ribonucleases or chemical agents, and determination of the secondary and tertiary structures of RNA.

Even though there has been a rapid increase in our knowledge of ribozyme action, there is still much to be learned. Ribozymes, like protein enzymes, require a specific shape and form for catalytic activity. A major current research interest is elucidation of the secondary and tertiary structure of ribozymes. Since ribozymes have not yet been crystallized, X-ray crystallography cannot be used for structure determination.

The discovery of catalytic RNA has interesting evolutionary implications. The fact that RNA can serve as both an informational molecule and a catalyst suggests that when life originated, RNA may have been one of the first biomolecules. During the evolutionary process, RNA's catalytic activity may have been replaced by more effective catalysts, protein enzymes, and the informational role of RNA taken over by the more chemically stable DNA.

Ribozymes have many potential biomedical applications in addition to their use as tools in specific cleavage of RNA. Many pathogenic viruses contain RNA, which could become a target for cleavage by carefully constructed ribozymes. There is the potential that ribozymes could be engineered to cleave RNA and DNA molecules only at a predetermined sequence of ribonucleotides. Infectious mammalian viruses, retroviruses, or cellular mRNAs involved in maintenance of the transformed state might be inactivated by such treatment. A ribozyme has been designed to cleave, sequence specifically, the RNA of the human immunodeficiency virus (HIV), the virus that causes AIDS (Sarver et al., 1990).

Overview of the Experiment

In this experiment you will investigate the action of a commercially available ribozyme. RNAzyme Tet 1.0, a shortened version of the pre-rRNA of *Tetrahymena,* catalyzes the sequence-specific cleavage of large or small, single-stranded RNA. The reaction catalyzed by the ribozyme is shown in Equation E22.1.

▶ 5'OH-GGGACUCUCUAAAAA + GTP

$$\updownarrow \text{ ribozyme} \qquad \text{(Equation E22.1)}$$

5'OH-GGGACUCUCU + 5'ppp-GAAAAA

The RNA substrate containing 15 ribonucleotide bases (a 15-mer) is cleaved to a 10-mer and a 6-mer. The substrate GTP is added to the new 5' end of one of the products. Since none of the products or substrates is colored (none absorbs light in the visible region) some method other than spectrophotometry must be used to measure the extent of reaction. Electrophoresis, the standard method for analyzing nucleic acids, may be used to separate the products; however, they must be labeled in some way for detection and quantification. Radioisotope labeling provides the most sensitive approach. The substrate will be labeled with ^{32}P at the 5' end (Equation E22.2). The reaction is catalyzed by polynucleotide kinase (PNK).

▶ 5'HO-GGGACUCUCUAAAAA + [$\gamma-^{32}$P]ATP

$$\updownarrow \text{ PNK} \qquad \text{(Equation E22.2)}$$

$_3^{2-}$O^{32}P-O-GGGACUCUCUAAAAA + ADP

Only one of the cleavage products from Equation E22.1, the 10-mer, will be labeled after hydrolysis. Since the products are relatively small, they can be separated on a 20% polyacrylamide gel under denaturing conditions (7 M urea).

The experiment consists of three parts: (A) radiolabeling the RNA substrate with ^{32}P, (B) performing the cleavage reaction, and (C) analyzing the product by electrophoresis. Part A requires 3 hours, part B 1 hour, and part C 3 hours. Since this experiment requires the use of radioisotopes, students must be aware of safety rules and special precautions (Chapter 6). In addition, it is necessary to avoid contamination of reagents and reaction vessels by ribonucleases. The enzymes, which are ubiquitous, will cleave the ribozyme and render it inactive. The best ways to avoid nonspecific degradation by nucleases are to wear gloves throughout the experiment, store reagents on ice during the experiment, use clean equipment, and work rapidly and carefully.

II. Materials and Supplies

RNAzyme Tet 1.0 kit from USBC. Consists of:

RNAzyme Tet 1.0 (activity varies with lot numbers)

Substrate : 5′GGGACUCUCUAAAAA

TMN buffer: 0.5 M Tris, 0.1 M MgCl$_2$, 0.1 M NaCl, pH 7.5

GTP: 1 mM

Urea: 7.5 M

Stop solution: 95% formamide, 0.02 M EDTA, 0.05% xylene cyanol, 0.05% bromphenol blue, pH 8.0

T4 Polynucleotide kinase (PNK)

T4 PNK reaction buffer: 0.5 M Tris, 0.05 M dithiothreitol, 0.1 M MgCl$_2$, 1mM spermidine pH 7.5

RNA diluent: sterile H$_2$O

Radiolabeled ATP, [γ-^{32}P]ATP, at least 1000 Ci/mmole

NaCl: 2.5 M

Ethanol: absolute or 95%

Microcentrifuge: capable of spinning 1.5-mL tubes at 12,000 $\times$ g

Polypropylene centrifuge tubes: 1.5 mL

Gel electrophoresis apparatus: glass plates (8.5 $\times$ 10 cm), 0.75-mm plastic spacers, comb to prepare at least 8 sample wells, power supply for 500 V, Scotch electrical 2-inch tape, apparatus for vertical electrophoresis.

Solutions for polyacrylamide gels:

TBE buffer: 0.1 M Tris, 0.08 M boric acid, 5 mM Na$_2$EDTA, pH 8.0

Acrylamide solution: 40% acrylamide, 2% bisacrylamide, 7 M urea in H$_2$O

TEMED: 0.02 M $N,N,N'N'$-tetramethylethylenediamine, 7 M urea in TBE buffer

Ammonium persulfate: 0.010 M ammonium persulfate, 7 M urea in H$_2$O

Temperature baths: 37°C and 50°C

Micropipettors and tips

Materials for autoradiography: film, film holder, developing solutions, and darkroom

III. Experimental Procedure

A. *Radiolabeling the Substrate with* ^{32}P

⚠ Caution

The following procedure requires the use of radioactive ^{32}P. Wear gloves and do all work behind a glass or Plexiglas shield. Read Chapter 6 for special guidelines in the use of ^{32}P.

If you are pregnant or may become pregnant during this course, special precautions in the use of radioisotopes must be considered. Since each institution has its own rules and regulations, consult your instructor. Some institutions may require the use of a personnel badge in order to monitor the exposure for each individual. Your instructor will provide this if necessary.

Acrylamide in the unpolymerized form is a skin irritant and a potential neurotoxin. Wear gloves and a mask while weighing the dry powder. Do not breathe the dust. Prepare all acrylamide solutions in the hood. Do not mouth pipet any solutions used for gel formation or staining.

Do not touch the electrophoresis chamber or wires while the electrophoretic operation is in progress. Voltages may reach as high as 300 volts and shocks may be fatal.

The 15-base RNA substrate must be 5'-end labeled before cleavage. Use the following protocol:

1. Combine all reagents in a 1.5-mL polypropylene microcentrifuge tube. Use the appropriate size micropipettor and tips. Mix well!

    10 μL PNK reaction buffer

    55 μL RNA diluent (sterile H$_2$O)

  10 μL ribonucleotide substrate (50 pmoles)

  20 μL [γ-^{32}P]ATP (60 pmole, 3000 Ci/mmole, 10 mCi/mL)

  5 μL polynucleotide kinase (30 units/μL or 150 units)

2. Incubate the kinase reaction mixture in a 37°C water bath for 30 minutes.

3. Precipitate the labeled RNA by adding 11 μL of 2.5 M NaCl and 330 μL of ethanol. Mix well and allow to sit at -20°C for 1 hour.

4. Centrifuge at 4°C for 30 minutes at 12,000 $\times$ g.

5. Pour off the supernatant and discard in a radioactive waste container. Add 1.0 mL of ice-cold 70% ethanol and rinse the precipitated RNA pellet by inverting four or five times. Centrifuge at 4°C for 30 minutes at 12,000 $\times$ g.

6. Pour off and discard the supernatant (may still be radioactive; test with a Geiger-Müller counter).

7. Dry the pellet at room temperature or under vacuum for 15 minutes. Dissolve the pellet in 20 μL of RNA diluent or sterile H_2O. The labeled substrate may be stored for a few days at -20°C if not used immediately for part B. The half-life of ^{32}P is 14 days. This procedure prepares enough labeled substrate for four student groups.

B. The Cleavage Reaction

 Caution

Continue working carefully, with gloves, behind a Plexiglas shield.

1. Prepare the following reaction mixtures in two separate 1.5-mL centrifuge tubes:

Complete reaction mixture:

  2.5 μL of ^{32}P-labeled RNA substrate (from part A)

  2.5 μL TMN buffer

  5.0 μL GTP (final concentration 1 mM)

  5.0 μL 7.5 M urea (final concentration 2.0 M)

  1 unit RNAzyme Tet 1.0 (amount will vary depending on concentration and activity of ribozyme)

  RNA diluent or sterile H_2O—add, if necessary, to bring final volume of reaction mixture to 25 μL

Control reaction mixture:

  2.5 μL of ^{32}P-labeled RNA substrate (from part A)

  2.5 μL TMN buffer

  5.0 μL GTP

5.0 μL 7.5 M urea

RNA diluent to bring final volume to 25 μL

2. Incubate the two reaction mixtures at 50°C for 30 minutes.
3. Add 25 μL of stop solution containing dyes to each mixture and heat at 80°C for 1 minute.
4. The reaction mixtures may be stored at -20°C for a few days or analyzed directly in part C.

C. Product Analysis by Electrophoresis

Prepare a 20% polyacrylamide gel (acrylamide-bisacrylamide = 20:1, 7 M urea) as described in Experiments 3 and 20. Prepare 0.5 to 0.75-mm-thick gels following the directions provided with the electrophoresis apparatus. Prepare 40 mL of acrylamide solution as follows:

20 mL acrylamide-bisacrylamide, 7 M urea solution

10 mL TEMED, 7 M urea solution

10 mL ammonium persulfate, 7 M urea solution

Mix well and immediately pour into the space between the glass plates until full. If some solution leaks from the edges, add more gel solution. Insert the plastic comb and allow the acrylamide to polymerize (about 30–45 minutes). Carefully remove the comb and rinse the sample wells with TBE buffer. Clamp the gel plate onto the electrophoresis chamber. Fill buffer reservoirs with TBE. Each student will need four sample wells, two for duplicate wells of the complete reaction mixture and two for duplicate application of the control reaction mixture. A 5- to 10-μL sample is loaded into each well. Run the electrophoresis at 300 volts for 30 minutes or until bromphenol blue, the faster-moving dye, has migrated about 3 cm. Prepare a permanent picture of the gel by autoradiography. A 3- to 5-minute exposure of the gel on the film should be sufficient. Dispose of the radioactive gel and reaction mixtures as directed by your instructor.

IV. Analysis of Results

A. Radiolabeling the Substrate with ^{32}P

Why is it necessary to use ^{32}P in this experiment? What is the purpose of 2.5 M NaCl and ethanol in preparing the labeled substrate? What other radioactive products in addition to substrate are formed?

B. The Cleavage Reaction

Describe the purpose of each of the reagents in the reaction mixtures. Why are two reaction mixtures (complete and control) run? Describe how the stop solution and heating step quench the reactions.

C. Product Analysis by Electrophoresis

Describe the pattern of bands displayed on the autoradiogram. Why is there not a band for the 6-mer product? In order to achieve better separation of products, why not run the electrophoresis longer (until both dyes are off the gel plate)?

V. Questions and Problems

1. Write the ribozyme-cleavage reaction if $[\gamma\text{-}^{32}P]GTP$ is used as substrate along with a "cold" strand of 15-mer RNA.

▶ 2. Explain on a molecular level how urea may affect the specificity of the ribozyme cleavage reaction.

▶ 3. Predict the site of cleavage by the action of RNAzyme Tet 1.0 on each of the following substrates.

    (a) 5'GGACCUCUAAAAA3'
    (b) 5'GGGACAUUUAAAAA3'
    (c) 5'GGGACGAAUAAAAA3'

▶ 4. What is the purpose of xylene cyanol and bromphenol blue in the stop solution?

▶ 5. How would you modify this experiment to determine K_M and V_{max} for the RNA substrate? For GTP?

▶ 6. Why would you not be able to store the ^{32}P-labeled RNA substrate in a frozen state and use it several weeks later?

7. The radioisotope ^{35}S is sometimes used in place of ^{32}P for labeling RNA and DNA substrates. ^{35}S has several advantages, including a longer half-life (87 days vs. 14 days for ^{32}P) and lower energy (0.167 MeV vs. 1.71 MeV for ^{32}P). The labeling procedure for ^{35}S involves the use of $[\gamma\text{-}^{35}S]ATP$ and PNK. Use structures to write a reaction showing the ^{35}S labeling of a short strand of RNA.

VI. References

R. Athota and T. Radhakrishnan, *Biochem. Educ.* **19,** 72–73 (1991). "Ribozyme: A Novel Enzyme."

T. Cech, *Sci. Am.* **255** (5), 64–75 (1986). "RNA As an Enzyme."

T. Cech, *Science* **236,** 1532–1537 (1987). "The Chemistry of Self-Splicing RNA and RNA Enzymes."

T. Cech, *J. Am. Med. Assoc.* **260,** 3030–3034 (1988). "Ribozymes and Their Medical Implications."

T. Cech, *Editorial Comments* **16** (No. 2), 1–5 (1989). United States Biochemical Corp. P.O. Box 22400, Cleveland, OH 44122. "Ribozymes: Tools for Sequence-Specific Cleavage of RNA."

T. Cech and B. Bass, *Annu. Rev. Biochem.* **55,** 599–629 (1986). "Biological Catalysis by RNA."

D. Herschlag and T. Cech, *Biochemistry* **29,** 10159–10171 (1990). "Catalysis of RNA Cleavage by the *Tetrahymena thermophila* Ribozyme."

D. Herschlag and T. Cech, *Nature* **344,** 405–409 (1990). "DNA Cleavage Catalyzed by the Ribozyme from *Tetrahymena.*"

C. Mathews and K. van Holde, *Biochemistry* (1990), Benjamin/Cummings (Redwood City, CA), pp. 373–375. Ribozymes.

G. McCorkle and S. Altman, *J. Chem. Educ.* **64,** 221–226 (1987). "RNAs As Catalysts: A New Class of Enzymes."

R. Padgett, P. Grabowski, M. Konarska, S. Seiler, and P. Sharp, *Annu. Rev. Biochem.* **55,** 1119–1150 (1986). "Splicing of Messenger RNA Precursors."

R. Padgett, P. Grabowski, M. Konarska, and P. Sharp, *Trends Biochem. Sci.* **10** (4), 154–157 (1985). "Splicing Messenger RNA Precursors: Branch Sites and Lariat RNAs."

J. Rawn, *Biochemistry,* 2nd ed. (1989), Neil Patterson Publishers (Burlington, NC), pp. 784–785, 792–799, 847–848. Catalytic RNA.

N. Sarver, E. Cantin, P. Chang, J. Zaia, P. Ladne, D. Stephens, and J. Rossi, *Science* **247,** 1222–1225 (1990). "Ribozymes as Potential Anti–HIV-1 Therapeutic Agents."

L. Stryer, *Biochemistry,* 3rd ed. (1988) W. H. Freeman (New York), pp. 113–114, 214–215. Ribozymes.

United States Biochemical Corporation (1989), P.O. Box 22400, Cleveland, OH 44122. "RNAzyme™ Tet 1.0 : Guidelines for Use."

D. Voet and J. Voet, *Biochemistry* (1990), John Wiley & Sons (New York), pp. 884–887. Ribozymes.

A. Zaug, M. Been, and T. Cech, *Nature* **324,** 429–433 (1986). "The *Tetrahymena* Ribozyme Acts Like an RNA Restriction Endonuclease."

APPENDIX I

Properties of Common Acids and Bases

Compound	Formula	Molecular Weight	Specific Gravity	% by Weight	Molarity (*M*)
Acetic acid, glacial	CH_3COOH	60.1	1.05	99.5	17.4
Ammonium hydroxide	NH_4OH	35.0	0.89	28	14.8
Formic acid	$HCOOH$	46.0	1.20	90	23.4
Hydrochloric acid	HCl	36.5	1.18	36	11.6
Nitric acid	HNO_3	63.0	1.42	71	16.0
Perchloric acid	$HClO_4$	100.5	1.67	70	11.6
Phosphoric acid	H_3PO_4	98.0	1.70	85	18.1
Sulfuric acid	H_2SO_4	98.1	1.84	96	18.0

APPENDIX II

Properties of Common Buffer Compounds

Compound	Abbreviation	Molecular Weight	pK_1	(20°C) pK_2	pK_3	pK_4
N-(2-Acetamido)-2-aminoethanesulfonic acid	ACES	182.2	6.9	—	—	—
N-(2-Acetamido)-2-iminodiacetic acid	ADA	212.2	6.60	—	—	—
Acetic acid		60.1	4.76	—	—	—
Arginine	Arg	174.2	2.17	9.04	12.48	—
Barbituric acid		128.1	3.79	—	—	—
N,N-bis(2-Hydroxyethyl)-2-aminoethane-sulfonic acid	BES	213.1	7.15	—	—	—
N,N-bis(2-Hydroxyethyl)glycine	Bicine	163.2	8.35	—	—	—
Boric acid		61.8	9.23	12.74	13.80	—
Citric acid		210.1	3.10	4.75	6.40	—
Ethylenediaminetetraacetic acid	EDTA	292.3	2.00	2.67	6.24	10.88
Formic acid		46.03	3.75	—	—	—
Fumaric acid		116.1	3.02	4.39	—	—
Glycine	Gly	75.1	2.45	9.60	—	—
Glycylglycine		132.1	3.15	8.13	—	—
N-2-Hydroxyethylpiperazine-N'-2-ethanesulfonic acid	HEPES	238.3	7.55	—	—	—
N-2-Hydroxyethylpiperazine-N'-3-propanesulfonic acid	HEPPS	252.3	8.0	—	—	—
Histidine	His	209.7	1.82	6.00	9.17	—
Imidazole		68.1	6.95	—	—	—
2-(N-Morpholino)ethanesulfonic acid	MES	195	6.15	—	—	—
3-(N-Morpholino)propanesulfonic acid	MOPS	209.3	7.20	—	—	—
Phosphoric acid		98.0	2.12	7.21	12.32	—
Succinic acid		118.1	4.18	5.60	—	—
3-Tris(hydroxymethyl)aminopropanesulfonic acid	TAPS	243.2	8.40	—	—	—
N-Tris(hydroxymethyl)methyl-2-aminoethanesulfonic acid	TES	229.2	7.50	—	—	—
N-Tris(hydroxymethyl)methylglycine	Tricine	179	8.15	—	—	—
Tris(hydroxymethyl)aminomethane	Tris	121.1	8.30	—	—	—

APPENDIX III

pK_a Values and pH$_I$ Values of Amino Acids

Name	Abbreviation	pK_1 (α-carboxyl)	pK_2 (α-amino)	pK_R (side chain)	pH$_I$
Alanine	ala	2.3	9.7	—	6.0
Arginine	arg	2.2	9.0	12.5	10.8
Asparagine	asn	2.0	8.8	—	5.4
Aspartate	asp	2.1	9.8	3.9	3.0
Cysteine	cys	1.7	10.8	8.3	5.0
Glutamate	glu	2.2	9.7	4.3	3.2
Glutamine	gln	2.2	9.1	—	5.7
Glycine	gly	2.3	9.6	—	6.0
Histidine	his	1.8	9.2	6.0	7.6
Isoleucine	ilu	2.4	9.7	—	6.1
Leucine	leu	2.4	9.6	—	6.0
Lysine	lys	2.2	9.0	10.5	9.8
Methionine	met	2.3	9.2	—	5.8
Phenylalanine	phe	1.8	9.1	—	5.5
Proline	pro	2.0	10.6	—	6.3
Serine	ser	2.2	9.2	—	5.7
Threonine	thr	2.6	10.4	—	6.5
Tryptophan	trp	2.4	9.4	—	5.9
Tyrosine	tyr	2.2	9.1	10.1	5.7
Valine	val	2.3	9.6	—	6.0

Molecular Weights of Some Common Proteins

Name (Source)	Molecular Weight
Albumin (bovine serum)	65,400
Albumin (egg white)	45,000
Carboxypeptidase A (bovine pancreas)	35,268
Carboxypeptidase B (porcine)	34,300
Catalase (bovine liver)	250,000
Chymotrypsinogen (bovine pancreas)	23,200
Cytochrome c	13,000
Hemoglobin (bovine)	64,500
Insulin (bovine)	5,700
α-Lactalbumin (bovine milk)	17,400
Lysozyme (egg white)	14,600
Malate dehydrogenase (pig heart mitochondria)	70,000
Myoglobin (horse heart)	16,900
Pepsin (porcine)	35,000
Peroxidase (horseradish)	40,000
Ribonuclease I (bovine pancreas)	12,600
Trypsinogen (bovine pancreas)	24,000
Trypsin (bovine pancreas)	23,800
Tyrosinase (mushroom)	128,000
Uricase (pig liver)	125,000

Common Abbreviations Used in This Text

A	adenine or absorbance
AMP, ADP, ATP	adenosine mono-, di-, or triphosphate
BCA	bicinchoninic acid
bp	base pairs
BPG	bisphosphoglycerate
Bq	Becquerel
BSA	bovine serum albumin
C	cytosine
CE	capillary electrophoresis
Ci	Curie
CM cellulose	carboxymethyl cellulose
CPM	counts per minute
DABITC	4-N,N-dimethylaminoazobenzene-4'-isothiocyanate
DA	dalton
DANSYL	1,1-dimethylaminonaphthalene-5-sulfonyl chloride
DCIP	dichlorophenolindophenol
DEAE cellulose	diethylaminoethyl cellulose
DNA	deoxyribonucleic acid
DNP	dinitrophenyl
DOPA	dihydroxyphenylalanine
E	absorption coefficient
EDTA	ethylenediaminetetraacetic acid
ELISA	enzyme linked immunosorbent assay
FAD(H$_2$)	flavin adenine dinucleotide (reduced form)
FAME	fatty acid methyl ester
FMN(H$_2$)	flavin mononucleotide (reduced form)
FPLC	fast protein liquid chromatography
G	guanine
GC	gas chromatography
GMP, GDP, GTP	guanosine mono-, di-, or triphosphate
Hb	hemoglobin
HDL	high-density lipoprotein
HMIS	Hazardous Materials Identification System
HPLC	high-performance liquid chromatography

IE	immunoelectrophoresis
IEF	isoelectric focusing
IHP	inositol hexaphosphate
IMAC	immobilized metal-ion affinity chromatography
IR	infrared
IVS	intervening sequence
kb	kilobase pairs
LDL	low-density lipoprotein
MAO	monoamine oxidase
MS	mass spectrometry
MSDS	material safety data sheet
NAD(H)	nicotinamide adenine dinucleotide (reduced form)
NADP(H)	nicotinamide adenine dinucleotide phosphate (reduced form)
NBT	nitroblue tetrazolium
NMR	nuclear magnetic resonance
OSHA	Occupational Safety and Health Administration
PAGE	polyacrylamide gel electrophoresis
PCR	polymerase chain reaction
PEG	polyethylene glycol
PFGE	pulsed field gel electrophoresis
PMS	phenazine methosulfate
PMT	photomultiplier tube
POPOP	1,4-bis[5-phenyl-2-oxazolyl]benzene
PPO	2,5-diphenyloxazole
PTH	phenylthiohydantoin
RCF	relative centrifugal force
RFLP	restriction fragment length polymorphism
RIA	radioimmunoassay
RNA	ribonucleic acid
s	sedimentation coefficient
S	Svedberg (10^{-13}s)
SDS	sodium dodecyl sulfate
SMP	submitochondrial particles
T	thymine
TAG	triacylglycerol
TEMED	N,N,N',N'-tetramethylethylenediamine
TLC	thin-layer chromatography
Tris	tris(hydroxymethyl)aminomethane
U	uracil
UV	ultraviolet
VIS	visible
VLDL	very low-density lipoprotein

APPENDIX VI

Units of Measurement

The International System of Units (SI)

Quantity	Unit	Abbreviation
length	meter	m
mass	kilogram	kg
time	second	s
temperature	kelvin	K
electric current	ampere	A
amount of substance	mole	mol
radioactivity	becquerel	Bq
volume	liter	L

Metric Prefixes

Name	Abbreviation	Multiplication Factor (relative to "1")
atto	a	10^{-18}
femto	f	10^{-5}
pico	p ($\mu\mu$)	10^{-12}
nano	n(mμ)	10^{-9}
micro	μ	10^{-6}
milli	m	10^{-3}
centi	c	10^{-2}
deci	d	10^{-1}
deca	da	10
kilo	k	10^{3}
mega	M	10^{6}
giga	G	10^{9}

Units of Length

Name	Abbreviation	Multiplication Factor (relative to meter)
kilometer	km	10^3
meter	m	1
centimeter	cm	10^{-2}
millimeter	mm	10^{-3}
micrometer	μm	10^{-6}
nanometer	nm	10^{-9}
Angstrom	Å	10^{-10}

Units of Mass

Name	Abbreviation	Multiplication Factor (relative to gram)
kilogram	kg	10^3
gram	g	1
milligram	mg	10^{-3}
microgram	μg	10^{-6}
nanogram	ng	10^{-9}

Units of Volume

Name	Abbreviation	Multiplication Factor (relative to liter)
liter	L	1
deciliter	dL	10^{-1}
milliliter	mL	10^{-3}
microliter	μL	10^{-6}

Table of the Elements*

	Symbol	Atomic No.	Atomic Mass		Symbol	Atomic No.	Atomic Mass
Actinium	Ac	89	227.0278	Mercury	Hg	80	200.59
Aluminum	Al	13	26.98154	Molybdenum	Mo	42	95.94
Americium	Am	95	[243]†	Neodymium	Nd	60	144.24
Antimony	Sb	51	121.75	Neon	Ne	10	20.179
Argon	Ar	18	39.948	Neptunium	Np	93	237.0482
Arsenic	As	33	74.9216	Nickel	Ni	28	58.70
Astatine	At	85	[210]	Niobium	Nb	41	92.9064
Barium	Ba	56	137.33	Nitrogen	N	7	14.0067
Berkelium	Bk	97	[247]	Nobelium	No	102	[259]
Beryllium	Be	4	9.01218	Osmium	Os	76	190.2
Bismuth	Bi	83	208.9804	Oxygen	O	8	15.9994
Boron	B	5	10.81	Palladium	Pd	46	106.4
Bromine	Br	35	79.904	Phosphorus	P	15	30.97376
Cadmium	Cd	48	112.41	Platinum	Pt	78	195.09
Calcium	Ca	20	40.08	Plutonium	Pu	94	[244]
Californium	Cf	98	[251]	Polonium	Po	84	[209]
Carbon	C	6	12.011	Potassium	K	19	39.0983
Cerium	Ce	58	140.12	Praseodymium	Pr	59	140.9077
Cesium	Cs	55	132.9054	Promethium	Pm	61	[145]
Chlorine	Cl	17	35.453	Protactinium	Pa	91	231.0359
Chromium	Cr	24	51.996	Radium	Ra	88	226.0254
Cobalt	Co	27	58.9332	Radon	Rn	86	[222]
Copper	Cu	29	63.546	Rhenium	Re	75	186.207
Curium	Cm	96	[247]	Rhodium	Rh	45	102.9055
Dysprosium	Dy	66	162.50	Rubidium	Rb	37	85.4678
Einsteinium	Es	99	[252]	Ruthenium	Ru	44	101.07
Erbium	Er	68	167.26	Samarium	Sm	62	150.4

* Atomic masses are based on Carbon-12.
† A value given in brackets denotes the mass number of the longest-lived or best-known isotope.

	Symbol	Atomic No.	Atomic Mass		Symbol	Atomic No.	Atomic Mass
Europium	Eu	63	151.96	Scandium	Sc	21	44.9559
Fermium	Fm	100	[257]	Selenium	Se	34	78.96
Fluorine	F	9	18.998403	Silicon	Si	14	28.0855
Francium	Fr	87	[223]	Silver	Ag	47	107.868
Gadolinium	Gd	64	157.25	Sodium	Na	11	22.98977
Gallium	Ga	31	69.72	Strontium	Sr	38	87.62
Germanium	Ge	32	72.59	Sulfur	S	16	32.06
Gold	Au	79	196.9665	Tantalum	Ta	73	180.9479
Hafnium	Hf	72	178.49	Technetium	Tc	43	[98]
Helium	He	2	4.00260	Tellurium	Te	52	127.60
Holmium	Ho	67	164.9304	Terbium	Tb	65	158.9254
Hydrogen	H	1	1.0079	Thallium	Tl	81	204.37
Indium	In	49	114.82	Thorium	Th	90	232.0381
Iodine	I	53	126.9045	Thulium	Tm	69	168.9342
Iridium	Ir	77	192.22	Tin	Sn	50	118.69
Iron	Fe	26	55.847	Titanium	Ti	22	47.90
Krypton	Kr	36	83.80	Tungsten	W	74	183.85
Lanthanum	La	57	138.9055	Uranium	U	92	238.029
Lawrencium	Lr	103	[260]	Vanadium	V	23	50.9415
Lead	Pb	82	207.2	Xenon	Xe	54	131.30
Lithium	Li	3	6.941	Ytterbium	Yb	70	173.04
Lutetium	Lu	71	174.967	Yttrium	Y	39	88.9059
Magnesium	Mg	12	24.305	Zinc	Zn	30	65.38
Manganese	Mn	25	54.9380	Zirconium	Zr	40	91.22
Mendelevium	Md	101	[258]				

APPENDIX VIII

Values of t for Analysis of Statistical Confidence Limits

d.f.	Probability of Larger Value of t, Sign Ignored			d.f.			
	0.05	0.02	0.01		0.05	0.02	0.01
1	12.706	31.821	63.657	14	2.145	2.624	2.977
2	4.303	6.0965	9.925	15	2.131	2.602	2.947
3	3.182	4.541	5.841	16	2.120	2.583	2.921
4	2.776	3.747	4.604	17	2.110	2.567	2.898
5	2.571	3.365	4.032	18	2.101	2.552	2.878
6	2.447	3.143	3.707	19	2.093	2.539	2.861
7	2.365	2.998	3.499	20	2.086	2.528	2.845
8	2.306	2.896	3.355	21	2.080	2.518	2.831
9	2.262	2.821	3.250	22	2.074	2.508	2.819
10	2.228	2.764	3.169	23	2.069	2.500	2.807
11	2.201	2.718	3.106	24	2.064	2.492	2.797
12	2.179	2.681	3.055	25	2.060	2.485	2.787
13	2.160	2.650	3.012				

From *Experimental Techniques in Biochemistry* by J. Brewer, A. Pesce, and R. Ashworth, p. 332. Copyright © 1974, Reprinted by permission of Prentice-Hall, Inc., Englewood Cliffs. NJ.

APPENDIX IX

Solutions to Selected Questions

Experiment 2

1.

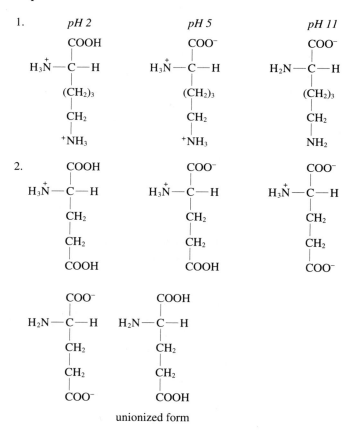

3. The separation of amino acids by paper chromatography is based on polarity. The less polar an amino acid, the greater the R_f. Arrangement in order of decreasing R_f is Leu > Val > Ala > Ser > Lys.

4. Alanine, pH 3.0: slightly +; aspartic acid, pH 4.0: very slightly − .

5. *Toward cathode*   *Toward anode*   *Remain stationary*
   Lys   Glu, Phe, Ser   Val

7. Order of elution: Ser > Glu > Pro > Ala > Arg

8. See Figure E2.2.

Experiment 3

2. Gel electrophoresis.

3. The mobility of bromphenol blue must be greater than that of any proteins in the sample.

4. (a) Slower mobility
   (b) Probably no effect
   (c) No effect
   (d) No effect
   (e) Slower mobility
   (f) No effect on mobility; however, actual distance moved would be less

5. Read Chapter 2, Section D and Experiment 5.

6. Read Chapter 5, Section A.

7. Read Chapter 2, Section D.

8. Read Chapter 2, Section D.

9. Read Chapter 2, Section D.

Experiment 4

1. Edman method vs. dansyl chloride. Because of the cyclization-cleavage reaction that occurs, the Edman method can be used for sequential degradation of a peptide or protein. NH_2-terminal residues can be removed, one at a time, and identified chromatographically. Dansyl chloride can be used only for single NH_2-terminal analysis. Dansyl chloride, however, has some advantages. Dansyl amino acids can be more readily separated by chromatographic techniques and detection by a UV lamp is more sensitive than for phenylthiohydantoin amino acids (PTH-amino acids). Also, work-up of the dansyl chloride reaction is less cumbersome and, hence, more rapid. Excess phenylisothiocyanate reagent and coupling products must be extracted from the Edman reaction mixture. This requires more work-up steps. In your experiment, this extraction was accomplished with toluene. In the dansyl chloride reaction, excess reagent is hydrolyzed to the corresponding sulfonic acid, which does not interfere with chromatographic analysis.

2. If the solvent level in a chromatography jar is above the application spots, the compounds spotted will be dissolved in the solvent and dif-

fuse throughout the chromatography solvent. The compounds will no longer be highly concentrated at a single area on the plate. Even if there is sufficient compound remaining on the plate for detection, the final developed spot will most likely be too large for an accurate R_f measurement.

3. At least three major factors must be considered in choosing a solvent for the Edman degradation. (1) The solvent must dissolve both the peptide and the phenylisothiocyanate reagent. This problem can best be solved by using an aqueous-organic system. (2) The coupling reaction requires an unprotonated α-amino group of the peptide for nucleophilic attack on the phenylisothiocyanate reagent:

$$R\overset{..}{N}H_2 + S{=}C{=}N{-}\bigcirc \longrightarrow RNH\underset{\underset{S}{\|}}{C}{-}NH{-}\bigcirc$$

This requires the use of basic solvent. (3) The solvent must be basic but not highly nucleophilic. A highly nucleophilic solvent, such as alcohol, would react with thiocyanate reagent, hence reducing the effective labeling of the peptide and producing side products that must be removed before analysis of the reaction mixture.

Another factor that should also be considered is the volatility of the solvent. Work-up of the reaction mixture calls for evaporation of the solvent; therefore, the lower boiling the solvent the more desirable, as long as the other factors are considered.

4. Analyzing the data given leads to the following conclusions.

Amino acids present in tripeptide: Phe, Leu, and Ala

NH$_2$-terminal amino acid: Ala

COOH-terminal amino acid: Phe

The most likely sequence: Ala-Leu-Phe

5. Total acid hydrolysis of a dipeptide would identify the amino acids present. Reaction of a dipeptide with carboxypeptidase would release both amino acids at the same time and it would be impossible to identify the COOH-terminal amino acid. COOH-terminal analysis of a dipeptide could be done using a chemical method—hydrazinolysis or reduction by lithium borohydride.

6. Dansic acid is very polar, in fact, it may be ionized during TLC analysis.

7. See Figures E4.1 and E4.3.

Experiment 5

1.

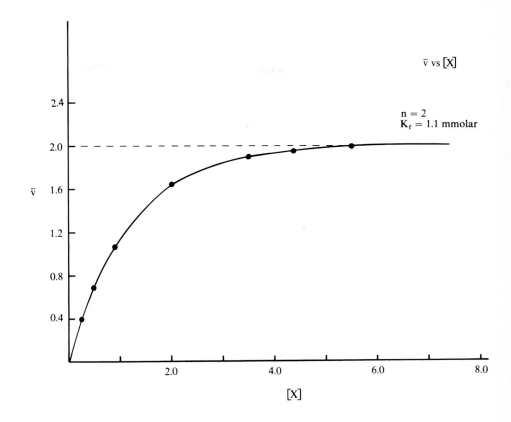

$\bar{v}$ vs $[X]$

$n = 2$
$K_f = 1.1$ mmolar

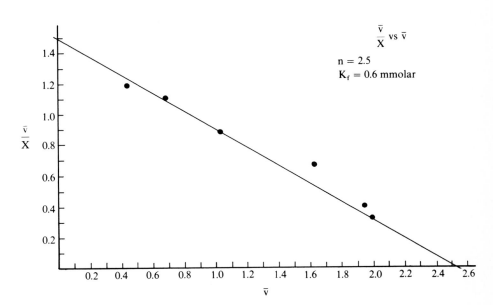

$\dfrac{\bar{v}}{X}$ vs $\bar{v}$

$n = 2.5$
$K_f = 0.6$ mmolar

2. The binding of the molecular probe must be reversible; the bound probe must have a different absorbance spectrum from the unbound probe.

3. A site for sugar binding to human albumin must be at or near the tryptophan residue.

4. Study the conversion of the Michaelis-Menten equation to the Lineweaver-Burk equation in your biochemistry textbook.

Experiment 6

1. Derivation of the Eadie-Hofstee equation:

$$\frac{1}{v_0} = \frac{K_M}{V_{max}} \frac{1}{[S]} + \frac{1}{V_{max}}$$

Multiply by V_{max}

$$\frac{V_{max}}{v_0} = \frac{K_M}{[S]} + 1$$

Multiply by v_0

$$V_{max} = \frac{K_M}{[S]} v_0 + v_0$$

Rearrange

$$v_0 = -\frac{K_M}{[S]} v_0 + V_{max}$$

A plot of v_0 vs. $v_0/[S]$ yields a straight line as shown in the graph. The intercept on the v_0 axis is V_{max}. The intercept on the $v_0/[S]$ axis is V_{max}/K_M.

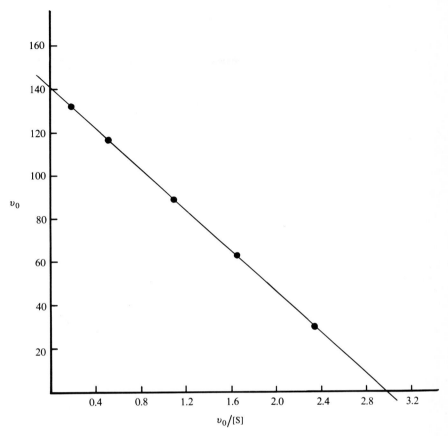

2. From Michaelis-Menten plot:

$$V_{max} = 140 \ \mu mole/min$$

$$K_M = 4.0{-}4.5 \times 10^{-5}M$$

3. The turnover number, k_3, is defined as follows:

$$k_3 = \frac{V_{max}}{[E_T]}$$

From Problem 2, $V_{max} = 140 \ \mu mole/min$ (see graph in Problem 1).

$$k_3 = \frac{140 \ \mu mole/min}{5 \times 10^{-7}M} = 280 \ min^{-1}$$

5.(a) sodium azide—noncompetitive
phenylalanine—competitive
tryptophan—competitive

cysteine—noncompetitive
4-chlororesorcinol—competitive
sodium cyanide—noncompetitive
8-hydroxyquinoline—noncompetitive
diethyldithiocarbamate—noncompetitive
thiourea—noncompetitive
phenylacetate—competitive

(b) The reverse of part a.

6. Type of inhibition: competitive

$$K_I = 5.9 \times 10^{-4}M$$

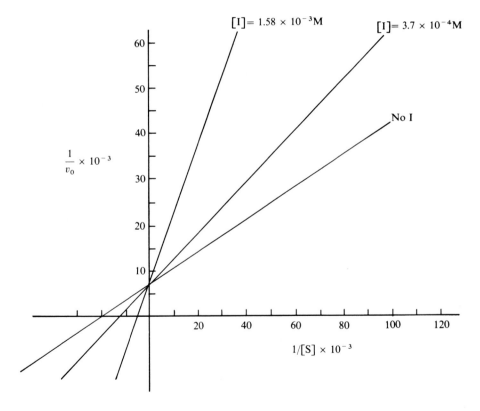

Experiment 7

1. After completing the experiment you should now recognize that the lipid is a triacylglycerol. The following experimental procedure could be followed to identify the structure.

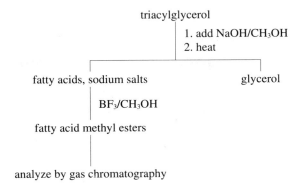

For a further discussion of this procedure, see Experiment 8.

2. Lipids that have both polar and nonpolar portions (fatty acids, etc.) tend to "streak" or "tail" during thin-layer chromatography development. This leads to long, narrow spots after iodine treatment. Addition of a small amount of a very polar solvent (acetic acid) greatly reduces this tailing and results in a more circular, better resolved spot.

3. The order of migration of the standard lipids during chromatography depends primarily on polarity. The less polar the lipid, the greater its affinity for the nonpolar solvent (compared to the polar silica gel) and hence the greater the migration.

4. Iodine interacts with many organic compounds to form pi complexes that are colored. This method of detection is especially useful for lipids containing double bonds. A lipid with several double bonds will give a darker spot with iodine. The darker spots may also be due to a higher concentration of lipid.

5. $R_f = \dfrac{63 \text{ mm}}{85 \text{ mm}} = 0.74$

Experiment 8

2. For chemical substances to be analyzed by gas chromatography, they must be relatively nonpolar and volatile. Fatty acids are not readily vaporized at temperatures attainable in a gas chromatograph (up to 200°C). FAMEs have much lower boiling points than fatty acids. Polar fatty acids also have very long retention times unless extremely nonpolar column packings are used.

3. Calcium ion stimulates the action of lipase by forming insoluble salts with liberated fatty acids. The fatty acids are no longer free in solution to inhibit the enzyme. Removal of fatty acid products from solution also retards the reverse reaction (resynthesis of acylglycerols), hence shifting the equilibrium toward hydrolysis.

5. Trimyristin

6. BF_3 is a Lewis acid that catalyzes ester formation.

8. (a) 12:0, 14:0, 16:0, 18:0, 20:0

Experiment 9

1.

Fischer projection
of D-ribose

Haworth structure
of D-ribose

2. Sucrose contains no anomeric carbons, because the anomeric carbons of the individual monosaccharides (glucose and fructose) are linked together.

3. Invertase catalyzes the hydrolysis of glycosidic bonds between fructose and glucose. Note that one of the glycosidic bonds in raffinose is between carbon 2 of fructose and carbon 1 of glucose as in sucrose. The hydrolysis of this bond is facilitated by invertase. The products are fructose plus O-α-D-galactosyl-(1 → 6)-O-α-D-glucose.

4. Specific optical rotation of pure sugar:

$$[\alpha]_D^{20} = \frac{[\alpha]_{obs}}{l \times c}$$

$$[\alpha]_D^{20} = \frac{-6.74}{0.05} = -135.2$$

Specific optical rotation of mutarotated sugar:

$$[\alpha]_D^{20} = \frac{-4.55 + 0.01}{0.05} = -91.2$$

The unknown carbohydrate is most likely β-D-fructose.

5. (a) Sample mean = +3.21°
   (b) Standard deviation = ± 0.043
   (c) 95% confidence limits = +3.21 ± 0.03 at a probability of 0.05.

Experiment 10

1. Bio-Rex is a cation exchange resin. (See Chapter 3, Section E.)
2. The proteins could be detected by monitoring column elution at a wavelength of 280 nm.
3. Glycohemoglobins elute faster than normal hemoglobin from ion-exchange columns.
5. The hemoglobins in this experiment have approximately the same molecular size, so Sephadex G-100 would not lead to effective separation.
6. See Chapter 5, Section A.
7. See Chapter 5, Section A.
8. Heme
9. Sorbitol, which is a sugar with a diol functional group, displaces the glycohemoglobin from the column.

Experiment 11

1. Methanol, ethanol, methyl ethyl ketone, 2-propanol.
2. Less polar pigments > more polar pigments. See Chapter 3, Section B.
3. See Chapter 3, Section B.
4. Aqueous-based buffers would be best for extracting intact protein-pigment complexes. Molecular weights could be determined by gel electrophoresis using appropriate standards. See Chapter 4, Section B.

Experiment 12

2. Chlorophyll is a nonpolar molecule so it is soluble in most organic solvents. Acetone disrupts protein-pigment complexes (see Experiment 11). A more efficient extraction could be achieved if several extractions were carried out and the extracts pooled.
3. ATP formation may also be measured by monitoring the uptake of inorganic phosphorus. This may be done with radioactive phosphorus (^{32}P) or by colorimetric measurement of P_i.
5. Ion-selective electrodes may be used to measure chloroplast exchange of K^+, Mg^{2+}, or Ca^{2+}. For other methods, see Dilley (1972).

Experiment 13

1. Mitochondria are most stable in an environment that is relatively polar, but not ionic. Although sucrose is polar, it is not ionic; therefore, it will not disrupt the structure of mitochondria and cause excessive loss of protein material.

2. Sodium deoxycholate is a detergent that releases proteins from membranes.

4. Electron transport and oxidative phosphorylation may be studied by incubating the mitochondrial fraction with various substrates including malate, succinate, and ascorbate. Oxidative phosphorylation is monitored by measuring the uptake of oxygen, using an oxygen electrode or Warburg manometer. Alternatively, the uptake of inorganic phosphate is measured with a colorimetric assay.

5. See Equation E13.2.

6. See Chapter 7, Section C.

8. Mitochondrial matrix.

Experiment 14

1.

Sodium urate

3. 5.4%

4. For the single standard method to be valid, there must be a linear relationship between absorbance and cholesterol concentration. The cholesterol assay is linear up to about 500 mg/100 mL.

5. Analyze a smaller serum sample, perhaps 50–75% of the standard amount.

6. Measure the amount of H_2O_2 produced as in the case of cholesterol analysis.

Experiment 15

2. 69 mg of vitamin C in 100 mL of orange juice.

3. Compounds containing the sulfhydryl group, shown as R—SH below, react with DCIP:

$$2R—SH + DCIP(oxidized) \longrightarrow R—S—S—R + DCIP(reduced)$$

p-Chloromercuribenzoic acid reacts with sulfhydryl groups, preventing reaction with the dye:

$$R-SH + \underset{\overset{\displaystyle |}{\underset{\displaystyle COOH}{Hg}}}{\underset{\overset{\displaystyle Cl}{|}}{\bigcirc}} \longrightarrow R-S-Hg-\bigcirc-COOH + HCl$$

4. 19 mg.

5. Carbohydrate

Experiment 16

1. To remove horseradish peroxidase that is not trapped in gel. Analyze rinse water for the presence of the enzyme.

2. See Experiment 6.

3. See Problem 1. Increase the percent of cross-linking agent, methylene bisacrylamide, in the gel.

4. Should be a linear graph.

5. Ammonium persulfate initiates free radical formation to begin gel polymerization. Riboflavin in the presence of light also produces radicals.

Experiment 17

1. DNA, which forms viscous solutions, is released from the cells to the medium.

2. See Experiment 19.

3. SDS is a detergent that dissociates protein-lipid complexes in cell membranes.

4. Highly purified DNA may be obtained by repeating the chloroform–isoamyl alcohol extraction several times. The alcohol precipitation step may also be carried out many times. Various chromatographic methods including ion exchange have been applied to the purification of nucleic acids.

5. 28 μg/mL

6. Electrophoresis on agarose with appropriate standards.

7. See Problem 6.

8. $A_{270} = 0.70$

9. (a) Lower T_m

   (b) Lower T_m

   (c) Lower T_m

   (d) Probably little or no effect.

10. Sample I, because of quick cooling, did not form complete double helix. Slow cooling of sample II allowed more complete double-helix formation.

11. % GC = $2.44(T_m - 69.3)$

    $40 = 2.44T_m - (2.44)(69.3)$

    $T_M = 85.7°C$

12. DNAase catalyzes the hydrolysis of phosphodiester bonds in DNA. This causes disruption of the helix formation, hence an increase in A_{260}.

Experiment 18

1. The following references provide an answer to this question: A. R. Morgan, J. S. Lee, D. E. Pulleyblank, N. L. Murray, and D. H. Evans, *Nucleic Acids. Res.* **7**, 547–569 (1979). G. J. Boer, *Anal. Biochem.* **65**, 225–231 (1975).

2. See: D. Kowalski, *Anal. Biochem.* **93**, 346 (1979).

3. Yes.

4. See the answer to Problem 1 above.

Experiment 19

1. Many plasmids contain genes that carry messages for the synthesis of proteins that protect microorganisms against antibiotics. The presence of certain antibiotics is a signal to the microorganism that more of these proteins are necessary. The microorganism responds by increasing the rate of production of plasmids.

2. See the section on protein synthesis in your biochemistry textbook.

4. Preparations of ribonuclease also contain deoxyribonuclease. If not removed or destroyed, DNAases would degrade the extracted plasmids. DNAases are not stable to high temperatures and are denatured during heating. Ribonucleases remain stable during the heat treatment.

Experiment 20

1. Isopropanol is added to precipitate plasmid DNA from solution. Most RNA remains in solution; hence, isopropanol precipitation provides an effective technique for removing contaminating RNA from DNA extracts.

Experiment 21

1. Restriction endonucleases require the presence of Mg^{2+} for activity. The "quench buffer" contains EDTA, which complexes transition metal ions in the solution. The metal ions are no longer available for binding by the nuclease molecules and enzyme activity is inhibited.

2. Glycerol is a dense, viscous chemical that aids in the application of samples to the gel. Bromphenol blue dye acts as a marker. It migrates very rapidly in electrophoresis.

4. Natural DNA molecules and restriction fragments are too large to penetrate polyacrylamide gels. Even gels with low percentage cross-linking are not useful.

Experiment 22

2. Urea acts as a mild denaturant for RNA. It may "open" the structure of the RNA so it can more selectively accept and bind substrate.

3. (a) 5′GGACCUCU ↓ AAAAA3′

   (b) No reaction

   (c) No reaction

4. These are marker dyes. See Chapter 4.

5. Do a series of reactions with constant amounts of catalytic RNA and increasing amounts of RNA substrate or GTP.

6. ^{32}P has a half-life of only 14 days.

Index